Progress in Nonlinear Differential Equations and Their Applications

Volume 42

Semigroups of Operators: Theory and Applications

International Conference in Newport Beach, December 14–18, 1998

A. V. Balakrishnan
Editor

Birkhäuser
Basel · Boston · Berlin

A. V. Balakrishnan
Department of Electrical Engineering
UCLA m/c 951594
Los Angeles, CA 90095
USA
e-mail: bal@ee.ucla.edu

2000 Mathematics Subject Classification 47Dxx

A CIP catalogue record for this book is available from the Library of Congress, Washington D.C., USA

Deutsche Bibliothek Cataloging-in-Publication Data

Semigroups of operators : theory and applications ; international conference in Newport Beach, December 14-18, 1998 / A. V. Balakrishnan, ed. – Basel ; Boston ; Berlin : Birkhäuser, 2000
(Progress in nonlinear differential equations and their applications ; Vol. 42)
ISBN 3-7643-6310-X

ISBN 3-7643-6310-X Birkhäuser Verlag, Basel – Boston – Berlin

Printed on acid-free paper produced of chlorine-free pulp. TCF ∞
Printed in Germany
ISBN 3-7643-6310-X

9 8 7 6 5 4 3 2 1

TABLE OF CONTENTS

PREFACE

These *Proceedings* comprise the bulk of the papers presented at the *International Conference on Semigroups of Operators: Theory and Control*, held 14–18 December 1998, Newport Beach, California, U.S.A.

The intent of the Conference was to highlight recent advances in the theory of Semigroups of Operators which provides the abstract framework for the time-domain solutions of time-invariant boundary-value/initial-value problems of partial differential equations. There is of course a firewall between the abstract theory and the applications and one of the Conference aims was to bring together both in the hope that it may be of value to both communities.

In these days when all scientific activity is judged by its value on "dot com" it is not surprising that mathematical analysis that holds no promise of an immediate commercial product-line, or even a software tool-box, is not high in research priority. We are particularly pleased therefore that the National Science Foundation provided generous financial support without which this Conference would have been impossible to organize. Our special thanks to Dr. Kishan Baheti, Program Manager.

It is my pleasant duty to thank the members of the International Program Committee: Q. Phong Vu (Chair), W. Arendt, C. J. K. Batty, R. Datko, H. Komatsu, C. S. Kubrusly, G. Lumer, Yu. I. Lyubich, J. M. Mazon, A. McIntosh, R. Nagel, S. Y. Shaw, B. Simon, R. Triggiani, for the outstanding technical program. I am also deeply indebted to Professors N. Levan and Q. Phong Vu for their unstinting help with the many tiresome tasks involved in organizing an international conference.

A. V. Balakrishnan
December 1999

Progress in Nonlinear Differential Equations
and Their Applications, Vol. 42

A SEMIGROUP APPROACH TO THE MAXIMUM LIKELIHOOD STATE ESTIMATION OF STOCHASTIC PARABOLIC SYSTEMS

SHIN ICHI AIHARA
THE SCIENCE UNIVERSITY OF TOKYO

ARUNABHA BAGCHI
TWENTE UNIVERSITY

Abstract. We consider a stochastic parabolic partial differential equation with unknown boundary conditions. We provide a state-space formulation of the problem using the semigroup approach. Introducing the Onsager-Machlup functional, we can formulate the maximum likelihood state estimation problem. The derived estimator has a recursive form.

1. Introduction

There is an enormous literature on identification of distributed parameter systems [5]. In practically all such models, the boundary conditions of the (state) partial differential equations are assumed known. There are ,however, significant practical situations where the boundary conditions are only available through noisy measurements. One case in point is modelling the salt concentration of a finite portion of a river with the purpose of predicting the quality of water downstream used for drinking purpose. This problem was first studied heuristically in [3] where the space-time dynamics of the salt concentration of the river Rhine was analyzed after it reached Holland. The state in this situation was described, as expected, by a stochastic parabolic partial differential equation.

We first reformulate the problem in the state-space form using the semigroup approach. We then use the maximum likelihood approach for this problem. By introducing the Onsager-Machlup functional, we derive a modified likelihood and construct a maximum likelihood state estimation algorithm. The derived state estimator is a recursive one and it is easy to implement this filter.

2. Semigroup Formulation

We consider the following simple heat diffusion equation:

$$u(t,x) - \int_0^t \{a(x)\frac{\partial^2 u(s,x)}{\partial x^2} + b(x)\frac{\partial u(s,x)}{\partial x}\}ds = u_o(x) + w(t,x) \ in\ [0,t_f]\times]0,1[\tag{2.1}$$

$$\int_0^t u(s,0)ds = \int_0^t g_0(s)ds \ on\ [0,t_f] \tag{2.2}$$

$$\int_0^t u(s,1)ds = \int_0^t g_1(s)ds \text{ on } [0,t_f] \tag{2.3}$$

where w is a cylindrical Brownian motion on $H = L^2(0,1)$, and g_0 and g_1 are the unknown conditions with

$$g_0, g_1 \in L^2(]0,t_f[) \tag{2.4}$$

We set the following observation mechanisms:

$$y(t) = \int_0^t \left(\int_{G_o} c(x)u(s,x)dx\right)ds + v(t), G_o \subset [0,1] \tag{2.5}$$

$$y^{b1}(t) = \int_0^t g_1(s)ds + v_1(t) \tag{2.6}$$

$$y^{b0}(t) = \int_0^t g_0(s)ds + v_0(t) \tag{2.7}$$

where $\tilde{v} = [v\ v_1\ v_0]'$ is a Brownian motion in R^3 which is independent of the system noise.

The stochastic system (2.1) has a cylindrical Brownian motion and non-homogeneous Dirichlet boundary conditions. Hence in order to formulate the considered system in the semigroup theory of stochastic parabolic systems, we introduce the Dirichlet map. See [6][Chap.2] for more details. We work in the following Sobolev spaces:

$$D = H^2 \cap H_0^1 \subset V = H_0^1 \subset H = L^2(0,1) \subset V' = H^{-1} \subset D' = \text{dual of } D$$

Define

$$-A\phi = a(x)\frac{\partial^2\phi(x)}{\partial x^2} + b(x)\frac{\partial\phi(x)}{\partial x} \quad \forall\phi \in D. \tag{2.8}$$

We assume that

$$a(x) \geq \alpha_o > 0, \exists\alpha_o, \forall x \in [0,1] \tag{2.9}$$

and

$$a \in C^1(0,1) \text{ and } b \in C(0,1). \tag{2.10}$$

Hence the domain of A becomes

$$\mathcal{D}(A) = D = H^2 \cap V \tag{2.11}$$

2.1 Riesz spectral operator

In order to obtain the semigroup formulation of the system (2.1), we consider the eigenvalue problem of $-A$, i.e., we find nonzero e_n and μ_n such that

$$\begin{cases} -Ae_n = \mu_n e_n \\ e_n(0) = e_n(1) = 0 \end{cases}. \tag{2.12}$$

The substitution

$$\text{(2.13)} \qquad \begin{cases} \lambda(x) = exp(\int_0^x \frac{b(\xi)}{a(\xi)} d\xi) \\ \rho(x) = \frac{1}{a(x)} exp(\int_0^x \frac{b(\xi)}{a(\xi)} d\xi) \end{cases}$$

transforms (2.12) into the weighted Sturm-Liouville problem

$$\text{(2.14)} \qquad \begin{cases} \frac{1}{\rho}\{\frac{\partial}{\partial x}(\lambda(x)\frac{\partial e_n}{\partial x}\} = \mu_n e_n(x) \\ e_n(0) = e_n(1) = 0 \end{cases} .$$

The space $H = L^2(0,1)$ is also a Hilbert space with the inner product

$$\text{(2.15)} \qquad (\phi, \psi)_\rho = \int_0^1 \phi(x)\psi(x)\rho(x)dx$$

It is well known that the transformed operator (2.14) is self adjoint with compact resolvent in H with this weighted inner product (2.15). Furthermore noting that $\{e_n\}_{n=0}^{\infty}$ forms a Riesz basis in H with the biorthogonal basis $\{\rho e_n\}_{n=0}^{\infty}$, we find that $-A$ is a generator of the following semigroup

$$\text{(2.16)} \qquad e^{tA} = \sum_{n=0}^{\infty} e^{-\mu_n t}(\cdot, e_n)_\rho e_n$$

The co-normal derivative with respect to the operator A is given by

$$\text{(2.17)} \qquad \frac{\partial \phi}{\partial \nu_A} = \lambda(x)\frac{\partial \phi}{\partial x}\nu(x) \; on \; \Gamma = \{x = 0, 1\}$$

with ν the unit external normal to Γ. In this case $\nu(0) = -1$ and $\nu(1) = 1$ and $(\phi_1, \phi_2)_{L^2(\Gamma)} = \phi_1(1)\phi_2(1) + \phi_1(0)\phi_2(0)$.

2.2 Dirichlet map

In order to transform the original non-homogeneous system to the homogeneous one, we need to construct the Dirichlet map. So we consider the following elliptic boundary value problem:

$$\text{(2.18)} \qquad \begin{cases} Av(t,x) = 0, \text{ in } (t,x) \in]0, t_f[\times]0,1[\\ v(t,0) = g_0(t), \; v(t,1) = g_1(t) \end{cases}$$

By using Green's formula, we obtain $\forall \phi \in D$,

$$(Av(t), \rho\phi) = (g(t), \frac{\partial \phi}{\partial \nu_A})_{L^2(\Gamma)} + (v(t), \rho A\phi),$$

where

$$\text{(2.19)} \qquad g(t,x) = \begin{cases} g_0(t), & x = 0 \\ g_1(t), & x = 1 \end{cases} .$$

This implies that

$$(v(t), -A\phi)_\rho = ((B^*)^* g(t), \phi),$$

where

$$B^*\phi = \frac{\partial \phi}{\partial \nu_A} \in \mathcal{L}(D : L^2(\Gamma)). \tag{2.20}$$

Setting $\phi = e_n$, we get

$$(v(t), \rho(-A)e_n) = ((B^*)^* g(t), e_n) \tag{2.21}$$

i.e.,

$$\mu_n (v(t), \rho e_n) = ((B^*)^* g(t), e_n),$$

where $(B^*)^*$ denotes the topological transpose of the linear map $B^* : \mathcal{D} \to L^2(\Gamma)$. Hence noting that

$$(-A)^{-1} = \sum_{n=0}^{\infty} \frac{1}{\mu_n} (\cdot, e_n)_\rho e_n, \tag{2.22}$$

we have

$$\begin{aligned} v(t) &= \sum_{n=0}^{\infty} \frac{1}{\mu_n} ((B^*)^* g(t), e_n) e_n \\ &= -\tilde{B} g(t), \end{aligned} \tag{2.23}$$

where

$$\tilde{B} = -(-A)^{-1} \frac{1}{\rho} (B^*)^* \in \mathcal{L}(L^2(\Gamma); H) \tag{2.24}$$

2.3 Transformation of system model

Following Balakrishnan [4, p.248], we can show that the original system (2.1-3) can be expressed in the following form:

$$u(t) = e^{tA} u_o + A \int_0^t e^{(t-s)A} \tilde{B} g(s) ds + \int_0^t e^{(t-s)A} dw(s) \tag{2.25}$$

Defining

$$\tilde{u}(t) = A^{-1} u(t), \tag{2.26}$$

we get

$$\tilde{u}(t) = e^{tA} \tilde{u}_o + \int_0^t e^{(t-s)A} \tilde{B} g(s) ds + \int_0^t e^{(t-s)A} d\tilde{w}(s) \tag{2.27}$$

where

$$\tilde{w} = A^{-1}w. \tag{2.28}$$

It is easy to show that

$$E\{(\tilde{w}(t),\phi_1)_\rho(\tilde{w}(t),\phi_2)_\rho\} = (\phi_1,\tilde{Q}\phi_2)_\rho t,\ \forall\phi_1,\phi_2 \in H \tag{2.29}$$

where

$$\begin{aligned}\tilde{Q} &= A^{-1}A^{-1}\\ &= \sum_{i=0}^{\infty}\frac{1}{\mu_i^2}(\cdot,\rho e_i)e_i\end{aligned} \tag{2.30}$$

with

$$Tr\{\tilde{Q}\} = \sum_{i=0}^{\infty}\frac{1}{\mu_i^2} < \infty, \tag{2.31}$$

because

$$\sqrt{\mu_i} = ci + O(\frac{1}{i}) \text{ as } i \rightarrow \infty. \tag{2.32}$$

It is also possible to represent (2.27) as the weak form:

$$(\tilde{u}(t),\phi)_\rho + \int_0^t(\tilde{u}(s),A\phi)_\rho ds = (\tilde{u}_o,\phi)_\rho + \int_0^t(\tilde{B}g(s),\phi)_\rho ds + \int_0^t(d\tilde{w}(s),\phi)_\rho, \forall\phi\in D. \tag{2.33}$$

Theorem 2.1 *Under (2.9), (2.10),(2.31) and*

$$\tilde{u}_o \in L^2(\Omega;H), \tag{2.34}$$

(2.27) has a unique continuous solution in $L^2(\Omega;L^2(T;V)\cap C(T;H))$.

Proof We introduce the following finite-dimensional system:

$$\left\{\begin{aligned}&\tilde{u}_m(t) = \textstyle\sum_{i=0}^m \tilde{u}_{im}(t)e_i\\ &d(\tilde{u}_m(t),e_j)_\rho + (\tilde{u}_m(t),Ae_j)_\rho dt = (\tilde{B}g(t),e_j)_\rho dt + d(\tilde{w}(t),e_j)_\rho,\\ &\qquad\qquad j=1,2,\cdots,m\\ &\tilde{u}_m(0) = \tilde{u}_{om}\end{aligned}\right. \tag{2.35}$$

where e_j is a sequence of linearly independent elements of D, which is complete in V. Equation (2.35) is equivalent to an initial-value problem for a linear m-dimensional ordinary stochastic differential equation for the $\tilde{u}_{im}(t)$. The existence and uniqueness is obvious. From [7], we obtain the following estimate:

$$\tilde{u}_m \to \tilde{u} \text{ in } L^2(\Omega \times T; V) \text{ weakly} \tag{2.36}$$

$$\tilde{u}_m \to \tilde{u} \text{ in } L^2(\Omega; L^\infty(T : H)) \text{ weakly star} \tag{2.37}$$

The proof of the continuity property can be also found in [7]. □

The observation mechanism is reformulated as follows:

$$y(t) = \int_0^t \int_{G_o} c(x) A A^{-1} u dx ds + v(t) \tag{2.38}$$

Assuming that

$$c \in C_0^2(G_o) \tag{2.39}$$

we have

$$\begin{aligned} y(t) &= \int_0^t \int_{G_o} c(x) A\tilde{u} dx ds + v(t) \\ &= \int_0^t C\tilde{u} ds + v(t) \end{aligned} \tag{2.40}$$

where

$$C(\cdot) = \int_{G_o} c(x) A(\cdot) dx. \tag{2.41}$$

Restricting the domain of A in V, A is also considered as the linear operator $\mathcal{L}(V; V')$. From (2.39), we find that

$$C \in \mathcal{L}(V'; R^3). \tag{2.42}$$

It is easy to show that $Y(t) = [y(t)\ y_0(t)\ y_1(t)]' \in C(T; R^3)$ a.s.

3. The Onsager-Machlup functional and finite dimensional MLE state estimation

In this section, we shall briefly indicate how to derive the ML state estimation algorithm for the system (2.27) with (2.38) as the observation process. The main advantage of the ML state estimate is that the derived algorithm is recursive and the unknown boundary inputs are just replaced by the boundary observations $dy_0(t)$ and $dy_1(t)$. The usual likelihood functional is not sufficient to formulate the ML state estimate as explained in [8] and we need to modify the likelihood functional by using the Onsager-Machlup functional. For the infinite-dimensional system , it is difficult to derive the Onsager-Machlup functional. Hence, we introduce the finite-dimensional approximated system for

(2.27) and derive the Onsager-Machlup functional. Modifying the likelihood functional by using this functional, we derive the ML state estimation algorithm for the approximated finite-dimensional system, resulting the so-called "Mortensen equation." The convergence property of the approximated finite-dimensional state estimation algorithm to the infinite-dimensional one is finally checked.

3.1. Finite-dimensional system

First we recall the finite-dimensional system in the proof of Theorem 2.1:

$$
\begin{cases} \tilde{u}_m(t) = \sum_{i=0}^{m} \tilde{u}_{im}(t)e_i \\ d(\tilde{u}_m(t), e_j)_\rho + (\tilde{u}_m(t), Ae_j)_\rho dt = (\tilde{B}g(t), e_j)_\rho dt + d(\tilde{w}(t), e_j)_\rho, \\ \qquad\qquad j = 1, 2, \cdots, m \\ \tilde{u}_m(0) = \tilde{u}_{om} \end{cases} \tag{3.1}
$$

The approximated observation is

$$
y_m(t) = \int_0^t C\tilde{u}_m(s)ds + v(t), \tag{3.2}
$$

with the actual boundary observations;

$$
y^{b0}(t) = \int_0^t g_0(s)ds + v_0(t) \tag{3.3}
$$

$$
y^{b1}(t) = \int_0^t g_1(s)ds + v_1(t) \tag{3.4}
$$

Let $\mathcal{F}_t^{Y_m}$ be a σ-algebra generated by $Y_m(s) = [y_m(s)\ y^{b0}(s)\ y^{b1}(s)], 0 \leq s \leq t$. The Onsager-Machlup functional under $\mathcal{F}_t^{Y_m}$ [8] is

(3.5)

$$
\begin{aligned}
\lim_{\epsilon \to 0} & \frac{Prob_{\tilde{u}_{om}}\{\sup_t |\tilde{u}_m(t) - \tilde{\phi}_m(t)|_\rho < \epsilon | \mathcal{F}_t^{Y_m}\}}{Prob\{\sup_t |\tilde{w}_m(t)|_\rho < \epsilon\}} \\
= \ & exp\{-\frac{1}{2}\int_0^t ((\dot{\tilde{\phi}}_m(s) + A_m\tilde{\phi}_m(s) - \tilde{B}_m g(s)), \\
& \tilde{Q}_m^{-1}(\dot{\tilde{\phi}}_m(s) + A_m\tilde{\phi}_m(s) - \tilde{B}_m g(s)))_\rho ds \\
& + \frac{1}{2}Tr\{A_m t\} + \int_0^t C\tilde{\phi}_m(s)dy_m(s) - \frac{1}{2}\int_0^t |C\tilde{\phi}_m(s)|^2 ds \\
& + \int_0^t g_0 dy^{b0}(s) - \frac{1}{2}\int_0^t |g_0(s)|^2 ds + \int_0^t g_1 dy^{b1}(s) - \frac{1}{2}\int_0^t |g_1(s)|^2 ds\}
\end{aligned}
$$

We now introduce here the piecewise linear approximation of the observation data y_m, y_0, y_1 . Let π_k be a partition of the interval $[0, t_f], 0 = t_0^k < t_1^k < \cdots < t_{K(k)}^k = t_f$. We assume that

$$\max_{1 \leq j \leq K(k)} |t_j^k - t_{j-1}^k| \leq \frac{c_1}{k} \tag{3.6}$$

$$K(k) \leq c_2 k \tag{3.7}$$

For a fixed partition π_n, we have

$$y_{\cdot\, n}(t) = y_\cdot(t_i^n) + \frac{t - t_i^n}{t_{i+1}^n - t_i^n}(y_\cdot(t_{i+1}^n) - y_\cdot(t_i^n)), t \in [t_i^n, t_{i+1}^n] \tag{3.8}$$

Replacing $dy_\cdot$ as $\frac{dy_{\cdot\, n}}{dt}dt$ in (2.8) , our ML-estimate is equivalent to minimizing

$$\begin{aligned} J(\tilde{\phi}_m, g, n) = & \frac{1}{2}\int_0^t ((\dot{\tilde{\phi}}_m(s) + A_m\tilde{\phi}_m(s) - \tilde{B}_m g(s)), \\ & \tilde{Q}_m^{-1}(\dot{\tilde{\phi}}_m(s) + A_m\tilde{\phi}_m(s) - \tilde{B}_m g(s)))_\rho ds \\ & - \int_0^t C\tilde{\phi}_m(s)\frac{dy_{mn}(s)}{ds}ds + \frac{1}{2}\int_0^t |C\tilde{\phi}_m(s)|^2 ds \\ & - \int_0^t g_0 \frac{dy_n^{b0}(s)}{ds}ds + \frac{1}{2}\int_0^t |g_0(s)|^2 ds \\ & - \int_0^t g_1 \frac{dy_n^{b1}(s)}{ds}ds + \frac{1}{2}\int_0^t |g_1(s)|^2 ds \end{aligned} \tag{3.9}$$

3.2. The Mortensen Equation

For the derivation of the likelihood cost $\tilde{J}$, the initial condition $\tilde{u}_o$ is assumed to be fixed. It is possible to consider the randomness of the initial condition as in [2]. Here we shall add the initial cost $\ell_o(\tilde{\phi}_m(0))$;

$$\ell_o(\tilde{\phi}_m(0)) = (\tilde{\phi}_m(0), P_o^{-1}\tilde{\phi}_m(0))_\rho$$

to $\tilde{J}$. The precise argument for the initial data can be found in [2]. In order to derive the recursive state estimator from the cost (3.9), we must perform the optimization procedure twice. First for a fixed t, the optimization problem (3.9) is rewritten as follows:
(System)

$$\dot{\tilde{\phi}}_m(t,n) + A_m\tilde{\phi}_m(t,n) - \tilde{B}_m g(t) = f(t), \ \tilde{\phi}_m(t,n) = \xi \tag{3.10}$$

with the cost

$$\tag{3.11} \begin{aligned} \tilde{J}(f,g,n) &= \ell_o(\tilde{\phi}_m(0)) + \frac{1}{2}\int_0^t (f(s), Q_m^{-1} f(s))_\rho ds \\ &+\frac{1}{2}\int_0^t |C\tilde{\phi}_m(s,n)|^2 ds - \int_0^t C\tilde{\phi}_m(s,n)\frac{dy_{mn}(s)}{ds}ds \\ &+\frac{1}{2}\int_0^t |g(s) - \frac{dy_n^b}{ds}|^2_{L^2(\Gamma)} ds \end{aligned}$$

where

$$\tag{3.12} y_n^b(t,x) = \begin{cases} y_n^{b0}(t), & x = 0 \\ y_n^{b1}(t), & x = 1 \end{cases}$$

Define

$$\tag{3.13} S(t,\xi,n) = \inf_{f,g} \tilde{J}(f,g,n)$$

The second optimization step is to find $\hat{\phi}_m^o$ such that

$$\tag{3.14} S(t,\hat{\phi}_m^o,n) \le S(t,\xi,n), \forall \xi \in H$$

at each time t.

By using the dynamic programming approach, the Mortensen equation becomes

$$\tag{3.15} \begin{cases} \frac{\partial S(t,\xi,n)}{\partial t} - \inf\{\ell(f,g,Y_n^m) - (D_\xi S(t,\xi,n), (-A_m\xi + \tilde{B}_m g + f))_\rho\} = 0 \\ S(0,\xi) = \ell_o(\xi), \end{cases}$$

where

$$\tag{3.16} \ell(t,g,Y_n^m) = \frac{1}{2}(f,\tilde{Q}_m^{-1}f)_\rho + \frac{1}{2}|C\xi|^2 - C\xi\cdot\frac{dy_{mn}}{dt} + \frac{1}{2}|g - \frac{dy_n^b}{dt}|^2_{L^2(\Gamma)}$$

and D_ξ denotes the Frechet derivative.

This leads to

$$\tag{3.17} \dot{\hat{\phi}}_m^o + A_m\hat{\phi}_m^o - \tilde{B}_m\frac{dy_n^b(t)}{dt} = P_m(t)C^*\{\frac{dy_{mn}(t)}{dt} - C\hat{\phi}_m^o\}, \hat{\phi}_m^o(0) = 0,$$

where $P_m(t)$ satisfies

$$\tag{3.18} \begin{aligned} &\dot{P}_m(t) + A_m P_m(t) + P_m(t)A_m + P_m(t)C^*CP_m(t) \\ &\qquad = \tilde{Q}_m + \tilde{B}_m\tilde{B}_m^*, P_m(0) = P_o \end{aligned}$$

The value function $S(t, \hat{\phi}^o_m, n)$ satisfies

$$S(t, \hat{\phi}^o_m, n) = \frac{1}{2}\int_0^t |C(\hat{\phi}^o_m(s,n)|^2 ds - \int_0^t C\hat{\phi}^o_m(s,n)\frac{dy_{mn}(s)}{ds}ds. \tag{3.19}$$

3.3. Convergence of the ML state estimation algorithm

First, we shall take a limit as $n \to \infty$ in (3.17). Noting that as $n \to \infty$

$$y_{mn}(t) \to y_m(t) \ a.s. \tag{3.20}$$

$$y^b_n(t) \to y^b(t) \ a.s., \tag{3.21}$$

(3.17) becomes

$$\begin{aligned} \hat{\phi}^o_m(t) + \int_0^t A_m\hat{\phi}^o_m(s)ds - \int_0^t \tilde{B}_m dy^b(s) \\ = \int_0^t P_m(s)C^*(dy_m(s) - C\hat{\phi}^o_m ds) \end{aligned} \tag{3.22}$$

It can be shown (See [1] for details.) that

$$\lim_{m\to\infty}(\hat{\phi}^o_m, \phi) = (\hat{\phi}^o(t), \phi) \ a.s. \tag{3.23}$$

where

$$\begin{aligned} (\hat{\phi}^o(t), \phi)_\rho + \int_0^t (\hat{\phi}^o(s), A^*\phi)_\rho ds - \int_0^t (\tilde{B}dy^b(s), \phi)_\rho \\ = \int_0^t (P(s)C^*(dy(s) - C\hat{\phi}^o ds), \phi)_\rho, \forall \phi \in D. \end{aligned} \tag{3.24}$$

and $\forall \phi_1, \phi_2 \in D$,

$$\begin{aligned} (\dot{P}(t)\phi_1, \phi_2)_\rho + (P(t)\phi_1, A\phi_2)_\rho + (A\phi_1, P(t)\phi_2)_\rho \\ +(P(t)C^*CP(t)\phi_1, \phi_2)_\rho = ((\tilde{Q} + \tilde{B}\tilde{B}^*)\phi_1, \phi_2)_\rho. \end{aligned} \tag{3.25}$$

Consequently we get the following optimal cost:

$$\begin{aligned} \lim_{n\to\infty, m\to\infty} S(t, \hat{\phi}^o_{mn}, n) = -\int_0^t C\hat{\phi}^o(s)dy(s) + \frac{1}{2}\int_0^t |C\hat{\phi}^o(s)|^2 ds \\ -\frac{1}{2}\int_0^t Tr\{CP(s)C^*\}ds \end{aligned} \tag{3.26}$$

4. PDE form of the derived filter

It is better to derive the filter equations in the form of the original system equations. Noting that the original state $u(t)$ is transformed to $\tilde{u}(t)$, we should transform the filter state $\hat{\phi}^o(t)$ to the original state. We set

$$\hat{u}(t) = A\hat{\phi}^o(t) \tag{4.1}$$

Hence (3.24) becomes

$$\begin{aligned}(A^{-1}\hat{u}(t), \phi)_\rho + \int_0^t (A^{-1}\hat{u}(s), A\phi)_\rho ds - \int_0^t (\tilde{B}dy^b(s), \phi)_\rho \\ = \int_0^t (P(s)C^*(dy(s) - CA_\lambda^{-1}\hat{u}(s)ds), \phi)_\rho\end{aligned} \tag{4.2}$$

and

$$\begin{aligned}(\hat{u}(t), A^{-1}\phi)_\rho + \int_0^t (\hat{u}(s), AA^{-1}\phi)_\rho ds + \int_0^t ((-\rho A)^{-1}(B^*)^* dy^b(s), \phi)_\rho \\ = \int_0^t (AP(s)C^*(dy(s) - C^o\hat{u}(s)ds), A^{-1}\phi)_\rho,\end{aligned} \tag{4.3}$$

where

$$C^o(\cdot) = \int_{G_o} c(x)(\cdot)dx. \tag{4.4}$$

From (2.22) and (2.24), we get

$$\begin{aligned}(\tilde{B}dy^b, \phi)_\rho = -((-\rho A)^{-1}(B^*)^* dy^b, \phi)_\rho &= ((B^*)^* dy^b, A^{-1}\phi) \\ &= (dy^b, B^* A^{-1}\phi)_{L^2(\Gamma)} \\ &= (dy^b, \frac{\partial}{\partial \nu_A} A^{-1}\phi)_{L^2(\Gamma)}.\end{aligned} \tag{4.5}$$

From (2.41) we also have

$$\begin{aligned}APC^* &= A(CP)^* \\ &= A(AP)^*(C^o)^*\end{aligned} \tag{4.6}$$

because the kernel c of C^o belongs to $C_0^2(G_o)$. Denoting

$$\begin{aligned}P^o &= A(AP)^* \\ &= \int_0^1 p^o(t,x,z)(\cdot)dz,\end{aligned} \tag{4.7}$$

we have for $\tilde{\phi} = A^{-1}\phi$

$$(4.8) \qquad (\hat{u}(t),\tilde{\phi})_\rho + \int_0^t (\hat{u}(s), A\tilde{\phi})_\rho ds - \int_0^t (dy^b(s), \frac{\partial}{\partial \nu_A}\tilde{\phi})_{L^2(\Gamma)} = \int_0^t (P^o(s)(C^o)^*(dy(s) - C^o\hat{u}(s)ds), \tilde{\phi})_\rho.$$

It is easy to show that

$$\begin{aligned}
&\int_0^t (\hat{u}(s), A\tilde{\phi})_\rho ds \\
&= \int_0^t\int_0^1 \rho(x)\hat{u}(s,x)(-a(x)\frac{\partial^2\tilde{\phi}}{\partial x^2} - b(x)\frac{\partial\tilde{\phi}}{\partial x})dxds \\
&= \int_0^t\int_0^1 \rho(x)\hat{u}(s,x)(-\frac{1}{\rho(x)}\frac{\partial}{\partial x}\{\lambda(x)\frac{\partial\tilde{\phi}}{\partial x}\})dxds \\
&\quad \text{(by using Green's formula)} \\
&= -\int_0^t [\hat{u}(s,x)\lambda(x)\frac{\partial\tilde{\phi}(x)}{\partial x}]_{x=0}^{x=1} ds + \int_0^t\int_0^1 \frac{\partial\hat{u}(s,x)}{\partial x}\lambda(x)\frac{\partial\tilde{\phi}(x)}{\partial x}dxds \\
&\quad \text{(noting that } \tilde{\phi} = 0 \text{ on } \Gamma) \\
&= \int_0^t (\hat{u}(s), \frac{\partial\tilde{\phi}}{\partial\nu_A})_{L^2(\Gamma)}ds - \int_0^t \frac{\partial}{\partial x}\{\lambda(x)\frac{\partial\hat{u}(s,x)}{\partial x}\}\tilde{\phi}(x)dxds \\
&= \int_0^t (\hat{u}(s), \frac{\partial\tilde{\phi}}{\partial\nu_A})_{L^2(\Gamma)}ds + \int_0^t\int_0^1 \rho(x)\{-a(x)\frac{\partial^2\hat{u}(s,x)}{\partial^2 x} - b(x)\frac{\partial\hat{u}(s,x)}{\partial x}\}\tilde{\phi}(x)dx \\
&= \int_0^t (\hat{u}(s), \frac{\partial\tilde{\phi}}{\partial\nu_A})_{L^2(\Gamma)}ds + \int_0^t (A\hat{u}(s), \tilde{\phi})_\rho ds.
\end{aligned}$$

This implies that

$$\begin{aligned}
&\int_0^t (\hat{u}(s), A\tilde{\phi})_\rho ds - \int_0^t (dy^b(s), \frac{\partial\tilde{\phi}}{\partial\nu_A})_{L^2(\Gamma)}ds \\
&= \int_0^t (\hat{u}(s)ds - dy^b(s), \frac{\partial\tilde{\phi}}{\partial\nu_A})_{L^2(\Gamma)}ds \\
&\quad + \int_0^t\int_0^1 \{-a(x)\frac{\partial^2\hat{u}(s,x)}{\partial^2 x} - b(x)\frac{\partial\hat{u}(s,x)}{\partial x}\}\rho(x)\tilde{\phi}(x)dxds.
\end{aligned}$$

Hence we can derive formally the original partial differential equation of the filtering equation:

$$\begin{aligned}
(4.9) \quad & d\hat{u}(t,x) - \{a(x)\frac{\partial^2\hat{u}(t,x)}{\partial x^2} + b(x)\frac{\partial\hat{u}(t,x)}{\partial x}\}dt \\
&= \int_{G_o} p^o(t,x,z)c(z)dx(dy(t) - \int_{G_o} c(x)\hat{u}(t,x)dxdt), x \in]0,1[\\
(4.10) \quad & d\hat{u}(t,0) = dy^{b0}(t) \\
(4.11) \quad & d\hat{u}(t,1) = dy^{b1}(t)
\end{aligned}$$

The remaining problem is to derive the equation of the kernel $p^o(t,x,z)$ of P^o from (3.25). Choosing $\phi_1 = A\varphi_1$ and $\phi_2 = A\varphi_2$ in (3.25) , we have

$$
\begin{aligned}
(4.12) \qquad & ((\tilde{Q}+\tilde{B}\tilde{B}^*)A\varphi_1, A\varphi_2)_\rho \\
&= (A^{-1}A^{-1}+\tilde{B}\tilde{B}^*)A\varphi_1, A\varphi_2)_\rho \\
&= (\varphi_1,\varphi_2)_\rho + (\frac{\partial\varphi_1}{\partial\nu_A}, \frac{\partial\varphi_2}{\partial\nu_A})_{L^2(\Gamma)}
\end{aligned}
$$

and

$$
\begin{aligned}
(4.13) \quad & (A(A\dot{P})^*\varphi_1,\varphi_2)_\rho + (A(AP)^*\varphi_1, A\varphi_2)_\rho + (A\varphi_1, A(AP)^*\varphi_2)_\rho \\
&+(A(AP)^*(C^o)^*C^o(A(AP)^*)\varphi_1,\varphi_2)_\rho \\
&= (\varphi_1,\varphi_2)_\rho + (\frac{\partial\varphi_1}{\partial\nu_A}, \frac{\partial\varphi_2}{\partial\nu_A})_{L^2(\Gamma)}
\end{aligned}
$$

Now we shall represent the term $(\frac{\partial\varphi_1}{\partial\nu_A}, \frac{\partial\varphi_2}{\partial\nu_A})_{L^2(\Gamma)}$ as

$$
(4.14) \quad (\varphi_1, \int_0^1 r(x,z)\varphi_2 dz)_\rho = \sum_{i,j=1}^{\infty} (\frac{\partial\phi_i}{\partial\nu_A}, \frac{\partial\phi_j}{\partial\nu_A})_{L^2(\Gamma)} (\phi_i,\varphi_1)_\rho (\phi_j,\varphi_2)_\rho
$$

where it should be noted that the above function $r(x,z)$ is not a usual function, i.e., this function contains the partial derivative at the boundary points (in practice we approximate this function as the finite sum) and

$$
(4.15) \qquad \phi_i(x) = \sqrt{2}sin(i\pi x).
$$

Hence the equation of the kernel p^o becomes

$$
\begin{aligned}
(4.16) \qquad & \frac{\partial p^o(t,x,z)}{\partial t} - a(x)\frac{\partial^2 p^o(t,x,z)}{\partial x^2} - b(x)\frac{\partial p^o(t,x,z)}{\partial x} \\
& -a(z)\frac{\partial^2 p^o(t,x,z)}{\partial z^2} - b(z)\frac{\partial p^o(t,x,z)}{\partial z} \\
& + \int_{G_o} p^o(t,x,z_1)c(z_1)dz_1 \int_{G_o} p^o(t,z_2,z)c(z_2)dz_2 \\
& = \delta(x-z) + r(x,z), for\ (t,x,z) \in]0,\infty[\times]0,1[\times]0,1[
\end{aligned}
$$

$$
(4.17) \qquad p^o(t,x,0) = p^o(t,x,1) = 0, for\ (t,x) \in]0,\infty[\times]0,1[
$$

$$
(4.18) \qquad p^o(t,0,z) = p^o(t,1,z) = 0, for\ (t,z) \in]0,\infty[\times]0,1[
$$

Conclusions

In this paper we used the semigroup approach to formulate a stochastic parabolic partial differential equation with unknown boundary conditions in the state space form. This formulation is essential for developing a maximum likelihood state estimation algorithm which we discuss in the remainder of the paper. The algorithm is recursive and easy to implement in practice. It is interesting to see whether the method can be used for the state estimation problem in hyperbolic systems as well.

References

[1] S.I. AIHARA and A. BAGCHI. On maximum likelihood state estimation for stochastic parabolic systems with unknown boundary conditions, *submitted for publication.*

[2] S.I. AIHARA and A. BAGCHI. On the mortensen equation for maximum likelihood state estimation. *IEEE Trans on AC*, to appear.

[3] A. BAGCHI, R.C.W. STRIJBOS, and G. THE. Identification of a distributed-parameter systems with boundary noise. *INT. J. SYSTEM SCI.*, 11:49–56, 1980.

[4] A.V. BALAKRISHNAN. *Applied functional analysis.* Springer, Berlin, 1978.

[5] H.T. BANKS and K. KUNISCH. *Estimation techniques for distributed parameter systems.* Birkhäuser, Boston, 1989.

[6] A. BENSOUSSAN, G. Da PRATO, M.C. DELFOUR, and S.K. MITTER. *Representation and Control of Infinite Dimensional Systems Volume 1.* Birkhäuser, Boston, 1992.

[7] E. PARDOUX. *Equation aux dérivées partielles stochastiques non linéaires monotones.* Thèse, Université Paris XI, 1975.

[8] O. ZEITOUNI and A. DEMBO. An existence theorem and some properties of maximum a posteriori estimators of trajectories of diffusions. *Stochastics*, 23:197–218, 1988.

SHIN ICHI AIHARA, DEPARTMENT OF MANAGEMENT AND SYSTEMS SCIENCE, THE SCIENCE UNIVERSITY OF TOKYO, SUWA COLLEGE, 5000-1, TOYOHIRA, CHINO, NAGANO, 391-02, JAPAN
E-mail address: aihara@ss.suwa.sut.ac.jp

ARUNABHA BAGCHI, DEPARTMENT OF APPLIED MATHEMATICS, TWENTE UNIVERSITY, PO BOX 217, 7500AE, ENSCHEDE, THE NETHERLANDS
E-mail address: bagchi@math.utwente.nl

Progress in Nonlinear Differential Equations
and Their Applications, Vol. 42

SEMIGROUP THEORY IN AEROELASTICITY

A.V. BALAKRISHNAN
UCLA

Abstract. Aerolasticity mixes Structural Dynamics with Aerodynamics — a "Tale of Two Semigroups," so to speak. A fundamental problem — determining the bending-pitching wing-flutter speed in subsonic compressible flow — formulates as the asymptotic stability of the initial value problem for a Convolution-Semigroup equation in a Hilbert space of the form:

$$\dot{Y}(t) = \mathcal{A}Y(t) + \int_0^t L(t-\sigma)\dot{Y}(\sigma)\, d\sigma$$

where $\mathcal{A}$ is the infinitesimal generator of a C_0-semigroup with compact resolvent and $L(t)$ linear bounded for each t, and strongly continuous in $t \geq 0$, both to be deduced from the solution of a boundary value problem for subsonic compressible aerodynamics of interest in itself. The latter is the linearized version of the transonic inviscid compressible flow equation (we consider only the 2D version) for the velocity potential $\phi(x, z, t)$:

$$a_\infty^2(1-M^2)\frac{\partial^2\phi}{\partial t^2} + a_\infty^2\frac{\partial^2\phi}{\partial z^2} = \frac{\partial^2\phi}{\partial t^2} + 2Ma_\infty\frac{\partial^2\phi}{\partial t\partial x}$$

in the half-plane $-\infty < x < \infty$ and $0 < z < \infty$ with the (main) boundary condition

$$\frac{\partial\phi}{\partial z}(x, 0+, t) = w_a(t, x), \quad |x| < b < \infty$$

where the function $w_a(t, x)$ is specified, and

$$U = Ma_\infty$$

is the airspeed, M is the Mach number, assumed less than one. This is a mixed initial-value boundary-value problem in which the Semigroup solution of the homogeneous boundary value problem plays an essential role. We also show that the convolution-semigroup equation can be represented as a pure semigroup equation in the form

$$Y(t) = PZ(t)$$

$$\dot{Z}(t) = AZ(t).$$

1. Introduction

Applied problems do not unfortunately fit into a neat bundle of a cogent theorem statement and proof. Much of the effort is in the formulation, cutting through the special jargon of the application problem. The abstract formulation is however only the beginning. The heart of the work is often the hard analysis that follows. And however deep the theoretical solution, success has to be measured in terms of the impact back on the application problem, which is modest at best. The aeroelastic problem I want to discuss is no exception.

Aeroelasticity deals with the problem of determining the dynamic response of structures when subject to aerodynamic forces. The most interesting phenomenon is flutter — instability induced by the aerodynamics in an otherwise stable structure. A reasonably faithful mathematical model can be constructed for small disturbances — where both the structural dynamics and the aerodynamics can be described in terms of time-invariant linear partial differential equations. Each being individually a time-invariant initial-value/boundary-value problem, Semigroup Theory plays a fundamental role in both in the abstract formulation. Of great interest is whether a controller can be designed which would eliminate flutter — an open question still at the present time.

While there is a huge literature on structures and an equally huge literature on aerodynamics, the literature on aeroelastic control involving partial differential equations is relatively sparse. A basic reference is [1]. Of the two — the structures and the aerodynamics — the aerodynamic problems would appear to be more complex. We begin with the latter, limiting it further to the linearized model.

2. Linearized Subsonic Invsicid Compressible Flow Equations (2D): Mixed Initial Value-Boundary Value Problem

The (linearized) inviscid compressible subsonic flow equation for the velocity potential $\phi(x, z, t)$ in the half-plane (R^2_+), $-\infty < x < \infty,\ 0 < z < \infty$ is taken in the form:

$$a^2_\infty(1-M^2)\,\frac{\partial^2\phi}{\partial x^2} + a^2_\infty\,\frac{\partial^2\phi}{\partial z^2} = \frac{\partial^2\phi}{\partial t^2} + 2Ma_\infty\,\frac{\partial^2\phi}{\partial t\partial x} \tag{2.1}$$

(see [2] for more on the field equations), where a_∞ is the speed of sound at the given altitude, and M is Mach number, assumed less than one. The all-important boundary conditions to be imposed are:

(i) *Flow Tangency Condition*

$$\frac{\partial\phi}{\partial z}\,(x, 0+, t) = w_a(t, x), \quad |x| < b$$

where $w_a(t, x)$ is a given ("downwash") function (to be specified later).

(ii) *Kutta-Joukowski*

This condition is stated in terms of the acceleration potential $\psi(x, z, t)$ defined by

$$\psi(x, z, t) = \frac{\partial}{\partial t}\int_{-\infty}^{x} \frac{\partial\phi(y, z, t)}{\partial y}\,dy + U\,\frac{\partial\phi(x, z, t)}{\partial x}$$

where $U = a_\infty M$. We require that

$$\psi(x,0+,t) = 0, \quad x = b-, \ x \geq b \tag{2.1a}$$

($\equiv$ zero pressure jump at trailing edge)

(iii) *Far-Field Conditions*

$$\frac{\partial\phi}{\partial x} = 0 = \frac{\partial\phi}{\partial z} \qquad \text{as } |x| \to \infty \quad \text{or} \quad z \to \infty. \tag{2.1b}$$

We can formulate this more precisely, following [3]. The function space chosen is $L_2[R_+^2]^3$ with

$$f = \begin{vmatrix} f_1 \\ f_2 \\ f_3 \end{vmatrix}$$

with the norm-squared:

$$\|f\|^2 = \|f_1\|^2 + (1-M^2)a_\infty^2\|f_2\|^2 + a_\infty^2\|f_3\|^2$$

where

$$\|f_i\|^2 = \int_0^\infty \int_{-\infty}^\infty |f_i(x,z)|^2 \, dx \, dz.$$

We denote the corresponding Hilbert space by $\mathcal{H}$. The domain R_+^2 being unbounded, the usual trace definition for boundary values has to be eschewed. Here we define them in the L_2-sense. In this way we first look at the pure initial value problem, formulating it as a semigroup problem. The infinitesimal generator A is defined by:

$$\mathcal{D}(A) = \left[f = \begin{vmatrix} f_1 \\ f_2 \\ f_3 \end{vmatrix}, \quad \begin{matrix} \frac{\partial f_1}{\partial x}, \frac{\partial f_1}{\partial z} \in L_2[R_+^2] \\ \frac{\partial f_2}{\partial x}, \frac{\partial f_3}{\partial z} \in L_2[R_+^2] \end{matrix} \right]$$

with zero boundary values ($w_a(t,\cdot) = 0$), and

$$Af = \begin{vmatrix} -2U\frac{\partial}{\partial x} & (1-M^2)a_\infty^2 \frac{\partial f}{\partial x} & a_\infty^2 \frac{\partial}{\partial z} \\ \frac{\partial}{\partial x} & 0 & 0 \\ \frac{\partial}{\partial z} & 0 & 0 \end{vmatrix} \begin{vmatrix} f_1 \\ f_2 \\ f_3 \end{vmatrix}. \tag{2.2}$$

Note that the Kutta-Joukowski condition is *not* invoked. Thus defined A has a dense domain and is dissipative thereon, and A can be closed. The closure $\bar{A}$ ($= -A^*$) generates a C_0-semigroup $S(t)$, $t \geq 0$, (actually a group) such that

$$\|S(t)\| = 1.$$

The point spectrum of $\bar{A}$ contains only zero.

The subdomain where the Kutta-Joukowski condition

$$\begin{aligned} f_1(t,x,0+) + Uf_2(t,x,0+) &= 0 \quad \text{at } x = b- \\ &= 0 \quad x \geq b \end{aligned}$$

is satisfied is seen to be dense in $\mathcal{D}(A)$.

To obtain the solution to the combined initial value/boundary value problem, let A_L denote an extension of A where we drop the condition that the boundary value be zero:

$$A_L f = g$$

$$g = \begin{vmatrix} -2U\frac{\partial}{\partial x} & (1-M^2)a_\infty^2 \frac{\partial f}{\partial x} & a_\infty^2 \frac{\partial}{\partial z} \\ \frac{\partial}{\partial x} & 0 & 0 \\ \frac{\partial}{\partial z} & 0 & 0 \end{vmatrix} f.$$

Given $w(t,\cdot) \in L_2[-b,b]$, find $f(t)$ in $\mathcal{H}$ such that

(i) $A_L f(t) = 0$

(ii) $\displaystyle\int_{-b}^{b} |f_3(t,x,z) - w(t,x)|^2 \, dx \to 0 \quad \text{as } z \to 0+.$

Such a solution is obtained by taking

$$\begin{aligned} f_1(t,x,z) &= 0 \\ f_2(t,x,z) &= \frac{\partial}{\partial x}\phi(t,x,z) \\ f_3(t,x,z) &= \frac{\partial}{\partial z}\phi(t,x,z) \end{aligned}$$

where $\phi(t,x,z)$ is the solution of the potential equation

$$(1-M^2)\frac{\partial^2\phi}{\partial x^2} + \frac{\partial^2\phi}{\partial z^2} = 0, \quad |x| < \infty,\ 0 < z$$

with the boundary condition

$$\frac{\partial\phi}{\partial z}(t,x,0+) = w(t,x), \quad |x| < b,$$

and satisfying (2.1a) and (2.1b).

The combined solution then has the form:

$$x(t) = x_0(t) + x_M(t) \tag{2.3}$$

where assuming $w(t,\cdot)$ is strongly, continuously differentiable in $t > 0$, and

$$x_0(t) = S(t)(x(0) - x_M(0)) - \int_0^t S(t-\sigma)\,\dot{x}_M(\sigma)\,d\sigma \tag{2.4}$$

and

$$\phi(t,x,z) = \phi(0,x,z) + \int_0^t f_1(\sigma,x,z)\, d\sigma.$$

$$\frac{\partial\phi(0,x,z)}{\partial t} = f_2(0,x,z)$$

$$\frac{\partial\phi(0,x,z)}{\partial t} = f_3(0,x,z)$$

where

$$x(t) = \begin{vmatrix} f_1(t) \\ f_2(t) \\ f_3(t) \end{vmatrix}.$$

However it is required that $x_0(t)$ be in the domain of A, not in the domain of the closure $\bar{A}$. In addition we want the solution to satisfy the Kutta-Joukowsky condition (2.1a). It is not clear whether this is possible in general.

If we are interested only in the aeroelastic stability problem (Section 3) the aerodynamics initial conditions do not matter and any particular solution which satisfies the boundary conditions will do. There is another technique for obtaining such a particular solution which satisfies the boundary condition — flow tangency and Kutta-Joukowski — obtained via an integral equation in the frequency domain known as the Possio integral equation. This is what is generally used, not the semigroup solution above. This has the further advantage that it yields directly what is needed in the aeroelastic problem: the pressure-differential

$$\Delta P(t,x) = 2(-\rho) \lim_{z\to 0} \psi(t,x,z)$$

where ρ is the air density (see [1, 3] for more on this).

In fact for the aeroelastic problem we only need

$$F(t) = \int_{-b}^{b} \Delta P(t,x)\, dx \qquad \text{(Force per unit span length)} \tag{2.5}$$

and

$$M(t) = \int_{-b}^{b} (x-a)\, \Delta P(t,x)\, dx \quad \text{(Moment about the pitch axis).} \tag{2.6}$$

3. The Aeroelastic Stability Problem

Consistent with the 2D aerodynamic model we can use the Goland [1] structure model (long slender rectangular wing) for the aeroelastic problem. Thus we take the structure model with two degrees of freedom — "plunge" $h(\cdot)$ (displacement) and "pitch" $\alpha(\cdot)$ (torsion angle) yielding

$$x(t,s) \;=\; \begin{vmatrix} h(t,s) \\ \alpha(t,s) \end{vmatrix}, \quad 0 < t, \;\; 0 \le s \le L$$

(3.1) $$M_s \ddot{x}(t,s) \;+\; D_s\, \dot{x}\,(s,t) \;+\; K_s x(t,s) \;=\; \begin{vmatrix} F(t,s) \\ M(t,s) \end{vmatrix}$$

where

$$M_s \;=\; \begin{vmatrix} m & s \\ s & I_\alpha \end{vmatrix}$$

is the structure moment of inertia matrix assumed nonnegative definite, and nonsingular

$$D_s \;=\; 0$$

$$K_s \;=\; \begin{vmatrix} EI\,\frac{\partial^4}{\partial s^4} & 0 \\ 0 & -GJ\,\frac{\partial^2}{\partial s^2} \end{vmatrix}$$

with end conditions

$$h(t,0) \;=\; h'(t,0) \;=\; 0$$
$$h'''(t,L) \;=\; 0; \quad \alpha(t,0) \;=\; 0$$

and modeling self-straining controllers (if any) by

$$EIh''(t,L) \;+\; g_h h'(t,L) \;=\; 0$$

$$GJ\alpha'(t,L) \;+\; g_\alpha\, \dot{\alpha}(t,L) \;=\; 0$$

where g_h, g_α are the control gains and may be set to zero if there are no controls.

The aerodynamic "loads," $F(t,\cdot)$, $M(t,\cdot)$ depend now also on the spanwise coordinate, s, since the downwash function $w_a(t,\cdot)$ now depends also on s. In fact the latter is given by:

$$w_a(t,x,s) \;=\; -\,\dot{h}(t,s) \;-\; (bx-a)\,\dot{\alpha}(t,s) \;-\; U\alpha(t,s), \quad 0 < t,\; 0 < s < L$$

where a, b are given parameters.

The aerodynamical "load" functions $F(t,\cdot)$ and $M(t,\cdot)$ can be deduced from the aerodynamic analysis in Section 2:

$$\begin{vmatrix} F(t,s) \\ M(t,s) \end{vmatrix} \;=\; M_a \ddot{x}(t,s) + UD_a\, \dot{x}\,(t,s) + U^2 K_a x(t,s) + \int_0^t L_c(t-\sigma)\, \dot{y}\,(\sigma,s)\, d\sigma$$
(3.2)

where

$$y(t,s) = \left| \begin{matrix} x(t,s) \\ \dot{x}(t,s) \end{matrix} \right|$$

and M_a , D_a , K_a are matrices such that $(M_s - M_a)$ is nonsingular and positive definite.

The abstract formulation of the problem if we wish to accommodate the controls as well, becomes a little involved. Here, since we are not concerned with presenting numerical results, we shall concentrate on the abstract problem it leads to, referring to [4] for details. Thus we are led to the following problem. Let A denote the infinitesimal generator of a C_0-semigroup with compact resolvent. The aeroelastic equation translates to the Hilbert space Convolution-Semigroup equation:

$$\left.\begin{aligned} \dot{Y}(t) &= AY(t) + \int_0^t L(t-\sigma)\,\dot{Y}(\sigma)\,d\sigma, \quad t>0 \\ & Y(0) \text{ given} \end{aligned}\right\} \tag{3.3}$$

where $L(t)$ is a strongly continuous linear bounded operator function such that

$$\int_0^\infty e^{-\sigma t}\,\|L(t)\|\,dt < \infty \qquad \text{for every } \sigma > 0$$

with the characteristic feature that:

$$\hat{L}(\lambda)Y = \int_0^\infty e^{-\lambda t}\,L(t)Y\,dt, \quad \operatorname{Re}\lambda > 0$$

(can be extended to be) analytic in λ except for a branch cut along the negative real axis. Note that this is *not* an evolution equation. Our main interest (in addition to existence and uniqueness of solution) is in the asymptotic (strong) stability of the solution (behavior of $Y(t)$ as t goes to infinity for each $Y(0)$) depending on the air speed U. The key quantity here is the operator function of λ:

$$\lambda I - A - \lambda\hat{L}(\lambda)$$

and a key result is that it has a bounded inverse except for a sequence $\{\lambda_k\}$ such that

$$\left(\lambda_k I - A - \lambda_k \hat{L}(\lambda_k)\right) Y_k = 0$$

where the $\{\lambda_k\}$ (which depend on U) are called the aeroelastic modes (see [1]) and do not have an accumulation point in the finite part of the plane. It is natural to call

$$R(\lambda) = \left(\lambda I - A - \lambda\hat{L}(\lambda)\right)^{-1}$$

the "resolvent" of (3.3). It is analytic except for simple poles at the aeroelastic modes and a branch cut along the negative real axis. For each U, only a finite number of the modes can have positive real parts. In fact defining the inverse Laplace transform in the usual way:

$$W(t)Y = \frac{1}{2\pi i} \oint_{\gamma - i\infty}^{\gamma + i\infty} e^{\lambda t} \mathcal{R}(\lambda) Y \, d\lambda$$

we have

$$Y(t) = W(t) Y(0) - \int_0^t W(t-\sigma) L(\sigma) Y(0) \, d\sigma$$

solving (3.3). What is more important (since it answers the stability question) is that we have the expansion of the form:

$$W(t)Y = \sum_0^\infty e^{\lambda_k t} [Y, Z_k] Y_k + \frac{1}{\pi} \int_0^\infty e^{-rt} (\operatorname{Im} \mathcal{R}(-r) Y) \, dr. \tag{3.4}$$

It should be noted that in this representation the $\{Y_k\}$, $\{Z_k\}$ are not biorthogonal. Such a solution (especially including the integral term) for the finite-dimensional structure model (the $\{\lambda_k\}$ are finite in number) is given in [5].

4. Representation Theory

By keeping the aerodynamic problem (2.1) and the structures problem (3.1) in parallel, rather than in series as is traditional (as in [1]), resulting in (3.3) as a stand-alone convolution-semigroup problem, it is possible to obtain a pure semigroup problem. This has the benefit as well of yielding an interpretation of the aeroelastic modes $\{\lambda_k\}$ as eigenvalues of an infinitesimal generator corresponding to the combined problem — explaining further why $\{Y_k\}$, $\{Z_k\}$ are not biorthogonal.

However the procedure is a bit too long to explain in this limited space. Rather we shall only include here how to obtain such a representation more abstractly but directly from (3.3). Even in this because of the space limitation we shall need to be sketchy. Basically we follow [6] for handling the second term in (3.3). Assume for simplicity that $L(t)$ is Hilbert-Schmidt for each $t \geq 0$ and that

$$\int_0^\infty \|L(t)\|^2 \, dt < \infty$$

where $\|\cdot\|$ denotes the Hilbert-Schmidt norm, and consider the L_2-space $\mathcal{H}_L$ of such functions. Introduce the shift semigroup $S_L(t)$, $t > 0$, on $\mathcal{H}_L$ defined by

$$S_L(t)L = L(t + \cdot)$$

with generator A_L. Define the unbounded unclosable operator C on the subspace of functions continuous at zero in $\mathcal{H}_L$ by

$$CL = L(0).$$

Then we can write

$$\begin{aligned}\int_0^t L(t-\sigma)\,\dot{Y}(\sigma)\,d\sigma &= C\int_0^t S_L(t-\sigma)\,\dot{Y}(\sigma)\,d\sigma \\ &= Cx(t)\end{aligned}$$

where $x(t)$ is the generalized solution of

$$\begin{aligned}\dot{x}(t) &= A_L x(t) + L\dot{Y}(t) \\ x(t) &= 0.\end{aligned}$$

Hence (3.3) becomes

$$\left.\begin{aligned}\dot{Y}(t) &= \mathcal{A}Y(t) + Cx(t) \\ \dot{x}(t) &= A_L x(t) + L\dot{Y}(t)\end{aligned}\right\}. \tag{4.1}$$

Let

$$Z(t) = \left|\begin{matrix} x(t) \\ Y(t)\end{matrix}\right|$$

and define, as before:

$$BZ = \left|\begin{matrix} LY \\ 0\end{matrix}\right| \qquad \text{for } Z = \left|\begin{matrix} x \\ Y\end{matrix}\right|.$$

Then we can express (3.3) as:

$$\dot{Z}(t) = (I-B)\left|\begin{matrix} A_L & 0 \\ C & \mathcal{A}\end{matrix}\right| Z(t); \quad Y(t) = PZ(t)$$

which is our semigroup equation again, where the pregenerator can be shown to be closable. Of course the procedure for determining $\{\lambda_k\}$ is still the same, in both cases.

5. Conclusion

What is of primary engineering interest in this problem is to determine the "flutter boundary" [7]. As noted in the Introduction, this part of the problem, while it has to draw on our analysis, is of a different nature. One looks in vain for a precise mathematical definition of flutter — even plain "wing" flutter — in the engineering literature. For the *linear* model (as here) we can follow the definition given in [3]. The real part of λ_k is zero for zero U, and it can be shown that $\operatorname{Re}\lambda_k$ decreases as U increases from zero. The flutter speed may be defined as the minimal value of U at which $\operatorname{Re}\lambda_k$ again becomes zero for some (any) k. At present the flutter speed has to be determined by numerical calculations for given structure parameters; no "analytical" formulas are known. The rate controllers as in [4], while they serve to increase damping, do not seem to affect the flutter speed. Indeed control design for flutter boundary enhancement is largely open.

References

[1] R.L. Bisplinghoff, H. Ashley and R.L. Halfman. *Aeroelasticity.* Addison-Wesley Publishing Company, Inc., 1955.

[2] D. Nixon, "Basic Equations for Unsteady Transonic Flow." Chapter 2 in: *Unsteady Transonic Aerodynamics.* Edited by D. Nixon. American Institute of Aeronautics and Astronautics, Inc., 1989. Pp. 57–73.

[3] A.V. Balakrishnan. "Unsteady Aerodynamics — Subsonic Compressible Inviscid Case." NASA Contractor Report, NASA/CR-1999-206583. 1999.

[4] A.V. Balakrishnan. "Aeroelastic Control with Self-Straining Actuators." In: *Proceedings of the Second International Conference on Nonlinear Problems in Aviation and Aerospace.* Edited by S. Sivasundaram. Embry-Riddle Aeronautical University, 1999.

[5] J.W. Edwards, H. Ashley and J.V. Breakwell. "Unsteady Aerodynamic Modeling for Arbitrary Motions." *AIAA Journal*, Vol. 17, No. 4 (April 1979), pp. 365–374.

[6] A.V. Balakrishnan. "Stochastic Filtering and Control of Linear Systems: A General Theory." In: *Proceedings of ONR Conference on Control Theory of Systems Governed by Partial Differential Equations.* Academic Press, 1976.

[7] J.W. Edwards. "Computational Aeroelasticity." Chapter 7 in: *Unsteady Transonic Aerodynamics.* Edited by D. Nixon. American Institute of Aeronautics and Astronautics, Inc., 1989.

A.V. Balakrishnan, Department of Electrical Engineering, UCLA, 405 Hilgard Ave., Los Angeles, CA 90095, U.S.A.

E-mail address: bal@ee.ucla.edu

Progress in Nonlinear Differential Equations
and Their Applications, Vol. 42

B-BOUNDED SEMIGROUPS, EXISTENCE FAMILIES AND IMPLICIT EVOLUTION EQUATIONS

J. BANASIAK
UNIVERSITY OF NATAL

1. Introduction

Let us consider a standard abstract Cauchy problem in a Banach space X:

$$\frac{du}{dt} = Au, \quad \lim_{t\to 0^+} u(t) = \overset{\circ}{u} . \tag{1.1}$$

Very often the existence of the semigroup $(\exp(tA))_{t\geq 0}$ describing the evolution of the system is established in a non-constructive way. This is especially the case when the positivity methods are employed, see e.g. [1]. Then, very little quantitative information on the evolution is available. On the other hand, there may exist an operator B such that $t \to Be^{tA}$ can be calculated constructively yielding some information about the evolution (note the similarity of this reasoning with that leading to C-semigroups, e.g. [9]; the final result is, however, different, see Section 5). An interesting example of this type, pertaining to the transport equation with multiplying boundary conditions, was analysed in [13] and has prompted one of the authors to define a class of evolution families which behave well if looked at through the "lens" of another operator. Such families, called B-bounded semigroups, have been introduced in [6], and analysed and applied to various problems in a few papers [2, 3, 7, 8]. In this paper we present some developments of the theory, discuss the relations between B-bounded semigroups and C-existence families, and also sketch applicability of B-bounded semigroups to solving implicit evolution equations.

An extended version of this talk together with proofs will appear elsewhere (see [4, 5]).

2. B-Bounded semigroups revisited

The definition of B-quasi bounded semigroups [7] (with some modifications) reads as follows.

Definition 2.1 *Let $(A, D(A))$ be a linear operator in a Banach space X and $(B, D(B))$ be another linear operator from X to another Banach space Z such that for some $\omega \in \mathbf{R}$ the operator*

$$(\lambda I - A) : D_B(A) \to D(B) \tag{2.1}$$

where $D_B(A) = \{x \in D(A) \cap D(B); Ax \in D(B)\}$, is bijective for all $\lambda > \omega$.

A one-parameter family of operators $(Y(t))_{t\geq 0}$ from X to Z, which satisfies

1. $D(Y(t)) =: \Omega \supseteq D(B)$, and for any $t \geq 0$ and $f \in D(B)$

$$\|Y(t)f\|_Z \leq M\exp(\omega t)\|Bf\|_Z, \tag{2.2}$$

2. The function $t \to Y(t)f \in C([0,\infty[, Z)$ for any $f \in \Omega$,

3. For any $f \in \Omega_0 := \{f \in D(A) \cap D(B); Af \in \Omega\} \subset D(A) \cap D(B)$

$$Y(t)f = Bf + \int_0^t Y(s)Af\,ds, \quad t \geq 0, \tag{2.3}$$

is called a B-quasi bounded semigroup generated by A and B.

If the A generates a B-bounded semigroup, then we write $A \in B - \mathcal{G}(M,\omega, X,Z)$. We also use the notation $A \in \mathcal{G}(M,\omega,X)$ to express the fact that A is the generator of a C_0-semigroup in X with the Hille-Yoshida constants M and ω.

The main role in the considerations is played by the space X_B which is the completion of $D(B)$ with respect to the semi-norm $\|\cdot\|_B = \|B\cdot\|_Z$. The operator B can be extended by density to an isometry $\mathbf{B}\colon X_B \to Z$. It follows that if A generates a B-bounded semigroup, then A preserves co-sets of $D(B)/N(B)$ and therefore it can be defined to act in X_B. Moreover, in such a case A is closeable in X_B; we denote this closure of $\mathbf{A}$. Let us define $Z_B = \overline{ImB}$ (the closure of the range of B in Z).

The main result of [3] reads as follows.

Theorem 2.1 *If $A \in B - \mathcal{G}(M,\omega,X,Z)$ and $\overline{B[D_B(A)]}^Z = Z_B$, then $\mathbf{A} \in \mathcal{G}(M,\omega,X_B)$. Conversely, if there is $\mathcal{A} \supset A$ such that $\mathcal{A} \in \mathcal{G}(M,\omega,X_B)$, then $\mathcal{A} = \mathbf{A}$ and $A \in B - \mathcal{G}(M,\omega,X,Z)$. The B-bounded semigroup $(Y(t))_{t\geq 0}$ for $x \in D(B)$ is given by*

$$Y(t)x = \exp(t\mathbf{BAB}^{-1})Bx = \mathbf{B}\exp(t\mathbf{A})x. \tag{2.4}$$

The assumption that $B[D_B(A)]$ is dense in Z_B can be discarded if Z (and consequently Z_B) are reflexive spaces ([3], Corollary 4.1). Moreover it can be proved [2] that the density is equivalent to the condition

$$\forall_{x\in D(B)} Y(0) = Bx. \tag{2.5}$$

Since the space X_B is in many cases rather difficult to handle, Theorem 2.1 is most often used in the following version [3, 2].

Theorem 2.2 *Let operators A and B satisfy the conditions of Definition 2.1. Then A is the generator of a B-quasi bounded semigroup satisfying (2.5) if and only if the following conditions hold:*

1. $B[D_B(A)]$ is dense in Z_B,

2. There exist $M > 0$ and $\omega \in \mathbf{R}$ such that for any $x \in D(B)$, $\lambda > \omega$ and $n \in \mathbf{N}$:

$$\|B(\lambda I - A)^{-n}x\|_Z \leq \frac{M}{(\lambda - \omega)^n}\|Bx\|_Z. \tag{2.6}$$

If we don't require $(Y(t))_{t\geq 0}$ to satisfy (2.5), then condition (1.) is sufficient but not necessary.

3. Further improvements of the generation theorem

Example 3.1 Let us consider $X = L_2(\mathbf{R}, e^{x^2}dx)$, $Au = \partial_x u$ on the maximal domain, and $(Bu)(x) = e^{-\frac{x^2}{x}}u(x)$. Clearly, $B : X \to X$ is a continuous operator. Moreover, $\|Bu\|_X = \|u\|_{L_2(\mathbf{R})}$ and since $C_0^\infty(\mathbf{R}) \subset X$, we can identify X_B with $L_2(\mathbf{R})$. Let us consider the closure $\mathbf{A}$ of A, that is, we take a sequence $(u_n)_{n\in\mathbf{N}}$ of elements of $D(A)$ such that $u_n \to u$ and $\partial_x u_n \to g$ as $n \to \infty$ in $L_2(\mathbf{R})$. However, this is the same as the closure of $D(A)$ in $W_2^1(\mathbf{R})$, and as $C_0^\infty(\mathbf{R}) \subset D(Z)$ is dense in $W_2^1(\mathbf{R})$, we obtain that $\mathbf{A}u = \partial_x u$ for $u \in W_2^1(\mathbf{R})$. thus, $\mathbf{A}$ generates a semigroup of contractions in X_B and therefore $(Y(t)u)(x) = e^{-\frac{x^2}{2}}u(t+x)$ satisfies conditions (1.)–(3.) of Definition 2.1.

On the other hand, since the formally adjoint equation $\lambda\Phi + 2x\Phi + \partial_x\Phi = 0$ has a solution $\Phi(x) = e^{-x^2-\lambda x}$ which can be proved to belong to the domain of a^*, $(\lambda I - A) : D(A) \to X$ is not a bijective operator. □

Note that this example was used in [9] to motivate the introduction of C-existence and uniqueness families. Here we have seen an alternative way of regularizing this problem.

Example 3.1 shows that we should be able to replace the assumption 2.1 by one which requires only the bijectivity of a suitable extension of A. Thus, what we really need is that the Hille-Yoshida estimate holds on some dense subspace $\mathbf{X}$ of X_B. Since we are using the pseudo resolvent identity, we need that $D_\lambda = (\lambda I - A)^{-1}\mathbf{X} \subset \mathbf{X}$ for $\lambda > \omega$. Finally, as our starting point are X and the operators defined in it, the space $\mathbf{X}$ must be accessible from X in the sense of the operator closure in X_B. This also shows that we should be able to free ourselves from the topological structure of X.

We assume therefore that

(2.1′) X is a linear space and the operator A_B is closeable in X_B. Denoting $\mathbf{A} = \overline{A_B}^{X_B}$, we assume further that there exist subspaces: $\mathbf{X}$ satisfying $D(B) \subseteq \mathbf{X} \subseteq X_B$, and $D_B(A) \subseteq D \subseteq \mathbf{X} \cap D(\mathbf{A})$ such that $(\lambda - \mathbf{A}|_D) : D \to \mathbf{X}$ is bijective for all $\lambda > \omega$.

Theorem 3.1 *Let the operators A and B satisfy the conditions of Definition 2.1 with assumption (2.1) replaced by assumption (2.1′). Then $A \in B - \mathcal{G}(M, \omega, X, Z)$ and (2.5) holds if and only if the following conditions are satisfied:*

1. $B(D)$ is dense in Z_B,

2. There exist $M > 0$ and $\omega \in \mathbf{R}$ such that for any $\mathbf{y} \in \mathbf{X}$, $\lambda > \omega$ and $n \in \mathbf{N}$:

$$\|\mathbf{B}(\lambda I - \mathbf{A}|_D)^{-n}\mathbf{y}\|_Z \leq \frac{M}{(\lambda-\omega)^n}\|\mathbf{B}\mathbf{y}\|_Z. \tag{3.1}$$

The B-bounded semigroup is generated by $\mathbf{A}$ which is also equal to $\overline{\mathbf{A}|_D}^{X_B}$.

If we don't assume (2.5), then the assumption (1.) is sufficient but not necessary.

This theorem shows, in particular, that the bounded semigroup $(Y(t))_{t\geq 0}$ is uniquely determined by A and B, being defined by the semigroup generated by the B-closure of $A|_{D_B(A)}$.

4. A special case

In various applications it is important to know when X_B can be treated as a subspace of X.

Theorem 4.1 *Let X,Z be Banach spaces and let $B : X \to Z$ be an injective operator. The following conditions are equivalent:*

(i) X_B has the following properties:

(i') each co-set $\tilde{x} \in X_B$ contains a sequence $(x_N)_{n\in\mathbf{N}}$ converging in the norm of X to some $x \in X$, and x is the limit of any other X-Cauchy sequence in $\tilde{x}$,

(i'') if $(x_n)_{n\in\mathbf{N}} \in \tilde{x}$, $(y_n)_{n\in\mathbf{N}}$ satisfy $\|x_n - y_n\|_X \to 0$ as $n \to \infty$ and $(y_n)_{n\in\mathbf{N}} \in \tilde{y}$ for some $\tilde{y} \to X_B$, then $\tilde{x} = \tilde{y}$,

(ii) the operator B is closeable and B^{-1} is bounded,

(iii) there is an isomorphism $T : X_B \to X'_B \hookrightarrow X$ which satisfies $T|_{D(B)} = Id$.

In the discussed case the operator $\mathbf{A}$ also becomes much simpler.

Theorem 4.2 *If B is a closeable operator such that B^{-1} is bounded, and A is closeable in X wigh $\lambda I - \overline{A}$ injective for some $\lambda > \omega$, and moreover $A \in B - \mathcal{G}(M,\omega,X,Z)$, then*

$$\mathbf{A} = \overline{A}\,|_{D_{\overline{B}}(\overline{A})} \tag{4.2}$$

where

$$D_{\overline{B}}(\overline{A}) = \{x \in D(\overline{A}) \cap D(\overline{B}); Ax \in D(\overline{B})\}.$$

Example 4.1 Let us consider $X = L_2(\mathbf{R}, e^{-x^2}dx)$, $Au = \partial_x u$ on the maximal domain, and $(Bu)(x) = e^{\frac{x^2}{x}}u(x)$. $B : X \to X$ is an unbounded operator and since $\|Bu\|_X = \|u\|_{L_2(\mathbf{R})}$, we see that $D(B) = L_2(\mathbf{R})$. Since $B(D(B)) = X$, we obtain that $X_B = D(B) = L_2(\mathbf{R})$ by Theorem 4.1. Then $D_B(A) = W_2^1(\mathbf{R})$ and A generates a contraction semigroup, say $(T(t))_{t\geq 0}$, in $L_2(\mathbf{R})$. Thus $Y(t)u = BT(t)u = e^{\frac{x^2}{2}}u(t+x)$ is the B-bounded semigroup generated by (A, B).

Note that here neither $D(A) \subset D(B)$, nor $\rho(A) \supset [\omega, \infty[$ (see similar considerations in Example 3.1), but the assumption (2.1) is satisfied.

5. Relation to C-regularized semigroups and similar objects

The philosophy and appearance of B-bounded semigroups are similar to C-semigroups and related objects, like C-existence and uniqueness families. It follows however that these objects are quite different, as we shall see, to the extent that the only objects which can be simultaneously C-semigroups and B-bounded semigroups in the same space X are necessarily defined by C_0-semigroups in X.

Theorem 5.1 *Let us assume that $A : X \to X$ is a closed operator, such that $[\omega, \infty[$ does not contain its eigenvalues, $B\colon X \to X$ is a bounded operator with the range ImB dense in X and $(W(t))_{t\geq 0}$ is a mild B-existence family for A. Then $(W(t))_{t\geq 0}$ is a B-bounded semigroups $(Y(t))_{t\geq 0}$ satisfying (2.5), generated by some operator $\mathcal{D}$ if and only if A generates a semigroup in X. In such a case*

$$\forall_{t\geq 0, x\in X} W(t)x = e^{tA}Bx = e^{t\mathbf{BDB}^{-1}}Bx = Y(t)x, \tag{5.3}$$

where $\mathbf{B}, \mathbf{D}$ are the closures of B and $\mathcal{D}$, respectively, in X_B.

A little more intrinsic link between C-existence families and B-bounded semigroups can be found by noting that any B-bounded semigroup gives rise to some evolution governed by a semigroup in X_B. Since the very concept of C-existence families is that they provide solutions for the initial values taken from a subspace of the original space X, we are placed in the very restricted situation described in Theorem 4.1 with $C = \overline{B}^{-1}$. In any case this suggests that evolution families are related to C^{-1}-bounded semigroups rather than C-bounded semigroups and with little algebra one can find that this relation must be of the form (5.4). The following theorem shows however that this places a stiff restriction on A.

Theorem 5.2 *Assume that $A : D(A) \to X$ is a closed operator with no eigenvalues in $]\omega, \infty[$ for some $\omega \in \mathbf{R}$, B^{-1} is densely defined and $X_B \hookrightarrow X$. Let $(W(t))_{t\geq 0}$ be a mild $\overline{B}^{-1}$-existence family for A. The formula*

$$Y(t)x = \overline{B}W(t)\overline{B}x \tag{5.4}$$

defines a $\overline{B}$-bounded semigroup if and only if $\overline{B}A\overline{B}^{-1}$ generates a C_0-semigroup in X and $(W(t))_{t\geq 0}$ is exponentially bounded. Then $\overline{BAB^{-1}} = \overline{B}A\overline{B}^{-1}$ and $(Y(t))_{t\geq 0}$ is generated by A.

If moreover $(W(t))_{t\geq 0}$ is a $\overline{B}^{-1}$ regularized semigroup generated by A, then the formula (5.4) defines a $\overline{B}$-bounded semigroup if and only if $(\overline{B}W(t))_{t\geq 0}$ is a semigroup in X generated by A.

6. B-bounded semigroups and implicit evolution equations

In this section we shall see that B-bounded semigroups are solutions to implicit evolution equations.

Let us consider the Cauchy problem for the implicit evolution equation

$$\frac{d}{dt}(Ku) = Lu, \quad \lim_{t\to 0+}(Ku)(t) = \mathring{u} \tag{6.1}$$

where $K: Z \to X$, $L: Z \to X$, Z, X are Banach spaces, and K, L are linear operators. There are a number of approaches to solving such problems. We do not aim at the comprehensive comparison of these methods, but rather we describe how B-bounded semigroups can be used to solve (6.1) and what are the connections between B-bounded semigroups and both the empathy and Showalter's method, which seem to be similar in spirit.

A "natural" way of approaching (6.1) would be to factor out K and, provided it is invertible, to consider a standard Cauchy problem with the operator $K^{-1}L$ on the right-hand side. In some cases, however, the operator K is not closeable and therefore there is no way the "time derivative" and the limit at $t = 0^+$ can commute with K. Thus, it is reasonable to study (6.1) as it stands. In [11] this problem is solved by introducing a pair of evolution families. As we shall see, the theory of B-bounded semigroups provides another convenient way of performing this "impossible" commutation by passing to a specially constructed space related to K. In this it is similar to the approach of [12] but some assumptions on K seem to be less restrictive.

6.1. X_B-solutions of implicit evolution equations

It is often the case that the original spaces X and Z are not the most convenient spaces from the mathematical point of view. We are usually interested to keep the values of the solution in the original space which may be related to some physical properties, like the finite total energy space, finite mass, etc., but for (6.1) to hld in the strict sense may be too restrictive and it is enough for it to hold for extensions, say $\tilde{K}$ and $\tilde{L}$, of the operators K and L. However, to have some link between K and L and these extensions, $D(K)$ and $D(L)$ should be cores for $\tilde{K}$ and $\tilde{L}$. This concept is related to so-called Haraux extrapolation in the theory of semigroups.

Moreover, thanks to Theorem 3.1, there is no need to assume any topological structure in X and therefore the operatrors K and L need not have any properties related to X, like closedness, etc.

We introduce the following definition.

Definition 6.1 *Let $X \hookrightarrow \tilde{X}$ and $\tilde{L} - \overline{L}^{\tilde{X}}$, $\tilde{K} = \overline{K}^{\tilde{X}}$. A Z-valued function $t \to u(t)$ is called an $\tilde{X}$-solution of the problem (6.1) if it is a classical solution of the problem*

$$\frac{d}{dt}(\tilde{K}u) = \tilde{L}u, \quad \lim_{t\to 0^+}(\tilde{K})(t) = \overset{\circ}{u}, \tag{6.2}$$

that is, $t \to \tilde{K}u(t)$ is continuously differentiable in $\tilde{X}$, the equality holds for all $t > 0$ in $\tilde{X}$, and the initial condition holds as a limit in the topology of $\tilde{X}$.

With this definition we can formulate the following theorem.

Theorem 6.1 *Let us suppose that we are given operators $K\colon D(K) \to X$ and $L\colon D(L) \to X$ with $D(L)$, $D(K) \subset Z$, Z, X-Banach spaces. Assume that K is a densely defined, one-to-one operator. Define $A = LK^{-1}$ with the natural domain $D(A) = K(D(L))$ and $B = K^{-1}$. If $A \in B - \mathcal{G}(M, \omega, X, Z)$, then, for any $x \in D_B(A) = \{x \in K(D(L) \cap D(K)); LK^{-1}x \in ImK\}$, the function $t \to Y(t)x$, where $(Y(t))_{t\geq 0}$ is the B-bounded semigroup generated by A, is an X_B-solution of the problem (6.1).*

Combining this theorem with Theorem 3.1 we obtain the following result.

Theorem 6.2 *Assume that*

1. $D(K^{-1}L) = \{y \in D(L) \cap D(K); Ly \in ImK\}$ is dense in $\overline{D(K)}^{Z}$,

2. The operator $(K^{-1}L, D(K^{-1}L))$ is closeable in Z and there exist spaces: $\mathbf{X}$ satisfying for $D(K) \subset \mathbf{X} \subset Z$, and D satisfying $D(K^{-1}L) \subset D \subset D(\overline{K^{-1}L})$ such that for $x \in \mathcal{X}$ either

$$\left\| \left(\lambda I - \overline{K^{-1}L}\,|_D\right)^{-1} x \right\|_Z \leq \lambda^{-1}\|x\|_Z, \quad \lambda > 0, \tag{6.3}$$

or

$$\left\| \left(\lambda I - \overline{K^{-1}L}\,|_D\right)^{-1} x \right\|_Z \leq \frac{M}{|\lambda - \omega|}\|x\|_Z, \quad \lambda \in S_\theta, \tag{6.4}$$

then, for any $x \in D(LK^{-1})$, the function $t \to Y(t)x$ is an X_B-solution to (6.1). For $x \in D(LK^{-1})$ the classical solution is given by

$$u(t,x) = e^{t\overline{K^{-1}L}}K^{-1}x. \tag{6.5}$$

In reflexive spaces assumption (1.) is superfluous.

6.2. B-bounded semigroups and empathy

Let X and Z be Banach spaces and consider two families of operators $\mathcal{E} = \{E(t) : X \to X\}_{t>0}$ and $\mathcal{S} = \{S(t) : X \to Z\}_{t>0}$ such that the Laplace transforms: $R(\lambda)y = \mathcal{L}(E(t)y)(\lambda)$ and $P(\lambda)y = \mathcal{L}(S(t)y)(\lambda)$ exist for any $y \in X$ and $\lambda > 0$. The pair $<\mathcal{S}, \mathcal{E}>$ is called an *empathy* if for any $s, t > 0$

$$S(t+s) = S(t)E(s)$$

and $P(\xi)$ is invertible for some $\xi > 0$. It can be proved that then $P(\lambda)$ is invertible for any $\lambda > 0$, and the same is valid for $R(\lambda)$. Also, the subspaces $\mathcal{D}_E = R(\lambda)X$ and $\mathcal{D} = P(\lambda)Z$ are independent of λ. Let us define the operators: $K: \mathcal{D} \to \mathcal{D}_E$ by $K = R(\lambda)P^{-1}(\lambda)$ and $L: \mathcal{D} \to X$ by $L = (\lambda R(\lambda) - I)P^{-1}(\lambda)$; both can be proved to be independent of λ and $P(\lambda) = (\lambda K - L)^{-1}$. The pair $<L,K>$ is called the generator of empathy $<\mathcal{S},\mathcal{E}>$. In this case, for any $y \in \mathcal{D}_E$, the function $t \to S(t)y$ is a solution of the Cauchy problem (6.1).

We call the empathy $<\mathcal{S},\mathcal{E}>$ is uniformly bounded if for some N', M', $\sup_{t>0} \|S(t)\| \leq M'$, $\sup_{t>0} \|E(t)\| \leq N'$. Then the following theorem is valid.

Theorem 6.3 *Suppose that the space X has the Radon-Nikodym property. The operator pair $<L,K>$ is the generator of a uniformly bounded empathy $<\mathcal{S},\mathcal{E}>$ if and only if the operators $P(\lambda)$ and $R(\lambda)$ are bounded for every $\lambda > 0$ and there exist positive numbers M,N such that for every $\lambda > 0$ and $k = 1, 2, \ldots$*

$$\|(\lambda K - L)^{-1}\| \leq M\lambda^{-1}, \quad \|K^k(\lambda K - L)^{-k}\| \leq N\lambda^{-k}. \tag{6.6}$$

To compare empathy with B-bounded semigroups we note first that if $<L,K>$ is to be the generator of an empathy, K by definition must be an injective operator. Moreover, since by definition the solution family $(S(t))_{t\geq 0}$ is a family of bounded operator in X, and we should have $(S(t))_{t\geq 0} = (Y(t))_{t\geq 0}$, $B = K^{-1}$ must be a bounded operator.

We have the following theorem.

Theorem 6.4 *Let $(Y(t))_{t\geq 0}$ be a B-bounded semigroup of generated by (A, B) where $A: D(A) \to X$, $B: D(B) \to Z$, $D(A), D(B) \subset X$, $\rho(A) \supset]0, \infty[$, and B is a bounded, one-to-one operator. The pair $<(Y(t))_{t\geq 0}, (e^{t\mathbf{A}}|_X)_{t\geq 0}>$ is an empathy generated by $<AB^{-1}, B^{-1}>$ if and only if*

$$\forall_{t\geq 0}\, e^{t\mathbf{A}} X \subset X \tag{6.7}$$

and

$$\forall_{x\in X, \lambda>0}\, t \to e^{-\lambda t}(e^{t\mathbf{A}}|_X)x \in L_1(0, \infty, X). \tag{6.8}$$

It is easy to observe that the pair $<(Y(t))_{t\geq 0}, (\exp(t\mathbf{A}))_{t\geq 0}>$ is always an empathy in the pair of spaces Z, X_B. This suggests that if one could find a space $X' \subset X$ such that X is a completion of X' with respect to the norm $\|K^{-1} \cdot\|_Z$, then the notion of B-bounded semigroup in the pair X', Z would coincide with the notion of empathy in the pair of spaces X, Z. However, the following proposition holds.

Proposition 6.1 *The pair $<(Y(t))_{t\geq 0}, (\exp(t\mathbf{A}))_{t\geq 0}>$ is an empathy in Z,X if and only if K is a bounded operator, with a bounded densely defined inverse K^{-1}.*

Note, however, that this result is not as restrictive as it looks as we have some freedom of choice for the spaces X and Z.

6.3. Showalter's approach

The approach presented in [12] is very similar in spirit to the ideas leading to construction of a B-bounded semigroup. In fact, the theory presented there, if restricted to linear operators, can be treated as a special case of the approach presented in this Section, as follows from the discussin below.

It is assumed in [12] that $K : E \to E^*$ is a linear, symmetric and non-negative operator from a linear space E to its algebraic dual. This allows to introduce a semi-scalar product on E; the corresponding semi-normed space is denoted by E_b. Its continuous dual E_b' is a Hilbert space. The operator L is assumed to act from a domain of E_b into E_b'.

In a particular case when K generates a scalar product under which E_b is complete, K becomes the Riesz isomorphism onto E_b'. Thanks to this one can proceed as in Theorem 6.1 converting the problem (6.1) into the standard Cauchy problem

$$\frac{du}{dt} = K^{-1}Lu, \qquad \lim_{t\to 0^+} u(t) = K^{-1}\overset{\circ}{u}$$

in E_b. Hence, the question to be solved is whether $K^{-1}L$ generates a semigroup in E_b.

Using the B-bounded semigroup approach we obtain the same problem under the following particular assumptions. Let $D(K) \subset X$ and $(K, D(K))$ be a symmetric, positive operator in X. Assume that E_b is the completion of $D(K)$ under the scalar product (Kx, y); then its continuous dual E_b' is exactly the space X_B ($B = K^{-1}$) as K is the Riesz isomorphism between E_b and E_b'.

6.4. The Sobolev-Galpern equation

Let us consider a model problem which can be dealt with using all three approaches:

$$\partial_t(\Delta u) = \Delta^2 u,$$

$$u\,|_{\partial\Omega} = 0 \qquad \partial_\nu u\,|_{\partial\Omega} = 0,$$

$$\Delta u\,|_{t=0} = \overset{\circ}{u}, \tag{6.9}$$

where $\Omega \subset \mathbf{R}^n$ is a bounded sufficiently smooth set.

In the solution provided by the empathy theory [11], the solution operators $t \to S(t)$ act from $L_2(\Omega)$ into $W_2^2(\Omega) \cap \overset{\circ}{W}{}_2^1(\Omega)$, the related semigroup $(E(t))_{t\geq 0}$, generated by a suitable realization of $LK^{-1} = \Delta^2\Delta^{-1}$ acts in $L_2(\Omega)$ and the equation is also satisfied in $L^2(\Omega)$.

Using Showalter's approach we obtain $E_b = \overset{\circ}{W}{}_2^1(\Omega)$ with $E_b' = W_2^{-1}(\Omega)$; it follows that the solution is a $\overset{\circ}{W}{}_2^1(\Omega)$-valued function and the equation is satisfied in $W_2^{-1}(\Omega)$.

Using B-bounded semigroups we are able to construct even weaker solutions. Let $Z = X = L_2(\Omega)$, $K = \Delta$ with $D(K) = W_2^2(\Omega) \cap \overset{\circ}{W}{}_2^1(\Omega)$ and $L = \Delta^2$

with $D(L) = W_2^4(\Omega) \cap \overset{\circ}{W}{}_2^2(\Omega)$. From the general theory of PDE's we know that K is an isomorphism onto $L_2(\Omega)$. From the elliptic regularity it follows that $X_B = (W_2^2(\Omega) \cap \overset{\circ}{W}{}_2^1(\Omega))'$ (the continuous dual) and the equation is satisfied in this space.

References

[1] W. Arendt, Resolvent positive operators, *Proc. Lond. Math. Soc.* (3) 54, 1984, 321–349.

[2] L. Arlotti, On the B-bounded semigroups as a generalization of the C_0-semigroups, *Quaderno n. 15 del Dipartimento di Ingengneria Civile dell' Universitá di Udine*, 1998.

[3] J. Banasiak, Generation results for B-bounded semigroups, *Annali di matematica pura et applicata* (IV), Vol. CLXXV (1998), 307–326.

[4] J. Banasiak, B-bounded semigroups and implicit evolution equations, *University of Natal, Faculty of Science, School of Mathematical and Statistical Sciences, Internal Report No. 1/99.*

[5] J. Banasiak, V. Singh, B-bounded semigroups and C-existence families, *University of Natal, Faculty of Science, School of Mathematical and Statistical Sciences, Internal Report No. 2/99.*

[6] A. Belleni-Morante, B-bounded Semigroups and Applications, *Annali di Matematica pura et applicata* (IV), Vol. CLXX (1996), 359–376.

[7] A. Belleni-Morante, On some properties of B-quasi bounded semigroups and applications, *Quaderno sezione modelli matematici n^0 1/96, Dipartimento di Ingengneria Civile, Universitá di Firenze.*

[8] L. Bartoli, S. Totaro, Approximations of B-bounded semigroups, *Adv. Math. Sci. Appl.*, 7, No. 2 (1997), 579–600.

[9] R. deLaubenfels, *Existence families, functional calculi and evolution equations*, Lecture Notes in Mathematics 1570, Springer Verlag, Berlin, 1994.

[10] A. Favini, A. Yagi, Multivalued Linear Operators and Degenerate Evolution Equations, *Annali di Matematica pura et applicata* (IV), Vol. CLXIII (1993), 353–384.

[11] N. Sauer, Empathy theory and the Laplace transform, in: E. Janas *et al.* (Eds), *Linear Operators*, Banach Center Publications **38** (1997), 325–338.

[12] R. Showalter, *Monotone operators in Banach Spaces and Nonlinear Partial Differential Equations*, Mathematical Surveys and Monographs 49, American Mathematical Society, Rhode Island, 1997.

[13] S. Totaro, A. Belleni-Morante, The successive reflection method in three dimensional particle transport, *Journal of Mathematical Physics*, **37** (6) (1996), 2815–2823.

J. Banasiak, Department of Mathematics and Applied Mathematics, University of Natal, Durban 4041, SOUTH AFRICA
E-mail address: banasiak@scifs1.und.ac.za

Progress in Nonlinear Differential Equations
and Their Applications, Vol. 42

SEMIGROUPS AND ASYMPTOTIC MEAN PERIODICITY

VALENTINA CASARINO
POLITECNICO DI TORINO

Abstract. We investigate the spectral structure of the infinitesimal generator of an equicontinuous semigroup, giving rise to asymptotically mean periodic functions.

1. Introduction

By extending a previous definition due to J. Delsarte, L. Schwartz introduced in 1947 the notion of mean periodicity. A continuous function $f : \mathbb{R} \to \mathbb{C}$ is said to be *mean periodic* if there exists a not identically vanishing measure μ, with compact support on $\mathbb{R}$, such that

$$\int_{\mathbb{R}} f(t-u)\, d\mu(u) = 0 \text{ for every } t \in \mathbb{R}.$$

Let U be an equicontinuous group of linear bounded operators acting on a locally convex space $\mathcal{E}$, generated by X. In [3] we analyzed the spectral structure of X under the assumption that there exist an element $x \in \mathcal{E}$ and an element $\lambda \in \mathcal{E}'$ such that the map $f : \ t \mapsto < U(t)x, \lambda >$ is mean periodic. Taking into account the notion of spectrum of a mean periodic map introduced by L. Schwartz in [9], first of all we established a connection between the spectrum of f and the spectrum of the infinitesimal generator X of U. As a consequence of this relationship, we proved that the existence of even only a mean periodic map, generated by U, forces the spectral structure of X and we explicited some of the constraints imposed in this case on the spectrum of X.

In this paper we present an extension of our previous results to the case of maps generated, in a weak sense, by a semigroup. Given a semigroup T of linear bounded operators, we apply the notion of asymptotic mean periodicity (introduced by P. Koosis [7]) to maps of the form $t \mapsto < T(t)x, \lambda >$, defined only on a half-line. This problem is, in particular, interesting, since, differently from the case of periodicity and almost periodicity, the theory of mean periodic functions on a half-line strongly differs from the theory of mean periodic maps on the entire real line.

The case is also considered of maps, arising from a semigroup, which are asymptotically mean periodic with respect to a weak topology on $L^2(0, +\infty)$, as suggested by Beurling [2].

2. Notations and preliminaries

For the notion and the main properties of mean periodic maps the reader is referred to [9],[6],[3]. The definition of asymptotic mean periodicity will now be introduced.

Let $\mathcal{C}_+$ denote the space of all complex-valued functions, continuous on $\mathbb{R}_+$, endowed with the topology of uniform convergence on the compact subsets of $\mathbb{R}_+$; if $f \in \mathcal{C}_+$, the symbol $\tau_+(f)$ will denote the closure, with respect to the same topology, of the linear subspace generated by the positive translates of f, *i.e.* of

$$\bigvee \{f(\cdot + u) : u \in \mathbb{R}_+\}.$$

A function f in $\mathcal{C}_+$ is said to be *asymptotically mean periodic* if $\tau_+(f) \neq \mathcal{C}_+$.

As a consequence of the Hahn-Banach theorem, a function f in $\mathcal{C}_+$ is asymptotically mean periodic if, and only if, there exists a non-zero measure μ, with compact support in $(-\infty, 0]$, such that $\int_{-\infty}^{0} f(t-u)\, d\mu(u) = 0$ for every $t \geq 0$, or, equivalently, such that the function $g(t) := (f * \mu)(t)$, as well as μ, has compact support in $(-\infty, 0]$.

Since g is compactly supported, its Fourier transform $\mathcal{G}(\zeta) := \int e^{-it\zeta}\, g(t)dt$ is an entire function. Let $\mathcal{M}(\zeta)$ denote the Fourier transform of μ, *i.e.* $\mathcal{M}(\zeta) := \int e^{-it\zeta}\, d\mu(t)$. Define the Fourier transform of f by the meromorphic function

$$\mathcal{F}(\zeta) := \frac{\mathcal{G}(\zeta)}{\mathcal{M}(\zeta)}.$$

By adopting the same procedure followed by L.Schwartz and J.P.Kahane in order to define the spectrum of a mean periodic map, it is now possible to define the spectrum of the asymptotically mean periodic function f (denoted with the symbol $S(f)$) as the set of all poles of $\mathcal{F}(\zeta)$, each counted with its order of multeplicity.

It is also possible to prove that a complex number ζ is a pole of order at least $n+1$ for $\mathcal{F}$ if, and only if, the monomial-exponential $t^n\, e^{i\zeta t}$ belongs to $\tau_+(f)$.

In the following the symbol $\mathcal{E}$ will denote a space satisfying the following assumption:

(H1) $\mathcal{E}$ is a locally convex Hausdorff sequentially complete linear topological space, such that $\mathcal{E}'$ is sequentially complete,

where $\mathcal{E}'$ represents the topological dual of $\mathcal{E}$. If $x \in \mathcal{E}$ and $\lambda \in \mathcal{E}'$, the symbol $< x, \lambda >$ will denote the value of λ in x.

Throughout the paper T will denote an equicontinuous semigroup of class $\mathcal{C}_0$ on $\mathcal{E}$, with infinitesimal generator X. $X^{\odot}$ will denote the infinitesimal generator of the dual semigroup $\left(T^{\odot}(t)\right)_{t\geq 0}$. For the main definitions and the duality theory for equicontinuous semigroups in locally convex spaces the reader is referred to [10].

3. Asymptotically mean periodic maps

In order to analyze the constraints imposed on $\sigma(X)$ by the existence of at least one asymptotically mean periodic map f generated by T, we would like to connect the spectrum of f with the spectrum of the infinitesimal generator of T. A similar connection was obtained for the mean periodic case in [3], by showing the equivalence between different types of spectrum for a bounded, uniformly continuous function defined on $\mathbb{R}$. Almost all notions of spectrum introduced in [3] make, unfortunately, no sense for a map f defined only on a half-line, and, therefore, the basic Theorem 3.3 in [3] can only partially be recovered, when the equicontinuous group U is substituted by a semigroup T.

Proposition 3.1. *Let $T : \mathbb{R}_+ \to \mathcal{L}(\mathcal{E})$ be an equicontinuous semigroup of class $\mathcal{C}_0$ on a space $\mathcal{E}$ fulfilling $(H1)$. Let X be the infinitesimal generator of T.*

If $\zeta \in p\sigma(X^{\odot})$ for some $\zeta \in \mathbb{C}$, then there exist $x_0 \in \mathcal{D}(X)$ and $\lambda_0 \in \mathcal{D}(X^{\odot})$ such that the map, from $\mathbb{R}_+$ to $\mathbb{C}$, $t \mapsto < T(t)x_0, \lambda_0 >$ is asymptotically mean periodic, and $-i\zeta \in S(< U(\cdot)x_0, \lambda_0 >)$.

Proof. Let $\lambda_0 \in \mathcal{D}(X^{\odot})$ be such that $X^{\odot}\lambda_0 = \zeta\lambda_0$. Take $x_0 \in \mathcal{D}(X)$ such that $< x_0, \lambda_0 > \neq 0$. Then the function

$$f : t \mapsto < T(t)x_0, \lambda_0 > = e^{\zeta t} < x_0, \lambda_0 >$$

is asymptotically mean periodic, since obviously $\tau_+(f) \neq \mathcal{C}_+$. Moreover, $e^{\zeta t}$ belongs to $\tau_+(< T(\cdot)x_0, \lambda_0 >)$, whence $-i\,\zeta$ belongs to the Schwartz spectrum of $< T(\cdot)x_0, \lambda_0 >$. □

During the proof of Proposition 3.1 it has been observed that, if ζ is an eigenvalue of $X^{\odot}$, then the map

$$t \mapsto < T(t)x_0, \lambda_0 > = e^{\zeta t} < x_0, \lambda_0 >,$$

$\lambda_0 \in \mathcal{D}(X^{\odot})$ and $x_0 \in \mathcal{D}(X)$ being, respectively, such that $X^{\odot}\lambda_0 = \zeta\lambda_0$ and $< x_0, \lambda_0 > \neq 0$, is asymptotically mean periodic, and therefore there exists a measure μ, compactly supported in $(-\infty, 0]$, such that for every $t \geq 0$

$$0 = \int_{-\infty}^{0} < T(t-u)x_0, \lambda_0 > d\mu(u) = e^{\zeta t} < x_0, \lambda_0 > \int_{-\infty}^{0} e^{-\zeta u}\, d\mu(u)\,.$$

Since the support of μ is contained in $(-\infty, 0]$, this implies $\int_{-\infty}^{+\infty} e^{-\zeta u}\, d\mu(u) = 0$, *i.e.* $\mathcal{M}(-i\zeta) = 0$, $\mathcal{M}$ denoting the Fourier transform of μ. In particular, if $\zeta = i\theta$ for some $\theta \in \mathbb{R}$, then $\mathcal{M}(\theta) = 0$.

Suppose now the family

$$\{< T(\cdot)x, \lambda >: x \in \mathcal{D}(X), \lambda \in \mathcal{D}(X^{\odot}), < x, \lambda > \neq 0\} \tag{3.1}$$

be mean periodic: this means that there exists some non zero measure $\mu \in \mathcal{C}'$, with compact support on $(-\infty, 0]$, such that $\int_{-\infty}^{0} < T(t-u)x, \lambda > d\mu(u) = 0$

for all $t \geq 0$ and for every $x \in \mathcal{D}(X)$, $\lambda \in \mathcal{D}(X^{\odot})$, with $< x, \lambda > \neq 0$. Since μ is compactly supported, its Fourier transform $\mathcal{M}$ is an entire function of exponential type, bounded on the real line; thus the set of zeros of $\mathcal{M}$ admits no finite accumulation point. Since every eigenvalue of $X^{\odot}$ belongs to the set of zeros of $\mathcal{M}$, the following result can be stated:

Proposition 3.2. *Let T be an equicontinuous semigroup of class $\mathcal{C}_0$ on a space $\mathcal{E}$ satisfying $(H1)$. If the family of functions* (3.1) *is asymptotically mean periodic, then $p\sigma(X^{\odot})$ (and also $p\sigma(X)$) has no finite accumulation point. Moreover, if $\mathcal{M}$ has no real zeros, then $p\sigma(X^{\odot}) \cap i\mathbb{R} = \emptyset$.*

By combining proposition above and the stability theorem by Arendt and Batty ([1],[8]) holding in Banach spaces, one obtains the following:

Proposition 3.3. *Let T be a uniformly bounded semigroup of class $\mathcal{C}_0$ on a Banach space $\mathcal{E}$. If family* (3.1) *is asymptotically mean periodic, $\mathcal{M}$ has no real zeros and $\sigma(X) \cap i\mathbb{R}$ is countable, then T is uniformly stable, i.e. $\lim_{t \to +\infty} ||T(t)x|| = 0$ for every $x \in \mathcal{E}$.*

From properties of zeros of entire functions of exponential type it follows also that, if $\{\theta_n\}$ represents an ordering of $\frac{1}{i}\mathrm{p}\sigma(X^{\odot})$, (with $\theta_n \in \mathbb{R}$ for every $n \in \mathbb{N}$), then $\sum \frac{1}{|\theta_n|^2} < \infty$.

Recall that the continuous spectrum $c\sigma(X)$ of a linear operator X, acting on a Fréchet space $\mathcal{E}$, is the set of all complex numbers ζ for which $\zeta I - X$ is injective, has a dense range, but $(\zeta I - X)^{-1}$ is not continuous.

We recall the following lemma, which was proved in [3].

Lemma 3.4. *Let X be a linear closed operator on a Fréchet space $\mathcal{E}$.*

If a complex number ζ belongs to $c\sigma(X)$, then there exists a seminorm q_0 such that for every seminorm p there exists $\{{y_\nu}^{(p)}\} \subset \mathcal{D}(X)$, satisfying $q_0({y_\nu}^{(p)}) = 1$ for every ν and such that $\{p(\,(\zeta I - X){y_\nu}^{(p)})\}$ tends to 0 when $\nu \to \infty$.

A complex number ζ, such that for every seminorm p there exists a sequence $\{{y_\nu}^{(p)}\} \subset \mathcal{D}(X)$ satisfying $q_0({y_\nu}^{(p)}) = 1$ for every ν, for some seminorm q_0 (not depending on p), and such that $\{p(\,(\zeta I - X){y_\nu}^{(p)})\}$ tends to 0, will be called *approximate eigenvalue* of X, in analogy to terminology holding in normed spaces. The set of all approximate eigenvalues will be called *approximate point spectrum* of X $(ap\sigma(X))$. The sequence $\{{y_\nu}^{(p)}\}$ will be said to be an *approximate eigenvector* of X, with respect to p.

The set of all complex numbers ζ such that $\overline{\mathcal{R}(\zeta I - X)} \neq \mathcal{E}$ is called the *compression spectrum* of X and denoted by $k\sigma(X)$.

In [3] the following spectral mapping theorem, relating the approximate point spectrum of a semigroup with that of its infinitesimal generator, was proved.

Lemma 3.5. *Let T be a semigroup of class C_0 on a Fréchet space $\mathcal{E}$, with infinitesimal generator X. Let T be equicontinuous, i.e. for every seminorm p there is a seminorm q for which*

$$p(T(t)x) \leq q(x) \text{ for every } x \in \mathcal{E},\, t \geq 0.$$

If $\{y_\nu{}^{(p)}\}$ is an approximate eigenvector for X with respect to p with approximate eigenvalue ζ, then $\{y_\nu{}^{(q)}\}$ is an approximate eigenvector for $T(t)$, with respect to the same seminorm p, with approximate eigenvalue $e^{\zeta t}$, for every $t \geq 0$.

In the framework of semigroups acting on Fréchet spaces, we can explicite some constraints also on the continuous spectrum of X.

Proposition 3.6. *Let T be an equicontinuous semigroup on a Fréchet space $\mathcal{E}$. Suppose that the family $\{< T(\cdot)x, \lambda >: x \in \mathcal{E}, \lambda \in \mathcal{E}'\}$ is asymptotically mean periodic. Then $c\sigma(X)$ has no finite accumulation point.*

Proof. By hypothesis, there exists some non-zero measure $\mu \in \mathcal{C}'$, with compact support on $(-\infty, 0]$, for which $\int_{-\infty}^{0} < T(t-u)x, \lambda > d\mu(u) = 0$ for all $t \geq 0$ and for every $x \in \mathcal{E}$, $\lambda \in \mathcal{E}'$.

Let ζ belong to $c\sigma(X)$. As a consequence of Lemma 3.4, there exists a seminorm q_0, such that for every seminorm p there is a sequence $\{y_\nu{}^{(p)}\} \subset \mathcal{D}(X)$, satisfying $q_0(y_\nu{}^{(p)}) = 1$ for every ν and such that $\{p\big(((\zeta I - X)y_\nu{}^{(p)})\}$ tends to 0.

Since T is equicontinuous, there is a seminorm q_1 for which

$$q_0(T(t)x) \leq q_1(x) \text{ for every } x \in \mathcal{E},\, t \in \mathbb{R}.$$

There exists also a sequence $\{y_\nu{}^{(q_1)}\} \subset \mathcal{D}(X)$ satisfying $q_0(y_\nu{}^{(q_1)}) = 1$ for every ν, and such that $\{q_1\big(((\zeta I - Y)y_\nu{}^{(q_1)})\}$ tends to 0.

For every ν consider the subspace of $\mathcal{E}$ $\mathcal{M}_\nu{}^{(q_1)} = \mathbb{C} \cdot y_\nu{}^{(q_1)}$ and define the linear functional $f_\nu{}^{(q_1)}$ on $\mathcal{M}_\nu{}^{(q_1)}$ defined by $< \alpha y_\nu{}^{(q_1)}, f_\nu{}^{(q_1)} >= \alpha$, for all $\alpha \in \mathbb{C}$. Since it holds

$$| < \alpha y_\nu{}^{(q_1)}, f_\nu{}^{(q_1)} > | = |\alpha| = |\alpha| \cdot q_0(y_\nu{}^{(q_1)}) = q_0(\alpha y_\nu{}^{(q_1)}),$$

as a consequence of the Hahn-Banach theorem for every ν there exists a functional $\lambda_\nu{}^{(q_1)} \in \mathcal{E}'$ for which

$$< \alpha y_\nu{}^{(q_1)}, \lambda_\nu{}^{(q_1)} >= \alpha \text{ for every } \alpha \in \mathbb{C} \text{ and for every } \nu$$
$$\text{and } | < x, \lambda_\nu{}^{(q_1)} > | \leq q_0(x) \text{ for all } x \in \mathcal{E}.$$

Choose now t such that $t-u \in [0, t_0]$, for some $t_0 > 0$. It results:

$$\left|\int e^{\zeta(t-u)} d\mu(u)\right| = \left|\int e^{\zeta(t-u)} < y_\nu{}^{(q_1)}, \lambda_\nu{}^{(q_1)} > d\mu(u) - \right.$$
$$\left.\int < U(t-u) y_\nu{}^{(q_1)}, \lambda_\nu{}^{(q_1)} > d\mu(u)\right| \leq$$
$$\int \left| < e^{\zeta(t-u)} y_\nu{}^{(q_1)} - U(t-u) y_\nu{}^{(q_1)}, \lambda_\nu{}^{(q_1)} > \right| d\mu(u) \leq$$
$$\int q_0\left(e^{\zeta(t-u)} y_\nu{}^{(q_1)} - U(t-u) y_\nu{}^{(q_1)}\right) d\mu(u),$$

and the last term tends to zero by Lemma 3.5, when ν tends to ∞. Thus $\int e^{\zeta(t-u)} d\mu(u) = 0$ for every $t \geq 0$, implying that $\mathcal{M}(-i\zeta) = 0$. Paley-Wiener theorem yields finally the thesis. □

4. Asymptotically (B)-mean periodic maps

The same argumentations hold by substituting the topology of uniform convergence on compact subsets of $\mathbb{R}_+$ with a weak topology on $L^2(0, \infty)$, suggested by A.Beurling in [2]. For more details on this topology see [7].

For any $\varepsilon \geq 0$, let $L^{2,\varepsilon}$ denote the Hilbert space of measurable functions from $[0, +\infty)$ to $\mathbb{C}$, arising from the inner product

$$(f, g)_\varepsilon := \int_0^{+\infty} f(t)\, g(t)\, e^{-\varepsilon t}\, dt.$$

Set $L^{2,0} := L^2$. Let A be a subset of $L^2(0, +\infty)$. The symbol A_ε will denote the closure of A in $L^{2,\varepsilon}$ under $||\cdot||_\varepsilon$ (where $||f||_\varepsilon := \left(\int_0^\infty |f(t)|^2 e^{-\varepsilon t} dt\right)^{\frac{1}{2}}$ for every $f \in L^{2,\varepsilon}$). Set

$$A^B := \bigcap_{\varepsilon > 0} (A_\varepsilon \cap L^2).$$

Let now f belong to $L^2(0, +\infty)$. Let S_f denote the set of all finite linear combinations of positive translates of f. Set $E_f := S_f{}^B$. The function f is said to be *asymptotically (B)-mean periodic* if, and only if, $E_f \subsetneq L^2(0, +\infty)$.

As a consequence of Hahn-Banach theorem, a function $f \in L^2(0, +\infty)$ is asymptotically (B)-mean periodic if, and only if, there exist $\varepsilon > 0$ and a non identically vanishing function $k \in L^{2,\varepsilon}$ such that $(f(\cdot + t), k)_\varepsilon = 0$ for every $t \geq 0$.

In the following, any function $k \in L^{2,\varepsilon}$ will be extended to the entire real line, by setting $k(t) = 0$ for all $t < 0$. Define $K(t) = k(-t) \cdot e^{\varepsilon t}$. Then the function $g(t) = (f * K)(t)$ vanishes on $[0, +\infty)$. Since f belongs to $L^2(0, +\infty)$, one can define the Fourier transform $\mathcal{F}$ of f in the usual way. As a consequence of the relation

$$\mathcal{F}(\zeta) = \frac{\mathcal{G}(\zeta)}{\mathcal{K}(\zeta)},$$

Koosis proved, moreover, that $\mathcal{F}$ has a meromorphic extension to the complex plane and that its poles lie necessarily in the half-plane $\mathcal{I}m\zeta < 0$; also in this case, it is possible to define the Schwartz spectrum of an asymptotically (B)-mean periodic function f (denoted with the symbol $S(f)$) as the set of all poles of $\mathcal{F}(\zeta)$, each counted with its order of multiplicity.

Suppose now T be an equicontinuous semigroup on a locally convex space $\mathcal{E}$ fulfilling $(H1)$. Let $< T(\cdot)x_0, \lambda_0 >$ belong to $L^2(0,\infty)$, for some $x_0 \in \mathcal{D}(X)$ and $\lambda_0 \in \mathcal{D}(X^\odot)$. If λ_0 is an eigenvector of $X^\odot$ with eigenvalue ζ, and $x_0 \in \mathcal{D}(X)$ is such that $< x_0, \lambda_0 > \neq 0$, then

$$\int_0^{+\infty} | < T(t)x_0, \lambda_0 > |^2 dt = | < x_0, \lambda_0 > |^2 \int_0^{+\infty} e^{2\Re e\zeta \cdot t}\, dt < +\infty,$$

entailing $\Re e\zeta < 0$.

Under the same assumptions, the map $< T(\cdot)x_0, \lambda_0 >$ is asymptotically (B)-mean periodic, and therefore there exist $\varepsilon > 0$ and $g \in L^{2,\varepsilon}$ such that, by setting $K(t) = k(-t) \cdot e^{\varepsilon t}$, it results:

$$0 = \int_{-\infty}^{+\infty} < T(t-u)x_0, \lambda_0 > K(u)\, du = \int_{-\infty}^{+\infty} e^{-\zeta u} K(u)\, du \text{ for every } t \geq 0,$$

i.e. the Fourier transform $\mathcal{K}$ of K vanishes in the point $-i\zeta$, implying, as in [7], that the exponential monomial $e^{\zeta t}$ belongs to $E_{<T(\cdot)x_0,\lambda_0>}$, and therefore that $-i\zeta$ belongs to the Schwartz spectrum of $<T(\cdot)x_0, \lambda_0>$. Summing up, if $\zeta \in p\sigma(X^\odot)$ for some $\zeta \in \mathbb{C}$, then there exist $x_0 \in \mathcal{D}(X)$ and $\lambda_0 \in \mathcal{D}(X^\odot)$ with $<x_0, \lambda_0> \neq 0$ such that the function, belonging to $L^2(0,+\infty)$, $t \mapsto <T(t)x_0, \lambda_0>$ is asymptotically (B)-mean periodic, and $-i\zeta \in S(<T(\cdot)x_0, \lambda_0>)$.

In view of what has been observed above, zeros of $\mathcal{K}$, arising from eigenvalues either of X or of $X^\odot$, have positive imaginary parts.

We say that the family (3.1) is asymptotically (B)-mean periodic if there exist $\varepsilon > 0$ and $g \in L^{2,\varepsilon}$ such that $\big(< T(\cdot + t)x, \lambda >, g \big)_\varepsilon = 0$ for every $t \geq 0$, for all $x \in \mathcal{D}(X)$, $\lambda \in \mathcal{D}(X^\odot)$ with $< x, \lambda > \neq 0$.

As a consequence of Paley-Wiener theorem, the following result can be stated:

Proposition 4.1. *Let T be an equicontinuous semigroup of class C_0 on a space $\mathcal{E}$ satisfying $(H1)$. If family (3.1) is asymptotically (B)-mean periodic, then both $p\sigma(X)$ and $p\sigma(X^\odot)$ are contained in the left half-plane $\Pi_l = \{\zeta \in \mathbb{C} : \Re\zeta < 0\}$ and are discrete sets.*

References

[1] W. Arendt and C.J.K. Batty, *Almost periodic solutions of first and second order Cauchy problems*, J. Diff. Equations, vol. 137 (1997), pp. 363–383.

[2] A. Beurling, *A theorem on functions defined on semigroups*, Math. Scand., vol. 1 (1953), pp. 127–130.

[3] V. Casarino, *Equicontinuous families of operators generating mean periodic maps*, to appear in Rendiconti di Matematica dell'Accademia dei Lincei.

[4] J. Delsarte, *Les fonctions moyenne-périodiques*, Journal de Mathématiques pures et appliquées, vol. 14 (1935), pp. 403–453.

[5] J.P. Kahane, *Sur quelques problémes d'unicité et de prolongement, relatifs aux fonctions approchables par des sommes d'exponentielles*, Annales de l'Institut Fourier, vol. 5 (1954), pp. 39–130.

[6] J.P. Kahane, *Lectures on mean periodic functions*, Tata Institut for fundamental Research, Bombay, 1958.

[7] P. Koosis, *On functions which are mean periodic on a half-line*, Comm. Pure and Appl. Math., vol. 10 (1957), pp. 133–149.

[8] J. van Neerven, *The asymptotic behaviour of semigroups of linear operators*, Birkhäuser, Basel, 1996.

[9] L. Schwartz, *Theorie générale des fonctions moyenne-périodiques*, Annals of Mathematics, vol. 48 (1947), pp. 857–929.

[10] K. Yosida, *Functional Analysis*, Springer Verlag, Berlin/Heidelberg/New York, 1968.

Valentina Casarino, Dipartimento di Matematica, Politecnico di Torino, Corso Duca degli Abruzzi 24, Torino, ITALY
E-mail address: casarino@dm.unito.it

Progress in Nonlinear Differential Equations
and Their Applications, Vol. 42

SOME PROBLEMS IN STOCHASTIC ANALYSIS AND SEMIGROUP THEORY

JAN A. VAN CASTEREN
UNIVERSITY OF ANTWERP

Abstract. For locally compact spaces there is a one-to-one correspondence between Feller semigroups (those are semigroups which send continuous functions to continuous ones) and (strong) Markov processes. This leads to an interaction between stochastic analysis and classical semigroup theory. However, many interesting topological spaces are not locally compact. Nevertheless from the point of view of (stochastic) analysis and possible applications these more general topological spaces are important. Examples of such spaces are Wiener space, Loop space, Fock space. These spaces are Polish spaces or more general Lusin spaces. We like to develop an analysis which encompasses this kind of spaces. We also like to bring in the martingale problem. In the commutative setting this leads to the problems sketched in §1. In the non-commutative setting some other problems are described in §5 as well. In §1 and in §2 a relationship is established between reciprocal Markov processes and Euclidean quantum mechanics. Moreover there should be a connection with Feynman propagators. The problem proposed in §3 is closely related to work done by Demuth and van Casteren [8]. In order to discuss Neumann scattering the problems proposed in §4 are included. In particular cases these can certainly be solved. In our opinion the described problems are feasible.

1. Strong Markov processes

In the sequel E denotes a locally compact second countable Hausdorff space. The assertions (a), (b), (c), and (d) of the following theorem are well known.

Theorem 1.

(a) *(R.M. Blumenthal and R.K. Getoor* [3], *Let* $\{S(t) : t \geq\}$ *be a Feller semigroup in* $C_0(E)$. *Then there exists a strong Markov process (in fact a Hunt process)*

$$\{(\Omega, \mathcal{F}, \mathbf{P}_x), (X(t) : t \geq 0), (\vartheta_t : t \geq 0), (E, \mathcal{E})\},$$

such that $[S(t)f](x) = \mathbf{E}_x[f(X(t))]$, $f \in C_0(E)$, $t \geq 0$. *Moreover this Markov process is normal (i.e.* $\mathbf{P}_x[X(0) = x] = 1$*), is right continuous (i.e.* $\lim_{t\downarrow s} X(t) = X(s)$, $\mathbf{P}_x$*-almost surely), possesses left limits in* E *on its life time (i.e.* $\lim_{t\uparrow s} X(t)$ *exists in* E, *whenever* $\zeta > s$*), and is quasi-left-continuous (i.e. if* $(T_n : n \in \mathbf{N})$ *is an increasing sequence of* $(\mathcal{F}_t)$*-stopping times,* $X(T_n)$ *converges* $\mathbf{P}_x$*-almost surely to* $X(T)$ *on the event* $\{T < \infty\}$, *where* $T = \sup_{n\in\mathbf{N}} T_n$*).*

(b) *Conversely, let* $\{(\Omega, \mathcal{F}, \mathbf{P}_x), (X(t) : t \geq 0), (\vartheta_t : t \geq 0), (E, \mathcal{E})\}$ *be a strong Markov process which is normal, right continuous, and possesses left limits in* E *on its life time. Put* $[S(t)f](x) = \mathbf{E}_x[f(X(t))]$, *for* f *a bounded Borel function,* $t \geq 0$, $x \in E$. *Suppose that* $S(t)f$ *belongs to* $C_0(E)$ *for* f *belonging to* $\in C_0(E)$, $t \geq 0$. *Then* $\{S(t) : t \geq 0\}$ *is a Feller semigroup.*

(c) *Let* A *be the generator of a Feller semigroup in* $C_0(E)$ *and let*

$$\{(\Omega, \mathcal{F}, \mathbf{P}_x), (X(t) : t \geq 0), (\vartheta_t : t \geq 0), (E, \mathcal{E})\}$$

be the corresponding Markov process. For every $f \in D(A)$ *and for every* $x \in E$, *the process* $t \mapsto f(X(t)) - f(X(0)) - \int_0^t Af(X(s))ds$ *is a* $\mathbf{P}_x$*-martingale for the filtration* $(\mathcal{F}_t)_{t\geq 0}$, *where each* σ*-field* $\mathcal{F}_t$, $t \geq 0$, *is (some closure of)* $\sigma(X(u) : u \leq t)$. *In fact the* σ*-field* $\mathcal{F}_t$ *may taken as* $\mathcal{F}_t = \bigcap_{s>t} \sigma(X(u) : u \leq s)$. *It is also possible to complete* $\mathcal{F}_t$ *with respect to* $\mathbf{P}_\mu$, *given by* $\mathbf{P}_\mu(A) = \int \mathbf{P}_x(A)d\mu(x)$. *For* $\mathcal{F}_t$ *the following* σ*-field may be chosen:*

$$\mathcal{F}_t = \bigcap_{\mu \in P(E)} \bigcap_{s>t} \{\mathbf{P}_\mu\text{-completion of } \sigma(X(u) : u \leq s)\}.$$

(d) *Conversely, let* A *be a densely defined linear operator with domain* $D(A)$ *and range* $R(A)$ *in* $C_0(E)$. *Let* $(\mathbf{P}_x : x \in E)$ *be a unique family of probability measures, on an appropriate measure space* $(\Omega, \mathcal{F})$ *with an appropriate filtration* $(\mathcal{F}_t)_{t\geq 0}$, *such that, for all* $x \in E$, $\mathbf{P}_x[X(0) = x] = 1$, *and such that for all* $f \in D(A)$ *the process*

$$t \mapsto f(X(t)) - f(X(0)) - \int_0^t Af(X(s))ds$$

is a $\mathbf{P}_x$*-martingale with respect to the filtration* $(\mathcal{F}_t)_{t\geq 0}$. *Then the operator* A *possesses a unique extension* A_0, *which generates a Feller semigroup in* $C_0(E)$.

(e) *(Unique Markov extensions) Suppose that the densely defined linear operator* A *(with domain and range in* $C_0(E)$*) possesses the Korovkin property as well as the following one. For every* $\lambda > 0$ *(large) and for every* $h \in D(L)$, *the inequality*

$$\lambda \sup_{x\in E} h(x) \leq \sup_{x\in E} (\lambda I - A) h(x) \text{ or, equivalently} \tag{1.1}$$

$$\lambda \inf_{x\in E} h(x) \geq \inf_{x\in E} (\lambda I - A) h(x) \tag{1.2}$$

is valid. Then A *extends to a unique generator* A_0 *of a Feller semigroup, and the martingale problem is well posed for the operator* A. *Moreover, the Markov process associated with* A_0 *solves the martingale problem uniquely for* A.

Definition 2. The operator A possesses the *Korovkin property* in the sense that there exists a strictly positive real number $t_0 > 0$ such that for every $x_0 \in E \cup \{\Delta\}$ the equality

$$\inf_{h\in D(A)} \sup_{x\in E} \left\{h(x_0) + \left[g - (I - t_0A)\, h\right](x)\right\} \tag{1.3}$$

$$= \sup_{h\in D(A)} \inf_{x\in E} \left\{h(x_0) + \left[g - (I - t_0A)\, h\right](x)\right\} \tag{1.4}$$

is valid for all $g \in C_0(E)$. In the presence of (1.1), this definition is equivalent to the following one. There exists $t_0 > 0$ such that for every $x_0 \in E \cup \{\Delta\}$, and for each $g \in C_0(E)$, and for every $\varepsilon > 0$, there exist functions h_1 and h_2 in the domain of A such that

$$-\varepsilon + (I - t_0A)\, h_1 \le g \le \varepsilon + (I - t_0A)\, h_2, \quad \text{and } h_2(x_0) - h_1(x_0) \le 0. \tag{1.5}$$

Problem 1. Is a result like Theorem true if the locally compact space E is replaced with a Polish space, and if $C_b(E)$ (space of all bounded continuous functions on E) replaces $C_0(E)$? Instead of the topology of uniform convergence we consider the *strict topology*. This topology is generated by seminorms of the form: $f \mapsto \sup_{x\in E} |u(x)f(x)|$, $f \in C_b(E)$. The functions $u \ge 0$ have the property that for every $\alpha > 0$ the set $\{u \ge \alpha\}$ is compact (or is contained in a compact subset of E. The functions u need not be continuous. What about Markov uniqueness? Is there a relationship with work done by Eberle [11, 12]?

Is it possible to rephrase Theorem for reciprocal Markov processes and diffusions? Martingales should then replaced with differences of forward and backward martingales. A stochastic process $(M(t) : t \ge 0)$ on a probability space $(\Omega, \mathcal{F}, \mathbf{P})$ is called a *backward martingale* if $\mathbf{E}\left[M(t) \mid \mathcal{F}^s\right] = M(s)$, $\mathbf{P}$-almost surely, where $t < s$, and $\mathcal{F}^s$ is the σ-field generated by the information from the future: $\mathcal{F}^s = \sigma\left(X(u) : u \ge s\right\}$. Of course we assume that $M(t)$ belongs to $L^1(\Omega, \mathcal{F}, \mathbf{P})$, $t \ge 0$.

Let $(\Omega, \mathcal{F}, \mathbf{P})$ be a probability space. An E-valued process $(X(t) : 0 \le t \le 1)$ is called *reciprocal* if for any $0 \le s < t \le 1$ and every pair of events $A \in \sigma\left(X(\tau) : \tau \in (s,t)\right)$, $B \in \sigma\left(X(\tau) : \tau \in [0,s] \cup [t,1]\right)$ the equality

$$\mathbf{P}\left[A \cap B \mid X(s), X(t)\right] = \mathbf{P}\left[A \mid X(s), X(t)\right] \mathbf{P}\left[B \mid X(s), X(t)\right] \tag{1.6}$$

is valid. By $\mathcal{D}$ we denote the set

$$\mathcal{D} = \left\{(s,x,t,B,u,z) : (x,z) \in E \times E,\ 0 \le s < t < u \le 1,\ B \in \mathcal{E}\right\}. \tag{1.7}$$

A function $P : \mathcal{D} \to [0, \infty)$ is called a *reciprocal probability distribution* or a *Bernstein probability* if the following conditions are satisfied:

(i) the mapping $B \mapsto P(s,x,t,B,u,z)$ is probability measure on $\mathcal{E}$ for any $(x,z) \in E \times E$ and for any $0 \le s < t < u \le 1$;

(ii) the function $(x,z) \mapsto P(s,x,t,B,u,z)$ is $\mathcal{E} \otimes \mathcal{E}$-measurable for any $0 \le s < t < u \le 1$;

(iii) For every pair $(C, D) \in \mathcal{E}\otimes E$, $(x, y) \in E\times E$, and for all $0 \le s < t < u \le 1$ the equality

$$\int_D P(s,x,u,d\xi,v,y)\,P(s,x,t,C,u,\xi) = \int_C P(s,x,t,d\eta,v,y)\,P(t,\eta,u,C,v,y)$$

is valid.

Then the following theorem is valid for $E = \mathbf{R}^\nu$ (see Jamison [22]).

Theorem 3. *Let $P(s,x,t,B,u,y)$ be a reciprocal transition probability function and let μ be a probability measure on $\mathcal{E}\otimes\mathcal{E}$. Then there exists a unique probability measure $\mathbf{P}_\mu$ on $\mathcal{F}$ with the following properties:*

(1) With respect to $\mathbf{P}_\mu$ the process $(X(t) : 0 \le t \le 1)$ is reciprocal;

(2) For all $(A, B) \in \mathcal{E}\otimes\mathcal{E}$ the equality $\mathbf{P}_\mu[X_0 \in A,\ X_1 \in B] = \mu(A\times B)$ is valid;

(3) For every $0 \le s < t < u \le 1$ and for every $A \in \mathcal{E}$ the equality

$$\mathbf{P}_\mu\left[X(t)\in A \,\middle|\, X(s), X(u)\right] = P(s, X(s), t, A, u, X(u)) \quad \textit{is valid.}$$

For more details see Thieullen [31] and [32].

What is the equivalent of this in the non-commutative setting? Notice that it is possible to define a strict topology for C^*-algebras: see *Problem 12.*

We notice that a positive solution to Problem 1 would make a link with work by J.R. Dorroh and J.W. Neuberger [9].

Problem 2 How useful is the martingale result on operators with the Korovkin property? We don't have a good example or application. Is there a non-commutative version of the Korovkin property?

2. Feynman Propagators and Euclidean Quantum Mechanics

We begin by describung a problem. *Problem 3.* Let $M(t)$ be a multiplicative process. Define the semigroup

$$\exp(-tK_M) \quad \text{by} \quad [\exp(-tK_M)f](x) = \int \exp(-tK_M)(x,y)f(y)dy,$$

where $\exp(-tK_M)(x,y) = \int M(t)d\mu_{0,x}^{t,y}$. Suppose that the operators $\exp(-tK_M)$, $t > 0$, are self-adjoint. Then, formally,

$$\begin{aligned}&\int [\exp(-isK_M)\exp(-tK_M)(\cdot,y)](x)f(y)dy\\ &= \left[\exp(-isK_M)\int \exp(-tK_M)(\cdot,y)f(y)dy\right](x)\\ &= [\exp(-isK_M)\exp(-tK_M)f](x) = [\exp(-(t+is)K_M)f](x).\end{aligned}$$

In what sense do we have convergence of

$$[\exp(-isK_M)\exp(-tK_M)(\cdot, y)](x) \quad \text{to} \quad \exp(-isK_M)(x,y),$$

if t tends to 0 downward?

Definition 4. The *pinned measure* $\mu_{0,x}^{t,y}$ on $\mathfrak{F}_{t-}$ is defined as follows:

$$\mu_{0,x}^{t,y}(A) = \mathbf{E}_x\left[p_0(t-s, X(s), y)1_A\right],$$

where $A \in \mathfrak{F}_s$, $s < t$. Extend this pre-measure as a genuine measure on $\mathfrak{F}_{t-}$ and notice that the process $s \mapsto p_0(t-s, X(s), y)$ is a martingale. The measure $\mu_{0,x}^{t,y}$ lives on the event $\{X(0) = x,\ X(t-) = y\}$.

Remark. A remark on Feynman propagators and pinned Brownian motion. Let $(C([0,t], \mathfrak{F}, m)$ be Wiener measure. Put (Feynman's operational calculus)

$$[(K_\lambda F)\psi](x) = \int F\left(\lambda^{-\frac{1}{2}}y + x\right)\psi\left(\lambda^{-\frac{1}{2}}y(t) + x\right) dm(y) \tag{2.1}$$

$$= \mathbf{E}_{\lambda^{-\frac{1}{2}}}\left[y \mapsto F\left(\lambda^{-\frac{1}{2}}y\right)\psi\left(\lambda^{-\frac{1}{2}}y(t)\right)\right] \tag{2.2}$$

$$= \int\int F d\mu_{0,x}^{\lambda^{-1}t,y}\psi(y)dy = \mathbf{E}_x\left[y \mapsto F(y)\psi\left(\lambda^{-1}t\right)\right], \tag{2.3}$$

where $\mu_{0,x}^{\lambda^{-1}t,y}$ is the measure corresponding to non-normalized pinned Brownian motion. The problem which poses itself, is to extend the measure $\mu_{0.x}^{\lambda^{-1}t,y}$, $\lambda > 0$, (as) an infinite-dimensional distribution on the subspace $C([0,t], \mathbf{R}^\nu)$, if $\Re\lambda \geq 0$. For $\lambda = -i$ this should lead to the Feynman propagators.

In what follows the measure $\mathbf{P}_{x,t}^{M_{v,t}}$ is defined as follows. For all finite n-tuples $t_1, \dots, t_n$ in $(0,\infty)$ the identity (2.5) is valid:

(2.4)

$$\mathbf{E}_{x,t}^{M_{v,t}}\left[\prod_{j=1}^n f_j\left(X\left(t_j+t\right), t_j+t\right)\right]$$

$$= \mathbf{E}_{x,t}\left[\exp\left(-\frac{1}{2}\int_t^u \Gamma_1(v,v)(X(\tau),\tau)\,d\tau - M_{v,t}(u)\right)\prod_{j=1}^n f_j\left(X\left(t_j+t\right), t_j+t\right)\right]$$

(2.5)

$$= \mathbf{E}_{x,t}\left[\prod_{j=1}^n f_j\left(q_v\left(t_j+t\right), t_j+t\right)\right],$$

where the $\mathbf{E}_{x,t}$-martingale $M_{v,t}(s)$, $s \geq t$, is given by

$$M_{v,t}(s) = v(X(s),s) - v(X(t),t) + \int_t^s \left(-\frac{\partial}{\partial\tau} + K_0\right) v(X(\tau),\tau)\,d\tau. \tag{2.6}$$

Its quadratic variation part $\langle M_{v,t}\rangle(s) := \langle M_{v,t}, M_{v,t}\rangle(s)$ is given by

$$\langle M_{v,t}\rangle(s) = \int_t^s \Gamma_1(v,v)(X(\tau),\tau)\,d\tau. \tag{2.7}$$

Let the process $\{(\Omega, \mathcal{F}, \mathbf{P}_{x,t}), ((q_v(t),t) : t \geq 0), (\vartheta_t : t \geq 0), (E \times \mathbf{R}, \mathcal{E} \otimes \mathcal{B})\}$ be the Markov process generated by the operator $-K_v + \dfrac{\partial}{\partial t}$, where K_v is defined by $K_v(f)(x,t) = K_0 f(x,t) + \Gamma_1(v,f)(x,t)$. Here, $\mathcal{B}$ denotes the Borel field of $\mathbf{R}$, and by $\Gamma_1(v,f)(x,t)$ we mean

(2.8)

$$\Gamma_1(v,f)(x,t) = \lim_{s\downarrow 0} \frac{1}{s}\mathbf{E}_x\left[(v(X(s),t) - v(X(0),t))\,(f(X(s),t) - f(X(0),t))\right].$$

Theorem 5. *Let $\chi : E \times [t,u] \to [0,\infty]$ be a function such that*

$$\mathbf{E}_{x,t}^{M_{v,t}}\left[|\log \chi(X(u),u)|\right], \quad v \in D(K_0),$$

is finite. Here $u > 0$ is a fixed time and

$$\{(\Omega, \mathcal{F}, \mathbf{P}_x), (X(t) : t \geq 0), (\vartheta_t : t \geq 0), (E, \mathcal{E})\}$$

is the strong Markov process generated by $-K_0$. Let S_L be a solution to the following Riccati type equation. (This equation is called the Hamilton-Jacobi-Bellman equation.) For $t \leq s \leq u$ and $x \in E$ the following identity is true:

(2.9)

$$\begin{cases} -\dfrac{\partial S_L}{\partial s}(x,s) + \dfrac{1}{2}\Gamma_1(S_L,S_L)(x,s) + K_0 S_L(x,s) - V(x,s) = 0; & \\ S_L(x,u) = -\log \chi(x,u), & x \in E. \end{cases}$$

Then for any real valued $v \in D(K_0)$ the following inequality is valid:

(2.10)

$$S_L(x,t) \leq \mathbf{E}_{x,t}^{M_{v,t}}\left[\int_t^u \left(\frac{1}{2}\Gamma_1(v,v) + V\right)(X(\tau),\tau)d\tau\right] - \mathbf{E}_{x,t}^{M_{v,t}}\left[\log \chi(X(u),u)\right],$$

and equality is attained for the "Lagrangian action" $v = S_L$.

Problem 4. Prove Theorem for viscosity solutions of the equation

$$H\left(x, s_1, s_2, v(x,\cdot), (\Gamma_1(v,v)(x,\cdot))^{1/2}, -K_0 v(x,\cdot)\right) = 0, \tag{2.11}$$

where the function $H(x, s_1, s_2, v(\cdot), p(\cdot), M(\cdot))$ is defined by

$$\begin{aligned} &H(x, s_1, s_2, v(\cdot), p(\cdot), M(\cdot)) \\ &= v(s_1) - v(s_2) + \frac{1}{2}\int_{s_1}^{s_2} p(s)^2 ds - \int_{s_1}^{s_2} M(s)ds - \int_{s_1}^{s_2} V(x,s)ds. \end{aligned} \tag{2.12}$$

Here a *viscosity solution* is defined as a function S_L for which

$$H\left(x_0, s_1, s_2, \varphi\left(x_0, \cdot\right), \left(\Gamma_1\left(\varphi, \varphi\right)\left(x_0, \cdot\right)\right)^{1/2}, -K_0\varphi\left(x, \cdot\right)\right) \geq 0, \tag{2.13}$$

whenever φ belongs to $D\left(K_0\right)$ and possesses the property that

$$S_L\left(x_0, s\right) - \varphi\left(x_0, s\right) \leq S_L\left(x, s\right) - \varphi\left(x, s\right), \quad \text{for all } s_1 \leq s \leq s_2, \tag{2.14}$$

and for all x in a neighborhood of x_0 (or for all $x \in E$). Moreover, if the inequality in (2.14) is reversed, then the one in (2.13) should also be reversed. The use of viscosity solutions here should be compared with that of W.H. Fleming and H.M. Soner [17]. The reader should also consult S.J. Sheu [30]. Another approach could be the use of *travelling wave solutions*: see e.g. K.D. Elworthy and H.Z.Zhao [15, 16], K.D. Elworthy, A. Truman, H.Z. Zhao and J.G. Gaines [14], X.-M. Li and H.Z. Zhao [24]. In [24] the authors study the following initial value problem, a so-called *reaction-diffusion* equation:

(2.15)
$$\begin{cases} \dfrac{\partial u(t,x)}{\partial t} & = \dfrac{1}{2}\mu^2 \Delta u(t,x) + \langle Z(x), \nabla u(t,x)\rangle + \dfrac{1}{\mu^2} c(t,x,u(t,x))u(t,x) \\ u(0,x) & = u_0^\mu(x). \end{cases}$$

This equation is considered in $[0,\infty) \times \mathbf{R}^n \times \mathbf{R}$. The authors want to take singular limits as $\mu \downarrow 0$. They impose certain growth and regularity conditons on the function $c(t,x,y)$ and on the vector field $Z(x)$. Under appropriate conditions the solution $u(t,x) = u^\mu(t,x)$ converges to a travelling wave solution, in the sense that there exists a function $V(t,x)$ with the property that $\lim_{\mu\downarrow 0} u^\mu(t,x) = 0$, if $V(t,x) < 0$, and $\lim_{\mu\downarrow 0} u^\mu(t,x) = 1$, if $V(t,x) > 0$. The region $\{(t,x) : V(t,x) < 0\}$ is called the *trough* and the region $\{(t,x) : V(t,x) > 0\}$ is called the *crest*. If $u^\mu(0,x) = T_0(x)\exp\left(-S_0(x)/\mu^2\right)$, then $J^\mu(t,x)$, defined by $J^\mu(t,x) = -\mu^2 \log u^\mu(t,x)$ satisfies the non-linear Hamilton-Jacobi-Bellman equation:

$$\frac{\partial}{\partial t} J^\mu(t,x) + \frac{1}{2}\left|\nabla J^\mu(t,x)\right|^2 + c\left(u^\mu(t,x)\right) = \frac{1}{2}\mu^2 \Delta J^\mu(t,x),$$

with $J^\mu(t,x) = -\mu^2 \log T_0(x) + S_0(x)$. Here we assume that $c(\cdot)$ depends only on the position. In [5] the authors M.G. Crandall, L.C. Evans, and P.L. Lions study properties of viscosity solutions of Hamilton-Jacobi equations. In [27] E. Pardoux uses viscosity solutions in the study of backward stochastic differential equations and semi-linear parabolic equations.

Formulate and prove a stochastic Noether theorem in terms of the squared gradient operator: see J.C. Zambrini [36]. It should read something like what follows ($(x,s) \in E \times [t,u]$, $\dot{h}$ denotes the time derivative of h).

Theorem 6. *Suppose that the functions h, $S_L : E \times \mathbf{R} \to \mathbf{C}$, $T : [0,\infty) \to [0,\infty)$ satisfy*

$$\dot{h}(x.s) - \dot{h}(x,t) - K_0 h(x,s) + K_0 h(x,t)$$

$$= \int_t^s \Gamma_1(h,V)(x,\sigma)d\sigma + V(x,s)T(s) - V(x,t)T(t); \tag{2.16}$$

$$\dot{S}_L(x,s) = K_0 S_L(x,s) + \frac{1}{2}\Gamma_1\left(S_L, S_L\right)(x,s) - V(x,s). \tag{2.17}$$

Then the proccess

$$\left[\Gamma_1\left(S_L,h\right) + \dot{S}_L T + \dot{h}\right]\left(q_{S_L}(s),s\right), \quad t \le s \le u, \tag{2.18}$$

is a $\mathbf{E}_{x,t}$-martingale. Under suitable conditions on the function $v : E \times [t,u] \to \mathbf{C}$ the process

$$\left[\Gamma_1\left(v,h\right) + \left(K_0 v + \frac{1}{2}\Gamma_1(v,v) - V\right)T + \dot{h}\right]\left(q_v(s),s\right), \quad t \le s \le u, \tag{2.19}$$

is a $\mathbf{E}_{x,t}$-martingale as well.

3. More on Feynman-Kac semigroups

In the summary below, the following notation is in vogue. The symbols V and W stand for Kato-Feller potentials, Σ denotes a (large) open subset of E, $-K_0$ is the generator of a Feller semigroup with continuous symmetric transition densities $\exp\left(-tK_0\right)(x,y) = \exp\left(-tK_0\right)(y,x)$, the symbol $D(t)$ stands for

$$D(t) = \exp\left(-t\left(K_0 \dot{+} V\right)\right) - \exp\left(-t\left(K_0 \dot{+} W\right)\right),$$

$D_\Sigma(t)$ denotes the difference

$$D_\Sigma(t) = \exp\left(-t\left(K_0 \dot{+} V\right)\right) - \mathfrak{J}^* \exp\left(-t\left(K_0 \dot{+} V\right)_\Sigma\right)\mathfrak{J},$$

where $\mathfrak{J}(f) = f\restriction_\Sigma$, $f \in L^2(E,m)$. Here the semigroup $\left\{\exp\left(-t\left(K_0 \dot{+} V\right)_\Sigma\right) : t \ge 0\right\}$, is defined by

(3.1)

$$\exp\left(-t\left(K_0 \dot{+} V\right)_\Sigma\right) f(x) = \mathbf{E}_x\left[\exp\left(-\int_0^t V(X(s))ds\right) f(X(t)) : S > t\right],$$

where S is the terminal stopping time defined by

$$S = \inf\left\{t > 0 : \int_0^t 1_{E\setminus\Sigma}\left(X(s)\right)ds > 0\right\}.$$

The function h_Σ^{a+V} is defined by

$$h_\Sigma^{a+V}(x) = \mathbf{E}_x\left[\exp\left(-\int_0^S (a+V)\left(X(s)\right)ds\right) : S < \infty\right].$$

Summary.

1. $\exp\left(-\frac{t}{2}K_0\right)|W-V|^2 \in L^{1/2}(E,m)$ implies: $D(t)$, $t>0$, is a trace operator;
2. $W-V\in L^2(E,m)$ implies: $D(t)$, $t>0$, is a Hilbert-Schmidt operator;
3. $W-V\in L^1(E,m)$ implies that the wave operators exist and are complete;
4. $h_\Sigma^{a+V}\in L^{1/2}(E,m)$ implies: $D_\Sigma(t)$, $t>0$, is a trace operator;
5. $h_\Sigma^{a+V}\in L^2(E,m)$ implies: $D_\Sigma(t)$, $t>0$, is a Hilbert-Schmidt operator;
6. $h_\Sigma^{a+V}\in L^1(E,m)$ implies that the wave operators exist and are complete;
7. $W-V\in C_0(E)$ implies: $D(t)$, $t>0$, is compact;
8. $h_\Sigma^{a+V}\in C_0(E)$ implies $D_\Sigma(t)$, $t>0$, is compact;
9. In 3, 4 and 6 (extensions) of the classical Pearson estimates are available. We owe this result to Demuth and Eder (Clausthal): see [7].

Problem 5. What happens to the results in the above summary if the additive functional $\int_0^t V(X(u))du$ is replaced with more general additive processes, like stochastic integrals? The general result on the existence of Feynman-Kac semigroups can be formulated and proved for a Kato type class of additive processes. This result was obtained by L. Smits (Antwerp). The compactness properties for these more general processes are not yet investigated in its full generality.

Problem 6. What happens to the compactness results, if the Dirichlet type boundary condition is replaced with a Neumann type boundary condition?

4. Neumann semigroups

Next we come to some abstract problems, in which Neumann semigroups and Neumann boundary conditions are involved. A general motivation is the following one. Let

$$\{(\Omega,\mathcal{F},\mathbf{P}_x),(X(t),t\geq 0),(\vartheta_t:t\geq 0),(\mathbf{R}^\nu,\mathcal{B})\}$$

be the Markov process of ν-dimensional Brownian motion. Consider a Neumann initial value problem on an open domain Σ in $\mathbf{R}^\nu$. This means find a solution to the following Cauchy problem:

$$\frac{\partial}{\partial t}u(x,t)=\frac{1}{2}\Delta u(x,t)\quad\text{on}\quad\Sigma, \tag{4.1}$$

where its normal derivative $D_n u(x,t)=0$, $x\in\partial D$. Assume $u(x,0)=f(x)$, $x\in\Sigma$. Find a sequence of multiplicative processes $M_n(t)$ such that

$$u(x,t)=\lim_{n\to\infty}\mathbf{E}_x\left[M_n(t)f(X(t))\right].$$

If we consider Dirichlet semigroups, then such multiplicative functionals can be found. For instance, the semigroup in (3.1) can be written as

$$\exp\left(-t\left(K_0\dot{+}V\right)_{\Sigma}\right) f(x) = \lim_{n\to\infty} \mathbf{E}_x\left[M_n(t)f(X(t))\right], \tag{4.2}$$

$$\text{where} \quad M_n(t) = \exp\left(-n\int_0^t 1_{E\setminus\Sigma}\left(X(s)\right)ds\right)\exp\left(-\int_0^t V(X(s))ds\right).$$

We extend this kind of problem to a general domain in a locally compact space. Let $v : E \to [0,\infty)$ be a function in $D(K_0)$ with $e^{-v} \in D(K_0)$ as well. Put

(4.3)
$$M_v(t) = v(X(t)) - v(X(0)) + \int_0^t K_0 v(X(s))ds;$$

(4.4)
$$\langle M_v\rangle(t) = \langle M_v, M_v\rangle(t) = \int_0^t \Gamma_1(v,v)(X(s))ds;$$

$$Z_v(t) = M_{-v}(t) - \frac{1}{2}\langle M_v\rangle(t)$$

$$= v(X(0)) - v(X(t)) + \int_0^t e^{v(X(s))}\left[K_0\left(e^{-v}\right)\right](X(s))ds; \tag{4.5}$$

(4.6)
$$T_v(t)f(x) = \mathbf{E}_x\left[\exp\left(Z_v(t) - \int_0^t V(X(s))ds\right)f(X(t))\right]$$

$$= \mathbf{E}_x\left[\exp\left(v(X(0)) - v(X(t))\right)\exp\left(\int_0^t e^{v(X(s))}\left[K_0\left(e^{-v}\right)\right](X(s))ds\right)\right.$$

$$\left.\times\exp\left(-\int_0^t V(X(s))ds\right)f(X(t))\right] \tag{4.7}$$

$$= \left[\exp\left(-tK_v\right)f\right](x), \quad \text{where} \tag{4.8}$$

(4.9)
$$K_v f = e^v K_0\left(e^{-v}f\right) + \left(V - e^v K_0\left(e^{-v}\right)\right)f.$$

In particular $K_v 1 = V$. Here $\Gamma_1(f,g)$ is the carré du champ operator, introduced by Roth, but popularized by Bakry (see e.g. D. Bakry [1])

$$\Gamma_1(f,g)(x) = \lim_{s\downarrow 0}\frac{1}{s}\mathbf{E}_x\left[\left(f(X(s)) - f(X(0))\right)\left(g(X(s)) - g(X(0))\right)\right], \tag{4.10}$$

and $\langle M_v\rangle = \langle M_v, M_v\rangle$ is the variation process corresponding to the martingale M_v. The family $\{T_v(t) : t \geq 0\}$ is a strongly continuous semigroup in $L^2(E,m)$. Put $u(t,x) = T_v(t)f(x)$. Then $u(0,x) = f(x)$ and

$$\frac{\partial u}{\partial t} = -K_0 u - Vu - \Gamma_1(v,u).$$

So the expression $f \mapsto \Gamma_1(v, f)$ is sort of a drift in gradient form. The corresponding quadratic form is given by

(4.11)
$$\begin{aligned}&\mathcal{E}_v^V(f,g)\\&= -\int v\Gamma_1(f,\overline{g})(x)dx + \int (2v(x)+1)K_0f(x)\overline{g}(x)dx + \int V(x)f(x)\overline{g}(x)dx.\end{aligned}$$

Notice the identity $K_0(fg) + \Gamma_1(f,g) = (K_0f)g + f(K_0g)$. Put

$$\begin{aligned}\mathbf{P}_{v,x}(A) = \mathbf{E}_x\Big[&\exp\left(v(X(0)) - v(X(t))\right)\exp\left(\int_0^t e^{v(X(s))}\left[K_0\left(e^{-v}\right)\right](X(s))ds\right)\\&\times\exp\left(-\int_0^t V(X(s))ds\right)1_A\Big],\end{aligned} \tag{4.12}$$

where A belongs to $\mathcal{F}_t$.

In case K_0 is $-\frac{1}{2}\Delta$, we have $\Gamma_1(f,g) = \nabla f \cdot \nabla g$. Moreover, $M_v(t) = \int_0^t \nabla v(X(s))dX(s)$, (Itô integral) where X is Brownian motion. In addition,

$$\mathbf{E}_x\left[F\left(Y(s) : 0 \le s \le t\right)\exp\left(Z_v(t)\right)\right] = \mathbf{E}_x\left[F\left(X(s) : 0 \le s \le t\right)\right], \tag{4.13}$$

where $Y(s) = X(s) + \int_0^s \nabla v(X(\sigma))d\sigma$. This is a version of the Girsanov transformation. For an up-to-date account of Girsanov transformations, the reader is referred to Üstünel and Zakai [33]. It could be useful to look at Driver and Lohrenz [10], Elworthy and Ma [13], Norris [26], Léandre and Norris [23]. We want to take (singular) limits in the expressions for $T_v(t)$ and $\mathcal{E}_v^V$, for v tending to $a1_{E\setminus\Sigma}$, $0 < a \le \infty$. If $0 < a < \infty$, the quadratic form converges to

(4.14)
$$\begin{aligned}\mathcal{E}_{\Sigma,a}^V(f,g) :=& -a\int_{E\setminus\Sigma}\Gamma_1(f,\overline{g})(x)dx\\&+2a\int_{E\setminus\Sigma}K_0f(x)\overline{g}(x)dx + \int K_0f(x)\overline{g}(x)dx + \int V(x)f(x)\overline{g}(x)dx.\end{aligned}$$

So that, if $\int_{E\setminus\Sigma}\left(\Gamma_1(f,\overline{g})(x) - 2K_0f(x)\overline{g}(x)\right)dx = 0$, then

(4.15)
$$\mathcal{E}_{\Sigma,N}^V(f,g) := \lim_{a\to\infty}\mathcal{E}_{\Sigma,a}^V(f,g) = \int K_0f(x)\overline{g}(x)dx + \int V(x)f(x)\overline{g}(x)dx.$$

Here $\mathcal{E}_{\Sigma,N}^V$ stands for *Neumann quadratic form*. In the presence of the carré du champ operator we may define a distance on $E \times E$:

$$d(x,y) = \sup\left\{|\psi(y) - \psi(x)| : \Gamma_1(\psi,\psi) \le 1\right\}.$$

The *local time* (occupation) the process X (up to time t) spends on the (boundary of the) complement of Σ is then the bounded variation part of the process $d(X(t), E \setminus \Sigma)$. Suitable logarithmic Sobolev inequalities imply $d(x, y) < \infty$: see e.g. D. Bakry, Théorème 3.2, page 39. A proof of a pointwise estimate for a heat kernels based on a logarithmic Sobolev inequality may be based on § 3 of Bakry in combination with the proof of Lemma 3.2.1 in E.B. Davies [6].

Remark. The operator K_0 generates a *diffusion* in the following sense: for every C^∞-function $\Phi : \mathbf{R}^\nu \to \mathbf{R}$, with $\Phi(0, \dots, 0) = 0$, the following identity is valid:

$$\begin{aligned} (4.16)\quad & K_0\left(\Phi(f_1, \dots, f_n)\right) \\ &= \sum_{j=1}^{n} \frac{\partial \Phi}{\partial x_j}(f_1, \dots, f_n)\, K_0 f_j - \frac{1}{2} \sum_{j,k=1}^{n} \frac{\partial^2 \Phi}{\partial x_j \partial x_k}(f_1, \dots, f_n)\, \Gamma_1(f_j, f_k) \end{aligned}$$

for all functions $f_1, \dots, f_n$ in a rich enough algebra of functions $\mathcal{A}$, contained in the domain of the generator K_0, as described in Bakry's lecture notes.

Problem 7. What relations, if any, do exist between the following concepts:

1. singular limit of quadratic form;
2. singular limit of Feynman-Kac semigroup;
3. local time spent by the process X in $E \setminus \Sigma$;
4. Girsanov transformation (SDE);
5. reflected Markov process?

Problem 8. A related, not completely understood problem, is to formulate and prove the precise correspondence between convergence of semigroups and of the associated quadratic forms. One ought to consider Γ-convergence of quadratic forms. For the latter see e.g. Dal Maso [25]. A sequence of (symmetric) quadratic forms $(\mathcal{E}_n : n \in \mathbf{N})$, defined on subspace of a Hilbert space $\mathcal{H}$, is said to be Γ-convergent if for every $f \in \mathcal{H}$ the equality

$$\lim_{\delta \downarrow 0} \liminf_{n \to \infty} \inf_{\{g \in \mathcal{H},\ \|g - f\| \le \delta\}} \mathcal{E}_n(g, g) = \lim_{\delta \downarrow 0} \limsup_{n \to \infty} \inf_{\{g \in \mathcal{H},\ \|g - f\| \le \delta\}} \mathcal{E}_n(g, g)$$

is valid. This limit $\mathcal{E}(f, f)$ is called the Γ-limit of the sequence $(\mathcal{E}_n : n \in \mathbf{N})$: $\mathcal{E}(f, f) = \Gamma\text{-}\lim_{n \to \infty} \mathcal{E}_n(f)$. In [25] the author explains among other things how Γ-convergence of quadratic forms is related to resolvent and semigroup convergence of the corresponding semigroups and their resolvent families. If $\mathcal{E}_n \equiv \mathcal{E}_0$, then

$$\Gamma\text{-}\lim_{n \to} \mathcal{E}_n(f, f) = \liminf_{g \to f} \mathcal{E}_0(g, g), \quad f \in \mathcal{H}.$$

The Γ-limit is called the *relaxation* of $\mathcal{E}_0$. Moreover, $\mathcal{E}_0$ is closed if and only if $\mathcal{E} = \mathcal{E}_0$, $\mathcal{E}_0$ is closable if and only if $\mathcal{E} = \mathcal{E}_0$ on the subspace $\{\mathcal{E}_0 < \infty\}$. Please notice that $\mathcal{E}$ extends $\mathcal{E}_0$, but that, for f in the domain of $\mathcal{E}_0$, $\mathcal{E}(f, f) \le \mathcal{E}_0(f, f)$. The author learned these results from K.T. Sturm. In connection with Neumann type problems other references of interest might be: R.F. Bass and P. Hsu [2], P. Hsu [21], J.-F. Le Gall [20], Z.Q. Chen, R.J. Williams and Z. Zhao [4], E. Pardoux and R.J. Williams [28], M. Fukushima and M. Tomisaki [18], [19], R.J. Williams and W.A. Zheng [35] and Z.Q. Chen, P.J. Fitzsimmons and R.J. Williams [37].

Problem 9. If possible, incorpurate Neumann scattering in our discussion.

Problem 10. Suppose the sequence $\{v_n : n \in \mathbf{N}\}$ converges to $a1_{E\setminus\Sigma}$ in a reasonable way ($a = \frac{1}{2}$, or, more generally, $\infty \geq a > 0$). Does it follow that the sequence $(\mathbf{P}_{v_n,x})$ converges or is tight?

5. Completely Positive Semigroups

Problem 11. As a mathematical problem R. Rebolledo (private communiation) proposed to find a link between the following concepts. The symbols H_0 and H_1 stand for self-adjoint Hamiltonians in the respective Hilbert spaces $\mathcal{H}_0$ and $\mathcal{H}_1$, and $\mathcal{J} : \mathcal{H}_1 \to \mathcal{H}_0$ is a bounded *identification operator*.

(i) *Quantum mechanical scattering theory*:

$$\text{s-}\lim_{t\to\infty} \exp(\pm itH_0)\,\mathcal{J}\exp(\mp itH_1)\,.$$

(ii) *Kullback information*, which now is defined by

$$K((\psi_t),\varphi) = \sup_{0<u<\infty,\ u\in C_b(\mathbf{R})} \left\{ \limsup_{t,s\to\infty} \Re\langle \log u(H_0)\psi_t, \psi_s\rangle - \log\langle u(H_1)\varphi,\varphi\rangle \right\}.$$

Here ψ_t denotes $\psi_t = \exp(-itH_0)\,\mathcal{J}\exp(itH_1)\,\varphi$. This Kullback information is inspired by the classical Kullback (Shannon) information. Let μ and ν be probability measures on a Polish space E. The expression

$$K(\mu,\nu) = \sup_{0<u<\infty, u\in C_b(E)} \left\{ \int \log u\,d\mu - \log \int u\,d\nu \right\} = \int \left(\log \frac{d\mu}{d\nu}\right) \frac{d\mu}{d\nu} d\nu$$

then defines the mutual information between μ and ν. Then the following assertions are true: $K(\mu,\nu) = 0$ if and only if $\mu = \nu$, and $K(\mu,\nu) < \infty$ if and only if μ is absolutely continuous with respect to ν.

(iii) Let $(X,\mathbf{P})$ be a real valued Markov process: for example Brownian motion. The following expression connects (for $t \to \infty$) "quantum probability" and "scattering theory":

$$\mathbf{E}\begin{pmatrix} \exp(iX(t)H_0) & 0 \\ 0 & \exp(iX(t)H_1) \end{pmatrix} \begin{pmatrix} I & \mathcal{J} \\ 0 & I \end{pmatrix} \begin{pmatrix} \exp(-iX(t)H_0) & 0 \\ 0 & \exp(-iX(t)H_1) \end{pmatrix}.$$

(iv) Contiguity of the Hamiltonian H_0 with respect to H_1. A possible definition of this notion could be

$$\text{w-}\lim_{t\to\infty} \exp\left(-tH_0\right)\pi_t \exp\left(-tH_1\right) = 0,$$

whenever $\pi_t^*\pi_t$ and $\pi_t\pi_t^*$ are orthogonal projections which converge weakly to 0. In [29] Rebolledo employs a similar notion for quantum dynamical semigroups in Fock space. Another definition, which is closer to wave operators and scattering theory is $\text{w-}\lim_{t\to\pm\infty} \exp\left(-itH_0\right)\pi_t \exp\left(itH_1\right) = 0$, whenever $\pi_t^*\pi_t$ and $\pi_t\pi_t^*$ are orthogonal projections which converge weakly to 0.

Remark. Another problem might be to define *contiguity* of one (sequence) spectral measure(s) $E_n^0(\cdot)$ with respect another $E_n^1(\cdot)$. For two Hamiltonians H_0 with corresponding respective spectral measures E_0 and E_1 two notions are suggested in Problem 11. The first one is based on the semigroups $\{\exp\left(-tH_j\right) : t \geq 0\}$, $j = 0, 1$, and the second one is based on the wave oerators $\{\exp\left(itH_j\right) : t \in \mathbf{R}\}$, $j = 0, 1$. In the classical commutative context we have a sequence of probability measures $\left(\mathbf{Q}_n\right)_n$ is contiguous to the sequence $\left(\mathbf{P}_n\right)_n$ if $\lim_{n\to\infty} \mathbf{Q}_n\left(A_n\right) = 0$, whenever $\lim_{n\to\infty} \mathbf{P}_n\left(A_n\right) = 0$. We notice that the sequence $\left(\mathbf{Q}_n\right)_n$ is contiguous to the sequence $\left(\mathbf{P}_n\right)_n$ if and only the following two conditions are satisfied:

(i) $\inf_N \sup_n \int_{\{Z_n > N\}} Z_n d\mathbf{P}_n = 0$;

(ii) $\lim_n \mathbf{Q}_n\left[Z_n = \infty\right] = 0$.

Problem 12. Let A be a C^*-algebra, and let $\pi : A \to \mathcal{L}\left(\mathcal{H}\right)$ be a faithful representation (obtained e.g. via a Gelfand-Naimark-Segal construction). Define, for every compact operator $T : \mathcal{H} \to \mathcal{H}$ the semi-norm $p_T : A \to [0, \infty)$ by $p_T(x) = \|T\pi(x)\|$, $x \in A$. The topology τ_β induced by these seminorms may be called the *strict topology*. This topology does not dpend on the particular choice of the Gelfand-Naimark-Segal representation. Is the following theorem true?

Theorem 7. *Suppose that the C^*-algebra A is equipped with the strict topology. Let $\mathcal{L}$ be a linear operator with domain and range in A. The following assertions are equivalent:*

(1) the operator $\mathcal{L}$ possesses the following properties:

(i) For every $\lambda > 0$ the operator $\left(\lambda I - \mathcal{L}\right)^{-1}$ exists as a bounded, everywhere defined linear operator and is completely positive;

(ii) The domain of $\mathcal{L}$ is dense for the strict topology on A.

(2) The operator $\mathcal{L}$ generates a τ_β-contiuous semigroup consisting of completely positive operators.

Here an operator $\mathcal{T}: A \to A$ is said to be completely positive, if for every finite choice $x_1, \dots, x_n$ and $y_1, \dots, y_n$ in A the element $\sum_{j,k=1}^{n} x_j^* \mathcal{T}\left[y_j^* y_k\right] x_k$ is positive in A. Let $\mathfrak{L}$ be as in Theorem (2). Suppose $\mathfrak{L}(I) = I$. Let $\exp\left(\tau\widetilde{\mathfrak{L}}\right)$ be its extension (dilation) of $\exp(\tau\mathfrak{L})$ to (symmetric) Fock space. Make a systematic study of the following limits:

$$\begin{aligned} &\text{s-}\lim_{\tau\to\pm\infty} \text{s-}\lim_{s\to\infty} \frac{1}{s}\int_0^s \exp\left((\sigma+i\tau)\widetilde{\mathfrak{L}}\right) d\sigma, \quad \text{and} \\ &\text{s-}\lim_{s\to\infty} \text{s-}\lim_{\tau\to\pm\infty} \frac{1}{s}\int_0^s \exp\left((\sigma+i\tau)\widetilde{\mathfrak{L}}\right) d\sigma. \end{aligned} \tag{5.1}$$

Problem 11 and 12 should be compared to the following result, which can be found in [34].

Theorem 8. *Suppose that for all bounded intervals A_0 and A_1 the following conditions are satisfied:*

(i) the wave operators $\Omega_\pm\left(H_0, E_0(A_0)\mathfrak{J}E_1(A_1), H_1\right)$ exist;

(ii) the operator $E_0(A_0)\left(H_0\mathfrak{J} - \mathfrak{J}H_1\right)E_1(A_1)$ is compact.

(iii) The operators $E_0(A_0)\left(\mathfrak{J}\mathfrak{J}^* - I\right)E_0(A_0)$ and $E_1(A_1)\left(\mathfrak{J}^*\mathfrak{J} - I\right)E_1(A_1)$ are compact.

Then the wave operators $\Omega_\pm\left(H_0, \mathfrak{J}_\varphi, H_1\right)$, $\varphi \in L^1(\mathbf{R})$, exist. Here

$$\mathfrak{J}_\varphi = \int_{-\infty}^{\infty} \varphi(\sigma)\exp\left(-i\sigma H_0\right)\mathfrak{J}\exp\left(i\sigma H_1\right)d\sigma.$$

Moreover, for φ_1 and $\varphi_2 \in L^1(\mathbf{R})$,

$$\int \varphi_1(\sigma)d\sigma\, \Omega_\pm\left(H_0, \mathfrak{J}_{\varphi_2}, H_1\right) = \int \varphi_2(\sigma)d\sigma\, \Omega_\pm\left(H_0, \mathfrak{J}_{\varphi_1}, H_1\right).$$

If, in addition to (i) and (ii), $\lim_{n\to\infty}\left\|E_0[n,\infty)\mathfrak{J}E_1(A_1)\right\| = 0$ for all bounded intervals A_1, then the wave operators $\Omega_\pm\left(H_0, \mathfrak{J}, H_1\right)$ exist.

If moreover, $\lim_{n\to\infty}\left\|E_0(A_0)\mathfrak{J}E_1[n,\infty)\right\| = 0$, then $D(t_0) = V_0(t_0)\mathfrak{J} - \mathfrak{J}V_1(t_0)$ is compact and the scattering system $\left(H_0, \mathfrak{J}, H_1\right)$ is complete in the sense that the wave operators $\Omega_\pm\left(H_0, \mathfrak{J}, H_1\right)$ are unitary operators from $P_{\mathrm{ac}}(H_1)\mathcal{H}_1$ onto $P_{\mathrm{ac}}(H_0)\mathcal{H}_0$.

Thanks. Thanks are due to the University of Antwerp (UIA) and the Flemish Fund for Scientific Research (FWO, Fanders) for their material support. The author is indebted to K.T. Sturm (Universität Bonn) and Xue-Mei Li (University of Connecticut, Storrs) for interesting discussions on the subject and some references.

References

[1] D. Bakry, L'hypercontractivité et son utilisation en théorie de semigroups, *Lectures on probability theory: Ecole d'été de Saint-Flour XXII-1992*, Lecture notes in Math., vol.1581, Springer Verlag, Berlin, 1994, pp.1–114, P. Bernard (editor).

[2] R.F. Bass and P. Hsu, *Some potential theory for reflecting brownian motion in hölder and lipschitz domains*, Ann. Probab. **19** (1991), no. 2, 486–508.

[3] R.M. Blumenthal and R.K. Getoor, *Markov processes and potential theory*, Pure and Applied Mathematics: A series of monographs and textbooks, vol. 29, Academic Press, New York, 1986.

[4] Z.Q. Chen, R.J. Williams, and Z. Zhao, *On the existence of positive linear solutions for semilinear elliptic equations with Neumann boundary conditions*, Probab. Theory Related Fields **101** (1995), 251–276.

[5] M.G. Crandall, L.C. Evans, and P.L. Lions, *Some properties of viscosity solutions of Hamilton-Jacobi equations*, Transactions Amer. Math. Soc. **282** (1984), no. 2, 487–502.

[6] E.B. Davies, *Heat kernels and spectral thoery*, Cambridge University Press, Cambridge, 1989.

[7] M. Demuth and S. Eder, *A trace class estimate for two-space wave operators*, Tech. report, Technische Universität Clausthal-Zellerfeld, Clausthal, 1999, Preprint.

[8] M. Demuth and J.A. van Casteren, *Stochastic spectral theory of Feller operators*, Birkhäuser Verlag, Basel, 1999, book in preparation (to appear).

[9] J.R. Dorroh and J.W. Neuberger, *A theory of strongly continuous semigroups in terms of Lie generators*, J. Functional Analysis **136** (1996), no. 1, 114–126.

[10] B.K. Driver and T. Lohrenz, *Logarithmic sobolev inequalities for pinned loop groups*, J. Functional Analysis **140** (1996), no. 2, 381–448.

[11] A. Eberle, *Weak sobolev spaces and markov uniqueness of operators*, C.-R.-Acad.-Sci.-Paris-Ser.-I-Math. **320** (1995), no. 10, 1249–1254.

[12] _______, *Girsanov-type transformations of local dirichlet forms: an analytic approach*, Osaka Journal of Mathematics **33** (1996), no. 2, 497–531.

[13] D. Elworthy and Z. Ma, *Vector fields on mapping spaces and related dirichlet forms and diffusions*, Osaka Journal of Mathematics **34** (1997), no. 3, 629–651.

[14] K.D. Elworthy, A. Truman, H.Z. Zhao, and J.G. Gaines, *Approximate travelling waves for generalized kpp equations and classical mechanics*, Proc.-Roy.-Soc.-London-Ser.-A **446** (1994), no. 1924, 529–554.

[15] K.E. Elworthy and H.Z. Zhao, The travelling wave solutions of scalar generalized KPP equations via classical mechanics and stochastic approaches, *Collection: Stochastics and quantum mechanics (Swansea, 1990)*, World Scientific, River Edge, NJ, 1992, pp. 298–316.

[16] _______, Approximate travelling waves for generalized and stochastic KPP equations, *Collection: Probability theory and mathematical statistics, (St. Petersburg, 1993)*, Gordon and Breach, Amsterdam, 1966, 141–154.

[17] W.H. Fleming and H.M. Soner, *Controlled markov processes and viscosity solutions*, Applications of Mathematics, vol. 25, Springer Verlag, Berlin, 1993.

[18] M. Fukushima and M. Tomisaki, *Reflecting diffusions on Lipschitz domains with cusps - Analytic construction and Skorohod representation*, Potential Anal. **4** (1995), no. 4, 377–408.

[19] _______, *Construction and decomposition of reflecting diffusions on lipschitz domains with hölder cusps - Analytic construction and Skorohod representation*, Probab. Theory Related Fields **106** (1996), no. 4, 521–557.

[20] J.-F. Le Gall, One-dimensional stochastic differential equations involving the local times of the unknown process, *Collection: Stochastic analysis and applications (Swansea 1983)*, Lecture Notes in Math., vol. 1095, Springer Verlag, Berlin, 1984, pp. 51–82.

[21] P. Shu, *Probabilistic approach to the Neumann problem*, Communications on Pure and Applied Mathematics **XXXVIII** (1985), 445–472.

[22] B. Jamison, *Reciprocal processes*, Z. Wahrscheinlichkeitstheor. Verw. Gebiete **30** (1974), 65–86.

[23] R. Léandre and J.R. Norris, Integration by parts and Cameron-Martin formulas for the free path space of a compact Riemannian manifold, *Collection: Séminaire de Probabilités*, Lecture Notes in Math., vol. 1655, Springer Verlag, Berlin, 1997, pp. 16–23.

[24] X.M. Li and H.Z. Zhao, *Gradient estimates and the smooth convergence of approximate travelling waves for reaction-diffusion equations*, Nonlinearity **9** (1996), no. 2, 459–477.

[25] Gianno Dal Maso, *An introduction to* Γ*-convergence*, Birkhäuser Verlag, Basel, 1993.

[26] J.R. Norris, *Twisted sheets*, J. Functional Analysis **132** (1995), no. 2, 273–334.

[27] E. Pardoux, Backward stochastic differential equations and viscosity solutions of systems of semilinear parabolic and elliptic PDEs of second order, *Collection: Stochastic analysis and related topics, VI (Geilo, 1996)*, Progress in Probability, vol. 42, Birkhäuser Verlag, Basel, 1998, pp. 79–127.

[28] E. Pardoux and R.J. Williams, *Symmetric reflected diffusions*, Ann. Inst. Henri Poincaré Probab. Statist. **30** (1994), no. 1, 13–62.

[29] R. Rebolledo, *Limit problems for quantum dynamical semigroups inspired from scattering theory*, to appear, 1999.

[30] S.J. Sheu, *Stochastic control and principal eigenvalue*, Stochastics **11** (1984), no. 3–4, 191–211.

[31] M. Thieullen, *Second order stochastic differential equations and non-gaussian reciprocal diffusions*, Probab. Theory and Related Fields **97** (1993), 231–257.

[32] _______, *Reciprocal diffusions and symmetries*, Stochastics and Stochastics Reports **65** (1998), 41–77.

[33] A.S. Üstünel and M. Zakai, *Transformation of measure on Wiener space*, Tech. report, Technion, Haifa, Israel, 1999, 302 pages.

[34] J.A. van Casteren, *Compactness properties of differences of self-adjoint semigroups*, Uia preprints, University of Antwepr (UIA), 1999, submitted.

[35] R.J. williams and W.A. Zheng, *On reflecting Brownian motion—a weak convergence approach*, Ann. Inst. Henri Poincaré Probab. Statist. **26** (1990), no. 3, 461–488.

[36] J.-C. Zambrini, *A special time-dependent quantum invariant and a general theorem on quantum symmetries*, Proceedings of the second International Workshop Stochastic Analysis and Mathematical Physics: ANESTOC'96 (Singapore) (R. Rebolledo, ed.), World Scientific, 1998, Workshop: Vina del Mar Chile 16–20 December 1996, pp. 197–210.

[37] P.J. Fitzsimmons, Z.Q. Chen and R.J. Williams, *Reflecting Brownian motion: quasimartingales and strong Caccippoli sets*, Potential Analysis **2** (1993), 219–243.

Jan van Casteren, Department of Mathematics and Computer Science, University of Antwerp (UIA), Universiteitsplein 1, 2610 Antwerp/Wilrijk, BELGIUM

E-mail address: vcaster@uia.ua.ac.be

Progress in Nonlinear Differential Equations
and Their Applications, Vol. 42

ON DIRAC OBSERVABLES

H.O. CORDES
UNIVERSITY OF CALIFORNIA, BERKELEY

Abstract. The Dirac equation admits a natural time-invariant algebra $\mathbf{Q}$ of unbounded operators of the Hilbert space $\mathbf{H} = L^2(\mathbb{R}^3, \mathbb{C}^4)$. Time-invariance means that $e^{iHt}\mathbf{Q}e^{-iHt} = \mathbf{Q}$, with the unitary $e^{iHt} : \mathbf{H} \to \mathbf{H}$. This is not a specific feature of the Dirac equation; any 1-st order strictly hyperbolic system (under mild additional assumptions) admits such an algebra. We propose to introduce $\mathbf{Q}$ as algebra of "really predictable observables" into v. Neumann's theory of physical states and observables: Only the self-adjoint operators $A \in \mathbf{Q}$ can be really predicted. For other observables — especially for most *dynamical observables* — one must find an observable $A_0 \in \mathbf{Q}$ "close" to A, perhaps only for a given physical state ψ. Energy H belongs to $\mathbf{Q}$, but location (angular) momentum, spin, ..., do not. This seems to explain at a glance the triumphs and difficulties Dirac's theory encountered.

0. Introduction

The concept of quantum mechanical observable occurs in 2 slightly conflicting ways. Abstractly, unit vectors ψ of a Hilbert space $\mathbf{H}$ are physical states; observables are defined as unbounded self-adjoint operators A. Predicting A in a state $\psi \in \mathbf{H}$ (simplified for A with discrete spectrum) means expanding ψ along the eigenvectors φ_j of A. The square $|\psi_j|^2$ of the expansion coefficient is the probability of measuring the eigenvalue λ_j to φ_j. The *expectation value* for this measurement is

$$(0.1) \qquad \bar{a} = \sum \lambda_j |\psi_j|^2 = (\psi, A\psi), \text{ with inner product } (\cdot,\cdot) \text{ of } \mathbf{H}.$$

To calculate $\bar{a}$ one does not need the spectral theory of A.

This concept received its rigorous mathematical formulation by J. v. Neumann [Nm1]. It is the result of discussions on interpretation of quantum mechanics among leading physicists of the 1920's. [Well known: A. Einstein strongly opposed it. More philosophical discussions extend up to present time (cf. [DV], meeting, June 1998).] At the time of [Nm1] only the Schroedinger equation was known, and the definitions seemed to fit very well.

Later on, for the Dirac equation

$$(0.2) \qquad \partial\psi/\partial t + ih(x, D)\psi = 0,\ (t, x) \in \mathbb{R}^4,\ h(x, \xi) = \alpha(\xi - \mathbf{A}) + \mathbf{V} + \beta,$$

with 4×4-Dirac matrices $\alpha_1, \alpha_2, \alpha_3, \beta$, and $D_j = -i\partial/\partial x_j$, and electromagnetic potentials $\mathbf{A}(x), \mathbf{V}(x)$, v. Neumann's concept was just adapted, choosing $\mathbf{H} = L^2(\mathbb{R}^3, \mathbb{C}^4)$ as Hilbert space.

The Dirac equation describes *relativistic* quantum mechanics. We must think about the change of the concept in a different space time frame — say the coordinate system (t', x') given by the 'boost'

$$
\begin{aligned}
t' &= tc - sx_1,\ x_1' = x_1 c - st,\ x_2' = x_2,\ x_3' = x_3, \\
c &= \cosh\theta,\ s = \sinh\theta,\ \theta \in \mathbb{R}.
\end{aligned}
\tag{0.3}
$$

The new Hilbert space $\mathbf{H}'$ would be that of L^2-vector functions along the hyperplane $t' = 0$, in coordinates x'. A state $\psi \in \mathbf{H}$ trivially extends to $(t,x) \in \mathbb{R}^4$, as solution of (0.2): $\psi(t,x) = e^{-iHt}\psi$, $H = h(x,D)$, and $\psi(t_0,\cdot)$ is a unit vector for all t_0, since e^{-iHt} is unitary. $\psi(t,x)$ for $t' = 0$ in new coordinates x' no longer has norm 1, but a simultaneous transform of dependent variable — left multiplication by $\kappa = \cosh(\theta/2) - \alpha_1 \sinh(\theta/2)$ — will cure this. This defines a unitary $R = R_\theta$, $\theta \in \mathbb{R}$, $R : \mathbf{H} \to \mathbf{H}'$ such that (i) under both transforms $(t,x) \to (t',x')$ and $\psi \to \kappa\psi = \psi'(t',x')$ the Dirac equation goes to itself — same α, β but covariantly changed potentials (now possibly time dependent); (ii) $\psi' = R\psi$ is the state corresponding to $\psi \in \mathbf{H}$; $A' = RAR^*$ the observable A in new coordinates (details in [C4]).

Apart from abstract observables we have *dynamical observables*, with concrete physical meaning. Many are derived from the variational principle of Hamiltonian mechanics. Using Noether's lemma, and the "correspondence principle" one obtains self-adjoint differential expressions of order 0, 1 or 2 representing energy, momentum, angular momentum, spin, current, ... They define unbounded self-adjoint operators of $\mathbf{H}$ qualifying as abstract observables, sometimes with small adjustments (of domain, etc....).

For a Schroedinger equation this gives a mathematically perfect theory, perhaps with philosophical objections? But from the start Dirac's theory, accepted mainly due to its very precise prediction of *energy states* — i.e., spectral lines — was plagued with paradoxes, relating to time-propagation of other observables: "Zitterbewegung, Klein's paradox, noncommutativity of velocity components,...". During the past 70 years, Physics learned to live with these, has "stepped around them", but they occur in standard texts.

Our aim here is to suggest that the above may come to order if we give up the postulate that *all* v. Neumann observables necessarily can be "predicted" as above. One should allow for some *uncertainty*, similar as in the well known *Heisenberg principle.* In fact *some observables* — in certain physical states *cannot be predicted at all*, we claim.

More precisely: Sometimes *even a complete knowledge of A and ψ does not allow a precise prediction* of the expectation value $\bar{a}$ (or the probabilities $|\psi_j|^2$) — may be only with a (calculable) error, or not at all.

Such thoughts were suggested to us by the evidence that the Dirac equation has a (time)-invariant algebra **P** of global pseudodifferential operators ([C1], 1983). Likely this was not known to Dirac or his contemporaries.

"Prediction" should be modified as follows: Only for the unbounded self-adjoint operators $A \in \mathbf{P}$ (or possibly a slightly larger algebra?) a *precise* prediction, in every physical state, is possible. Let $\mathbf{Q} \supset \mathbf{P}$ be such an algebra. Predicting $A \notin \mathbf{Q}$ should require finding $A_0 \in \mathbf{Q}$ with A_0 "close" to $\mathbf{A}$, and then predicting A_0, while the possible error is limited by the difference $A - A_0$.

For a given physical state $\psi \in \mathbf{H}$ one just might find $A_0 \in \mathbf{Q}$ with $\|(A - A_0)\psi\| = \epsilon$ — or even only $|(\psi, (A - A_0)\psi)| = \epsilon$. *Then one predicts that* $\bar{a} = (\psi, A_0\psi)$, *with a possible error of* ϵ.

Of course we know from perturbation theory that even a "uniformly close" A_0 might have a very different spectral theory. Such consequences we leave without an answer, at present.

An example: Consider x_1, the first coordinate of the location of the particle. Multiplication by x_1 defines a self-adjoint operator of **H** — an observable A. We find that $A \notin \mathbf{Q}$: Position *cannot* be predicted with *absolute* accuracy. Experimentally, this is clear. Often, an electron acts like a wave. That wave's length should give an idea of a lower limit of accuracy in measuring position.

It should be desirable to describe a *maximal* algebra **Q** of precisely predictable observables. Or else, we should characterize **Q** by physically natural conditions. We will discuss the latter in section 2. Surprisingly, this may be done, to some extent. For technical reasons we offer **P**, an algebra of *global pseudodifferential operators* ($\psi do - s$). We assume the potentials $\mathbf{A}, \mathbf{V}$ as smooth functions, with some growth conditions at ∞.

P is a subalgebra of $Op\psi\mathbf{c}$ defined with the symbols $a \in \psi c$, with $\psi c = \{a(x,\xi) \in C^\infty(\mathbb{R}^6)$: a satisfies (0.4)$\}$, where

$$a^{(\alpha)}_{(\beta)}(x,\xi) = \partial_\xi^\alpha \partial_x^\beta a(x,\xi) = O((1+|x|)^{m_2-|\beta|}(1+|\xi|)^{m_1-|\alpha|}), \tag{0.4}$$

$x, \xi \in \mathbb{R}^3$, with some "order" $m = (m_1, m_2) \in \mathbb{R}^2$. Then of course

$$Au(x) = a(x,D)u(x) = \frac{1}{(2\pi)^3}\int d\xi \int dy e^{i\xi(x-y)} a(x,\xi)u(y), \quad u \in \mathbf{S}. \tag{0.5}$$

The algebra **P** then consists of all $A = a(x,D) \in Op\psi\mathbf{c}_m$ with

$$A_t = e^{iHt}Ae^{-iHt}, \ \partial_t^j A_t \in C^\infty(\mathbb{R}, Op\psi\mathbf{c}_{m-je^2}), \ j = 0,1,\ldots, \tag{0.6}$$

where $e^2 = (0,1)$, and $Op\psi\mathbf{c}_m \subset Op\psi\mathbf{c}$ is the subset of order m.

For a more detailed description of **P** we mention three criteria, discussed in detail in [C1], [C2], [C5]. Assume *here* our strongest growth condition: For $a(x) = \mathbf{V}(x), \mathbf{A}_j(x)$ require $\lim_{|x|\to\infty} a(x) = 0$, and

$$a^{(\alpha)}(x) = \partial_x^\alpha a(x) = O((1+|x|)^{-|\alpha|}), \ x \in \mathbb{R}^3, \text{ all } \alpha. \tag{0.7}$$

Under such conditions the Dirac Hamiltonian H has the well-known gap in its spectrum: For $|\lambda| < 1$ the spectrum is discrete. Let $\lambda_0 \in (-1,1)$ not be an eigenvalue, and $1 = P + Q$ be the decomposition of 1 with projections P, Q pertaining to $Sp(H)$ above and below λ_0. One finds that both P and Q are ψdo's in $Op\psi\mathbf{c}_0$ ([C5]). The decomposition is unique modulo order $-\infty$ (mod $\mathbf{O}(-\infty)$), in the sense that $\varphi\rangle\langle\varphi \in \mathbf{O}(-\infty)$ for any eigenvector φ, $H\varphi = \lambda\varphi$, $|\lambda| < 1$.

(A) *The algebra* $\mathbf{P}$ *consists precisely of all* $A \in Op\psi\mathbf{c}$ *such that* PAQ *and* QAP *have order* $-\infty$ — they have all orders $m = (m_1, m_2) \in \mathbb{R}^2$. (Note, we clearly get $PAQ, QAP \in Op\psi\mathbf{c}$, by calculus of $\psi do - s$.)

Another description of $\mathbf{P}$ using the "Foldy–Wouthuysen transform": In [C2] we showed that there exists a unitary $M = m(x, D) \in Op\psi\mathbf{c}_0$, decoupling the Dirac Hamiltonian into a pair of 2×2-matrix-operators, again only (mod $\mathbf{O}(-\infty)$), under assumptions (0.7):

$$M^*HM = \begin{pmatrix} H_+ & 0 \\ 0 & H_- \end{pmatrix} + \Gamma,\ \Gamma \in \mathbf{O}(-\infty), \tag{0.8}$$

with (2×2-matrices) $H_\pm \in Op\psi\mathbf{c}_{e^1}$. Refer to M as the Foldy–Wouthuysen transform. We have the criterion, below:

(B) $\mathbf{P}$ *consists precisely of all* $A \in Op\psi\mathbf{c}$ *with* $M^*AM = \begin{pmatrix} V & W \\ X & Y \end{pmatrix}$ *where the* (2×2-blocks) X, W *belong to* $\mathbf{O}(-\infty)$.

A third description of $\mathbf{P}$:

(C) *For every symbol* $a(x,\xi) \in \psi\mathbf{c}_m$ *commuting with* $h(x,\xi)$ *of* (0.2), *for all* (x,ξ), *there exists a symbol* $z(x,\xi) \in \psi\mathbf{c}_{m-e}$, $e = (1,1)$, *such that* $a_0(x,D) = a(x,D) + z(x,D) \in \mathbf{P}$. *Vice versa for any* $A_0 \in \mathbf{P}$ *the symbol* $a_0(x,\xi)$ *allows a split* $a_0 = a + z$, *where* $z \in \psi\mathbf{c}_{m-e}$ *and* $a \in \psi\mathbf{c}_m$ *commutes with* $h(x,\xi)$, *for all* (x,ξ).

To make our point with the proposal, regarding v. Neumann's observable concept we mainly use criterion (C): *Energy* is given by the self-adjoint operator H. Trivially we get $H \in \mathbf{P}_{e^1}$, $e^1 = (1,0)$, since $e^{iHt}He^{-iHt} = H$: *H is precisely predictable. Momentum* components belong to $Op\psi\mathbf{c}_{e^1}$, with symbols ξ_1, ξ_2, ξ_3. *Location* components x_1, x_2, x_3 belong to $Op\psi\mathbf{c}_{e^2}$, and *angular momentum* components $(x \times D)_j$ to $Op\psi\mathbf{c}_e$. These observables are *not* precisely predictable — they do not belong to $\mathbf{P}$.

But their symbols $\xi_j, x_j, (x \times \xi)_j$ *commute* with $h(x,\xi)$ of (0.2). Thus criterion (C) supplies a lower order *correction symbol* $z(x,\xi)$. We are tempted to use $A_0 = a(x,D) + z(x,D)$ for predictions. We had the physical constants $\hbar, c, e, m = 1$. If they are reintroduced, corrections seem of proper magnitude to reflect experimental uncertainties.

There are dynamical observables (like spin and current) still belonging to $Op\psi\mathbf{c}$, but with symbol *not* commuting with $h(x,\xi)$, so that criterion (C) does not apply. For such observables physical states exist, in which they *cannot be predicted at all.* More details on spin and current in section 3. Here we want to make a *general point*, regarding observables with "non-commuting symbol":

Given $A = a(x,D) \in Op\psi\mathbf{c}_m$, *where* (the 4×4-matrix) $a(x,\xi)$ *does not necessarily commute with* $h(x,\xi)$. Then A still is predictable (with more or less accuracy) in all physical states where it is known with *certainty* that the particle is an electron. [Or similarly, if we know that it is a positron (with certainty).] Mathematically, the fact that we have an electron and not a positron, is interpreted to mean that, with P,Q of (A), we have $P\psi = \psi$ [or, $Q\psi = \psi$, for the positron]. We will use this language, although there clearly might be electrons of energy $< \lambda$ or positrons of energy $> \lambda$, with λ of criterion (A). [Ignore "charge conjugation", at present, so normal positrons have $\lambda < 0$.] At any rate, this pertains to a split

$$\mathbf{H} = \mathbf{H}_+ \oplus \mathbf{H}_-, \ \mathbf{H}_+ = \operatorname{im} P, \ \mathbf{H}_- = \operatorname{im} Q \tag{0.9}$$

of $\mathbf{H}$ into an orthogonal direct sum, and asking either $\psi \in \mathbf{H}_+$ or $\psi \in \mathbf{H}_-$.

If $\psi \in \mathbf{H}_\pm$ for "+" or for "−", then let $A_1 = PAP + QAQ$. Clearly

$$\bar{a} = (\psi, A\psi) = (\psi, A_1\psi). \tag{0.10}$$

By calculus of $\psi do - s$, if $p(x,\xi)$, $q(x,\xi)$ are the symbols of P,Q, and p_0, q_0 are the projections of $h(x,\xi)$ (of (0.2)) to the eigenvalues $\lambda_\pm(x,\xi) = \mathbf{V}(x) \pm \sqrt{1+(\xi-\mathbf{A}(x))^2}$, we first have $p - p_0$, $q - q_0 \in \psi\mathbf{c}_{-e}$, and also, $\operatorname{symb}(A_1) = pap + qaq (\operatorname{mod} \psi\mathbf{c}_{m-e})$. It follows that

$$\operatorname{symb}(A_1) = p_0 a p_0 + q_0 a q_0 = a_2(x,\xi) (\operatorname{mod} \psi\mathbf{c}_{m-e}).$$

Here $a_2(x,\xi)$ commutes with $h(x,\xi)$. Hence $z \in \psi\mathbf{c}_{m-e}$ exists such that $A_0 = a_2(x,D) + z(x,D) \in \mathbf{P}_m$ using criterion (C). Therefore also A_1 differs from A_0 by an operator B of lower order, and we get

$$\bar{a} = (\psi, A_0\psi) + (\psi, B\psi), \text{ where } A_0 \in \mathbf{P}, \ B \in Op\psi\mathbf{c}_{m-e}. \tag{0.11}$$

The usefulness of (0.11) depends on the question whether "B of lower order" means that $|(\psi, B\psi)|$ is small. On the other hand, the general principle we used seems to be a perfect motivation for the construction of *fermionic Fock space*, used in quantum field theory. There one practically *rejects* superpositions ψ of $\psi_\pm \in \mathbf{H}_\pm$. At least in a state $\psi = \psi_+ + \psi_-$ the two components are treated very differently. This is offered as a final cure for the maladies of Dirac theory.

Interesting also: Every precisely predictable observable $A \in \mathbf{P}$ may be implemented in the Fock space

$$\begin{aligned} \mathbf{F} &= \bigoplus_{m,n=0}^{\infty} \mathbf{F}^{m,n}, \ \mathbf{F}^{m,n} = \mathbf{F}_+^m \oplus \mathbf{F}_-^n, \\ \mathbf{F}_\pm^j &= \mathbf{H}_\pm \wedge \cdots \wedge \mathbf{H}_\pm \ (j \text{ copies}), \ \mathbf{F}^0 = \mathbb{C}. \end{aligned} \tag{0.12}$$

By 'observable' we here mean a formally self-adjoint $A \in Op\psi\mathbf{c}$ with $A : \mathbf{S} \to \mathbf{S}$ having a *unique* self-adjoint extension. This is a direct consequence of the Shale–Stinespring criterion applied to the unitary e^{iAt}, using criterion (A) above.

Note, on the other hand that the spin s, while being an operator of $Op\psi\mathbf{c}$, cannot be implemented in $\mathbf{F}$, again by Shale–Stinespring.

A final example of a nonpredictable observable. Set $\mathbf{A}(x) \equiv 0$, $\mathbf{V}(x) = \frac{c}{|x|}$, with a constant c. This Coulomb potential $\mathbf{V}(x)$ must be modified ["capped"] near 0, where it is not C^∞, but then it will satisfy our general assumptions. Its symbol commutes with h, hence criterion (C) will supply a correction $\mathbf{V}_{\text{corr}}$ for the observable $\mathbf{V}(x)$ — while $\mathbf{V}(x)$ also serves as potential $\mathbf{V}$ in (0.2). For details on $\mathbf{V}_{\text{corr}}$ see section 4. An inspection shows that $\mathbf{V}_{\text{corr}}$ may be predicted with acceptable accuracy as long as $\psi(x)$ vanishes near 0. However, this no longer remains true if the "cap" on $\mathbf{V}$ is applied high enough, and supp ψ comes very close to 0 (cf. section 4).

Note, an electron very close to 0 might amount to a beta ray emitted by a nucleus, in radio-active decay. One even might ask whether the neutrino emitted simultaneously will amount to a breakdown of energy conservation, under these circumstances.

While proofs of above criteria and of other facts mentioned above were given in [C1] through [C5], we will use the sections below to fill in further detail and make additional points. Section 1: Invariant algebra and classical orbits. Section 2: Invariant algebra and Poincare smoothness. Section 3: Spin and current. Section 4: Nearly singular potentials.

We again express our thanks and appreciation to G. Lumer and A. Unterberger for their interest in these problems. Without them we might not have found the incentive to continue.

1. Invariant algebra and classical orbits

In [C1] we achieved a simple connection between classical and quantum mechanics by looking at the time propagation of the symbol of $A \in \mathbf{P}_m$. By definition of $\mathbf{P}$, $A_t = e^{iHt}Ae^{-iHt}$ again is a *ψdo* of $Op\psi\mathbf{c}_m$; we must have $A_t = a_t(x, D)$, with some $a_t(x,\xi) \in \psi\mathbf{c}_m$. In [C1], Theorem 4.1, we showed that, modulo lower order, a_t may be obtained by letting $a(x,\xi)$ propagate along the Hamiltonian flow of the two eigenvalues of $h(x,\xi)$. More specifically, we rather focus on the decomposition $a = q + z$ of criterion (C), written as $a = q(\text{mod } \psi\mathbf{c}_{m-e})$, where $q(x,\xi)$ commutes with $h(x,\xi)$. Here $[h,q] = 0$ (for the 4×4-matrices q and h) means that $q(x,\xi)$ transforms each of the two eigenspaces $\mathbf{S}_\pm(x,\xi)$ of $h(x,\xi)$ to its eigenvectors

$$\lambda_\pm(x,\xi) = \mathbf{V}(x) \pm \sqrt{1 + (\xi - \mathbf{A}(x))^2} \tag{1.1}$$

to itself. q induces Hermitian operators $q_\pm(x,\xi) : \mathbf{S}_\pm(x,\xi) \to \mathbf{S}_\pm(x,\xi)$.

$h(x,\xi)$ is diagonalized by the unitary matrix (with $\langle\kappa\rangle = \sqrt{1+\kappa^2}$)

$$\begin{aligned} \chi(x,\xi) &= \sqrt{\frac{1}{2}\left(1+\frac{1}{\langle\kappa\rangle}\right)}\left\{1+\frac{\alpha\beta\kappa}{1+\langle\kappa\rangle}\right\}, \\ \kappa &= \xi - \mathbf{A}(x), \\ \chi^* h\chi &= \begin{pmatrix} \lambda_+ & 0 \\ 0 & \lambda_- \end{pmatrix}, \end{aligned} \tag{1.2}$$

where $\chi(x,\xi)$ also is a symbol of $\psi\mathbf{c}_0$. Thus the first two (last two) columns of $\chi(x,\xi)$ form orthonormal bases of $\mathbf{S}_+(x,\xi)$ (of $\mathbf{S}_-(x,\xi)$). Then the operators $q_\pm(x,\xi)$ define Hermitian 2×2-matrices $q_\pm(x,\xi)$. Now we use the *Wightmann-representation*: The Hermitian 2×2-matrix $q_\pm(x,\xi)$ is uniquely represented as

$$q_\pm(x,\xi) = v_{0\pm}(x,\xi) + \sigma\cdot v_\pm(x,\xi), \tag{1.3}$$

with a real-valued symbol $v_{0\pm}(x,\xi) \in \psi\mathbf{c}_m$, and a 3-vector $v_\pm(x,\xi)$, components $\in \psi\mathbf{c}_m$, and with Pauli-matrices $\sigma = (\sigma_1,\sigma_2,\sigma_3)$. We use

$$\sigma_1 = \begin{pmatrix} 0 & i \\ -i & 0 \end{pmatrix},\ \sigma_2 = \begin{pmatrix} 0 & 1 \\ 1 & 0 \end{pmatrix},\ \sigma_3 = \begin{pmatrix} 1 & 0 \\ 0 & -1 \end{pmatrix}, \tag{1.4}$$

also for our Dirac matrices

$$\alpha_j = \begin{pmatrix} 0 & i\sigma_j \\ -i\sigma_j & 0 \end{pmatrix},\ j=1,2,3,\ \beta = \begin{pmatrix} I & 0 \\ 0 & -I \end{pmatrix},\ I = \begin{pmatrix} 1 & 0 \\ 0 & 1 \end{pmatrix}. \tag{1.5}$$

The *classical motion — of orbit and spin of the particle* — then is recovered as follows: The operator $A_t = a_t(x,D) = e^{-Ht}Ae^{-iHt}$ allows the same decomposition: $a_t = q_t + z_t$, $[q_t,h] = 0$, $z_t \in \psi\mathbf{c}_{m-e}$. Repeating the above with q_t instead of q, we obtain time-dependent symbols $v_{0\pm} = v_{0\pm,t}(x,\xi)$, and 3-vectors $v_{\pm,t}(x,\xi)$.

In [C1] it was seen that *the propagation* $(v_{0\pm}, v_\pm) \to (v_{0\pm,t}, v_{\pm,t})$ *is governed by the Hamiltonian flow* $\nu_\pm$ *of the eigenvalue* $\lambda_\pm$:

The Hamiltonian system of ordinary differential equations

$$\frac{dx}{dt} = \lambda_{\pm|\xi}(x,\xi),\ \frac{d\xi}{dt} = -\lambda_{\pm|x}(x,\xi) \tag{1.6}$$

for the 6 unknown functions $x(t),\xi(t)$ generates a flow

$$\nu^\pm(t,x^0,\xi^0) = (x^\pm(t,x^0,\xi^0),\xi^\pm(t,x^0,\xi^0)),\ t\in\mathbb{R},\ x^0,\xi^0\in\mathbb{R}^3, \tag{1.7}$$

where $x(t,x^0,\xi^0)$, $\xi(t,x^0,\xi^0)$ just is the solution of (1.6) with initial values x^0,ξ^0. (1.7) indeed defines a smooth family of diffeomorphisms $\mathbb{R}^6 \to \mathbb{R}^6$,

$t \in \mathbb{R}$, and it preserves symbols of $\psi\mathbf{c}_m$, as seen in [C3], 6.6: The composition $a \circ \nu_t = a(x_t(x,\xi), \xi_t(x,\xi))$, where $\nu_t(x,\xi) = \nu(-t,x,\xi)$ again is $\psi\mathbf{c}_m$, for each $a \in \psi\mathbf{c}_m$.

Note (1.7) is the characteristic flow of the first order PDE

$$\partial\theta/\partial t - \langle\lambda_\pm, \theta\rangle = 0 \text{ where } \langle\lambda,\theta\rangle = \lambda_{|\xi}\theta_{|x} - \theta_{|\xi}\lambda_{|x} \tag{1.8}$$

is the *Poisson bracket* of λ and θ. In particular, this implies

(i) (1.8) with initial values $\theta(0,x,\xi) = \theta_0(x,\xi)$ is solved by $\theta(t,x,\xi) = \theta_0 \circ \nu_t$;

(ii) differentiation along the flow, for any $\theta(t,x,\xi)$ is given by the expression

$$\nabla_t\theta = \frac{d}{dt}(\theta(t,\xi_\tau(x,\xi)) = \partial\theta/\partial t - \langle\lambda_\pm,\theta\rangle. \tag{1.9}$$

In [C3] it was shown:

(a) *The flow ψ_t precisely describes the classical orbits of (the spinless) electron* (for "+") *and positron* (for "−").

(b) *The real-valued symbol $\upsilon_{0\pm,t}$ solves the PDE* (1.8), *so that*

$$\upsilon_{0\pm,t} = \upsilon_{0\pm} \circ \nu_t. \tag{1.10}$$

That is, $\upsilon_{0\pm}$ just flows along the particle flow.

(c) *The 3-vector-valued symbol $\upsilon_{\pm,t}$ satisfies a differential equation, exactly as if it were a magnetic moment vector, influenced by the (relativistically corrected) magnetic field*

(1.11)

$$\mathbf{B}^\sim = \mathbf{B} + (1 + \sqrt{1 + x^{\cdot 2}})^{-1}(x^\cdot \times \mathbf{E}), \text{ with } \mathbf{E} = \text{grad } \mathbf{V}, \ \mathbf{B} = \text{curl } \mathbf{A},$$

with equations of motion (1.6) *used to express ξ by $x^\cdot$ and x:*

$$\nabla_t \upsilon_{\pm,t} \pm \sqrt{1 - x^{\cdot 2}}\mathbf{B}^\sim \times \upsilon_{\pm,t} = 0, \ \upsilon_{\pm,0} = \upsilon_\pm, \tag{1.12}$$

with ∇_t of (1.8). [Remember, we have $e = m = c = \hbar = 1$ here.]

Notice, this relation between classical and quantum theory does *not* reflect the *Stern–Gerlach effect*: Our orbits are that of a spinless electron, while it is known that the experimentally observed orbits depend on the spin of the electron.

Here our present improvement will fit in: First it must be recalled, that the equations for the orbits are 'asymptotic': If a real-valued symbol $\epsilon_\pm(x,\xi)$ of order $-e^2 = (0,-1)$ is added to $\lambda_\pm(x,\xi)$ we will get slightly different orbits (only asymptotically agreeing with the above ones), but the same theory.

We will get orbits incorporating the Stern–Gerlach effect, if we first apply the Foldy–Wouthuysen transform of (B), section 0. The Dirac equations then will assume the form

$$\partial\psi/\partial t + i\begin{pmatrix} \Lambda_+ & P \\ Q & \Lambda_- \end{pmatrix}\psi, \text{ with } P = Q^* \in O(-\infty). \tag{1.13}$$

In [C2], (3.4), we found, effectively, for the upper left corner (κ of (1.2))

$$\mathrm{symb}(\Lambda_+) = \lambda_+ - \frac{1}{\langle\kappa\rangle}\left\{\mathbf{B} + \frac{1}{2}\frac{1}{1+\langle\kappa\rangle}\mathbf{E}\times\kappa\right\}\cdot\sigma(\mathrm{mod}\ \psi\mathbf{c}_{-e^2-e}). \tag{1.14}$$

Now, the invariant algebra of (1.13) is of the form $\left\{\begin{pmatrix} V & W \\ X & Y \end{pmatrix}\right\}$, $V, W \in Op\psi\mathbf{c}$, $W, X \in O(-\infty)$, (cf. [C4]). For propagation of symbols we may look at the two hyperbolic systems

$$\partial\psi/\partial t + i\Lambda_\pm\psi = 0. \tag{1.15}$$

Both systems fit our assumptions of [C3], but not with the highest order symbols of (1.14), since the originally 2-fold eigenvalues $\lambda_\pm$ now split into 2 eigenvalues each, which do not spread far enough to make it semi-strictly hyperbolic, with that highest order symbol. Still, one may use the 2 eigenvalues of the 2×2-matrix (1.14) instead of λ_+ to generate the classical orbits of the electron. That indeed will generate a Stern–Gerlach effect.

2. Invariant algebra and Poincare smoothness

In this section we examine (slightly?) larger invariant algebras, having essentially the same properties as $\mathbf{P}$, but with physically more natural description.

The time-smoothness (0.6) might be a natural condition. But why "pseudodifferential operator"?

Note that "translation-smoothness" of $\mathbf{A} \in \mathbf{L}(\mathbf{H})$ — both in the configuration and momentum representation — make A a ψdo with symbol $a(x,\xi) \in CB^\infty(\mathbb{R}^6)$ — all derivatives exist and are $O(1)$. This is a necessary and sufficient condition ([C3],8,[Be1]). What we mean precisely: Let $T_z u(x) = u(x+z) = e^{iDz}u(x)$ and $A^\wedge = F^{-1}AF$, with the Fourier transform F, then

$$T_{-z}AT_z,\ T_{-z}A^\wedge T_z \in C^\infty(\mathbb{R}^6, \mathbf{L}(\mathbf{H})), \tag{2.1}$$

in norm convergence of $\mathbf{L}(\mathbf{H})$. The norm-smoothness (2.1) of an $A \in \mathbf{L}(\mathbf{H})$ is necessary and sufficient for "$A = a(x,D)$, $a \in CB^\infty(\mathbb{R}^6)$".

Similarly we may define *rotation and dilation smoothness*, using $S_{\tau o}u(x) = u(\tau o x)$ instead of T_z, with a rotation $o : \mathbb{R}^3 \to \mathbb{R}^3$, and $\tau > 0$. Here $S^*_{\tau o}AS_{\tau o} \in C^\infty$ is enough, since it implies the same for $A^\wedge$. The translation-rotation-dilation smooth operators $A \in \mathbf{L}(\mathbf{H})$ form an algebra $Op\psi\mathbf{s}_0$ of (0-order) ψdo –

s, and the polynomials

$$(2.2)\qquad Op\psi\mathbf{s} = \left\{\sum x^\alpha A_{\alpha\beta} D_x^\beta : A_{\alpha\beta} \in Op\psi\mathbf{s}_0\right\},\ D_x = -i\partial_x,$$

form an algebra $Op\psi\mathbf{s} \supset Op\psi\mathbf{c}$. The symbols of $Op\psi\mathbf{s}$ are characterized with the differential expressions η_{jl}, $j,l = 0,\dots,3$, where

$$(2.3)\qquad \begin{aligned} \eta_{00} &= \sum(\xi_j\partial_{\xi_j} - x_j\partial_{x_j}),\ \eta_{j0} = \partial_{x_j},\ \eta_{0l} = \partial_{\xi_l},\ j,l > 0, \\ \eta_{jl} &= (\xi_j\partial_{\xi_l} - \xi_l\partial_{\xi_j}) + (x_j\partial_{x_l} - x_l\partial_{x_j}). \end{aligned}$$

A function $a(x,\xi)$ belongs to $\psi\mathbf{s}_m$ if and only if any finite application of the expressions η_{jl} (in any order) *always produces a function which is $O((1+|x|)^{m_2}(1+|\xi|)^{m_1})$* ([C3],VIII). A *$\psi do$*-calculus still works in $Op\psi\mathbf{s}$, insofar as products AB with one factor in $Op\psi\mathbf{c}$ still allow an asymptotic Leibniz–Taylor expansion. This is sufficient for definition of an invariant algebra $\mathbf{P}^s$ of (0.2) ([C3],X). We have $Op\psi\mathbf{s} \supset \mathbf{P}^s \supset \mathbf{P}$. A criterion similar to (C) of section 0 still holds for $\mathbf{P}^s$. All dynamical observables belong to $\mathbf{P}$ already, so the enlargement $\mathbf{P} \to \mathbf{P}^s$ seems to have little significance.

But the requirement of translation-rotation-dilation smoothness, imposed on an *observable* might be more natural, physically, as the condition that A be a *ψdo*.

Here we want to reflect on the following question: The spatial translations $x \to x+z$, rotations $x \to ox$, as well as time-translation $t \to t+t_0$ all belong to the Poincare group $\mathbf{p}$ of linear maps of Minkowski-space $\mathbb{R}_4 = \{(t,x)\}$ leaving the Lorentz-metric $t^2 - |x|^2$ invariant, but the dilations $x \to \tau x$ do not belong to $\mathbf{p}$. If dilation smoothness is omitted, from the above, we just lose the differential expression η_{00} from (2.3), producing a larger algebra of symbols.

Suppose we assume *boost-smoothness* instead of dilation-smoothness — where we want to involve the coordinate transform (0.3). Note, the boost (0.3) is a map $\mathbb{R}^4 \to \mathbb{R}^4$, not $\mathbb{R}^3 \to \mathbb{R}^3$, so it cannot directly be used to transform a *ψdo* over $\mathbb{R}^3$. But in section 0 we introduced the unitary maps $R = R_\theta$ transforming the space $\mathbf{H}$ of physical states. By boost-smoothness we mean that, for all $o \in SO(3,\mathbb{R})$,

$$(2.4)\qquad R_\theta^* S_o A S_o^* R_\theta \in C^\infty(\{-\infty < \theta < \infty\}, \mathbf{L}(\mathbf{H})).$$

Clearly this condition depends on the Dirac equation, and its potentials $\mathbf{V}, \mathbf{A}$. At the moment we want to focus on (2.4), for $\mathbf{V} = \mathbf{A} = 0$.

It will be possible to obtain the operator R_θ explicitly, in that case. [Note, this also will strongly simplify the procedures of [C4] in proving the relativistic covariance of the algebra $\mathbf{P}$.]

In [C4], section 1, R was obtained as composition $R = \kappa S_c G$, with

$$\begin{aligned} \kappa &= c_{\theta/2} - \alpha_1 s_{\theta/2},\ c_{\theta/2} = \cosh(\theta/2),\ s_{\theta/2} = \sinh(\theta/2), \\ S_c u(\xi) &= u(c\xi_1, \xi_2, \xi_3) = u(c\xi_1, \xi^\sim), \\ Gu(x) &= g(x, D),\ g(x,\xi) = e^{-i\eta x_1 h(\xi)} \\ &= p(\xi)e^{-i\eta x_1 \langle\xi\rangle} + q(\xi)e^{i\eta x_1\langle\xi\rangle}. \end{aligned} \tag{2.5}$$

Recall, we have (with vanishing potentials)

$$\begin{aligned} H &= h(D),\ h(\xi) = \sum \alpha_j \xi_j + \beta,\ p(\xi) = \frac{1}{2}(1 + h/\langle\xi\rangle), \\ q(\xi) &= \frac{1}{2}(1 - h/\langle\xi\rangle). \end{aligned} \tag{2.6}$$

Here we start with showing that — according to [C4], section 6, the operator R assumes the simple form of a substitution operator, in momentum space — i.e., after applying a Fourier transform. Write

$$G = E_\theta P + E_{-\theta} Q,\ P = p(D),\ Q = q(D),\ E_\theta = e^{-i\eta x_1\langle\xi\rangle}|_{\xi=D}. \tag{2.7}$$

Explicitly we have

$$E_\theta u(x) = \int d\!\!\!/\xi_1 d\!\!\!/\xi^\sim e^{i(x_1(\xi_1 - \eta\langle\xi\rangle) + x^\sim \xi^\sim)} u^\wedge(\xi),\ u \in \mathbf{S}. \tag{2.8}$$

Accordingly,

$$S_c E_\theta u(x) = \int d\!\!\!/\xi_1 d\!\!\!/\xi^\sim e^{i(x_1(\xi_1 c - s\langle\xi\rangle) + x^\sim \xi^\sim)} u^\wedge(\xi),\ u \in \mathbf{S}. \tag{2.9}$$

Here we introduce the new integration variable $\kappa = \xi_1 c - \langle\xi\rangle s$ instead of ξ_1: Get $(\kappa - \xi_1 c)^2 = s^2(\xi_1^2 + \sigma^2)$, $\sigma^2 = \langle\xi^\sim\rangle^2 = 1 + \xi^{\sim 2}$, $\Rightarrow \kappa^2 + \xi_1^2 c^2 - 2\kappa\xi_1 c = s^2\xi_1^2 + s^2\sigma^2 \Rightarrow \kappa^2 + \xi_1^2 - 2\kappa\xi_1 c = s^2\sigma^2 \Rightarrow (\xi_1 - \kappa c)^2 = s^2\sigma^2 + s^2\kappa^2 \Rightarrow \xi_1 = \kappa c \pm s\sqrt{\sigma^2 + \kappa^2}$. Considerations at large and small ξ_1 show that we must choose "+". Accordingly,

$$\kappa = \xi_1 c - \langle\xi\rangle s \Leftrightarrow \xi_1 = \kappa c + s\sqrt{\sigma^2 + \kappa^2},\ d\xi_1 = \left(c + \frac{\kappa}{\sigma^2 + \kappa^2}\right) d\kappa. \tag{2.10}$$

Carrying this into (2.6) and renaming κ to ξ_1 again, we get

$$S_c E_\theta u(x) = \int d\!\!\!/\xi e^{ix\xi}(c + s s_1(\xi)) u^\wedge(\xi_1 c + \langle\xi\rangle s, \xi^\sim) = F^{-1}\Phi T_\theta F u, \tag{2.11}$$

with the Fourier transform F, the multiplication by $\Phi = \Phi_\theta = c + s s_1(\xi)$, and the substitution operator $T_\theta u(\xi) = u(\xi_1 c + \langle\xi\rangle s, \xi^\sim)$. This result suggests to transfer all consideration of R into momentum space:

Proposition 2.1. *Let* $a(M)$ *be defined by* $(a(M)u)(x) = a(x)u(x)$. *Then*

$$a(D) = F^{-1}a(M)F,\ \kappa S_c E_\theta = F^{-1}(\kappa\Phi(M)T_\theta)F,\ \text{for functions } a. \tag{2.12}$$

Hence,

$$R_\theta = F^{-1}\{\kappa\Phi_\theta(M)T_\theta p(M) + \kappa\Phi_{-\theta}T_\theta q(M)\}F. \tag{2.13}$$

Or, in momentum space, if we write $A^\wedge = FAF^{-1}$, *for operators* A,

$$\begin{aligned} H^\wedge &= h(M),\ P^\wedge = p(M),\ Q^\wedge = q(M), \\ R_\theta^\wedge &= \kappa\Phi_\theta(M)T_\theta p(M) + \kappa\Phi_{-\theta}T_{-\theta}q(M). \end{aligned} \tag{2.14}$$

Formula (2.14) still may be improved, as follows: Write $T_\theta p(M) = (T_\theta p(M)T_{-\theta})T_\theta p(M)$, using that $\varphi_\theta(\kappa) = \kappa c + s\sqrt{\sigma^2+\kappa^2}$ has inverse function $\varphi_{-\theta}(\kappa)$, by (2.10), so that also $T_\theta T_{-\theta} = T_{-\theta}T_\theta = 1$, and also using that the matrix $p(\xi)$ is a projection — i.e., $p^2(\xi) = p(\xi)$. Here let us calculate the projection $P' = T_\theta p(M)T_{-\theta}$: Get $p(\xi) = \frac{1}{2\langle\xi\rangle}(\langle\xi\rangle + \sum\alpha_j\xi_j + \beta)$, and $P' = p'(M)$, where $p'(\xi) = p(\varphi_\theta(\xi_1),\xi^\sim)$. But $\langle(\varphi_\theta(\xi_1),\xi^\sim)\rangle^2 = (c\xi_1 - s\sqrt{\sigma^2+\xi_1^2})^2 + \sigma^2 = c^2\xi_1^2 + s^2\sigma^2 + s^2\xi_1^2 + \sigma^2 + 2cs\xi_1\sqrt{\sigma^2+\xi_1^2} = c^2(\xi_1^2+\sigma^2) + s^2\xi_1^2 + 2cs\xi_1\sqrt{\xi_1^2+\sigma^2} = (c\sqrt{\xi_1^2+\sigma^2} + s\xi_1)^2 \Rightarrow \langle(\varphi_\theta(\xi_1),\xi^\sim)\rangle = c\langle\xi\rangle + s\xi_1$. Accordingly,

$$\begin{aligned} p'(\xi) &= \frac{1}{2}\frac{1}{c\langle\xi\rangle + s\xi_1}\{(c\langle\xi\rangle + s\xi_1) + \alpha_1(c\xi_1 + s\langle\xi\rangle) + \alpha^\sim\xi^\sim + \beta\} \\ &= \frac{1}{2}\frac{1}{c\langle\xi\rangle + s\xi_1}\{(c+\alpha_1 s)(\langle\xi\rangle + \alpha_1\xi_1) + \alpha^\sim\xi^\sim + \beta\}. \end{aligned}$$

We get $c + \alpha_1 s = \kappa^{-2}$. Hence $\kappa p'(\xi) = \frac{1}{2}\frac{1}{c+ss_1(\xi)}\{(\langle\xi\rangle + \alpha_1\xi_1)\kappa^{-1} + \kappa(\alpha^\sim\xi^\sim + \beta)\}$. Here we use that $\alpha_1(\alpha^\sim\xi^\sim + \beta) = -(\alpha^\sim\xi^\sim + \beta)\alpha_1$, by the commutation rules for α,β. Hence, $\kappa(\alpha^\sim\xi^\sim + \beta) = (a^\sim\xi^\sim + \beta)\kappa^{-1}$. Conclusion:

$$\kappa p'(\xi) = \frac{1}{c + ss_1(\xi)}p(\xi)\kappa^{-1},\ s_1(\xi) = \xi_1/\langle\xi\rangle. \tag{2.15}$$

Or also, with the definition of the function Φ_θ:

$$\kappa\Phi_\theta(M)T_\theta p(M) = \kappa\Phi_\theta(M)p'(M)T_\theta p(M) = p(M)\kappa^{-1}T_\theta p(M). \tag{2.16}$$

With a similar argument for $\kappa\Phi_{-\theta}T_{-\theta}q(M)$ we arrive at the following conclusion:

Proposition 2.2. *We have*

$$\begin{aligned} R_\theta^\wedge &= p(M)\kappa^{-1}T_\theta p(M) + q(M)\kappa^{-1}T_{-\theta}q(M), \\ R_\tau^{*\wedge} &= R_{-\theta}^\wedge = p(M)\kappa^{-1}T_{-\theta}p(M) + q(M)\kappa^{-1}T_\theta q(M). \end{aligned} \tag{2.17}$$

Since $\{R_\theta\}$ is a Lie-group, it suffices to evaluate the derivatives $(\partial_\theta)^j A_\theta$, $A_\theta = R_\theta^* A R_\theta = R_{-\theta} A R_\theta$ at $\theta = 0$ for any $A \in \mathbf{L}(\mathbf{H})$. Also, we may look at smoothness of $A_\theta{}^\wedge$ instead of A_θ.

Note: $\kappa = \kappa_\theta$ forms a group under matrix multiplication. Also the substitutions T_θ form a group: If $\kappa = \xi_1 c_\theta - s_\theta\sqrt{\xi_1^2+\sigma^2}$, $\lambda = \kappa c_\varphi - s_\varphi\sqrt{\kappa^2+\sigma^2}$, then $\lambda = c_{\theta+\varphi}\xi_1 - s_{\theta+\varphi}\sqrt{\xi_1^2+\sigma^2}$. Clearly κ_θ and T_φ commute, hence also $\Psi_\theta = \kappa_{-\theta}T_\theta$, $W_\theta = \kappa_{-\theta}T_{-\theta}$ are groups. By differentiation we get $\frac{d}{d\theta}V_\theta = \left(\frac{1}{2}\alpha_1 + (\sigma\xi_1 + c\langle\xi\rangle)\partial_{\xi_1}\right)V_\theta$, hence $\frac{d}{d\theta}V_\theta|_{\theta=0} = \frac{1}{2}\alpha_1 + \langle\xi\rangle\partial_{\xi_1}$. Thus,

$$\text{(2.18)}\qquad \begin{aligned} \frac{d}{d\theta}V_\theta &= \left(\frac{1}{2}\alpha_1 + \langle\xi\rangle\partial_{\xi_1}\right)V_\theta, \\ \frac{d}{d\theta}W_\theta &= \left(\frac{1}{2}\alpha_1 - \langle\xi\rangle\partial_{\xi_1}\right)W_\theta, \end{aligned}\qquad \theta\in\mathbb{R}.$$

By a calculation (or using [C3],X,(7.5)) get

$$\text{(2.19)}\qquad p\alpha_1 p = p(\xi)s_1(\xi),\ q\alpha_1 q = -q(\xi)s_1(\xi),\ s_1(\xi) = \xi_1/\langle\xi\rangle.$$

It follows that

$$\text{(2.20)}\qquad Z = \frac{d}{d\theta}R_\theta^\wedge|_{\theta=0} = pL_+p + qL_-q,\ L_\pm = \frac{1}{2}s_1(\xi) \pm \langle\xi\rangle\partial_{\xi_1}.$$

Note pL_+p and qL_-q both are skew symmetric first order differential operators, mapping im $p(\xi)$ and im $q(\xi)$ to itself each. We get

$$\text{(2.21)}\quad \left(\frac{d}{d\theta}\right)^j A_\theta^\wedge|_{\theta=0} = (-1)^j(\mathrm{ad}\ Z)^j A^\wedge,\ (\mathrm{ad}\ Z)X = ZX - XZ,\ j = 1,2,\dots$$

Summarizing we get the following

Theorem 2.3. *For vanishing potentials $\mathbf{V}, \mathbf{A}_j$ boost-smoothness (i.e. condition (2.4)) is equivalent to the following: For $A \in \mathbf{L}(\mathbf{H})$ write*

$$\text{(2.22)}\quad A^\wedge = p(M)A^\wedge p(M) + p(M)A^\wedge q(M) + q(M)A^\wedge p(M) + q(M)A^\wedge q(M).$$

Or, matrix-wise, write $A^\wedge = \begin{pmatrix} G & H \\ J & K \end{pmatrix}$, $G = (pA^\wedge p)|\mathbf{H}_+^\wedge$, $H = (pA^\wedge q)|\mathbf{H}_-^\wedge$, $J = (qA^\wedge p)|\mathbf{H}_+^\wedge$, $K = (qA^\wedge q)|\mathbf{H}_-^\wedge$, *according to the orthogonal direct decomposition* $\mathbf{H} = \mathbf{H}_+^\wedge \oplus \mathbf{H}_-^\wedge = \{\mathrm{im}\ p(\xi)\} \oplus \{\mathrm{im}\ q(\xi)\}$, *with operators* $G : \mathbf{H}_+^\wedge \to \mathbf{H}_+^\wedge$, $H : \mathbf{H}_-^\wedge \to \mathbf{H}_+^\wedge$, $J : \mathbf{H}_+^\wedge \to \mathbf{H}_-^\wedge$, $K : \mathbf{H}_-^\wedge \to \mathbf{H}_-^\wedge$. *Then let $L_{\pm,j}$, $j = 1,2,3$, be defined like $L_\pm$ of (2.20), with ξ_1, ∂_{ξ_1} replaced by ξ_j, ∂_{ξ_j}.*

(i) *Any finite application of* ad $L_{+,j}$ *to G and any finite application of* ad $L_{-,j}$ *to K must be $\mathbf{L}(H_\pm^\wedge, \mathbf{H}_\pm^\wedge)$, respectively;*

(ii) *Any finite application of* $ch_{+,j}(X) = L_{+,j}X + XL_{-,j}$ *to* H *must be* $\mathbf{L}(\mathbf{H}_-^\wedge, \mathbf{H}_+^\wedge)$, *and any finite application of* $ch_{-,j}(X) = L_{-,j}X + XL_{+,j}$ *to* J *must be* $\mathbf{L}(\mathbf{H}_+^\wedge, \mathbf{H}_-^\wedge)$.

Notice that special translations, rotations, together with boosts and time-translations generate a unitary representation of the connected component of 1 of the Poincare group $\mathbf{p}$ on $\mathbf{L}(\mathbf{H})$. Accordingly we introduce the algebra $\mathbf{P}^p$ of *Poincare-smooth* operators by requiring all above smoothness — (spatial) translation, rotation, time translation and boost (and spatial translation in momentum space — i.e., smoothness in $\mathbf{L}(\mathbf{H})$ of the map $A \to e^{izx}Ae^{-izx}$, $z \in \mathbb{R}^3$). Clearly $\mathbf{P}^p \supset \mathbf{P} \cdot \mathbf{P}^p$ consists of $\psi do - s$ with symbol in CB^∞, both in configuration space and in momentum space. Moreover, for any angular derivative ∂_θ we have $(\partial_{\theta_x} - \partial_{\theta_\xi})a(x,\xi) \in CB^\infty$, and the same even after a finite number of applications of these operators. Again this holds also for the symbol of $A^\wedge$. In addition we must have the conditions of Theorem 2.3 satisfied. Their evaluation is left open here, but it is expected that condition (ii) should give something like order $-\infty$ for H and J, and perhaps some calculus of $\psi do{-}s$ for G and K, to obtain parallels of criteria (A), (B), and (C) of section 0.

3. Spin and current

The spin observable normally is represented by the matrices

$$s_1 = \frac{1}{2}\begin{pmatrix} \sigma_1 & 0 \\ 0 & \sigma_1 \end{pmatrix}, \quad s_2 = \frac{1}{2}\begin{pmatrix} \sigma_2 & 0 \\ 0 & \sigma_2 \end{pmatrix}, \quad s_3 = \frac{1}{2}\begin{pmatrix} \sigma_3 & 0 \\ 0 & \sigma_3 \end{pmatrix} \tag{3.1}$$

with Pauli-matrices σ of (1.4). For the current we have components

$$c_j(x^0) = \alpha_j \delta(x - x^0), \quad \alpha \text{ of } (1.5). \tag{3.2}$$

Actually, $c_j(x^0)$ represents *current density*; with expectation value $\bar{c}_j(x^0) = (\psi, \chi_j(x^0)\psi) = \psi^*(x^0)\alpha_j\psi(x^0)$. Together with the particle density $\rho(x^0) = |\psi(x_0)|^2$ — expectation value of the observable $\rho(x^0) = \delta(x - x^0)$ (with Dirac measure δ) the Dirac equation implies

$$\partial\bar{\rho}/\partial t + \operatorname{div}\bar{c} = 0, \tag{3.3}$$

in analogy to the classically expected relation between particle density and current.

Note, the c_j and ρ are not proper observables; due to $\delta(x)$. As densities they are (distribution-) limits of c_j^ϵ and ρ^ϵ, as $\epsilon \to 0$, with δ replaced by $\delta_\epsilon(x) = \delta_1(x/\epsilon)$, $\epsilon > 0$, $\delta_1 \in C_0^\infty(\mathbb{R}^3)$, $\int \delta_1 dx = 1$, where $c_j^\epsilon, \rho^\epsilon \in Op\psi\mathbf{c}_0$. Similarly for the spin components $s_j \in Op\psi\mathbf{c}_0$.

Note, the matrix-multiplication operators s_j, ρ^ϵ, c_j^ϵ do *not* belong to **P**. But now the symbols of s_j and c_j^ϵ do *not* commute with $h(x,\xi)$, so that our criterion (C) cannot be used for construction of corrections. For $A = s_j$ or $A = c_j^\epsilon(x^0)$ we may apply the procedure around (0.10), with a split (0.9) of **H**, yielding approximate expectation values (0.11) for physical states $\psi \in \mathbf{H}_\pm$ only.

The spectral theory of A itself seems to lose its meaning here unless we are prepared to actually replace A by the corresponding A_0 of (0.11). Note, s_1 — for example — has eigenvalues $\pm\frac{1}{2}$, in agreement with physical observations.

In [C1] we obtained a corrected spin s_{corr} (only the first correction calculated). Assuming $\mathbf{A}(x) \equiv 0$, its symbol is

(3.4)
$$\begin{aligned} \mathrm{symb}(s_{\mathrm{corr}}) &= \frac{1}{2}\frac{1}{\langle\xi\rangle^2}\{\mu\times\xi+\rho+(\rho\xi)\xi\}, \\ \mu &= \begin{pmatrix} 0 & \sigma \\ \sigma & 0 \end{pmatrix}, \ \rho = \begin{pmatrix} \sigma & 0 \\ 0 & \sigma \end{pmatrix}. \end{aligned}$$

In diagonal representation of $h(x,\xi)$ this yields the matrix

(3.5)
$$\tau = \frac{1}{2}\sigma + \frac{1}{2}\frac{1}{\langle\xi\rangle^2}\frac{1}{1+a_0}\{(\sigma\xi)\xi - \sigma|\xi|^2\}, \ a_0 = \frac{1}{\langle\xi\rangle},$$

in the upper left corner. This will agree with the traditional $s = \frac{1}{2}\sigma$ if ξ is not too large (for relativistically small velocities).

If $\mathbf{A}(x) = 0$ and $\mathbf{V}(x) = V(|x|)$ is rotationally symmetric, then it is known that the sum $j = x\times\xi + s$ (the total angular momentum) commutes with H. Thus evidently, $j \in \mathbf{P}$ is a precisely predictable observable. But the summands $x\times\xi$ and s do not belong to **P**. For $x\times\xi$ we found a correction $(x\times\xi)_{\mathrm{corr}} = x\times\xi+Z$, $Z \in Op\psi c_0$, using criterion (C). Surprisingly it is found that s_{corr} above coincides with $s - Z$, so that also $j = (x\times\xi)_{\mathrm{corr}} + s_{\mathrm{corr}}$, a possible vote for the replacement $(x\times\xi) \to (x\times\xi)_{\mathrm{corr}}$, $s \to s_{\mathrm{corr}}$.

For a possible replacement $c_j^\epsilon(x^0) \to c_j^\epsilon(x^0)_{\mathrm{corr}}$ it might be crucial to preserve the divergence relation (3.3) in some form.

4. Nearly singular potentials

Let us finally look at the (capped) Coulomb potential mentioned in section 0. Again assume $\mathbf{A}(x) \equiv 0$. We may use criterion (C) to obtain a correction $\mathbf{V}_{\mathrm{corr}} \in Op\psi c_0$ of the observable $\mathbf{V}(x)$. In [C1] we obtained the first approximation of $\mathbf{V}_{\mathrm{corr}}$ as (with $\mathbf{E} = \mathrm{grad}\,\mathbf{V}$)

(4.1)
$$\mathrm{symb}(\mathbf{V}_{\mathrm{corr}}) = \mathbf{V} - \mathbf{E}\cdot\lambda_c, \ \lambda_c = \frac{1}{2(m^2c^2+\xi^2)}(mc\mu+\rho\times\xi), \ \mu,\rho \text{ of } (3.4),$$

where we now have left the physical constants e, m, c in place, but still assume $\hbar = 1$. This is for a general $\mathbf{V}(x)$ satisfying (0.7). In section 0 we mentioned the possible importance of this observable. Below we report the result of a lengthy calculation also incorporating the second approximation. The symbol then obtained (from the iteration of [C1]) is

$$
\begin{aligned}
v(x,\xi) &= \mathrm{symb}(\mathbf{V}_{\mathrm{corr}}(x,\xi)) \\
&= \mathbf{V}(x) - \mathbf{E}(x)\lambda_C(\xi) \\
&\quad -\frac{3}{4c}\frac{1}{(m^2c^2+\xi^2)^2}\{(\mathbf{E}(x)\xi)(\mathbf{E}(x)\alpha) \\
&\quad -\frac{(E(x)\xi)^2}{m^2c^2+\xi^2}(mc\beta+\alpha\xi)\Big\}.
\end{aligned} \tag{4.2}
$$

However, the operator $v(x,D)$ will not be self-adjoint — thus will not qualify as "observable". Instead we must take

$$
\mathbf{V}_{\mathrm{corr}} = \frac{1}{2}\{v(x,D) + (v(x,D))^*\}, \tag{4.3}
$$

with Hilbert space adjoint "$*$". This $\mathbf{V}_{\mathrm{corr}}$ still will belong to $\mathbf{P}_0-$ modulo order $-3e$. It will be self-adjoint — in the proper domain. Interestingly, one finds $\mathbf{V}_{\mathrm{corr}}$ to belong to the algebra finitely generated by multiplications $\mathbf{E}(x)$ and convolutions $D^\alpha(m^2c^2-\Delta)^{-1}$, $|\alpha|\le 2$, an algebra investigated by this author.

Perhaps it might be of interest to reintroduce the Planck constant $\hbar$ at this time: We were setting $\hbar = 1$, so far, but note that $H_\hbar = \mathbf{V} + mc^2\beta + c\alpha\hbar D$ appears if we just replace the constants m, c of our H (with $\hbar = 1$) by $m/\hbar^2$, $c\hbar$, respectively. With this reorientation we find that

$$
\begin{aligned}
\lambda_C(D) &= \hbar(mc\mu + \hbar\rho\times D)(m^2c^2 - \hbar^2\Delta)^{-1}, \\
\mathbf{V}_{\mathrm{corr}} &= \frac{1}{2}(v(x,D) + v(x,D)^*), \\
v(x,D) &= \mathbf{V}(x) - \mathbf{E}(x)\lambda_C(D) \\
&\quad -\frac{3}{4c}\hbar^2(m^2c^2-\hbar^2\Delta)^{-1}\{(\mathbf{E}(x)\hbar D)(\mathbf{E}(x)\alpha) \\
&\quad -(\mathbf{E}(x)\hbar D)^2(mc\beta+\hbar\alpha D)(m^2c^2-\hbar^2\Delta)^{-1}\}
\end{aligned} \tag{4.4}
$$

but in the last two lines all products must be reordered such that all multiplications $\mathbf{E}_j(x)$ are to the left of all convolutions $\varphi(D)$. [Note, with $\hbar \neq 1$ we also have $e^{iHt/\hbar}$, i.e., must replace t by $t/\hbar$.] The approximation of symbols is useful only for large ξ (that is, for velocities close to c). For the capped Coulomb potential it will degenerate for x close to 0, if the cap is applied very close to 0.

Clearly $V_{\rm corr}$ has the dimension of $\mathbf{V}$ — i.e., that of an energy. For a very high cap it is clear that $\mathbf{E}(x)$ will be much more singular than $\mathbf{V}(x)$, so the usefulness of (4.4) may be in doubt. Perhaps a different representation, not using our asymptotics might be of use.

References

[Be1] R. Beals, *Characterization of pseudodifferential operators and applications*, Duke Math. J. **4** (1977), 45–57.

[Bg1] N. N. Bogoliubov and D. V. Shirkov, "Quantum Fields," Benjamin, Reading, MA, 1983.

[C1] H. O. Cordes, *A pseudoalgebra of observables for the Dirac equation*, Manuscripta Math. **45** (1983), 77–105.

[C2] H. O. Cordes, *A pseudodifferential Foldy–Wouthuysen transform*, Communications in PDE **8**(13) (1983), 1475–1485.

[C3] H. O. Cordes, "The Technique of Pseudodifferential Operators," London Math. Soc. Lecture Notes 202, Cambridge University Press, 1995.

[C4] H. O. Cordes, *Lorentz transform of the invariant Dirac algebra*, J. Integr. Equ. Oper. Theory **34** (1999), 9–27.

[C5] H. O. Cordes, *Dirac algebra and Foldy–Wouthuysen transform*, to appear in "Evolution Equations and Their Applications," Bad Herrenalb, Marcel Dekker, 1998.

[DV] D. Dieks and P. Vermaas, "The Modal Interpretation of Quantum Mechanics," Kluver Akad. Pub., Dordrecht, Boston, London, 1998.

[Nm1] J. v. Neumann, "Grundlagen der Quantenmechanik," Dover Publishing, 1943.

[Th] B. Thaller, "The Dirac Equation," Springer Verlag, Berlin, Heidelberg, New York, 1992.

[Un1] A. Unterberger, *A calculus of observables on a Dirac particle*, Ann. Inst. Henri Poincare (Physique Theor.) (1998).

[Un2] A. Unterberger, *Quantization, symmetries and relativity*, Contemporary Math. **214** AMS (1998), 169–187.

H.O. Cordes, Department of Mathematics, University of California, Berkeley, California, 94720, U.S.A.

E-mail address: cordes@math.berkeley.edu

Progress in Nonlinear Differential Equations
and Their Applications, Vol. 42

SOLUTIONS OF THE OPERATOR EQUATION $A^*K + KA + KRK = -W$

RICHARD DATKO
GEORGETOWN UNIVERSITY

Introduction

The operator equation in the title appears in a number of control and stability problems. For example the solutions of Linear Quadratic Regulator (LQR) problems in both finite- and infinite-dimensional settings (e.g. [1]) lead to equations of this type. Then $R = 0$ and W is positive semidefinite the solution, if it exists and is positive semidefinite, determines that the abstract linear differential equation

$$\dot{x} = Ax$$

is Lyapunov stable (see e.g. [1]). This operator equation also arises in linear quadratic differential games (e.g. [1]). The purpose of this note is to show that there is a general abstract setting for the solution which is independent of dimensionality and also the specific context in which the equation arises. For example in both LQR and differential games problems certain a priori assumptions are made on the corresponding linear dynamical systems, such as stabilizability or detectability, which are often difficult to verify in practice (see e.g. [1]). This is particularly true in the case of detectability (e.g. [1]). Another virtue of our approach is that it presents an explicit method for solving these operator equations, whereas the more traditional approaches in infinite dimensional settings are usually only concerned with the existence of solutions and perhaps some of their properties (see e.g. [1]). This is not to infer that our method is some kind of magic bullet, since its major component is the computation of a line integral over the imaginary axis. However this integral can often be used to initiate numerical approximations to solutions, and in a finite-dimensional setting depends only on the ability to compute the eigenvalues of a specific Hamiltonian matrix. The presentation in this note is limited to a relatively narrow class of problems. The main reason for this is to expose our method to as broad a group as possible and not overburden it with excessive technical details. Indeed one may treat the operators involved as matrices. However the methods of proof in many instances must make allowance for infinite-dimensionality and unbounded operators. In the matrix case the method may be succinctly outlined as follows.

Consider the $2n$-dimensional matrix

$$H = \begin{pmatrix} A & R \\ -W & -A^* \end{pmatrix} \tag{0.1}$$

where A is a real $n \times n$ matrix, A^* is its adjoint and R and W are real symmetric matrices. Assume H has no eigenvalues on the imaginary axis. Let Γ represent a Jordan curve in the negative complex plane whose interior contains all the eigenvalues of H with real past negative (there must be n of them). Let

$$\mathcal{P} = \frac{1}{2\pi i}\int_\Gamma (\lambda I - H)^{-1} d\lambda = \begin{pmatrix} P_1 & P_2 \\ P_3 & P_4 \end{pmatrix}, \tag{0.2}$$

where P_j, $1 \le j \le n$, are $n \times n$ matrices. If there exists an $n \times n$ matrix K such that

$$P_3 = KP_1, \quad P_4 = KP_2, \tag{0.3}$$

then K satisfies the matrix equation

$$A^*K + KA + KRK = -W. \tag{0.4}$$

One condition for this to occur is that

$$(I - P_4) \tag{0.5}$$

is invertible (I is the identity matrix). Then

$$K = (I - P_4)^{-1}P_3. \tag{0.6}$$

This method is a variation of a procedure found in [2].

The idea there was to find n-linearly independent solutions of the equation

$$\begin{pmatrix} \dot{x} \\ \dot{q} \end{pmatrix} = H \begin{pmatrix} x \\ q \end{pmatrix} \tag{0.7}$$

in the matrix form $\begin{pmatrix} \underline{\overline{X}}(t) \\ Q(t) \end{pmatrix}$, where $\underline{\overline{X}}(t)$ and $Q(t)$ are $n \times n$ matrices such that $\underline{\overline{X}}(t)$ and $Q(t)$ decay exponentially. Then if $\underline{\overline{X}}^{-1}(t)$ exists

$$Q(t)\underline{\overline{X}}^{-1}(t) = K$$

solves (0.4). However this approach will not work in infinite dimensions since in general the exponential operator e^{Ht} does not exist. Thus one must modify the finite-dimensional approach, which is the content of the next sections.

1. Solutions of $A^*K + KA + KRK = -W$

Assumption 1.1.

(i) A generates a C_0-semigroup in a complex Hilbert space $\mathcal{H}$ (A^* denotes the adjoint of A). $S(t)$ is the associated semigroup.

(ii) R and W are bounded linear operators from $\mathcal{H}$ into itself, i.e., R and W are in $[\mathcal{H}]$.

(iii) There exists a finite subspace $\mathcal{H}_1 \subset \mathcal{H}$ and a subspace $\mathcal{H}_2 \subset \mathcal{H}$ such that $\mathcal{H} = \mathcal{H}_1 \oplus \mathcal{H}_2$ and on $\mathcal{H}_2$

$$|S(t)x| \leq Me^{\alpha t}, \ \alpha > 0$$

for all $x \in \mathcal{H}_1$.

(iv) For some $\sigma > 0$

$$[\lambda I - (A - \sigma I)]^{-1}$$

is L_2-square integrable over $Re\lambda = 0$ and analytic in $Re\lambda > -\epsilon$, $\epsilon > 0$.

(v) If

$$H = \begin{pmatrix} A & R \\ -W & A^* \end{pmatrix}, \tag{1.1}$$

then there exists $\epsilon > 0$ such that

$$(\lambda I - H)^{-1}$$

is square integrable over $Re\lambda = c = \text{const}$ for all $|c| < \epsilon$.

Definition 1.1. For all $\begin{pmatrix} x \\ q \end{pmatrix}$ in $\mathcal{H} \times \mathcal{H}$ and $t > 0$ define

$$\tau(t)\begin{pmatrix} x \\ q \end{pmatrix} = \frac{1}{2\pi}\int_{-i\infty}^{i\infty} e^{iWt}(iWI - H)^{-1}\begin{pmatrix} x \\ q \end{pmatrix} dW. \tag{1.2}$$

Proposition 1.1.

(i)

$$\lim_{t \to 0+} \tau(t) = \begin{pmatrix} P_1 & P_2 \\ P_3 & P_4 \end{pmatrix} = \mathcal{P}, \tag{1.3}$$

where P_j, $1 \leq j \leq 4$, are in $[\mathcal{H}]$.

(ii)

$$\mathcal{P}^2 = \mathcal{P}, \quad P\tau(t) = \tau(t)\mathcal{P} = \tau(t), \tag{1.4}$$

for $t > 0$.

(iii)

(1.5)
$$\begin{bmatrix} P_1 = P_1^2 + P_2P_3, & P_2 = P_1P_2 + P_2P_4 \\ P_3 = P_3P_1 + P_4P_3, & P_4 = P_3P_2 + P_4^2 \end{bmatrix}$$

Proof. From the definition it is easy to show that if $t_1 + t_2 > 0$

(1.6)
$$\tau(t_1 + t_2) = \tau(t_1)\tau(t_2).$$

The proof is identical to the same property for semigroups (see e.g. [3]).

Thus, if one can establish the limit (1.3) the other properties follow immediately. Notice that

(1.7)
$$(\lambda I - H)^{-1} = \begin{pmatrix} S_1(\lambda) & S_1(\lambda)R(\lambda I + A^*)^{-1} \\ -S_2(\lambda)W(\lambda I + A)^{-1} & S_2(\lambda) \end{pmatrix},$$

where

(1.8)
$$S_1(\lambda) = [(\lambda I - A) + R(\lambda I + A^*)^{-1}W]^{-1}$$

and

(1.9)
$$S_2(\lambda) = [(\lambda I + A^*) + W(\lambda I - A)^{-1}R]^{-1}.$$

We can further manipulate S_1 and S_2 as

(1.10)
$$S_1(\lambda) = [(\lambda + \sigma)I - A]^{-1}[I + (\sigma I - R(\lambda I + A^*)^{-1}W)S_1(\lambda)]$$

and

(1.11)
$$S_2(\lambda) = [(\lambda - \sigma)I + A^*]^{-1}[I - (\sigma I + W(\lambda I - A)^{-1}R)S_2(\lambda)].$$

Since by assumption

$$((\lambda + \sigma)I - A)^{-1}$$

is analytic in $Re\lambda > -\epsilon$ and $((\lambda - \sigma)I + A^*)^{-1}$ is analytic in $Re\lambda < \epsilon$,

(1.12)
$$\lim_{t \to 0^+} \tau(t) = \begin{pmatrix} I + R_1 & P_2 \\ P_3 & P_4 \end{pmatrix}$$

where

(1.13)
$$R_1 = \frac{1}{2\pi}\int_{-i\infty}^{i\infty} ((iW + \sigma)I - A)^{-1}[(\sigma I + R(iWI - A)^{-1}W)S_1(iW)]\, dW$$

(1.14)
$$P_2 = \frac{1}{2\pi}\int_{-i\infty}^{i\infty} S_1(iW)R(iWI + A^*)^{-1}dW$$

(1.15)
$$P_3 = \frac{1}{2\pi}\int_{-i\infty}^{i\infty} -S_2(iW)W(iWI - A)^{-1}dW$$

(1.16)
$$P_4 = -\frac{1}{2\pi}\int_{-i\infty}^{i\infty} ((iW - \sigma)I + A^*)^{-1}[(\sigma I + W(iWI - A)^{-1})S_2(iW)]\, dW.$$

This establishes the proposition.

Remark 1.1. $\tau(t)$ may be partitioned as

$$\tau(t) = \begin{pmatrix} T_1(t) & T_2(t) \\ T_3(t) & T_4(t) \end{pmatrix} \tag{1.17}$$

where T_j, $1 \le j \le 4$, are strongly continuous in $[\mathcal{H}]$.

Proposition 1.2.

(i) If $(I - P_4)^{-1} \in [\mathcal{H}]$, then there exists $K \in [\mathcal{H}]$ such that

$$P_3 = KP_1, \quad P_4 = KP_2, \tag{1.18}$$

$$K = (I - P_4)^{-1} P_3. \tag{1.19}$$

(ii) If K is one-to-one, then

$$P_1 + P_2 K = I. \tag{1.20}$$

Proof.

(i) The proof of (i) is immediate from (1.5) if $(I - P_4)^{-1} \in [\mathcal{H}]$.

(ii) Assume K is one-to-one. Then, since

$$K = P_3 + P_4 K = K[P_1 + P_2 K],$$

it follows that (1.20) must hold.

Proposition 1.3. Assume $P_3 = KP_1$, $P_4 = KP_2$ where $K \in [\mathcal{H}]$ is one-to-one. Then

$$T(t) = T_1(t) + T_2(t)K \tag{1.21}$$

is a C_0-semigroup on $[\mathcal{H}]$ and there exists $\delta > 0$ such that

$$|T(t)| \le Me^{-\delta t}, \quad t > 0, \tag{1.22}$$

where $|\cdot|$ is any suitable norm on $[\mathcal{H}]$.

Proof. Clearly for some $\delta > 0$

$$|\tau(t)| \le M_1 e^{-\delta t} \tag{1.23}$$

where $|\cdot|$ is some norm on $[\mathcal{H} \times \mathcal{H}]$. The semigroup property of $T(t)$ follows from the identities

$$\tau(t)\mathcal{P} = \mathcal{P}\tau(t) = \tau(t)$$

$$\mathcal{P} = \begin{pmatrix} P_1 & P_2 \\ KP_1 & KP_2 \end{pmatrix}$$

and

$$P_1 + P_2K = I,$$

since these imply that

$$T_3 = KT_1, \quad T_4 = KT_2, \quad T(0) = I.$$

In some circumstances the operator R in H is analytically dependent on a parameter α. Thus assume

$$R = R(\alpha), \quad \alpha \in U \tag{1.24}$$

is analytic in an open neighborhood U of the origin.

Assumption 1.2. Let Assumption 1.1 and the hypotheses of Proposition 1.2 be satisfied when $\alpha = 0$. Also assume $(\lambda I - H)^{-1}$ is compact in some neighborhood of the origin.

Definition 1.2. $\tau(t,0)$, $\mathcal{P}(0)$ and $T(t,0)$ will respectively denote the operators $\tau(t)$, $\mathcal{P}$ and $T(t)$ when $R = R(0)$.

Proposition 1.4. let Assumption 1.2 be satisfied.

(i) Then there exists an open neighborhood, $\mathcal{V}$, of the origin such that the following operators exist:

$$\tau(t,\alpha) = \frac{1}{2\pi i}\int_\Gamma (\lambda I - H(\alpha))^{-1} e^{\lambda t} d\lambda, \tag{1.25}$$

where $\Gamma = \{\lambda : Re\lambda = 0\}$ is oriented in the positive direction, and

$$\mathcal{P}(\alpha) = \lim_{t\to 0^+} \tau(t,\alpha) = \begin{pmatrix} P_1(\alpha) & P_2(\alpha) \\ P_3(\alpha) & P_4(\alpha) \end{pmatrix}. \tag{1.26}$$

(ii) There exists $K(\alpha) \in [\mathcal{H}]$ such that

$$P_3(\alpha) = K(\alpha)P_1(\alpha), \quad P_4(\alpha) = K(\alpha)P_4(\alpha), \tag{1.27}$$

$$I = P_1(\alpha) + P_2(\alpha)K(\alpha), \tag{1.28}$$

$$\tau(t,\alpha) = \begin{pmatrix} T_1(t,\alpha) & T_2(t,\alpha) \\ K(\alpha)T_1(t,\alpha) & K(\alpha)T_2(t,\alpha) \end{pmatrix} \tag{1.29}$$

and

$$T(t,\alpha) = T_1(t,\alpha) + T_2(t,\alpha)K(\alpha) \tag{1.30}$$

is a C_0-semigroup on $[\mathcal{H}]$.

Proof.

(i) We first show that (1.25) and (1.26) are valid. Thus notice that

$$\begin{aligned}(\lambda I - H(\alpha))^{-1} &= (\lambda I - H(0))^{-1} \\ &\quad + (\lambda I - H(\alpha))^{-1}(H(0) - H(\alpha))(\lambda I - H(0))^{-1}.\end{aligned} \tag{1.31}$$

Thus, by the analyticity of $(\lambda I - H(\alpha))^{-1}$ and $(H(0) - H(\alpha))$, Assumption 1.1 holds in some open neighborhood of the origin for $(\lambda I - H(\alpha))^{-1}$.

Consequently Proposition 1.1 is satisfied and (1.25) and (1.26) hold.

(ii) Let $L(\alpha)$ and $Q(\alpha)$ be defined as follows

$$\begin{bmatrix} P_1(\alpha) + P_2(\alpha)K(0) &= L(\alpha) \\ P_3(\alpha) + P_4(\alpha)K(0) &= Q(\alpha) \end{bmatrix}. \tag{1.32}$$

Clearly by analyticity

$$\lim_{\alpha\to 0} L(\alpha) = I, \quad \lim_{\alpha\to 0} Q(\alpha) = K(0).$$

Hence

$$P_1(\alpha)L^{-1}(\alpha) + P_2(\alpha)K(0)L^{-1}(\alpha) = I \tag{1.33}$$

$$P_3(\alpha)L^{-1}(\alpha) + P_4(\alpha)K(0)L^{-1}(\alpha) = Q(\alpha)L^{-1}(\alpha). \tag{1.34}$$

Let

$$K(\alpha) = Q(\alpha)L^{-1}(\alpha). \tag{1.35}$$

Then, since

$$\mathcal{P}(\alpha) = \mathcal{P}^2(\alpha)$$

$$P_1(\alpha) + P_2(\alpha)K(\alpha) = I \tag{1.36}$$

$$P_3(\alpha) + P_4(\alpha) = K(\alpha) = Q(\alpha)L^{-1}(\alpha). \tag{1.37}$$

Since $P_4(0)$ is compact $K(\alpha)$ also must satisfy

$$K(\alpha) = (I - P_4(\alpha))^{-1}P_3(\alpha).$$

Hence (1.27) is also satisfied.

Proposition 1.5. Let Assumption 1.1 be satisfied and the hypotheses of Proposition 1.2. Then the operator K given by (1.19) satisfies the equation

$$A^*K + KA + KRK = -W \tag{1.38}$$

for all $x \in \mathcal{D}(A)$.

Proof. Let $x_0 \in \mathcal{D}(A)$. Then if $x(t) = T(t)x_0$ and $q(t) = KT(t)x_0$ the pair $x(t)$, $q(t)$ must satisfy

$$\left[\begin{array}{rcl} \dot{x}(t) & = & Ax(t) + Rq(t) \\ \dot{q}(t) & = & -Wx(t) - A^*q(t) \end{array}\right]. \tag{1.39}$$

Thus

$$KAx_0 + KRKx_0 = -Wx_0 - A^*Kx_0.$$

Remark 1.2. Proposition 1.2 gives a method for solving (1.38). However it is clear that more general quadratic equations of the same type can be solved by the projection method. The essential requirement is a suitable contour in the complex plane where conditions similar to Assumption 1.1 and the hypotheses of Proposition 1.2 are satisfied.

Remark 1.3. Let Γ be any Jordan curve in $Re\lambda < 0$ such that $S_1(\alpha)$ (Eq.(1.10)) exists on Γ. Assume $K \in [\mathcal{H}]$ exists and satisfies $P_3 = KP_1$. Then

$$\lim_{t\to 0+} \frac{K}{2\pi i}\int_\Gamma e^{\lambda t} S_1(\lambda)\, d\lambda = -\frac{1}{2\pi i}\int_\Gamma (\lambda I + A^*)^{-1} W S_1(\lambda)\, d\lambda. \tag{1.40}$$

(see e.g. [4], Eqs. (3.4) and (3.5)).

Furthermore if there exists a sequence, $\{\Gamma_j\}$, of such curves and

$$\lim_{t\to 0+} \sum_{j=1}^{\infty} \frac{1}{2\pi i}\int_{\Gamma_j} e^{\lambda t} S_1(\lambda)\, d\lambda = P_1, \tag{1.41}$$

then, if $P_1 \in [\mathcal{H}]$,

$$P_3 P_1^{-1} = K = \left(\sum_{j=1}^{\infty} -\frac{1}{2\pi i}\int_{\Gamma_j} (\lambda I + A^*)^{-1} W S_1(\lambda)\, d\lambda\right) P_1^{-1}. \tag{1.42}$$

The relation (1.42) is a Lyapunov operator, if

$$\dot{x} = Ax$$

is uniformly exponentially stable and $\cup\Gamma_j = \{Re\lambda = 0\}$.

2. Some Applications

Assume

$$\dot{x}(t) = Ax(t) \tag{2.1}$$

is a uniformly exponentially stable system in $\mathcal{H}$ which generates a C_0-semigroup, $S(t)$. Let $W > 0$ (i.e., a positive symmetric operator in $[\mathcal{H}]$). Let

$$H = \begin{pmatrix} A & 0 \\ -W & -A^* \end{pmatrix}. \tag{2.2}$$

Then

$$(\lambda I - H)^{-1} = \begin{bmatrix} (\lambda I - A)^{-1} & 0 \\ -(\lambda I + A^*)^{-1}W(\lambda I - A)^{-1} & (\lambda I + A^*)^{-1} \end{bmatrix}$$

and

$$P_1 = I, \quad P_4 = 0, \quad P_2 = 0$$

$$P_3 = -\frac{1}{2\pi}\int_{-i\infty}^{i\infty} (iWI + A^*)^{-1}W(iWI - A)^{-1}dW$$

$$= \int_0^\infty S^*(t)WS(t)\,dt = K. \tag{2.3}$$

Equation (2.3) is well known and represents a Lyapuov operator (see e.g.[1]).

A Quadratic Game

Let

$$\dot{x}(t) = Ax(t) + Bu(t) + Lv(t) \tag{2.4}$$

where

$$\dot{x}(t) = Ax(t) \tag{2.5}$$

generates a C_0-semigroup on $[\mathcal{H}]$, B is a bounded linear mapping from a Hilbert space H_1 into $\mathcal{H}$ and L is a bounded linear mapping from a Hilbert space H_2 into $\mathcal{H}$.

Let $x_0 \in \mathcal{H}$ and $u : R^+ \to H_1$, $v : R^+ \to H_2$, and $x(t)$ the mild solution of (2.4) such that $x(0) = x_0$. Define the quadratic functional

$$C(u, v, \alpha) = \int_0^\infty \left[(Wx(t), x(t)) + |w(t)|^2 - \frac{|v(t)|^2}{\alpha}\right] dt, \tag{2.6}$$

where $\alpha > 0$. The object is to find

$$m(x_0, \alpha) = \min_u \left(\max_v C(u, v, \alpha)\right). \tag{2.7}$$

The solution, if it exists for all $x_0 \in \mathcal{H}$, reduces to mild solutions of

$$\begin{bmatrix} \dot{x}(t) = Ax(t) + (-BB^* + \alpha LL^*)q(t) \\ \dot{q}(t) = -Wx(t) - A^*q(t), \;\; q(t) = Kx(t) \end{bmatrix}, \tag{2.8}$$

i.e., if $x_0 \in \mathcal{D}(A)$ and a solution exists then (2.8) is satisfied.

Notice that the operator K is unknown. If (2.7) has a solution for all $x_0 \in \mathcal{H}$, then

$$m(x_0, \alpha) = (K(\alpha)x_0, x_0) \geq 0,$$

where $K(\alpha) \in [\mathcal{H}]$ is symmetric.

Thus if

$$R(\alpha) = (-BB^* + \alpha LL^*) \tag{2.9}$$

the problem reduces to finding $K = K(\alpha)$. It is known (see e.g. [1]) that as $\alpha \to 0$ the solution tends to

$$\min_u C(u, 0, 0) = m_0(x_0) \tag{2.10}$$

and that for all $x_0 \in \mathcal{H}$

$$m_0(x_0) \leq m(x_0, \alpha), \quad \alpha > 0 \tag{2.11}$$

and

$$m(x_0, \alpha_1) \leq m(x_0, \alpha_0) \quad \text{if } \alpha_1 < \alpha_0. \tag{2.12}$$

Furthermore there exists $\alpha_0 > 0$ such that no solution exists for $\alpha \geq \alpha_0$ if (2.12) is strictly monotonic. (If monotonicity is not strict α_0 need not exist.)

Some Examples

We present two finite-dimensional examples which what may occur as the paramter α tends to α_0 in a quadratic game.

Example 2.1. Let

$$\dot{x}(t) = \begin{pmatrix} 0 & 1 \\ -1 & -1 \end{pmatrix} x(t) + \begin{pmatrix} 0 \\ 1 \end{pmatrix} u + \begin{pmatrix} 1 \\ 0 \end{pmatrix} v$$

and

$$c(u, v, \alpha) = \int_0^\infty \left[|x(t)|^2 + |u(t)|^2 - \frac{1}{\alpha} |v(t)|^2 \right] dt.$$

Then

$$H = \begin{pmatrix} 0 & 1 & \alpha & 0 \\ -1 & -1 & 0 & -1 \\ -1 & 0 & 0 & 1 \\ 0 & -1 & -1 & 1 \end{pmatrix}.$$

If

$$\alpha < \alpha_0 \doteq .604572948585$$

$K(\alpha)$ exists and $K(\alpha) > 0$. At α_0, $H(\alpha_0)$ has no eigenvalues on $Re\lambda = 0$, but $K(\alpha)$ does not exist. However for $\alpha > \alpha_0$ but near α_0 $K(\alpha) < 0$. For example when $\alpha = 0$

$$K(0) = \begin{pmatrix} 1.378 & .414 \\ - & .682 \end{pmatrix}.$$

When $\alpha = \alpha_0 - .0001$

$$K(\alpha) = \begin{pmatrix} 2107.397 & 1637.457 \\ - & 1273.374 \end{pmatrix}.$$

When $\alpha = \alpha_0 + .0001$

$$H(\alpha) \doteq \begin{pmatrix} -2102.213 & -1635.697 \\ - & -1271.644 \end{pmatrix}.$$

Example 2.2. If we modify Example 2.1 and let

$$\dot{x}(t) = \begin{pmatrix} 0 & 1 \\ -1 & -1 \end{pmatrix} x + \begin{pmatrix} 0 \\ 1 \end{pmatrix} u + \begin{pmatrix} 0 \\ 1 \end{pmatrix} v,$$

the critical α_0 is

$$\alpha_0 = 2\sqrt{3} - 2.$$

At this value $H(\alpha_0)$ has roots on the imaginary axis, so Assumption 1.1 fails. When

$$\alpha = \alpha_0 - .001$$

$$K(\alpha) \doteq \begin{pmatrix} 2.0824 & .577 \\ - & 2.054 \end{pmatrix}.$$

Moreover

$$\lim_{\alpha \to \alpha_0^-} K(\alpha) = K_0 \doteq \begin{pmatrix} 2.155 & .57735 \\ - & 2.1547 \end{pmatrix}.$$

References

[1] R.F. Curtain and H.J. Zwart, *An Introduction to Infinite Dimensional Linear Systems Theory*, Texts in Applied Mathematics, Vol. 21, Springer Verlag, New York, 1995.

[2] J. Doyle, K. Glover, P. Khargonekar and B. Francis, *State space sollutions to standard* $\mathcal{H}_2/\mathcal{H}_\infty$ *control problems*, IEEE Trans. Autom. Control., vol. 34 (1989), pp. 831–847.

[3] E. Hille and R.S. Phillips, *Functional Analysis and Semi-groups*, American Math. Soc., Providence, Rhode Island, 1957.

[4] R. Datko, *Some spectral solutions for a class of LQR problems*, Control and Cybernetics, vol. 19 (1990), pp. 109–129.

Richard Datko, Department of Mathematics, Georgetown University, Washington, DC 20057, U.S.A.

E-mail address: datkoa@erols.com

Progress in Nonlinear Differential Equations
and Their Applications, Vol. 42

FLUTTER AS A PERTURBATION PROBLEM FOR SEMIGROUPS

P. DIEROLF[1]
D. SCHRÖDER
UNIVERSITÄT TRIER

J. VOIGT
TECHNISCHE UNIVERSITÄT DRESDEN

Flutter is an instability phenomenon which may occur when the following three types of forces are present:

inertial forces,

elastic forces, and

aerodynamic forces.

For certain values of the parameters the mechanical system (e.g. an airplane) absorbs enough energy from the surrounding airstream to destroy itself. This type of dynamic instability is observed at bridges (Tacoma bridge, 1940), propellers, turbine blades, aircraft wings, ailerons, rudders, helicopter rotors, empennages ... According to the terminology of vibration theory, flutter is a *self excited vibration.* (We therefore expect homogeneous equations.)

In this talk we use a simple (and well known) model of a bending–torsion wing to show how stability against flutter amounts mathematically to a perturbation problem for semigroups. Figure 1 shows the clamped wing, and Figure 2 shows the cross section of the wing. For simplicity we assume the wing to have constant cross sections.

We describe the deformation of the wing by means of the functions

$$(1)\qquad \begin{aligned} &(y,t) \mapsto w(y,t) && \text{(lateral deflection)},\\ &(y,t) \mapsto \alpha(y,t) && \text{(torsion angle)}. \end{aligned}$$

Balance of forces and balance of moments yield the following coupled system

[1]The talk was given by P. Dierolf

of equations

$$(2)\quad \begin{cases} m \cdot \partial_t^2 w(y,t) - m \cdot s \cdot \partial_t^2 \alpha(y,t) + E \cdot I \cdot \partial_y^4 w(y,t) = L(y,t), \\ (\Theta + m \cdot s^2)\partial_t^2 \alpha(y,t) - m \cdot s \cdot \partial_t^2 w(y,t) - G \cdot J \cdot \partial_y^2 \alpha(y,t) = M_L(y,t). \end{cases}$$

(Our notation is in accordance with the literature; see for instance Bisplinghoff et al. [1955], p. 102 ff.)

At this stage of development, the aerodynamic forces lift density L and moment density M_L act as external forces. But they are functions of the deformation of the wing:

$$(3)\qquad L = \frac{\rho}{2} v^2 \cdot \ell \cdot (c_L' \cdot \beta).$$

Here β is the effective angle of attack, and c_L' is the derivative of the lift coefficient. The velocity $\partial_t w$ of the lateral deflection reduces the angle of attack by γ:

$$(4)\qquad \gamma \approx \tan \gamma = \frac{\partial_t w}{v}.$$

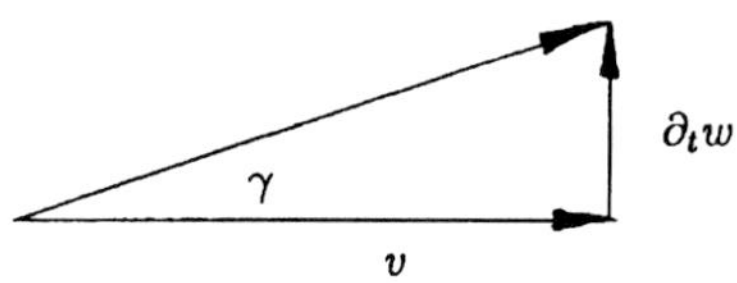

We thus obtain

$$(5)\qquad L = \frac{\rho}{2} v^2 \cdot \ell \cdot c_L' \left(\alpha - \frac{\partial_t w}{v} \right),$$

$$(6)\qquad M_L = -(\ell \cdot a) \cdot L = -\frac{\rho}{2} v^2 \cdot \ell \cdot (\ell \cdot a) \cdot c_L' \cdot \left(\alpha - \frac{\partial_t w}{v} \right).$$

We now consider w and α as functions of the time $t \in \mathbb{R}_0^+$ with values in a function space and introduce the generalized coordinate

$$(7)\qquad q = \begin{pmatrix} w \\ \alpha \end{pmatrix} : \mathbb{R}_0^+ \longrightarrow X([0,b]; \mathbb{R}^2),$$

where the space $X([0,b]; \mathbb{R}^2)$ of functions defined on $[0,b]$ with values in $\mathbb{R}^2$ will be chosen later. The flutter equation then takes the form:

$$(8)\qquad M \cdot \ddot{q} + D(v) \cdot \dot{q} + C(v, \partial_y) q = 0.$$

M is the *mass matrix* with the elements

$$(9)\qquad \begin{cases} m_{11} = 1, & m_{12} = -s, \\ m_{21} = -s, & m_{22} = \frac{\Theta}{m} + s^2. \end{cases}$$

$D(v)$ is the *damping matrix*

$$(10)\qquad \begin{cases} d_{11} = \mu \cdot v \cdot c'_L, & d_{12} = 0, \\ d_{21} = -\mu \cdot \ell \cdot a \cdot v \cdot c'_L, & d_{22} = 0. \end{cases}$$

(Here we have set $\mu := \frac{\rho \cdot \ell}{2m}$.)

$C(v, \partial_y)$ is a *matrix differential operator* which describes the elastic forces and also aerodynamic forces:

$$(11)\qquad \begin{cases} c_{11} = \frac{E \cdot I}{m} \partial_y^4, & c_{12} = -\mu \cdot v^2 \cdot c'_L, \\ c_{21} = 0, & c_{22} = -\frac{G \cdot J}{m} \partial_y^2 + \mu \cdot \ell \cdot a \cdot v^2 \cdot c'_L. \end{cases}$$

We note

$$(12)\qquad D(0) = 0, \quad \text{and} \quad C(0, \partial_y) = \begin{pmatrix} \frac{E \cdot I}{m} \partial_y^4 & 0 \\ 0 & -\frac{G \cdot J}{m} \partial_y^2 \end{pmatrix}.$$

In addition to the equation (8) we have the following time–independent boundary conditions:

Kinematic (or essential) boundary conditions at $y = 0$ for all $t \geq 0$:

$$(13)\qquad \begin{cases} q_1(0,t) = w(0,t) = 0, \\ \partial_y q_1(0,t) = \partial_y w(0,t) = 0, \\ q_2(0,t) = \alpha(0,t) = 0. \end{cases}$$

(The wing is clamped at $y = 0$.)

Dynamic (or remaining) boundary conditions at $y = b$ for all $t \geq 0$:

$$(14)\qquad \begin{cases} \partial_y^2 q_1(b,t) = \partial_y^2 w(b,t) = 0, & \text{no bending moment,} \\ \partial_y^3 q_1(b,t) = \partial_y^3 w(b,t) = 0, & \text{no force,} \\ \partial_y q_2(b,t) = \partial_y \alpha(b,t) = 0, & \text{no torsion moment.} \end{cases}$$

We now transform (8) into a system of first order in time:

$$(15)\qquad u := \begin{pmatrix} q \\ \dot{q} \end{pmatrix} = \begin{pmatrix} w \\ \alpha \\ \dot{w} \\ \dot{\alpha} \end{pmatrix} : \mathbb{R}_0^+ \longrightarrow X([0,b]; \mathbb{R}^4),$$

$$(16)\quad \begin{cases} \dot{u} = Au, \\ A = A(v,\partial_y) = \begin{pmatrix} 0 & I \\ -M^{-1}C(v,\partial_y) & -M^{-1}D(v) \end{pmatrix}, \\ A(v,\partial_y) = A_0(\partial_y) + v\cdot A_1 + v^2\cdot A_2. \end{cases}$$

Well-posedness of the evolution problem formulated above means that A generates a C_0–semigroup $(T(t); t\geq 0)$ on X. Then "stability" of the motion would mean boundedness of the semigroup.

How to choose the space X? The space X should be small enough to allow A to generate a C_0–semigroup on X, and at the same time large enough to contain all "reasonable" initial conditions for our evolution. At the same time, the space X determines the notion of boundedness for our stability concept.

The energy of the mechanical system

$$(17)\quad E(q,\dot{q}) = \frac{m}{2}\int_0^b \dot{q}^T M\dot{q}dy + \frac{1}{2}\int_0^b E\cdot I\cdot (q_1'')^2 dy + \frac{1}{2}\int_0^b G\cdot J\cdot (q_2')^2 dy$$

suggests the Hilbert space

$$(18)\quad \begin{cases} H = H([0,b];\mathbb{C}^4) := \{f\in (L^2)^4; f_1'', f_2' \in L^2, f_1(0)=0, f_1'(0)=0, f_2(0)=0\}, \\ (f|g)_H := \frac{m}{2}\left[\int_0^b \begin{pmatrix} f_3 \\ f_4 \end{pmatrix}^T M \begin{pmatrix} g_3 \\ g_4\end{pmatrix} dy + \int_0^b \frac{E\cdot I}{m}\overline{f_1''}\cdot g_1'' dy + \int_0^b \frac{G\cdot J}{m}\overline{f_2'}\cdot g_2' dy\right] \end{cases}$$

For $v=0$ the mechanical system is *conservative.* Therefore, the evolution is given by a *group of isometries* on the above Hilbert space. The operator

$$A_0 = \begin{pmatrix} 0 & I \\ -M^{-1}C(0,\partial_y) & 0 \end{pmatrix}$$

with domain

$$(19)\quad \begin{cases} D(A_0) = & \{g\in H([0,b];\mathbb{C}^4); \\ & g_1^{(4)} \in L^2([0,b]), g_1''(b)=0, g_1'''(b)=0, \\ & g_2'' \in L^2([0,b]), g_2'(b) = 0, \\ & g_3'' \in L^2([0,b]), g_3(0)=0, g_3'(0)=0, \\ & g_4' \in L^2([0,b]), g_4(0) = 0\} \end{cases}$$

is skew-adjoint (as it should be). Observe that the kinematic boundary conditions are used in the definition of the space H whereas the dynamic boundary conditions determine the domain $D(A_0)$.

The operator $A = A_0 + vA_1 + v^2A_2$ is a perturbation of A_0 by bounded operators. Therefore, A with domain $D(A) := D(A_0)$ generates a C_0–semigroup $(T(t); t \geq 0)$ on H (cf. Pazy [1983], p. 76). The resolvent of A is a Hilbert–Schmidt operator.

We have now reformulated our stability problem as the following perturbation problem:

$$\text{(20)}\quad \left\{ \begin{array}{l} \text{Given a semigroup } (T(t);\ t \geq 0) \text{ of isometries on a Hilbert space } H, \\ \text{determine all bounded perturbations } B \text{ of the generator } A \text{ of } (T(t);\ t \geq 0) \\ \text{such that } (A+B) \text{ generates a bounded semigroup.} \end{array} \right.$$

In our context the evolution is stable if the energy of the system remains bounded.

Engineers expect stability as long as

$$\text{(21)}\qquad \sup\{\mathrm{Re}\ \lambda;\ \lambda \in \sigma(A)\} < 0.$$

The following result by Zabczyk shows that this condition is not sufficient for stability.

Theorem (Zabczyk [1975]). *For all $\omega_s, \omega_0 \in \mathbb{R}$, $\omega_s < \omega_0$, there is a C_0–semigroup $(T(t); t \geq 0)$ on a Hilbert space H with generator A such that*

$$\omega_s = \sup\{Re\ \lambda;\ \lambda \in \sigma(A)\} < \omega_0 = \lim_{t \to \infty} \frac{1}{t} \log ||T(t)||.$$

Zabczyk's construction can be adapted to yield that the resolvent $R(\lambda, A)$ is a Hilbert–Schmidt operator for all $\lambda \in \rho(A)$.

We therefore need additional information about the semigroup to determine stability.

A formulation of such additional conditions was given by Renardy. We are grateful to K.–J. Engel (Università di L'Aquila) for calling our attention to Renardy's result. Let $B[z, r]$ denote the closed ball in $\mathbb{C}$ of radius r centered at z.

Theorem (Renardy [1993]). *Let H be a Hilbert space, A_0 a normal operator, B a bounded operator such that $A = A_0 + B$ generates a C_0–semigroup $(T(t); t \geq 0)$ of bounded operators on H. Assume further that there exist $M > 0$ and $n \in \mathbb{N}$ such that*

(a) $\sigma(A_0) \setminus B[0, M-1]$ consists of isolated eigenvalues λ of finite multiplicity $\mu(\lambda)$.

(b) $\forall\ z \in \mathbb{C}, |z| > M:\ \sum_{\lambda \in \sigma(A_0) \cap B[z,1]} \mu(\lambda) \leq n.$

Then

$$\sup\{Re\ \lambda; \lambda \in \sigma(A)\} = \lim_{t\to\infty}\left(\frac{1}{t}\log ||T(t)||\right),$$

i.e. the spectral bound of the generator coincides with the type of the semigroup.

The known estimates for elliptic problems,

$$\exists\, C_1, C_2 > 0, \mu > 0 : \forall\, k \in \mathbb{N} : C_1 k^\mu \leq 1 + |\lambda_k| \leq C_2 k^\mu$$

(cf., e.g., Triebel [1978], p. 395 ff.), do not imply (b).

Acknowledgement. Research supported by the Stiftung Rheinland–Pfalz für Innovation.

References

[1] Bisplinghoff, R.L., Ashley, H., Halfman, R.L. [1955]: Aeroelasticity. Addison–Wesley Publishing Company, Reading, Mass.

[2] Pazy, A. [1983]: Semigroups of linear operators and applications to partial differential equations. Springer–Verlag, New York.

[3] Renardy, M. [1993]: On the type of certain C_0–semigroups. Commun. Partial Differential Equations **18**, 1299–1307.

[4] Triebel, H. [1978]: Interpolation theory, function spaces, differential operators. North-Holland Publ. Comp., Amsterdam.

[5] Zabczyk, J. [1975]: A note on C_0–semigroups. Bulletin de l'Acad. Polon. Sci. **33**, 895–898.

P. Dierolf, D. Schröder, FB IV – Mathematik, Universität Trier, D–54286 Trier, GERMANY

E-mail address: `dierolf@uni-trier.de`

J. Voigt, Institut für Analysis, Technische Universität Dresden D–01062 Dresden, GERMANY

E-mail address: `voigt@math.tu-dresden.de`

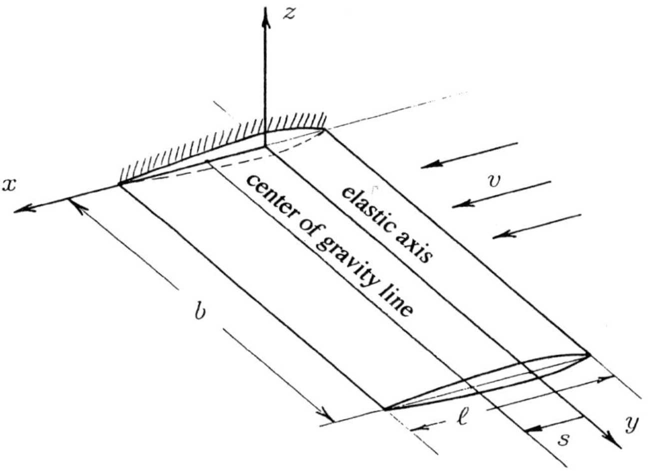

Figure 1

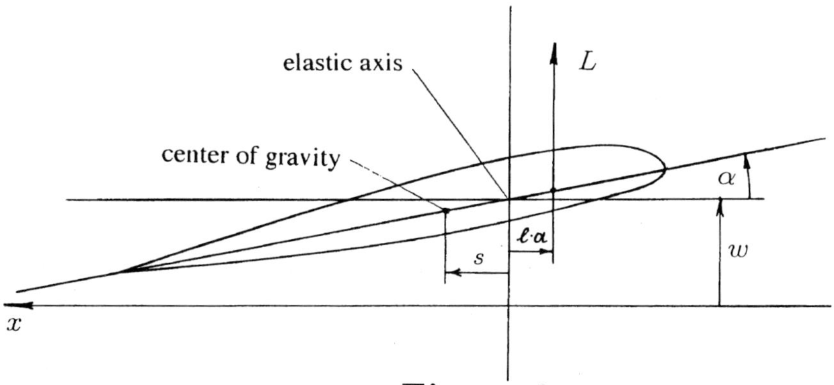

Figure 2

Progress in Nonlinear Differential Equations
and Their Applications, Vol. 42

LINEAR EXTENSIONS OF NONLINEAR SEMIGROUPS

J.R. DORROH
LOUISIANA STATE UNIVERSITY

J.W. NEUBERGER
UNIVERSITY OF NORTH TEXAS

Abstract. This note concerns extending a nonlinear semigroup to a linear one after identifying points with corresponding Dirac measures. Generators of such linear extensions are characterized. A linear extension and hence the original semigroup is recovered from this generator by means of an exponential formula.

1. Introduction

Suppose X is a complete separable metric space and T is a jointly continuous one-parameter semigroup of transformations from X to X. Starting, perhaps, in [7], there have been attempts to assign a generator to classes of such semigroups. Attempting to imitate established linear semigroup theory, (cf [5]), one tried to differentiate T at 0 (in some sense) to obtain a generator B and then show that T could be recovered from B by an exponential formula. The hope was, for substantial classes of semigroups T, to characterize the set of all such generators, a goal being to uncover classes of stationary evolution equations which could be solved. While this program met with some success (for summeries see [1], [2]), theories developed fell far short of early expectations for a nonliner theory comparable in strength to the linear one.

In the early 1970's, work appeared ([3],[8]) which drew inspiraton from work of vonNeumann and Koopman, [10], and most importantly, on seminal work of Sophus Lie (cf [6]) on linear representations of nonlinear semigroups. These last four references seek to *apply* (as distinct from emulating) linear semigroup theory to study nonlinear semigroups. More recently in [4] a complete characterization of generators of such linear representations was obtained for jointly continuous semigroups T on a complete separable metric space X. The present work is based on this last reference. Here we, in effect, extend a nonlinear semigroup to a linear one on a larger space. On the larger space, the extended semigroup has a generator. The extended semigroup, and hence the original one, is recovered from the generator by means of an exponential formula. We characterize the set of all such generators. This gives a version of a long-sought pairing of a wide class of nonlinear semigroups with corresponding generators.

2. A generator and a recovery from that generator

Denote by X a complete separable metric space and denote by $C(X)$ the normed ring (using sup norm) of bounded continuous real-valued functions on X. In addition to the topology induced by the sup-norm on $C(X)$, we follow [4] in using a second topology, the β topology. This topology can be defined by the notion of convergence given by the following (see [4] for more details):

Definition. The statement that the sequence $\{f_n\}_{n=1}^{\infty}$ in $C(X)$ is β convergent to f in $C(X)$ means that there is $M > 0$ so that

$$\|f_n\| \leq M, \ n = 1, 2, \ldots,$$

and

$$\{f_n\}_{n=1}^{\infty}$$

converges to f uniformly on compact subsets of X.

We denote by $C(X)^{*\beta}$ the dual space of $C(X)$ under the β topology, that is, $C(X)^{*\beta}$ is the vector space of all β continuous linear functions from $C(X)$ to R. Note that if $g \in C(X)^{*\beta}$, then g is also in $C(X)^*$, the dual of $C(X)$ derived from $\|\cdot\|$. Hence for $g \in C(X)^{*\beta}$, there is a smallest number $|g|$ so that $\|gf\| \leq |g|\|f\|$, $f \in C(X)$.

Suppose T is a jointly continuous semigroup on X and U denotes the function from $[0, \infty)$ to $C(X)^{*\beta}$ so that

$$(U(t)g)f = g(fT(t)), \ g \in C(X)^{*\beta}, \ f \in C(X), \ t \geq 0 \tag{1}$$

where $fT(t)$ denotes the composition of f with $T(t)$. It was shown in [4] that if $f \in C(X)$ and $h(t) = fT(t)$, $t \geq 0$, then h has range in $C(X)$ and is β continuous. Define

$$C = \{(g, h) \in (C(X)^{*\beta})^2 : h(f) = \lim_{t \to 0+} (1/t)(g(fT(t)) - g(f)), \ f \in C(X)\}. \tag{2}$$

Theorem 1. *If $\lambda \geq 0$ then $(I - \lambda C)^{-1}$ exists, has domain all of $C(X)^{*\beta}$ and*

$$|(I - \lambda C)^{-1} g| \leq |g|, \ g \in C(X)^{*\beta}. \tag{3}$$

Theorem 2. *The domain of C is dense in $C(X)^{*\beta}$ in the sense that if $g \in C(X)^{*\beta}$ then there is $\{g_n\}_{n=1}^{\infty}$ in $D(C)$ so that*

$$\lim_{n \to \infty} g_n(f) = g(f), \ f \in C(X).$$

Theorem 3. *If $\lambda \geq 0$ and $f \in C(X)$, $g \in C^{*\beta}$, then*

$$\lim_{n\to\infty}((I-(\lambda/n)C)^{-n}g)f = g(fT(\lambda)). \tag{4}$$

Note that (2) may be rewritten as

$$C = \{(g,h) \in (C^{*\beta})^2 : h(f) = \lim_{t\to 0+}(1/t)(U(t)g)(f) - g(f)),\ f \in C(X) \tag{5}$$

and (4) may be rewritten

$$\lim_{n\to\infty}(I-(\lambda/n)C)^{-n}g)(f) = (U(\lambda)g)(f), \tag{6}$$

to emphasize that U is differentiated to obtain C and that the resolvent for C generates U.

3. Characterization of generators

Following [4], for T as in the previous section, define A (called the Lie generator of T and also the generator of the linear representaton S on $C(X)$ of T) by

$$A = \{(f,k) \in C(X)^2 : k(x) = \lim_{t\to 0+}(1/t)(f(T(t)x) - f(x)),\ x \in X\}. \tag{7}$$

The following was demonstrated in [4]:

Theorem 4. *Suppose A is a linear operator with domain and range in $C(X)$. Then there is a jointly continuous semigroup T on X so that (7) holds if and only if*
(i) A is a derivation
(ii) the domain of A is β dense in $C(X)$
(iii) for each $\lambda > 0, (I-\lambda A)^{-1}$ has a norm nonexpansive and β continuous inverse defined on all of $C(X)$,
and
(iv) if $\eta > 0$, then the collection

$$\{(I-(\lambda/n)A)^{-n} : 0 < \lambda \leq \eta, n \in N\}$$

is β equicontinuous.

Moreover,

$$\beta - \lim_{n\to\infty}(I-(\lambda/n)A)^{-n}f = fT(\lambda),\ \lambda \geq 0.$$

This leads us to our promised characterization:

Theorem 5. *In order that a linear transformation C on $C(X)^{*\beta}$ arise, using (2) from a jointly continuous semigroup T on X it is necessary and sufficient that*

$$C = A^*$$

for some transformation A satisfying (i)-(iv) of Theorem 4.

We explain precisely our intended meaning for A^* in the above even though we follow conventional usage. The domain of A^* consists of all $g \in C(X)^{*\beta}$ such that the transformation v

$$vf = g(Af), \; f \in D(A)$$

is continuous. For such $g \in C(X)^{*\beta}$, $A^*g = v$.

4. Proofs

If $\lambda > 0$ denote by I_λ the tranformation on $C(X)^{*\beta}$ so that

$$(8) \; (I_\lambda g)(f) = (1/\lambda) \int_0^\infty \exp(-s/\lambda) \, g(fT(s)) \, ds, \; g \in C(X)^{*\beta}, f \in C(X).$$

It is an exercise to note that if $g \in C(X)^{*\beta}, \lambda > 0$, then $I_\lambda g \in C(X)^{*\beta}$.

Proof. [Theorem 1] Suppose $\lambda > 0$, $g \in C(X)^{*\beta}$ and $t > 0$. Note that if $f \in C(X)$, then

$$U(t)(I_\lambda g)f = (I_\lambda g)(fT(t)) = \int_0^\infty \exp(-s/\lambda)g(fT(t+s)) \, ds =$$

$$\int_t^\infty \exp(-(s-t)/\lambda)g(fT(s)) \, ds =$$

$$\exp(t/\lambda) \int_0^\infty \exp(-s/\lambda)g(fT(s)) \, ds - \exp(t/\lambda) \int_0^t \exp(-s/\lambda)g(fT(s)) \, ds.$$

Accordingly,

$$\lambda(1/t)(U(t)(I_\lambda g) - I_\lambda g))f =$$

$$((\exp(t/\lambda) - 1)/t)(I_\lambda g)f - (\exp(t/\lambda/(t\lambda)) \int_0^t \exp(-s/\lambda)g(fT(s))$$

$$\to (1/\lambda)((I_\lambda g)f - gf) \text{ as } t \to 0.$$

Thus $I_\lambda \in D(C)$ and

$$C(I_\lambda g) = (1/\lambda)((I_\lambda g)f - g(f)),$$

i.e.,

$$(I - \lambda C)I_\lambda g = g, \; g \in C(X)^{*\beta}.$$

Now suppose that $g \in D(C)$. Then

$$\lim_{t \to \infty} (1/t)((U(t)g)f - gf) = (Cg)f, \; f \in C(X).$$

In particular,

$$(Cg)(fT(s)) = \lim_{t\to\infty}(1/t)((U(t)g)(fT(s)) - g(fT(s)) =$$

$$\lim_{t\to\infty}(1/t)(g(fT(t+s)) - g(fT(s))),\ f\in C(X),\ s\geq 0.$$

Thus the function $g(fT(\cdot))$ is right differentiable on $[0,\infty)$. Since this right derivative function is also continuous, it follows that that $g(fT(\cdot))$ is continuously differentiable on $[0,\infty)$ and

$$(g(fT(\cdot)))' = (Cg)(fT(\cdot)).$$

Thus if $\lambda > 0, g\in C(X)^{*\beta}, f\in C(X)$,

$$(I_\lambda(Cg))f = (1/\lambda)\int_0^\infty \exp(-s/\lambda)(Cg)(fT(s))\,ds =$$

$$(1/\lambda)\int_0^\infty \exp(-s/\lambda)(g(fT(\cdot)))'(s)\,ds =$$

$$\lim_{s\to\infty}(1/\lambda)\exp(-s/\lambda)g(fT(s)) - (1/\lambda)gf + (1/\lambda^2)\int_0^\infty \exp(-s/\lambda)g(fT(s))\,ds =$$

$$(1/\lambda)(-gf + (I_\lambda g)f).$$

Hence

$$\lambda I_\lambda Cg = -g + I_\lambda g,$$

i.e.,

$$I_\lambda(I-\lambda C)g = g,\ g\in C(X)^{*\beta}.$$

This together with the previous result in this argument gives that $(I-\lambda C)^{-1}$ has domain all of $C(X)^{*\beta}$ and that

$$((I-\lambda C)^{-1}g)f = (1/\lambda)\int_0^\infty \exp(-s/\lambda)\,g(fT(s))\,ds.$$

To show that (3) holds, suppose $g\in C(X)^{*\beta},\ f\in C(X),\ \lambda > 0$. Then

$$|(I_\lambda g)f| \leq 1/\lambda)\int_0^\infty \exp(-s/\lambda)\,|g(fT(s))|\,ds \leq |g|\|f\|$$

and hence $|(I_\lambda g)| \leq |g|,\ g\in C(X)^{*\beta}$. □

Proof. [Theorem 2] To see this observe that if $g\in C(X)^{*\beta}, f\in C(X)$, then

$$\lim_{\lambda\to 0+}(I_\lambda g)f = gf. \qquad \square$$

Proof. [Theorem 3] We follow a similar argument in [4],[8]. For $\lambda > 0, g \in C(X)^{*\beta}, f \in C(X), n$ a positive integer , use (8) to inductively calculate

$$(I^m_{\lambda/n}g)f,\ m = 1, 2, \ldots, n$$

as an m – fold multiple integral. Reduce this multiple integral to a single integral and then proceed as in [4],[8] to deduce the conclusion. □

Proof. [Theorem 5] Suppose T is a jointly continuous semigroup on X. Define A using (7), define U from (1) and define C from (2). Note that from Theorem 4, it follows that A satisfies $(i) - (iv)$ of this theorem. We want to show that $C = A^*$. To this end, suppose that $g \in D(C)$. Then

$$(Cg)f = \lim_{t\to 0+} (1/t)((U(t) - I)g)f = \lim_{t\to 0+} g((1/t)(fT(t) - f)) = g(Af)$$

provided $f \in D(A)$. But this implies that

$$(Cg)f = g(Af) \text{ for all } f \in D(A)$$

and hence $g \in D(A^*)$. Thus A^* is an extension of C.

Observe now that $(I - \lambda A^*)^{-1}$ exists for otherwise there would be $g \in D(A^*)$, $g \neq 0$, so that $(I - \lambda A^*)g = 0$ and so for all $f \in D(A)$,

$$0 = ((I - \lambda A^*)g)f = g((I - \lambda A)f), f \in D(A)$$

and so $g = 0$ since the range of $(I - \lambda A)$ is β dense in $C(X)$, a contradiction. Thus $(I - \lambda A^*)^{-1}$ exists and is an extension of $(I - \lambda C)^{-1}$. But this extension can not be proper since the domain of $(I - \lambda C)^{-1}$ is all of $C(X)^{*\beta}$. Thus we have that $C = A^*$.

Now suppose that C is a linear transformation on $C(X)^{*\beta}$ to itself and $C = A^*$ for some A^* satisfying $(i) - (iv)$ of Theorem **??**. By this theorem, there is a jointly continuous semigroup T from which A arises by means of (7). Let U be the semigroup on $C(X)^{*\beta}$ connected with T by (1). Then U has a generator Q defined by

$$Q = \{(g, h) \in (C(X)^{*\beta})^2 \ :\ h(f) = \lim_{t\to 0+} (1/t)(g(fT(t)) - g(f)),\ f \in C(X)\}. \tag{9}$$

But as in the first part of this argument, it follows that $Q = A^*$ so it must be that $C = Q$. □

If $x \in X$, denote by g_x the member of $C(X)^{*\beta}$ so that $g_x f = f(x)$, $f \in C(X)$. We say that g_x is the Dirac measure associated with x. Note that Theorem 3 gives in this case that

$$\lim_{n\to\infty} ((I - (\lambda/n)C)^{-1} g_x)f = (U(\lambda)g_x)f,\ x \in X,\ f \in C(X),\ \lambda \geq 0.$$

In particular we have that

$$(U(\lambda)g_x)f = g_x(fT(\lambda)) = f(T(\lambda)x) = g_{T(\lambda)x}f, \ \lambda \geq 0,$$

so

$$U(\lambda)g_x = g_{T(\lambda)x}, \ x \in X, \lambda \geq 0$$

and hence we have that if T is a jointly continuous semigroup on X, U its corresponding linear extension and C the generator of U, then T itself is recovered from C in the sense that if $x \in X, \lambda \geq 0$, then the Dirac measure $g_{T(\lambda)x}$ associated with $T(\lambda)x$ is recovered from C.

5. Closing Comments

For any jointly continuous semigroup T on a complete separable metric space we have identified a generator C for a linear extension of T and have shown how to recover our linear extension of T (and hence T itself) from C. We have characterized the set of all such generators C. We hope that this will be useful in dealing with nonlinear semigroups for which no adequate generator theory has been available.

References

[1] H. Brezis, *Operateurs maximaux monotones*, North Holland, 1973.

[2] G. DaPrato, *Applications croissantes et équations d'évolutions dans les espacé de Banach*, Adademic Press, 1976.

[3] J.R. Dorroh, *Semigroups of maps in a locally compact space*, Canad. J. Math 19 (1967).

[4] J.R. Dorroh and J.W. Neuberger, *A theory of strongly continuous semigroups in terms of Lie generators*, J. Func. Analysis 136 (1996), 114-126.

[5] J. Goldstein, *Semigroups of linear operators and applications*, Oxford, 1985.

[6] E. L. Ince, *Ordinary Differential Equations*, Dover 1956.

[7] J.W. Neuberger, *An exponential formula for one-paramter semi-groups of non-linear transformations*, J. Math. Soc. Japan 18 (1966), 154-157.

[8] J.W. Neuberger, *Lie generators for one parameter semigroups of transformations*, J. reine angewandte Mathematik 258 (1973), 133-136.

[9] F.D. Sentilles, *Bounded continuous functions on a completely regular space*, Trans. Amer. Math. Soc. 168 (1972), 311-336.

[10] J. von Neumann , *Dynamical systems of continuous spectra*, Proc. Nat. Acad. Sci. 18 (1932), 278-286.

J.R. Dorroh, Department of Mathematics, Louisiana State University, Baton Rouge, Louisiana 70803, U.S.A.

J.W. Neuberger, Department of Mathematics, University of North Texas, P.O. Box 305118, Denton, TX 76203-5118, U.S.A

E-mail address: jwn@unt.edu

Progress in Nonlinear Differential Equations
and Their Applications, Vol. 42

SECOND-ORDER SUBELLIPTIC OPERATORS ON LIE GROUPS II: REAL MEASURABLE PRINCIPAL COEFFICIENTS

A.F.M. TER ELST
EINDHOVEN UNIVERSITY OF TECHNOLOGY

DEREK W. ROBINSON
AUSTRALIAN NATIONAL UNIVERSITY

Abstract. Let G be a connected Lie group with Lie algebra $\mathfrak{g}$ and $a_1, \ldots, a_{d'}$ an algebraic basis of $\mathfrak{g}$. Further let A_i denote the generators of left translations, acting on the L_p-spaces $L_p(G\,;dg)$ formed with left Haar measure dg, in the directions a_i. We consider second-order operators

$$H = -\sum_{i,j=1}^{d'} A_i\, c_{ij}\, A_j + \sum_{i=1}^{d'} (c_i\, A_i + A_i\, c'_i) + c_0\, I$$

corresponding to a quadratic form with real measurable coefficients c_{ij} and complex c_i, c'_i, $c_0 \in L_\infty$. The matrix $C = (c_{ij})$ of principal coefficients, which is not necessarily symmetric, is assumed to satisfy the subellipticity condition

$$\Re C = 2^{-1}\Big(C + C^*\Big) \geq \mu I > 0$$

uniformly over G.

We prove that H generates a strongly continuous holomorphic semigroup S on L_2 with a kernel K which satisfies Gaussian bounds

$$|K_z(g\,;h)| \leq a\,|z|^{-D'/2} e^{\omega|z|} e^{-b(|gh^{-1}|')^2|z|^{-1}}$$

for $g, h \in G$ and z in a subsector $\Lambda(\theta)$ of the sector of holomorphy. Moreover, the kernel is Hölder continuous and there is a $\nu \in \langle 0,1\rangle$ such that for all $\kappa > 0$ one has estimates

$$|K_z(k^{-1}g\,;l^{-1}h) - K_z(g\,;h)|$$
$$\leq\ a\,|z|^{-D'/2} e^{\omega|z|}\left(\frac{|k|' + |l|'}{|z|^{1/2} + |gh^{-1}|'}\right)^{\nu} e^{-b(|gh^{-1}|')^2|z|^{-1}}$$

for $g, h, k, l \in G$ and z in the subsector with $|k|' + |l|' \leq \kappa\,|z|^{1/2} + 2^{-1}|gh^{-1}|'$.

In addition, if all the coefficients of H are real-valued then

$$K_t(g\,;h) \geq a'\, t^{-D'/2} e^{-\omega' t} e^{-b'(|gh^{-1}|')^2 t^{-1}}$$

for some $a', b' > 0$ and $\omega' \geq 0$ uniformly for $g, h \in G$ and $t > 0$.

1. Introduction

We analyze the semigroup kernels associated with second-order subelliptic operators with variable coefficients acting on the L_p-spaces over a d-dimensional Lie group G begun in [11]. In [11] we considered operators

$$H = -\sum_{i,j=1}^{d'} A_i c_{ij} A_j + \sum_{i=1}^{d'} (c_i A_i + A_i c'_i) + c_0 \tag{1}$$

with complex coefficients c_{ij}, c_i, c'_i, $c_0 \in L_\infty$ where $A_i = dL(a_i)$ denotes the generator of left translations L, acting on one of the classical function spaces, in the direction a_i of the Lie algebra $\mathfrak{g}$ of G, and $a_1, \ldots, a_{d'}$ is an algebraic basis of $\mathfrak{g}$. Subellipticity corresponds to the condition

$$\Re C = 2^{-1}(C + C^*) \geq \mu I > 0 \quad ,$$

in the sense of $d' \times d'$-matrices, uniformly over G. In this paper we analyze operators for which the principal coefficients c_{ij} are real-valued but the lower-order coefficients can still be complex. Moreover, we do not assume that the matrix $C = (c_{ij})$ of principal coefficients is symmetric.

Our main result establishes Hölder continuity estimates comparable to the classic results of Morrey [18], Nash [19] and De Giorgi [16] for strongly elliptic operators with real measurable coefficients on $\mathbf{R}^d$ and Gaussian upper bounds of the type first obtained by Aronson [3]. In addition we derive Gaussian lower bounds on the kernels associated with operators for which all coefficients are real. The proofs are again a combination of parabolic and elliptic techniques based on a mixture of the methods introduced by Nash [19], as refined by Fabes, Saloff-Coste and Stroock [12] [21], together with the method of De Giorgi [16] as influenced by the exposition of Giaquinta [14] and recent work of Auscher [4]. In particular De Giorgi estimates are combined with arguments involving Morrey–Campanato spaces.

Throughout the paper we adopt the notation and definitions of [11]. In particular the operator H formally given by (1) is defined as a closed sectorial operator on the L_2-space, $L_2(G\,;dg)$, where dg denotes left Haar measure. Then H is the generator of a strongly continuous semigroup S on L_2 with a holomorphic extension to a sector $\Lambda(\theta) = \{z \in \mathbf{C}\backslash\{0\} : |\arg z| < \theta\}$ in the complex plane with $\theta \geq \theta_C = \operatorname{arccot}(\gamma_C/\mu_C)$ where

$$\begin{aligned} \mu_C &= \sup\{\, \mu : 2^{-1}(C + C^*) \geq \mu I \,\} \quad , \\ \gamma_C &= \inf\{\, \gamma : \gamma I \geq (2i)^{-1}\Big(C - C^*\Big) \geq -\gamma I \,\} \quad . \end{aligned}$$

(The value of μ_C corresponds to the ellipticity constant.) Moreover, S has a distribution kernel $K_t \in \mathcal{D}'(G \times G)$ such that

$$(\psi, S_t\varphi) = \int_G dg\, \overline{\psi(g)} \int_G d\hat{h}\, K_t(g\,;h)\, \varphi(h)$$

for all $\varphi, \psi \in C_c^\infty(G)$ and $t > 0$ where $d\hat{h}$ denotes right Haar measure. We prove that the kernel has a holomorphic extension K_z to a sector $\Lambda(\theta) \supseteq \Lambda(\theta_C)$ satisfying

$$K_{z_1+z_2}(g\,;h) = \int_G d\hat{k}\, K_{z_1}(g\,;k)\, K_{z_2}(k\,;h) \tag{2}$$

for all $z_1, z_2 \in \Lambda(\theta_C)$. In addition, we derive bounds on K in terms of the right invariant distance $d'(\,\cdot\,;\,\cdot\,)$, the control distance, canonically associated with the algebraic basis $a_1, \ldots, a_{d'}$ (see, for example, [20] Sections IV.2 and IV.4c). This distance has the characterization

$$d'(g\,;h) = \sup\{\, |\psi(g) - \psi(h)| \,:\, \psi \in C_c^\infty(G)\,,\, \sum_{i=1}^{d'} |(A_i\psi)|^2 \le 1,\ \psi \text{ real}\,\}$$

(see, for example, [20], Lemma IV.2.3, or [8], Lemma 4.2). Other parameters which enter the estimates are the subelliptic modulus $g \mapsto |g|' = d'(g\,;e)$, where e is the identity of G, and the local dimension D', i.e., the integer for which the left Haar measure $|B'(g\,;r)|$ of the ball $B'(g\,;r) = \{\, h \in G \,:\, d'(g\,;h) < r\,\}$ satisfies bounds

$$c^{-1} r^{D'} \le |B'(e\,;r)| \le c\, r^{D'} \tag{3}$$

for some $c > 0$ and all $r \le 1$.

Theorem 1.1 *Let H be a subelliptic operator of the form (1) with real measurable principal coefficients c_{ij} and complex measurable lower order coefficients c_i, c_i' and c_0. Then, for each $\theta \in \langle 0, \theta_C\rangle$, there exist $a, b > 0$ and $\omega \ge 0$ such that the kernel K satisfies*

$$|K_z(g\,;h)| \le a\,|z|^{-D'/2} e^{\omega|z|} e^{-b(|gh^{-1}|')^2|z|^{-1}} \tag{4}$$

uniformly for $g, h \in G$ and $z \in \Lambda(\theta)$. The kernel has the semigroup property (2) and is Hölder continuous. In particular there is a $\nu \in \langle 0, 1\rangle$ such that for each $\theta \in \langle 0, \theta_C\rangle$ and $\kappa > 0$ there exist $a, b > 0$ and $\omega \ge 0$ such that

$$\begin{aligned} &|K_z(k^{-1}g\,;l^{-1}h) - K_z(g\,;h)| \\ &\quad\le\ a\,|z|^{-D'/2} e^{\omega|z|} \left(\frac{|k|' + |l|'}{|z|^{1/2} + |gh^{-1}|'} \right)^{\nu} e^{-b(|gh^{-1}|')^2|z|^{-1}} \end{aligned} \tag{5}$$

for all $g, h, k, l \in G$ and $z \in \Lambda(\theta)$ with $|k|' + |l|' \le \kappa\,|z|^{1/2} + 2^{-1}|gh^{-1}|'$.

Moreover, if all the coefficients of H are real-valued then

$$K_t(g\,;h) \ge a'\, t^{-D'/2} e^{-\omega' t} e^{-b'(|gh^{-1}|')^2 t^{-1}} \tag{6}$$

for some $a', b' > 0$ and $\omega' \ge 0$ uniformly for $g, h \in G$ and $t > 0$.

The theorem gives a satisfactory improvement of earlier results [5], [21], for second-order subelliptic operators on Lie groups, insofar it removes all unnecessary regularity, symmetry and reality restrictions on the coefficients and is valid for all groups, modular or unimodular, with polynomial or exponential growth. Bounds such as (4) and (5) are well known for subelliptic operators of all orders if the coefficients are smooth (see, for example, [7]). It is also known that positivity and bounds such as (6) are only possible for second-order operators with real coefficients (see [1] and [20]). Note that the theorem establishes the bounds (4) and (5) throughout any subsector of the sector $\Lambda(\theta_C)$ and consequently if the matrix of principal coefficients $C = (c_{ij})$ is symmetric the results are valid in any subsector of the open right half-plane. In fact the bounds (4) follow for complex z from their real z counterparts together with the L_2-holomorphy of the semigroup S (see [6], Theorem 3.4.8) and similar reasoning could be applied to the continuity estimates (5). These bounds allow one to extend the semigroup S to a holomorphic semigroup on all the L_p-spaces with a holomorphy sector containing the p-independent subsector $\Lambda(\theta_C)$.

The proof of the upper bounds and the Hölder continuity uses the strategy developed in [11] following ideas of [4]. It suffices to prove De Giorgi estimates for the principal part of the operator. These estimates are established by a refinement of the Nash–De Giorgi techniques based on the general reasoning given in Chapter 5 of Giaquinta's book [14]. For the proof of the lower bounds we combine methods of [5], [22] and [21]

2. De Giorgi estimates

The key to bounds on the semigroup kernel is the derivation of De Giorgi estimates for weak solutions of the elliptic equation $H\varphi = 0$ on small balls $B'(g\,;r) = \{h \in G : d'(g\,;h) < r\}$. In [11] these estimates were derived for operators with complex uniformly continuous coefficients by a series of L_2-estimates based on the Poincaré and Caccioppoli inequalities. In the present context we need more refined L_p-estimates with $p \in [1,2\rangle$. A principal new ingredient is the Poincaré–Sobolev inequality.

First, for each bounded open subset $\Omega \subseteq G$, introduce the spaces

$$H'_{p;1}(\Omega) = \{\varphi \in L_p(\Omega\,;dg) : A_i\varphi \in L_p(\Omega\,;dg) \text{ for all } i \in \{1,\ldots,d'\}\} \quad ,$$

where $A_i\varphi$ denotes the distributional derivative in $\mathcal{D}'(\Omega)$. We use the notation $\nabla'\varphi = (A_1\varphi,\ldots,A_{d'}\varphi)$ and equip the spaces with the norms $\varphi \mapsto (\|\varphi\|_{p,\Omega}^p + \|\nabla'\varphi\|_{p,\Omega}^p)^{1/p}$ where

$$\|\varphi\|_{p,\Omega} = \Big(\int_\Omega dh\,|\varphi(h)|^p\Big)^{1/p}$$

and

$$\|\nabla'\varphi\|_{p,\Omega} = \Big(\int_\Omega dh\,\Big(\sum_{i=1}^{d'}|(A_i\varphi)(h)|^2\Big)^{p/2}\Big)^{1/p} \quad .$$

Moreover, for $\varphi \in L_{1,\mathrm{loc}}$ we denote by $\langle\varphi\rangle_\Omega$ the average of φ over Ω. Finally, if $\Omega = B'(g\,;r)$ we simplify notation by setting $\|\cdot\|_{p,g,r} = \|\cdot\|_{p,\Omega}$ and for $g = e$ we drop the e, e.g., $B'(r) = B'(e\,;r)$, $\|\nabla'\varphi\|_{2,r} = \|\nabla'\varphi\|_{2,e,r}$ etc.

Proposition 2.1 *Let $p \in [1, D'\rangle$ and $1/q = 1/p - 1/D'$. Then there exist $c_p > 0$ and $R_p \in \langle 0, 1]$ such that*

$$\|\varphi - \langle\varphi\rangle_{g,r}\|_{q,g,r} \leq c_p \|\nabla'\varphi\|_{p,g,r} \tag{7}$$

for all $g \in G$, $r \in \langle 0, R_p]$, $\varepsilon > 0$ and $\varphi \in H'_{q;1}(B'(g\,; r(1+\varepsilon)))$.

Proof. The proof of the estimates (7) for $g = e$ and $\varphi \in C^\infty(\overline{B'(r)})$ follows from [13], Theorem 1.1. Their extension to $\varphi \in H'_{q;1}(B'(r(1+\varepsilon)))$ then follows by a density argument, [7] Lemma 2.4, and for general g by right translations. □

Our aim is to establish that the principal part H_P of the subelliptic operator H, and the principal part H_P^* of its adjoint H^*, satisfy De Giorgi estimates. These estimates then yield the upper bounds and continuity estimates for the semigroup kernel stated in Theorem 1. Since these results were already established for dimensions $d = 1$ and $d = 2$ in [9], even if the principal coefficients are complex, it suffices to assume $d \geq 3$. In particular $D' \geq 2$.

The De Giorgi estimates for $G = \mathbf{R}^d$ are established in Section 5.1 of [14]. The proof contains two technical ingredients. The first is the classical Sobolev–Poincaré inequality on balls of radius R and Proposition 2.1 provides the Lie group replacement. The second is the existence of a sequence of cut-off functions η with $\eta = 1$ on the ball of radius r, with $\operatorname{supp}\eta$ contained in the ball of radius R and $\|\nabla'\eta\|_\infty \leq c(R-r)^{-1}$ for all small $r < R$, where c is independent of r and R. It is particularly important that these cut-off functions exist for all r arbitrarily close to R. These functions exist by the following lemma.

Lemma 2.2 *There exists a $b_1 > 0$ and, for all $r, R \in \langle 0, 1]$ with $r < R$, an $\eta_{r,R} \in C_c^\infty(B'_R)$ with $0 \leq \eta_{r,R} \leq 1$ and $\eta_{r,R} = 1$ on B'_r such that $\|A_i\eta_{r,R}\|_\infty \leq b_1(R-r)^{-1}$ for all $i \in \{1, \ldots, d'\}$.*

Proof. By Lemma 2.6 of [11] there exist $\sigma \in \langle 0, 1\rangle$, $c_1 > 0$ and for all $r \in \langle 0, 1]$ there exists an $\eta_r \in C_c^\infty(B'_r)$ such that $0 \leq \eta_r \leq 1$, $\eta_r = 0$ on $B'_{\sigma r}$ and $\|A_i\eta_r\|_\infty \leq c_1 r^{-1}$ for all $i \in \{1, \ldots, d'\}$ and $r \in \langle 0, 1]$. Then there is a $c_2 > 0$ such that $\|\eta_r\|_1^{-1}|B'_r| \leq c_2$ uniformly for all $r \in \langle 0, 1]$. For all $s, t \in \langle 0, 1]$ with $s < t$ define $\varphi_{s,t} = \|\eta_s\|_1^{-1}\eta_s * 1_{B'_t}$. Then $\varphi_{s,t} \in C_c^\infty(G)$, $\operatorname{supp}\varphi_{s,t} \subset B'_{s+t}$ and $\varphi_{s,t}(g) = 1$ for all $g \in B'_{t-s}$. Moreover,

$$\begin{aligned} |(A_i\varphi_{s,t})(g)| &= \|\eta_s\|_1^{-1}|((A_i\eta_s) * 1_{B'_t})(g)| \leq \Big|\Big(\|\eta_s\|_1^{-1}\|A_i\eta_s\|_\infty 1_{B'_s}\Big) * 1_{B'_t})(g)\Big| \\ &\leq \Big\|\|\eta_s\|_1^{-1}\|A_i\eta_s\|_\infty 1_{B'_s}\Big\|_1 \|1_{B'_t}\|_\infty \leq c_1\, c_2\, s^{-1} \end{aligned}$$

for all $i \in \{1, \ldots, d'\}$ and $g \in G$. Finally, for all $0 < r < R \leq 1$ choose $s = 2^{-1}(R-r)$ and $t = 2^{-1}(R+r)$. □

For $M, \mu > 0$ let $\mathcal{E}^{\rm div}_{r,P}(\mu, M)$ be the set of all pure second-order subelliptic operators

$$H = -\sum_{i,j=1}^{d'} A_i \, c_{ij} \, A_j$$

with real-valued coefficients c_{ij} such that $\mu_C \geq \mu$ and $\|C\|_\infty = \sup_{g\in G} \|C(g)\| \leq M$, where $\|C(g)\|$ denotes the l_2-norm of the matrix $C(g) = (c_{ij}(g))$. We aim to prove estimates uniform for all operators in $\mathcal{E}^{\rm div}_{r,P}(\mu, M)$. We first prove a version of the Caccioppoli inequalities, as in [11], Lemma 2.7.

Lemma 2.3 *For all $M, \mu > 0$ there exists a $b_2 > 0$ such that*

$$\int_{B'_r} |\nabla'((\varphi - k)^+)|^2 \leq b_2 (R-r)^{-2} \int_{B'_R} |(\varphi - k)^+|^2$$

uniformly for all $H \in \mathcal{E}^{\rm div}_{r,P}(\mu, M)$, all $r, R \in \mathbf{R}$ with $0 < r < R \leq 1$, all $k \in \mathbf{R}$ and all real-valued $\varphi \in H'_{2;1}(B'_R)$ satisfying $H\varphi = 0$ weakly on B'_R.

Proof. We may assume that $k = 0$, since one also has $H(\varphi - k) = 0$ weakly on B'_R. Next, let $\eta_{r,R}$ and b_1 be as in Lemma 2.2. Note that $\varphi^+ \in H'_{2;1}(B'_R)$ and $A_i(\varphi^+) = 1_{[\varphi>0]} A_i \varphi$ by an argument similar to the proof of Lemma 7.6 in [15] but using the estimates underlying the proof of Lemma 2.4 in [7]. Therefore $A_i(\varphi^+) \, A_j(\varphi^+) = 1_{[\varphi>0]} A_i\varphi \, A_j\varphi = A_i(\varphi^+) \, A_j\varphi$ and $\varphi^+ A_i \varphi = \varphi^+ A_i(\varphi^+)$. Then

$$\begin{aligned}
\mu_C \, \|\eta_{r,R} \nabla'(\varphi^+)\|_2^2 &\leq (\eta_{r,R}\nabla'(\varphi^+), C \eta_{r,R} \nabla'(\varphi^+)) \\
&= (\eta_{r,R}^2 \nabla'(\varphi^+), C\nabla'\varphi) = \sum_{i,j=1}^{d'} ([\eta_{r,R}^2, A_i](\varphi^+), c_{ij} A_j \varphi)
\end{aligned}$$

where the last step uses $H\varphi = 0$ weakly on B'_R. But $[\eta_{r,R}^2, A_i] = -2\eta_{r,R}(A_i \eta_{r,R})$ and hence

$$\begin{aligned}
\mu_C \, \|\eta_{r,R} \nabla'(\varphi^+)\|_2^2 &\leq -2((\nabla'\eta_{r,R})(\varphi^+), C\eta_{r,R} \nabla'\varphi) \\
&= -2((\nabla'\eta_{r,R})\varphi^+, C\eta_{r,R}\nabla'(\varphi^+)) \\
&\leq 2\,\|C\|_\infty \, \|(\nabla'\eta_{r,R})\varphi^+\|_2 \, \|\eta_{r,R}\nabla'(\varphi^+)\|_2 \quad .
\end{aligned}$$

Therefore

$$\begin{aligned}
\int_{B'_r} |\nabla'((\varphi)^+)|^2 &\leq \|\eta_{r,R}\nabla'(\varphi^+)\|_2^2 \leq 4\,\mu_C^{-2} \|C\|_\infty^2 \, \|(\nabla'\eta_{r,R})\varphi^+\|_2^2 \\
&\leq 4\, b_1^2 \, \mu_C^{-2} \|C\|_\infty^2 \, (R-r)^{-2} \int_{B'_R} |\varphi^+|^2 \quad ,
\end{aligned}$$

by the estimates of Lemma 2.2. □

Set $\theta = 2^{-1}+(4^{-1}+2(D')^{-1})^{1/2} > 1$. Then $\theta^2-\theta-2(D')^{-1} = 0$. Moreover, if $k \in \mathbf{R}$, $0 < r \leq R < \infty$ and $\varphi \in H'_{2;1}(B'_R)$ set

$$A(k,r) = [\varphi > k] \cap B'_r \quad .$$

It will be clear from the context which φ is used.

Proposition 2.4 *For all $M, \mu > 0$ there exists a $b_3 > 0$ such that*

$$\operatorname*{ess\,sup}_{g\in B'_{R/2}} \varphi(g) \leq k + b_3\Big(R^{-D'}\int_{A(k,R)} |\varphi-k|^2\Big)^{1/2}\Big(R^{-D'}|A(k,R)|\Big)^{(\theta-1)/2}$$

uniformly for all $H \in \mathcal{E}^{\mathrm{div}}_{r,P}(\mu,M)$, $R \in \langle 0,1]$, $k \in \mathbf{R}$ and real-valued $\varphi \in H'_{2;1}(B'_R)$ satisfying $H\varphi = 0$ weakly on B'_R.

Proof. The proof is almost the same as the proof of Theorem 5.1 in [14]. The only problem arises with the application of the Sobolev inequality.

Let $\eta_{r,R}$ and b_1 be as in Lemma 2.2 and b_2 be as in Lemma 2.3. Let $H \in \mathcal{E}^{\mathrm{div}}_{r,P}(\mu,M)$, $S \in \langle 0,1]$ and $\varphi \in H'_{2;1}(B'_S)$ be a real-valued function satisfying $H\varphi = 0$ weakly on B'_S. Then for all $0 < r < R \leq S$ and $k \in \mathbf{R}$ one has

$$\begin{aligned}
&\int_{B'_{(r+R)/2}} |\nabla'(\eta_{r,(r+R)/2}(\varphi-k)^+)|^2 \\
&\quad\leq 2\int_{B'_{(r+R)/2}} |\nabla'\eta_{r,(r+R)/2}|^2\,|((\varphi-k)^+)|^2 \\
&\qquad + 2\int_{B'_{(r+R)/2}} |\eta_{r,(r+R)/2}|^2\,|\nabla'((\varphi-k)^+)|^2 \\
&\quad\leq (8d'\,b_1^2+8b_2)(R-r)^{-2}\int_{B'_R} |(\varphi-k)^+|^2 \quad .
\end{aligned}$$

Next, by the subelliptic Sobolev inequalities for Lie groups, [20], Theorem IV.5.6, there exists a $b > 0$ such that

$$\Big(\int_G |\psi|^{2D'/(D'-2)}\Big)^{(D'-2)/D'} \leq b\int_G |\nabla'\psi|^2 + b\int_G |\psi|^2$$

uniformly for all $\psi \in L'_{2;1}(G)$. Therefore

$$\begin{aligned}
&\Big(\int_{B'_r} |(\varphi-k)^+|^{2D'/(D'-2)}\Big)^{(D'-2)/D'} \\
&\quad\leq \Big(\int_G |\eta_{r,(r+R)/2}(\varphi-k)^+|^{2D'/(D'-2)}\Big)^{(D'-2)/D'}
\end{aligned}$$

$$
\begin{aligned}
&\le b\int_G |\nabla'(\eta_{r,(r+R)/2}(\varphi-k)^+)|^2 + b\int_G |\eta_{r,(r+R)/2}(\varphi-k)^+|^2 \\
&= b\int_{B'_{(r+R)/2}} |\nabla'(\eta_{r,(r+R)/2}(\varphi-k)^+)|^2 + b\int_{B'_{(r+R)/2}} |\eta_{r,(r+R)/2}(\varphi-k)^+|^2 \\
&\le b(8d'\,b_1^2+8b_2+1)(R-r)^{-2}\int_{B'_R} |(\varphi-k)^+|^2 \quad .
\end{aligned}
$$

The rest of the proof is as in [14]. The uniformity of the constant follows from the proof. Actually, one can take $b_3 = 2^{(\theta-1)^{-1}2^{D'\theta/2}\theta(1+D'\theta/2)}(b(8d'b_1^2+8b_2+1))^{D'\theta/4}c^{(\theta-1)/2}$, where the constant c is as in (3). □

Corollary 2.5 *For all $M,\mu>0$ there exists a $b_4>0$ such that*

$$
\operatorname*{ess\,sup}_{g\in B'(R/2)} |\varphi(g)| \le b_4\Big(R^{-D'}\int_{B'(R)} |\varphi|^2\Big)^{1/2}
$$

uniformly for all $H\in\mathcal{E}^{\rm div}_{r,P}(\mu,M)$, $R\in\langle 0,1]$, $k\in\mathbf{R}$ and all real-valued $\varphi\in H'_{2;1}(B'(R))$ satisfying $H\varphi=0$ weakly on $B'(R)$.

Proof. Let b_3 be as in Proposition 2.4. Then applying Proposition 2.4 with $k=0$ one deduces that

$$
\operatorname*{ess\,sup}_{g\in B'(R/2)} \varphi(g) \le b_3\Big(R^{-D'}\int_{B'(R)} |\varphi|^2\Big)^{1/2}\Big(R^{-D'}|B'(R)|\Big)^{(\theta-1)/2} \quad .
$$

But these estimates are also valid for φ replaced by $-\varphi$. Now the corollary follows from the volume estimates (3). □

Before we can prove the De Giorgi estimates for subelliptic operators with real measurable coefficients we need one more technical lemma.

Lemma 2.6 *For all $M,\mu>0$ there exists a $b_5>0$ such that*

$$
R^{-D'}|A(k_n,R)| \le b_5\, n^{-\beta}
$$

uniformly for all $H\in\mathcal{E}^{\rm div}_{r,P}(\mu,M)$, $R\in\langle 0,R_1/4]$, $n\in\mathbf{N}$ and real-valued $\varphi\in H'_{2;1}(B'(4R))$ satisfying $H\varphi=0$ weakly on $B'(4R)$ and $|A(k_0,R)|\le 2^{-1}|B'(R)|$ where

$$
k_n = \operatorname*{ess\,sup}_{g\in B'(2R)} \varphi(g) - 2^{-(n+1)}\Big(\operatorname*{ess\,sup}_{g\in B'(2R)} \varphi(g) - \operatorname*{ess\,inf}_{g\in B'(2R)} \varphi(g)\Big) \quad ,
$$

R_1 is as in Proposition 2.1 and $\beta=(2D'-2)^{-1}D'$.

Proof. Note that the essential suprema and infima are finite by Corollary 2.5. It follows from the proof of [14], Proposition 5.2, that

$$n\Big(R^{-D'}|A(k_n,R)|\Big)^{1/\beta} \le b\,R^{-D'}|A(k_0,R)|$$

for all $n \in \mathbf{N}$, where $b = 8(1-2^{-1/D'})^{-4}b_2^2c^2c_1^4\sigma^{-2D'}$ and the constants b_2, c and c_1 are as in Lemma 2.3, (3) and Proposition 2.1. But $|A(k_0,R)| \le |B'(R)| \le c\,R^{D'}$. Therefore $R^{-D'}|A(k_n,R)| \le (b\,c)^\beta\, n^{-\beta}$ and the lemma follows. □

At this point we can state and prove the uniform De Giorgi estimates which we require.

Proposition 2.7 *For all $M,\mu > 0$ there exist $\nu > 0$ and $c_{16} > 0$ such that*

$$\int_{B'(g;r)} |\nabla'\varphi|^2 \le c_{16}\,(r/R)^{D'-2+2\nu} \int_{B'(g;R)} |\nabla'\varphi|^2$$

uniformly for all $H \in \mathcal{E}^{\rm div}_{r,P}(\mu,M)$, $r,R \in \mathbf{R}$ with $0 < r \le R \le 1$, $g \in G$ and all (complex-valued) $\varphi \in H'_{2;1}(B'(R))$ satisfying $H\varphi = 0$ weakly on $B'(g\,;R)$.

Proof. As in the proof of Proposition 3.3 in [11] it suffices to prove the estimates for $g = e$, $R \in \langle 0, R_1 \wedge R_N]$, $r \le R/4$ and $\varphi \in H'_{2;1}(B'(2R))$ satisfying $H\varphi = 0$ weakly on $B'(2R)$ and $\langle\varphi\rangle_R = 0$, where R_N is the constant in the Poincaré inequality (11) of Proposition 2.4 in [11] and R_1 is as in Proposition 2.1. Moreover, since H is a real operator, it suffices to prove the inequality for real-valued φ.

Let b_2, b_3, b_4, b_5 and β be as in Lemma 2.3, Proposition 2.4, Corollary 2.5 and Lemma 2.6. Moreover, let c be as in (3) and c_N be as in the Neumann-type Poincaré inequality of Proposition 2.4 in [11]. Let R and φ be as above. We follow the ideas of [14] Section 5.1. For all $r \in \langle 0, R/2]$ the function φ is essentially bounded on $B'(r)$ by Corollary 2.5 and we set

$$m(r) = \operatorname*{ess\,inf}_{g\in B'(r)} \varphi(g) \quad , \quad M(r) = \operatorname*{ess\,sup}_{g\in B'(r)} \varphi(g) \quad .$$

Now suppose $r \in \langle 0, R/4]$. Set $k_0 = 2^{-1}\Big(M(2r)+m(2r)\Big)$. We may assume that $|A(k_0,r)| \le 2^{-1}|B'(r)|$, otherwise replace φ by $-\varphi$. Next, for all $n \in \mathbf{N}$ we set $k_n = M(2r) - 2^{-(n+1)}\Big(M(2r) - m(2r)\Big)$. Then it follows from Proposition 2.4 that

$$\begin{aligned} M(r/2) &\le k_n + b_3\Big(r^{-D'}\int_{A(k_n,r)} |M(r)-k_n|^2\Big)^{1/2}\Big(r^{-D'}|A(k_n,r)|\Big)^{(\theta-1)/2} \\ &\le k_n + b_3\,c^{1/2}\Big(M(2r)-k_n\Big)(b_5\,n^{-\beta})^{(\theta-1)/2} \end{aligned}$$

uniformly for all $n \in \mathbf{N}$, where Lemma 2.6 is used in the last inequality. Next fix $N \in \mathbf{N}$ such that

$$b_3\, c^{1/2} (b_5\, N^{-\beta})^{(\theta-1)/2} \leq 2^{-1} \quad .$$

Note that N depends only on M and μ. Then

$$\begin{aligned} M(r/2) &\leq M(2r) - 2^{-(N+1)}\Big(M(2r) - m(2r)\Big) + 2^{-1}2^{-(N+1)}\Big(M(2r) - m(2r)\Big) \\ &= M(2r) - 2^{-(N+2)}\Big(M(2r) - m(2r)\Big) \end{aligned}$$

and hence

$$\begin{aligned} M(r/2) - m(r/2) &\leq M(2r) - m(2r) - 2^{-(N+2)}\Big(M(2r) - m(2r)\Big) \\ &= (1 - 2^{-(N+2)})\Big(M(2r) - m(2r)\Big) \quad . \end{aligned}$$

This is valid for all $r \in \langle 0, R/4]$. Therefore one deduces by induction that

$$M(2^{-(2n+1)}r) - m(2^{-(2n+1)}r) \leq (1 - 2^{-(N+2)})^n \Big(M(R/2) - m(R/2)\Big)$$

for all $n \in \mathbf{N}_0$ and

$$M(r) - m(r) \leq 4^\nu\, (r/R)^\nu \Big(M(R/2) - m(R/2)\Big)$$

for all $r \in \langle 0, R/2]$, where $\nu = -(2\log 2)^{-1} \log(1 - 2^{-(N+2)}) > 0$. Finally, applying Lemma 2.3 to φ with $k = \langle\varphi\rangle_{2r}$ and to $-\varphi$ with $k = -\langle\varphi\rangle_{2r}$ one deduces that for all $r \in \langle 0, R/4]$

$$\begin{aligned} \int_{B'(r)} |\nabla'\varphi|^2 &\leq 4b_2\,(2r)^{-2} \int_{B'(2r)} |\varphi - \langle\varphi\rangle_{2r}|^2 \\ &\leq 4b_2\,(2r)^{-2} \int_{B'(2r)} |M(2r) - m(2r)|^2 \\ &\leq 4a^2\, b_2\, c\,(2r)^{D'-2}(2r/R)^{2\nu}\Big(M(R/2) - m(R/2)\Big)^2 \\ &\leq 4a^2\, b_2\, c\,(2r)^{D'+2\nu-2} R^{-2\nu}\Big(2 \operatorname*{ess\,sup}_{g\in B'(R/2)} |\varphi(g)|\Big)^2 \\ &\leq 16a^2\, b_2\, b_4^2\, c\,(2r)^{D'+2\nu-2} R^{-(D'+2\nu)} \int_{B'(R)} |\varphi|^2 \\ &\leq 16a^2\, b_2\, b_4^2\, c\, c_N\, 2^{D'+2\nu-2} (r/R)^{D'+2\nu-2} \int_{B'(R)} |\nabla'\varphi|^2 \quad , \end{aligned}$$

where we have used $\langle\varphi\rangle_R = 0$ and the Neumann-type Poincaré inequality in the last step. This completes the proof of the De Giorgi estimates for operators with real measurable coefficients. □

3. Kernel estimates

In this section we consider the bounds on the semigroup kernel stated in Theorem 1.1.

The upper bounds and continuity estimates were established for dimensions one and two in [2] [9] and these results do not require the principal coefficients of H to be real-valued. But if $D' \geq 2$ then the bounds and continuity are a direct corollary of Theorem 4.1 of [11] and the De Giorgi estimates of Proposition 2.7. Note that this latter proposition can be applied both to H and its adjoint H^*.

It remains to prove the Gaussian lower bounds in the last statement of Theorem 1.1. There is a large literature on lower bounds of semigroup kernels associated with real second-order elliptic operators. (References directly relevant to the current paper can be found in the books [6], [20], [24].) Most of this work is based on the arguments of Nash [19] for pure second-order operators on $\mathbf{R}^d$ and the subsequent discussion is largely composed of arguments already contained in the literature. The only essential new feature arises from the lack of symmetry of the principal coefficients. Previous Lie group results also require some smoothness of the coefficients or place restrictions on the growth properties of the group. Therefore we have to rearrange the reasoning to take care of these difficulties.

First we reduce to the case of smooth coefficients by regularization as in [5], [22], [11]. If $\tau_n \in C_c^\infty(G)$, $\tau_n \geq 0$, $\|\tau_n\|_1 = 1$ is an approximation to the identity and $c \in L_\infty$ the regularization $c^{(n)}$ is defined by

$$c^{(n)}(g) = (R(\tau_n)c)(g) = \int_G dh\, \tau_n(h)\, c(gh) \quad ,$$

where R is the right regular representation. Then let H_n denote the operators constructed from H by regularization of the coefficients as in [11], Section 2.1, i.e., one replaces the c_{ij} in H by $c_{ij}^{(n)}$, etc.. Since $\tau_n \geq 0$ and $\|\tau_n\|_1 = 1$ it follows that $\mu_{C^{(n)}} \geq \mu_C$, $\|C^{(n)}\|_\infty \leq \|C\|_\infty$, $\|c^{(n)}\|_\infty \leq \|c\|_\infty$ and $\|c_0^{(n)}\|_\infty \leq \|c_0\|_\infty$. Moreover, if $K^{(n)}$ denotes the kernel of the semigroup generated by H_n it follows from the uniform bounds of Proposition 2.7 together with Propositions 4.5 and 2.3 of [11] that $\lim_{n\to\infty} K_t^{(n)} = K_t$ uniformly on compacta of $G \times G$. Thus we can effectively assume the coefficients of H are smooth as long as the final estimates are independent of the smoothness. For operators with smooth coefficients it follows from [7], Corollary 3.5, that the kernel belongs to $C_{b;\infty}(G \times G)$.

Next we observe that the kernel K_t associated with an operator H with all real coefficients is pointwise positive, i.e., $K_t(g\,;h) \geq 0$ for all $g, h \in G$. If the principal coefficients are symmetric this is a standard result but it also follows for non-symmetric c_{ij} by the dispersivity of H (see [1], Proposition 2.7). It is not a priori evident, however, that K_t is strictly positive and this complicates the ensuing arguments. To avoid this difficulty we replace K_t by $K_t^{(\delta)} = K_t + \delta$ with $\delta \in \langle 0, 1]$ and eventually take the limit $\delta \to 0$. Now we turn to Nash's arguments. The first important step is to deduce local lower bounds on the kernel.

A basic ingredient in Nash's method is an L_1-lower bound

$$\inf_{t\in\langle 0,1]}\inf_{g\in G}\int_G d\hat{h}\,K_t(g\,;h)\geq c>0\quad. \tag{8}$$

This is straightforward for pure second-order operators since

$$\int_G d\hat{h}\,K_t(g\,;h)=1\quad.$$

The lower-order terms present a difficulty, however, which can be circumvented by an argument of Stroock [22] which we apply to the modified kernels $K^{(\delta)}$.

Choose a real $\chi\in C^\infty(G)$ such that $\int_G d\hat{h}\,e^{-\chi(h)}=1$ and $A_i\chi\in L_\infty$ for all $i\in\{1,\ldots,d'\}$. If G is not compact a function with these properties can be constructed by regularization of the modulus $g\mapsto|g|'$ with a C^∞-function. (See, for example, [20], pages 201–202.) Next fix $g\in G$ and define ψ by setting $\psi(h)=\chi(hg^{-1})$. Then ψ satisfies the properties required of χ and in addition $\|A_i\psi\|_\infty=\|A_i\chi\|_\infty$. Further define k_t by setting $k_t(h)=K_t^{(\delta)}(g\,;h)=K_t(g\,;h)+\delta$. Next for each $\gamma\in\langle 0,1]$ introduce $H_\gamma\colon\langle 0,\infty\rangle\to\langle 0,\infty\rangle$ by

$$H_\gamma(t)=\int_G d\hat{h}\,e^{-\psi(h)}k_t^\gamma(h)\quad,$$

where $k_t^\gamma(h)=\big(k_t(h)\big)^\gamma$. Note that the factor $e^{-\psi}$ is necessary since $k_t^\gamma\geq\delta^\gamma$. Then

$$H_\gamma'(t)=-\gamma\int_G d\hat{h}\,e^{-\psi(h)}k_t^{\gamma-1}(h)\,(\widetilde{H}_hK_t)(g\,;h)$$

where $\widetilde{H}_h$ denotes the adjoint of H with respect to right Haar measure acting on the h variable. (Note that $k_t^{\gamma-1}$ is bounded for $\gamma\in\langle 0,1\rangle$ because $k_t\geq\delta$. This together with the boundedness of $h\mapsto(\widetilde{H}_hK_t)(g\,;h)$ ensures that the integral exists.) Using the explicit form of $\widetilde{H}$ we can now integrate by parts, i.e., evaluate $\widetilde{H}$ in terms of the associated form. This operation causes no difficulty because we are assuming the coefficients are smooth and hence the kernel is a C^∞-function over $G\times G$. One finds

$$\begin{aligned}H_\gamma'(t)&=\gamma(1-\gamma)\sum_{i,j=1}^{d'}\int_G d\hat{h}\,e^{-\psi(h)}k_t^{\gamma-2}(h)\,(A_ik_t)(h)\,c_{ji}(h)\,(A_jk_t)(h)\\&\quad+\gamma\sum_{i,j=1}^{d'}\int_G d\hat{h}\,e^{-\psi(h)}k_t^{\gamma-1}(h)\,(A_i\psi)(h)\,c_{ji}(h)\,(A_jk_t)(h)+R\end{aligned}$$

where the first terms on the right hand side denote the contribution to the differential equation of the terms in $\widetilde{H}$ with two derivatives and R indicates the contribution of the lower order terms. Note that we have identified $A_j k_t$ with the derivative in the j-th direction, with respect to the second variable, of K_t. Simple rearrangement then gives

$$\begin{aligned} H'_\gamma(t) &= 4\gamma^{-1}(1-\gamma)\sum_{i,j=1}^{d'}\int_G d\hat{h}\, e^{-\psi(h)}(A_i k_t^{\gamma/2})(h)\, c_{ji}(h)\,(A_j k_t^{\gamma/2})(h) \\ &\quad + 2\sum_{i,j=1}^{d'}\int_G d\hat{h}\, e^{-\psi(h)} k_t^{\gamma/2}(h)\,(A_i\psi)(h)\, c_{ji}(h)\,(A_j k_t^{\gamma/2})(h) + R\ . \end{aligned}$$

The first term on the right hand side, which only depends on the hermitian part of the matrix $C = (c_{ij})$ of principal coefficients, can now be bounded below using the subellipticity condition,

$$\begin{aligned} &\sum_{i,j=1}^{d'}\int_G d\hat{h}\, e^{-\psi(h)}(A_i k_t^{\gamma/2})(h)\, c_{ji}(h)\,(A_j k_t^{\gamma/2})(h) \\ &\qquad\qquad \geq \mu_C \sum_{i=1}^{d'}\int_G d\hat{h}\, e^{-\psi(h)}|(A_i k_t^{\gamma/2})(h)|^2\ , \end{aligned}$$

and the second term can be bounded by an $\varepsilon, \varepsilon^{-1}$ argument, e.g.,

$$\begin{aligned} &\Big|\sum_{i,j=1}^{d'}\int_G d\hat{h}\, e^{-\psi(h)} k_t^{\gamma/2}(h)\,(A_i\psi)(h)\, c_{ji}(h)\,(A_j k_t^{\gamma/2})(h)\Big| \\ &\qquad \leq \|C\|_\infty \sum_{i=1}^{d'}\int_G d\hat{h}\, e^{-\psi(h)}\Big(\varepsilon\,|(A_i k_t^{\gamma/2})(h)|^2 + (4\varepsilon)^{-1} k_t^{\gamma}(h)\|A_i\chi\|_\infty^2\Big) \end{aligned}$$

for all $\varepsilon > 0$.

The lower order terms contained in the remainder R can be bounded in a similar manner with an $\varepsilon, \varepsilon^{-1}$ argument. (Care has to be taken with the terms with no derivatives on the right hand side. These contain a factor $K_t(g\,;h)$ but this can be replaced by $k_t(h)$ at the cost of introducing an extra factor $K_t(g\,;h)/K_t^{(\delta)}(g\,;h)$. But this factor satisfies $0 \leq K_t(g\,;h)/K_t^{(\delta)}(g\,;h) \leq 1$ and consequently plays no role in the estimate.) Therefore one obtains a differential inequality

$$H'_\gamma(t) \geq 2\gamma^{-1}(1-\gamma)\mu_C \sum_{i=1}^{d'}\int_G d\hat{h}\, e^{-\psi(h)}|(A_i k_t^{\gamma/2})(h)|^2 - \nu_\gamma H_\gamma(t) \tag{9}$$

where ν_γ is a constant which depends only on γ, μ_C, $\|C\|_\infty$, $\|c\|_\infty$, $\|c_0\|_\infty$ and $\|A_i\chi\|_\infty$.

Similarly,

$$H_1'(t) = \sum_{i,j=1}^{d'} \int_G d\hat{h}\, e^{-\psi(h)} (A_i\psi)(h)\, c_{ji}(h)\, (A_j k_t)(h) + R \quad .$$

So there exists a $c > 0$ such that

$$H_1'(t) \geq -c \sum_{i=1}^{d'} \int_G d\hat{h}\, e^{-\psi(h)} |(A_i k_t)(h)| - c \int_G d\hat{h}\, e^{-\psi(h)} k_t(h)$$

where the value of c depends only on μ_C, $\|C\|_\infty$, $\|c\|_\infty$, $\|c_0\|_\infty$ and $\|A_i\chi\|_\infty$. But

$$|(A_i k_t)(h)| \leq \varepsilon\, k_t^{\gamma-2}(h)\, |(A_i k_t)(h)|^2 + (4\varepsilon)^{-1} k_t^{2-\gamma}(h)$$

for any $\varepsilon > 0$. Therefore

$$\begin{aligned} H_1'(t) \;\; \geq \;\; & -4c\gamma^{-2}\varepsilon \sum_{i=1}^{d'} \int_G d\hat{h}\, e^{-\psi(h)} |(A_i k_t^{\gamma/2})(h)|^2 \\ & - (4\varepsilon)^{-1} c\, d' \int_G d\hat{h}\, e^{-\psi(h)} k_t^{2-\gamma}(h) - c \int_G d\hat{h}\, e^{-\psi(h)} k_t(h) \end{aligned}$$

for all $t > 0$.

Next using the upper bounds for K_t of Theorem 1.1, which have already been established, and making the choice $\varepsilon = (2c)^{-1}\gamma\,(1-\gamma)\,\mu_C$ one deduces that there exists $\tau_\gamma > 0$, whose value depends on γ, μ_C, $\|C\|_\infty$, $\|c\|_\infty$, $\|c_0\|_\infty$ and $\|A_i\chi\|_\infty$ such that

$$H_1'(t) \geq -2\gamma^{-1}(1-\gamma)\,\mu_C \sum_{i=1}^{d'} \int_G d\hat{h}\, e^{-\psi(h)} |(A_i k_t^{\gamma/2})(h)|^2 - \tau_\gamma t^{-(1-\gamma)D'/2} H_1(t) \tag{10}$$

uniformly for all $t \in \langle 0, 1]$. Hence

$$(H_\gamma + H_1)'(t) \geq -(\nu_\gamma + \tau_\gamma t^{-(1-\gamma)D'/2})(H_\gamma + H_1)(t)$$

by (9) and (10). Thus with $\gamma = 1 - (D')^{-1}$ and $\omega_\gamma = \nu_\gamma + \tau_\gamma$ one has

$$(H_\gamma + H_1)'(t) \geq -\omega_\gamma t^{-1/2} (H_\gamma + H_1)(t)$$

and since this inequality is valid for all $t \in \langle 0, 1]$ one may integrate between t_0 and t and conclude that

$$\log(H_\gamma(t) + H_1(t)) \geq -2\omega_\gamma + \log(H_\gamma(t_0) + H_1(t_0)) \geq -2\omega_\gamma + \log H_1(t_0) \quad .$$

But $\lim_{t_0 \to 0} H_1(t_0) = e^{-\psi(g)} + \delta \geq e^{-\chi(e)}$. Therefore

$$H_\gamma(t) + H_1(t) \geq c_\gamma$$

for some $c_\gamma > 0$, whose value depends on the coefficients only through μ_C, $\|C\|_\infty$, $\|c\|_\infty$ and $\|c_0\|_\infty$, uniformly for $g \in G$ and $t \in \langle 0, 1]$. Note that c_γ is independent of the parameter δ used to modify the kernel. Since $H_\gamma(t) \leq (H_1(t))^\gamma$, it follows that $H_1(t) \geq c$ for a suitable $c > 0$, which is again independent of δ, uniformly for all $t \in \langle 0, 1]$ and $g \in G$. Therefore

$$\int_G d\hat{h}\, K_t(g\,;h) \geq \int_G d\hat{h}\, e^{-\psi(h)} K_t(g\,;h) = H_1(t) - \delta \geq c - \delta \quad .$$

Finally taking the limit $\delta \to 0$ one concludes that (8) is valid.

Combination of the Gaussian upper bounds of of Theorem 1.1 and the L_1-lower bounds of (8) now give bounds that indicate that K_t is localized near the diagonal $g = h$ uniformly for all small t.

Lemma 3.1 *There exist $c', \lambda > 0$ such that*

$$\int_{B'(g;\lambda t^{1/2})} d\hat{h}\, K_t(g\,;h) \geq c'$$

uniformly for all $t \in \langle 0, 1]$ and $g \in G$.

Proof. It follows from the Gaussian upper bounds on the kernel and a quadrature estimate (see, for example, [20] pages 223–224) that one has bounds

$$\int_{G \backslash B'(g;\lambda t^{1/2})} d\hat{h}\, K_t(g\,;h) \leq a\, e^{\omega t} \lambda^{-1}$$

for all $\lambda, t > 0$, uniform in $g \in G$. Therefore using (8) one deduces that

$$\int_{B'(g;\lambda t^{1/2})} d\hat{h}\, K_t(g\,;h) \geq c - a\, e^{\omega} \lambda^{-1}$$

for all $t \in \langle 0, 1]$ and $g \in G$. Consequently the statement of the lemma holds, with $c' = 2^{-1}c$, for all $\lambda \geq 2\, a\, c^{-1} e^{\omega}$. □

The next lemma is the key to pointwise local lower bounds. It is a version of an estimate first given by Nash [19] in his fundamental analysis of strongly elliptic operators on $\mathbf{R}^d$. Nash's idea was to use the relative entropy as a measure of localization. Nash considered the entropy of the semigroup kernel relative to a Gaussian measure and used its properties to obtain bounds

$$K_1(x\,;y) \geq c > 0 \tag{11}$$

for all x, y such that $|x - y| \leq \kappa$. Then a scaling argument gives the crucial local bounds

$$K_t(x\,;y) \geq c\, t^{-d/2} \tag{12}$$

for all $t \in \langle 0,1]$ and all x, y such that $|x-y| \leq \kappa t^{1/2}$.

One can apply Nash's arguments directly in the current context and deduce the Lie group equivalent of (11) but these bounds are difficult to exploit since one does not have scaling arguments. Therefore there is no obvious way of deducing the analogue of the bounds (12). But Saloff-Coste and Stroock [21] realized that this difficulty can be circumvented by considering the relative entropy with respect to a suitably chosen family of measures with compact support. We will closely follow their reasoning.

For $r > 0$ define $\rho_r, \sigma_r : G \to [0,\infty\rangle$ by $\rho_r = (\sigma_r)^2$ and $\sigma_r(g) = 1 - |g|' r^{-1}$ if $|g|' \leq r$ and $\sigma_r(g) = 0$ if $|g|' > r$. Both ρ_r and σ_r are weakly* differentiable and it follows from the triangle inequality that $\sum_{i=1}^{d'} \|A_i \sigma_r\|_\infty^2 \leq a\, r^{-2}$ for a suitable $a > 0$. Moreover, there exists a local weighted Poincaré inequality with density function ρ_r. Define the weighted average of φ by $\langle\varphi\rangle_{r,\rho} = \Big(\int_G d\hat{g}\, \rho_r(g)\Big)^{-1} \int_G d\hat{g}\, \rho_r(g)\, \varphi(g)$.

Proposition 3.2 *There exist $R_0 \in \langle 0,1]$ and $c > 0$ such that*

$$\int_G d\hat{g}\, \rho_r(g) |\varphi(g) - \langle\varphi\rangle_{r,\rho}|^2 \leq c\, r^2 \int_G d\hat{g}\, \rho_r(g) \sum_{i=1}^{d'} |(A_i\varphi)(g)|^2$$

uniformly for all $r \in \langle 0, R_0]$ and $\varphi \in L'_{2;1}(B'(2))$.

Proof. This has been proved essentially in the appendix of [21]. In [21] the group has polynomial growth, but for a general group all the estimates are valid locally, just as in Jerison [17]. □

This Poincaré inequality will be used in the derivation of the following result.

Lemma 3.3 *There exist $c, \lambda > 0$ and $t_0 \in \langle 0,1]$ such that*

$$\int_G d\hat{h}\, \rho_{\lambda t^{1/2}}(h) \log\Big(t^{D'/2} K_t(g\,;h)\Big) \geq -c \int_G d\hat{h}\, \rho_{\lambda t^{1/2}}(h)$$

uniformly for all $t \in \langle 0, t_0]$ and $g \in B'(4^{-1}\lambda t^{1/2})$.

Remark 3.4 It is not evident that the relative entropy defined by the integral on the left hand side of the above inequality is finite. But this will be established in the following proof.

Proof. Let $\lambda \geq 1$ and $R_0 \in \langle 0,1]$ be as in Lemma 3.1 and Proposition 3.2. Set $t_0 = \lambda^{-2} R_0^2$. Next fix $t \in \langle 0, t_0]$ and $g \in B'(4^{-1}\lambda t^{1/2})$ and for $s, \delta \in \langle 0,1]$ set $l_s^{(\delta)}(h) = t^{D'/2} K_{st}^{(\delta)}(g\,;h)$ where $K_t^{(\delta)} = K_t + \delta$ as above. We simplify notation by setting $\rho(h) = \rho_{\lambda t^{1/2}}(h)$ and $\sigma(h) = \sigma_{\lambda t^{1/2}}(h)$. Then $l_s^{(\delta)} \geq t^{D'/2}\delta > 0$ and we can introduce the functions $\Phi_\delta \colon \langle 0,1] \to \mathbf{R}$ by

$$\Phi_\delta(s) = \left(\int_G d\hat{h}\,\rho(h)\right)^{-1} \int_G d\hat{h}\,\rho(h) \log l_s^{(\delta)}(h) \quad .$$

One must prove that for suitable $c, \lambda > 0$ one has $\Phi_\delta(1) \geq -c$ uniformly for all $\delta \in \langle 0,1]$, $t \in \langle 0, t_0]$ and $g \in B'(4^{-1}\lambda t^{1/2})$. Once this is established it follows from the monotone convergence theorem, by setting $\delta = n^{-1}$ and taking the limit $n \to \infty$, that the relative entropy integral exists. Moreover, in the limit $\delta \to 0$, the required bounds are valid.

Differentiating Φ_δ and then integrating by parts one finds

$$\begin{aligned}
\int_G d\hat{h}\,\rho(h)\,\Phi'_\delta(s) &= -t \int_G d\hat{h}\,\rho(h)\, l_s^{(\delta)}(h)^{-1} (\widetilde{H} l_s^{(0)})(h) \\
&= t\Big\{ \int_G d\hat{h}\,\rho(h) \sum_{i,j=1}^{d'} (A_i \log l_s^{(\delta)})(h)\, c_{ji}(h)\, (A_j \log l_s^{(\delta)})(h) \\
&\quad - 2 \int_G d\hat{h}\,\rho(h) \sum_{i,j=1}^{d'} (A_i\sigma)(h)\, c_{ji}(h)\, \sigma(h) (A_j \log l_s^{(\delta)})(h) \Big\} + R
\end{aligned}$$

where we have again explicitly exhibited the contribution to the differential equation of the part of $\widetilde{H}$ containing two derivatives and used R to indicate the contribution of the lower order terms. Note that we have also used the identity $A_j l_s^{(0)} = A_j l_s^{(\delta)}$.

Therefore by using subellipticity on the first term on the right and an $\varepsilon, \varepsilon^{-1}$ estimate on the second term one obtains bounds

$$\begin{aligned}
\int_G d\hat{h}\,\rho(h)\,\Phi'_\delta(s) &\geq (1-\varepsilon)\mu_C\, t \int_G d\hat{h}\,\rho(h) \sum_{i=1}^{d'} |(A_i \log l_s^{(\delta)})(h)|^2 \\
&\quad - \varepsilon^{-1} \mu_C^{-1}\, \|C\|_\infty^2\, t \int_G d\hat{h}\,\rho(h) \sum_{i=1}^{d'} |(A_i\sigma)(h)|^2 + R
\end{aligned}$$

for all $\varepsilon \in \langle 0,1]$. Since $\sum_{i=1}^{d'} \|A_i\sigma\|_\infty^2 \leq a\lambda^{-2}t^{-1}$ one then finds

$$\begin{aligned}
\int_G d\hat{h}\,\rho(h)\,\Phi'_\delta(s) &\geq (1-\varepsilon)\,\mu_C\, t \int_G d\hat{h}\,\rho(h) \sum_{i=1}^{d'} |(A_i \log l_s^{(\delta)})(h)|^2 \\
&\quad - \varepsilon^{-1} \mu_C^{-1}\, \|C\|_\infty^2\, a\,\lambda^{-2} |B'_{\lambda t^{1/2}}| + R \quad .
\end{aligned}$$

The remainder R can, however, be dealt with by similar $\varepsilon, \varepsilon^{-1}$ estimates. (Care has to be taken again with the terms in $\widetilde{H}$ with no derivatives on the right hand side. These contain a factor $l_s^{(0)}$ but this can be replaced by $l_s^{(\delta)}$ at the cost of introducing an extra factor $l_s^{(0)}/l_s^{(\delta)}$. But this factor takes values in $[0,1]$ and plays no essential role in the estimate which is expressed totally in terms of $l^{(\delta)}$ with all constants independent of δ.) Then, choosing ε appropriately, one obtains bounds

$$\int_G d\hat{h}\,\rho(h)\,\Phi'_\delta(s) \geq 2^{-1}\mu_C\,t\int_G d\hat{h}\,\rho(h)\sum_{i=1}^{d'}|(A_i\log l_s^{(\delta)})(h)|^2 - \nu\,|B'_{\lambda t^{1/2}}|$$

with the value of ν independent of δ, t and g. Next we use Proposition 3.2 to deduce that

$$\int_G d\hat{h}\,\rho(h)\sum_{i=1}^{d'}|(A_i\log l_s^{(\delta)})(h)|^2 \geq a\,\lambda^{-2}\,t^{-1}\int_G d\hat{h}\,\rho(h)|\log l_s^{(\delta)}(h) - \Phi_\delta(s)|^2$$

where a is a constant independent of δ, t, g and the coefficients of H. Therefore combining these bounds and using a straightforward estimate on the integral of ρ one concludes that

$$\Phi'_\delta(s) \geq \mu\Big(\int_G d\hat{h}\,\rho(h)\Big)^{-1}\int_G d\hat{h}\,\rho(h)|\log l_s^{(\delta)}(h) - \Phi_\delta(s)|^2 - \nu'$$

with the values of ν' and μ dependent on μ_C, $\|C\|_\infty$, $\|c\|_\infty$ and $\|c_0\|_\infty$ but independent of the choice of δ, t and g. This is the Nash differential inequality, [19], Part II, which is 'solved' by repetition of his original arguments (see, for example, [6] p. 95). One needs bounds $\int_{B'(g;4^{-1}\lambda t^{1/2})} d\hat{h}\,l_s(h) \geq c\,t^{D'/2}$ uniformly for all $s \in \langle 0,1]$, $t \in \langle 0,t_0]$ and $g \in B'(4^{-1}\lambda t)$, but these have been proved under weaker restrictions in Lemma 3.1. The conclusion is that $\Phi_\delta(1) \geq -c$ uniformly for all $\delta \in \langle 0,1]$, $t \in \langle 0,t_0]$ and $g \in B'(4^{-1}\lambda t)$. Finally taking the limit $\delta \to 0$ gives the statement of the lemma. □

The required local lower bounds on the kernel follow straightforwardly from the entropy estimate.

Lemma 3.5 *There exist $a, \lambda > 0$ and $t_0 \in \langle 0,1]$ such that*

$$K_t(g\,;h) \geq a\,t^{-D'/2}$$

uniformly for all $t \in \langle 0,t_0]$ and $g,h \in G$ with $|gh^{-1}|' \leq \lambda t^{1/2}$.

Proof. First note that as $K_t(h\,;g) = \widetilde{K}_t(g\,;h)$ where $\widetilde{K}_t$ is the kernel associated with the adjoint semigroup on the $L_{\hat{p}}$-spaces it follows from Lemma 3.3 that one has estimates

$$\int_G d\hat{h}\,\rho_{\lambda t^{1/2}}(h)\log\Big(t^{D'/2}K_t(g\,;h)\Big) \;\geq\; -c\int_G d\hat{h}\,\rho_{\lambda t^{1/2}}(h)$$

$$\int_G d\hat{h}\,\rho_{\lambda t^{1/2}}(h)\log\Big(t^{D'/2}K_t(h\,;g)\Big) \;\geq\; -c\int_G d\hat{h}\,\rho_{\lambda t^{1/2}}(h)$$

uniformly for all $t \in \langle 0, t_0]$ and $g \in B'(\lambda t^{1/2})$, for suitable $c, \lambda > 0$ and $t_0 \in \langle 0, 1]$. Now using the convolution semigroup property one has

$$\begin{aligned} t^{D'} K_{2t}(g\,;k) &= \int_G d\hat{h}\, t^{D'/2} K_t(g\,;h) t^{D'/2} K_t(h\,;k) \\ &\geq \int_G d\hat{h}\, \rho_{\lambda t^{1/2}}(h)\, t^{D'/2} K_t(g\,;h)\, t^{D'/2} K_t(h\,;k) \quad . \end{aligned}$$

Therefore using concavity of the logarithm one deduces that

$$\begin{aligned} \log\Big(\Big(\int_G d\hat{h}\, \rho_{\lambda t^{1/2}}(h)\Big)^{-1} t^{D'} K_{2t}(g\,;k)\Big) &\geq \Big(\int_G d\hat{h}\, \rho_{\lambda t^{1/2}}(h)\Big)^{-1} \cdot \\ &\cdot \int_G d\hat{h}\, \rho_{\lambda t^{1/2}}(h)\Big(\log\big(t^{D'/2} K_t(g\,;h)\big) + \log\big(t^{D'/2} K_t(h\,;k)\big)\Big) \\ &\geq -2\,c \end{aligned}$$

uniformly for all $t \in \langle 0, t_0]$ and $g, k \in B'(\lambda t^{1/2})$. Hence

$$K_{2t}(g\,;k) \geq e^{-2c} t^{-D'} \int_G d\hat{h}\, \rho_{\lambda t^{1/2}}(h) \geq a\, t^{-D'/2}$$

under the same restrictions on g, k and t. Now the restrictions on g and k can be weakened by noting that $(g, k) \mapsto K_t(gh\,;kh)$ is the kernel associated with the subelliptic operator H conjugated with right translations by h. Since the coefficients of the conjugated operator are the right translates of the original coefficients the ellipticity constant μ_C and the parameters $\|C\|_\infty$ etc. are unchanged. Hence one has similar estimates

$$K_{2t}(gh\,;kh) \geq a\, t^{-D'/2}$$

valid under the previous restrictions but for all $h \in G$. The statement of the lemma follows immediately. □

The final step in the proof of the Gaussian lower bounds on the kernel is to convert the local lower bounds into the Gaussian bounds by the use of the convolution semigroup property. This is achieved by a now standard procedure (see, for example, [20], Proposition III.5.2).

Finally we note that if one can establish the local lower bounds for all $t > 0$ then the Gaussian lower bounds of Theorem 1.1 are valid with $\omega' = 0$.

4. Concluding remarks

There are several direction in which the bounds of Theorem 1.1 can be improved under more restrictive circumstances and this leaves several questions unanswered. First if the matrix of principal coefficients c_{ij} of H is symmetric and one introduces the distance

$$d_c(g\,;h) = \sup\{\,|\psi(g) - \psi(h)| \,:\, \psi \in C_c^\infty(G)\,,\,\sum_{i=1}^{d'}(A_i\psi)c_{ij}(A_j\psi) \leq 1,\ \psi \text{ real}\}$$

then there exists an $\omega \geq 0$ and for each $\varepsilon > 0$ an $a_\varepsilon > 0$ such that

$$|K_t(g\,;h)| \leq a_\varepsilon\, t^{-D'/2} e^{\omega t} e^{-d_c(g;h)^2(4(1+\varepsilon)t)^{-1}}$$

uniformly for $g, h \in G$ and $t > 0$. These upper bounds have been established by a variety of arguments (see, for example, [6] [12]) and the argument of the Gaussian is near optimal. If, however, the matrix of principal coefficients is non-symmetric none of the reasonings appear to allow optimization of the argument.

Secondly, assume that $a_1, \ldots, a_{d'}$ is a vector space basis of $\mathfrak{g}$ and all coefficients of H are real. Then if the matrix of principal coefficients is symmetric and the c_{ij} are (left) uniformly continuous one can establish lower bounds

$$K_t(g\,;h) \geq a_\varepsilon\, t^{-D'/2} e^{-\omega t} e^{-d_c(g;h)^2(4(1-\varepsilon)t)^{-1}}$$

for each $\varepsilon \in \langle 0, 1\rangle$ uniformly for $g, h \in G$ and $t > 0$. These are derived by a variation of Nash's original reasoning as developed by Stroock [23]. The arguments do not seem to extend to the subelliptic situation nor to the case that c_{ij} is non-symmetric.

Thirdly, it is possible to derive good large time bounds for groups of polynomial growth. If the operator H is pure second-order and c_{ij} is symmetric one has upper and lower bounds

$$a'\, V(t)^{-1/2} e^{-b'(|gh^{-1}|')^2 t^{-1}} \leq K_t(g\,;h) \leq a\, V(t)^{-1/2} e^{-d_c(g;h)^2(4(1+\varepsilon)t)^{-1}}$$

uniformly for $g, h \in G$ and $t > 0$ where $V(t) = |B'(e\,;t)|$. These follow from [21] together with a regularization argument (see, for example, [11], Section 2.1). The distance d_c is, however, not necessarily optimal for large time bounds. The situation of non-symmetric principal coefficients is again unclear.

Acknowledgements. Parts of this work were carried out whilst the first author was visiting the Centre of Mathematics and its Applications at the ANU and the second author was visiting the Technical University in Eindhoven. Both authors wish to thank the respective institutes for supporting these visits.

References

[1] Arendt, W., Batty, C.J.K. and Robinson, D.W., Positive semigroups generated by elliptic operators on Lie groups. *J. Operator Theory* **23** (1990), 369–407.

[2] AUSCHER, P., MCINTOSH, A. and TCHAMITCHIAN, P., Heat kernels of second order complex elliptic operators and their applications. *J. Funct. Anal.* **152** (1998), 22–73.

[3] ARONSON, D.G., Bounds for the fundamental solution of a parabolic equation. *Bull. Amer. Math. Soc.* **73** (1967), 890–896.

[4] AUSCHER, P., Regularity theorems and heat kernel for elliptic operators. *J. London Math. Soc.* **54** (1996), 284–296.

[5] BRATTELI, O. and ROBINSON, D.W., Subelliptic operators on Lie groups: variable coefficients. *Acta Appl. Math.* **42** (1996), 1–104.

[6] DAVIES, E.B., *Heat kernels and spectral theory.* Cambridge Tracts in Mathematics 92. Cambridge University Press, Cambridge etc., 1989.

[7] ELST, A.F.M. TER and ROBINSON, D.W., Subcoercive and subelliptic operators on Lie groups: variable coefficients. *Publ. RIMS. Kyoto Univ.* **29** (1993), 745–801.

[8] ——, Reduced heat kernels on nilpotent Lie groups. *Commun. Math. Phys.* **173** (1995), 475–511.

[9] ——, High order divergence-form elliptic operators on Lie groups. *Bull. Austral. Math. Soc.* **55** (1997), 335–348.

[10] ——, Weighted subcoercive operators on Lie groups. *J. Funct. Anal.* **157** (1998) 88–163.

[11] ——, Second-order subelliptic operators on Lie groups I: complex uniformly continuous principal coefficients. *Acta Appl. Math.* (2000). To appear.

[12] FABES, E.B. and STROOCK, D.W., A new proof of Moser's parabolic Harnack inequality using the old ideas of Nash. *Arch. Rat. Mech. and Anal.* **96** (1986), 327–338.

[13] FRANCHI, B., LU, G. and WHEEDEN, R.L., Weighted Poincaré inequalities for Hörmander vector fields and local regularity for a class of degenerate elliptic equations. *Potential Anal.* **4** (1995), 361–375.

[14] GIAQUINTA, M., *Introduction to regularity theory for nonlinear elliptic systems.* Lectures in Mathematics ETH Zürich. Birkhäuser Verlag, Basel etc., 1993.

[15] GILBARG, D. and TRUDINGER, N.S., *Elliptic partial differential equations of second order.* Second edition, Grundlehren der mathematischen Wissenschaften 224. Springer-Verlag, Berlin etc., 1983.

[16] GIORGI, E.D., Sulla differenziabilità e l'analiticità delle estremali degli integrali multipli regolari. *Mem. Accad. Sci. Torino cl. Sci. Fis. Mat. Nat.* **3** (1957), 25–43.

[17] JERISON, D., The Poincaré inequality for vector fields satisfying Hörmander's condition. *Duke Math. J.* **53** (1986), 503–523.

[18] MORREY, C.B., *Multiple integrals in the calculus of variations.* Springer-Verlag, Heidelberg etc., 1966.

[19] NASH, J., Continuity of solutions of parabolic and elliptic equations. *Amer. J. Math.* **80** (1958), 931–954.

[20] ROBINSON, D.W., *Elliptic operators and Lie groups.* Oxford Mathematical Monographs. Oxford University Press, Oxford etc., 1991.

[21] SALOFF-COSTE, L. and STROOCK, D.W., Opérateurs uniformément sous-elliptiques sur les groupes de Lie. *J. Funct. Anal.* **98** (1991), 97–121.

[22] STROOCK, D.W., Diffusion semigroups corresponding to uniformly elliptic divergence form operators. In AZÉMA, J., MEYER, P.A. and YOR, M., eds., *Séminaire de probabilités XXII*, Lecture Notes in Mathematics 1321. Springer-Verlag, Berlin etc., 1988, 316–347.

[23] ———, Estimates for the heat kernel of second order elliptic operators. In *Nonlinear partial differential equations and their applications. Collège de France Seminar, Vol. XII (Paris, 1991–1993)*, vol. 302 of Pitman Res. Notes Math. Ser., 226–235. Longman Sci. Tech., Harlow, 1994.

[24] VAROPOULOS, N.T., SALOFF-COSTE, L. and COULHON, T., *Analysis and geometry on groups.* Cambridge Tracts in Mathematics 100. Cambridge University Press, Cambridge, 1992.

A.F.M. TER ELST, DEPARTMENT OF MATHEMATICS AND COMPUTING SCIENCE, EINDHOVEN UNIVERSITY OF TECHNOLOGY, P.O. BOX 513, 5600 MB EINDHOVEN, THE NETHERLANDS
E-mail Address: `terelst@win.tue.nl`

DEREK W. ROBINSON, CENTRE FOR MATHEMATICS AND ITS APPLICATIONS, SCHOOL OF MATHEMATICAL SCIENCES, AUSTRALIAN NATIONAL UNIVERSITY, CANBERRA, ACT 0200, AUSTRALIA
E-mail Address: `derek.robinson@anu.edu.au`

Progress in Nonlinear Differential Equations
and Their Applications, Vol. 42

GENERALIZED WENTZELL BOUNDARY CONDITIONS AND ANALYTIC SEMIGROUPS IN $C[0,1]$

ANGELO FAVINI
UNIVERSITÁ DI BOLOGNA

GISÈLE R. GOLDSTEIN
JEROME A. GOLDSTEIN
UNIVERSITY OF MEMPHIS

SILVIA ROMANELLI
UNIVERSITÁ DI BARI

0. Introduction

In [4], we introduced for the first time the so-called generalized Wentzell boundary conditions for some classes of linear, or nonlinear, second order differential operator with domain in the space $C[0,1]$ of all real-valued continuous functions on $[0,1]$. There we proved generation results which extended substantially those referred to Dirichlet, Neumann, Robin and Wentzell boundary conditions.

Here we examine a particular case where, in addition, the analyticity holds for the generated semigroup. Our operator is defined as

$$Au := \alpha u'',$$

and its domain, included in $C[0,1]$, equipped with the supremum norm $\|\cdot\|_\infty$, contains suitable generalized Wentzell boundary conditions of the type

$$Au(j) + b_j u'(j) = 0, \qquad j = 0, 1,$$

where the coefficient α is assumed to satisfy

$$\alpha \in C[0,1], \qquad \alpha(x) > 0 \quad \text{in} \quad (0,1).1$$

as in [1], [2], [4], [5], and moreover

$$\alpha \in C^1(0,1), \qquad \frac{1}{\alpha} \in L^1(0,1), .2$$

as in [1] and [4]. The numbers b_o, b_1 are exactly the values at the endpoints of a suitable function $\beta \in C[0,1]$, which is Hölder continuous at 0 and 1, as in [1].

Indeed, we obtain an extension of the analyticity result due to Campiti and Metafune [1, Theorem 4.2].

Let us remark that the degeneration at the endpoints of the coefficient α (i.e. $\alpha(0) = 0 = \alpha(1)$) is allowed, provided that it is of low order.

1. Preliminary results

In order to make more clear the tools involved in the proof of the main theorem, we recall a result by Dorroh and Rieder [3, Lemma 3.1] and a theorem due to Campiti and Metafune [1, Theorem 4.2].

Lemma 1.1. *Let $u \in C^1[0,1] \cap C^2(0,1)$ with $u'' \in L^1[0,1]$.*

Then, for $0 < \delta \le \frac{1}{2}$, we have that

$$\|u'\|_\infty \le \frac{2}{\delta}\|u\|_\infty + \omega_L(u'', \delta),$$

where

$$\omega_L(g,\delta) := \sup\{\int_E |g| \ : \ E \text{ is a subinterval of } [0,1] \text{ with } |E| \le \delta\},$$

for $g \in L^1[0,1]$ and $0 \le \delta \le 1$.

Theorem 1.2. *Assume that $\alpha, \beta \in C[0,1]$, where α satisfies (0.1), (0.2) and β is Hölder continuous at $x = 0, 1$. Then the operator $(L, D(L))$ given by*

$$Lu := \alpha u'' + \beta u'$$

$$D(L) := \{u \in C[0,1] \cap C^2(0,1) \ : \ \lim_{x \to j} Lu(x) = 0, \text{ for } j = 0, 1\}$$

generates an analytic semigroup on $C[0,1]$.

2. Main Result

The main result of this paper reads as follows

Theorem 2.1. *Assume that $\alpha, \beta \in C[0,1]$ verify the following assumptions :*

(2.1) α *satisfies* (0.1) *and* (0.2), .1

(2.2) β *is Hölder continuous at* $0, 1$, .2

(2.3) $(\beta(0), \beta(1)) \ne (0,0)$..3

Then, denoting by $b_j := \beta(j)$ for $j = 0, 1$, the operator

$$Au := \alpha u''$$

with domain

$$D_b(A) := \{u \in C[0,1] \cap C^2(0,1) \ : \ Au \in C[0,1],\ Au(j) + b_j u'(j) = 0, \text{ for } j = 0, 1\},$$

generates an analytic semigroup on $C[0,1]$.

Proof. As a consequence of Theorem 1.2, the operator $(L, D(L))$ given by

$$\begin{aligned} Lu &:= \alpha u'' + \beta u' \\ D(L) &:= D_b(A), \end{aligned}$$

generates an analytic semigroup on $C[0,1]$. Now, let $Bu := -\beta u'$ with domain $D(B) := D(L)$; we will show that $(B, D(B))$ is L-bounded with L-bound equal to zero (see [7, p.190]).

Preliminarily, we show that every $u \in C[0,1] \cap C^2(0,1)$ with $\alpha u'' + \beta u' \in C[0,1]$ has its first derivative u' in $C[0,1]$.

Indeed, from $\dfrac{1}{\alpha} \in L^1(0,1)$ it follows that the function

$$W(x) := exp(-\int_{\frac{1}{2}}^{x} \frac{\beta(t)}{\alpha(t)}\, dt)$$

satisfies

$$0 < c_o \leq W(x) \leq c_1,$$

for suitable positive constants c_o, c_1. Since

$$f := \alpha W(\frac{u'}{W})' \in C[0,1],$$

we deduce

$$\frac{u'(x)}{W(x)} - \frac{u'(\frac{1}{2})}{W(\frac{1}{2})} = \int_{\frac{1}{2}}^{x} \frac{f(t)}{\alpha(t)W(t)}\, dt$$

and, consequently,

$$u'(x) = W(x)\left[\frac{u'(\frac{1}{2})}{W(\frac{1}{2})} + \int_{\frac{1}{2}}^{x} \frac{f(t)}{\alpha(t)W(t)}\, dt\right].$$

Now we observe that every $u \in D(L)$ verifies the inequality

$$|u''(x)| \leq \frac{1}{\alpha(x)}\|\alpha u'' + \beta u'\|_\infty + \frac{|\beta(x)|}{\alpha(x)}|u'(x)|, \quad x \in (0,1)$$

hence $u'' \in L^1(0,1)$ and

$$\|u''\|_1 \leq \|\frac{1}{\alpha}\|_1(\|Lu\|_\infty + \|\beta u'\|_\infty).$$

Thus, according to Lemma 1.1, each $u \in D(L)$ satisfies

$$\|u'\|_\infty \leq \frac{2}{\delta}\|u\|_\infty + \delta\|u''\|_1,$$

for all $0 < \delta \leq \frac{1}{2}$. Therefore

$$\begin{aligned} \|\beta u'\|_\infty &\leq \|\beta\|_\infty \left(\frac{2}{\delta}\|u\|_\infty + \delta\|u''\|_1\right) \\ &\leq \|\beta\|_\infty \left(\frac{2}{\delta}\|u\|_\infty + \delta\|Lu\|_\infty\|\frac{1}{\alpha}\|_1 + \right. \\ &\qquad \left. + \delta\|\beta u'\|_\infty\|\frac{1}{\alpha}\|_1\right) \end{aligned}$$

or, equivalently

$$\left(1 - \|\beta\|_\infty\|\frac{1}{\alpha}\|_1\delta\right)\|\beta u'\|_\infty \leq \frac{2\|\beta\|_\infty}{\delta}\|u\|_\infty + .4 \\ + \|\beta\|_\infty\delta\|\frac{1}{\alpha}\|_1\|Lu\|_\infty. \tag{2.4}$$

If

$$0 < \delta \leq \min\{\frac{1}{2}, \frac{1}{2\|\beta\|_\infty\|\frac{1}{\alpha}\|_1}\} := \delta_o \in \left(0, \frac{1}{2}\right],$$

from (2.4) we obtain

$$\| - \beta u'\|_\infty \leq \frac{2\|\beta\|_\infty}{\delta\left(1 - \|\beta\|_\infty\|\frac{1}{\alpha}\|_1\delta\right)}\|u\|_\infty + \frac{\delta\|\frac{1}{\alpha}\|_1\|\beta\|_\infty}{1 - \|\beta\|_\infty\|\frac{1}{\alpha}\|_1\delta}\|Lu\|_\infty..5$$

Define

$$\varepsilon_\delta := \frac{\|\beta\|_\infty\,\|\frac{1}{\alpha}\|_1\,\delta}{1 - \|\beta\|_\infty\,\|\frac{1}{\alpha}\|_1\,\delta}$$

and observe that

$$\delta = \frac{\varepsilon_\delta}{(\varepsilon_\delta + 1)\|\beta\|_\infty\,\|\frac{1}{\alpha}\|_1} \leq \frac{1}{2\|\beta\|_\infty\,\|\frac{1}{\alpha}\|_1} \quad \textit{if and only if} \quad 0 < \varepsilon_\delta < 1,$$

while

$$\begin{aligned} \delta \leq \tfrac{1}{2} \quad & \textit{if and only if} \quad 2\varepsilon_\delta\|\beta\|_\infty\|\frac{1}{\alpha}\|_1 \leq \varepsilon_\delta + 1 \\ & \textit{if and only if} \\ & \varepsilon_\delta\left(2\|\beta\|_\infty\,\|\frac{1}{\alpha}\|_1 - 1\right) \leq 1 \\ & \textit{if and only if} \quad \varepsilon_\delta \leq \frac{1}{2\|\beta\|_\infty\,\|\frac{1}{\alpha}\|_1 - 1}. \end{aligned}$$

Hence, if

$$\varepsilon_o := \min\{1, \frac{1}{2\|\beta\|_\infty\,\|\frac{1}{\alpha}\|_1 - 1}\}$$

we deduce that

$$\delta \in (0, \delta_o) \quad \textit{if and only if} \quad \varepsilon \in (0, \varepsilon_o).$$

Consequently, for all $\varepsilon \in (0, \varepsilon_o)$ there exists $\eta > 0$ such that

$$\|Bu\|_\infty \leq \eta \|u\|_\infty + \varepsilon \|Lu\|_\infty,$$

i.e. $(B, D(B))$ is L-bounded with L-bound equal to zero.

Taking into account [7, Chap. IX Theorem 2.4] (see also [6, Corollary 6.9 p.42]), we conclude that $A = L + B$ with domain $D(L)$ generates an analytic semigroup on $C[0,1]$. □

Example

Let $\alpha(x) := m(x)[x^i(1-x)^\lambda]$, for $x \in [0,1]$, and assume that

$$m \in C[0,1], \quad m(x) > 0, \qquad x \in [0,1]$$

and

$$0 \leq i < 1, \qquad 0 \leq \lambda < 1.$$

Then, for all $\beta \in C[0,1]$ which satisfies (2.2) and (2.3) the operator

$$Au := m(x)[x^i(1-x)^\lambda]u''$$

with domain

$$D_b(A) := \{u \in C[0,1] \cap C^2(0,1) | \, m(x)[x^i(1-x)^\lambda]u'' \in C[0,1],$$
$$Au(j) + \beta(j)u'(j) = 0, \text{ for } j = 0, 1\}$$

generates an analytic semigroup on $C[0,1]$.

References

[1] M. Campiti and G. Metafune, *Ventcel's boundary conditions and analytic semigroups,* Arch. Math. **70** (1998), 377–390.

[2] Ph. Clément and C. Timmermans, *On C_o-semigroups generated by differential operators satisfying Ventcel's boundary conditions,* Indag. Math. **89** (1986), 379–387.

[3] J.R. Dorroh and G.R. Rieder, *A singular quasilinear parabolic problem in one space dimension,* Pacific Journal of Diff. Equations **91**, No. 1 (1991), 1–23.

[4] A. Favini, G.R. Goldstein, J.A. Goldstein and S. Romanelli, *C_o-semigroups generated by second order differential operators with general Wentzell boundary conditions,* (preprint).

[5] A. Favini and S. Romanelli, *Analytic semigroups on $C[0,1]$ generated by some classes of second order differential operators*, Semigroup Forum **56** (1998) 367–372.

[6] J.A. Goldstein, *Semigroups of Linear Operators and Applications*, Oxford University Press, Oxford, New York, 1985.

[7] T. Kato, *Perturbations Theory for Linear Operators*, Springer-Verlag, Berlin, Heidelberg, New York, 1966.

Angelo Favini, Dipartimento di Matematica, Universitá di Bologna, Piazza di Porta S.Donato 5, 40127 Bologna, ITALY
E-mail address: favini@dm.unibo.it

Gisèle R. Goldstein, CERI & Department of Mathematical Sciences, University of Memphis, Memphis, Tennessee 38152, U.S.A.
E-mail address: gisele@ceri.memphis.edu

Jerome A. Goldstein, Department of Mathematical Sciences, University of Memphis, Memphis, Tennessee, 38152, U.S.A.
E-mail address: goldstej@msci.memphis.edu

Silvia Romanelli, Dipartimento Interuniversitario di Matematica, Universitá di Bari, Bari 70125, ITALY
E-mail address: romanspascal.dm.uniba.it

Progress in Nonlinear Differential Equations
and Their Applications, Vol. 42

SEMIGROUP THEORY AND OPERATOR TRIGONOMETRY

KARL GUSTAFSON
UNIVERSITY OF COLORADO

Abstract. The origins of an interesting but still little known operator trigonometry as it developed from perturbation questions in the Hille-Yosida-Phillips operator semigroup theory over 30 years ago, will be traced. This operator trigonometry, largely a creation of this author, features a notion of antieigenvalue which corresponds to the largest turning angle of an operator. As such, it constitutes an extension of the Rayleigh-Ritz theory, to now include both eigenvectors and antieigenvectors. Then the application of the operator trigonometry, during the last 10 years, to iterative solvers of linear systems in computational linear algebra, will be described. Roughly, the operator trigonometry provides the previously missing geometrical meanings of these algorithms. For example, the famous Kantorovich bound, known now for 50 years, is now clearly understood geometrically. Finally, a very interesting recent (1998) application to quantum probabilities will be exposed.

1. Introduction and Outline of This Paper

As indicated in the Abstract above, it is convenient to think of the development of the operator trigonometry, which one may regard as a new chapter in Linear Algebra, in three chronological stages. Of course there are some overlaps and some work in the intervening periods. However, I shall present the three stages in Sections 2, 3, and 4, respectively, as outlined in the Abstract. I shall also add a few brief remarks which include some recent results and observations not yet published elsewhere. For all the history and results that I just skip over or leave out, the reader will find most of them in the two recent books [1,2] and the survey [3].

2. Semigroup Theory and Operator Trigonometry

The origins of the operator trigonometry can be traced back to 1966 and three theorems characterizing contraction semigroup generators.

Theorem 1. (Hille-Yosida-Phillips-Lumer) *In a Banach space X, a densely defined operator A is the infinitesimal generator of a contraction semigroup e^{tA} iff it is dissipative and $\mathcal{R}(I - A) = X$.*

Theorem 2. (Rellich-Kato-Sz.Nagy-Gustafson) *If a generator and B is A-small, then $A + B$ is a generator iff $A + B$ is dissipative.*

Theorem 3. (Dorroh-Gustafson-Lumer-Sato) *If A is a generator and B is bounded and strongly accretive, then BA is a generator if BA is dissipative.*

Of course there are a number of generalizations of these results to bounded semigroups, left and right perturbations, etc., see the literature. Recall that A dissipative means that $\mathrm{Re}\,[Ax, x] \leqq 0$ for all x in the domain of A, for some given semi-inner-product, B strongly accretive means that $\mathrm{Re}\,[Bx, x] \geqq m\|x\|^2$ for some $m > 0$, and B is A-small means $\|Bx\| \leqq a\|x\| + b\|Ax\|$ for some $b < 1$.

Theorem 2 may be proved from Theorem 1, Theorem 3 from Theorem 2, see [1,2,3]. What is missing from Theorem 3 is a criteria for BA to be dissipative when A is dissipative and B is accretive. This question is the origin of the operator trigonometry. Consider in particular the question of when bounded operators A and B are both m-accretive and you want BA to be accretive. From that question in 1966–68 I found the following three results.

Theorem 4. *For bounded A and B which are both m-accretive, a sufficient condition for BA to be accretive is*

$$\sin\phi(B) \leqq \cos\phi(A). \tag{1}$$

In (1) $\phi(A)$ denotes the angle of an operator A, defined through the first antieigenvalue μ of A

$$\mu = \cos\phi(A) = \inf_{x\neq 0} \frac{\mathrm{Re}\,[Ax, x]}{\|Ax\|\,\|x\|}, \tag{2}$$

and $\sin\phi(B)$ denotes the entity

$$\nu = \sin\phi(B) = \inf_{\epsilon>0} \|\epsilon B - I\|. \tag{3}$$

That ν deserves to be called $\sin\phi(B)$ follows from

Theorem 5. *For any bounded strongly accretive operator B on a Hilbert space X,*

$$\nu^2(B) + \mu^2(B) = 1. \tag{4}$$

Theorem 6. *For A a symmetric positive definite (SPD) $n \times n$ matrix,*

$$\cos\phi(A) = \frac{2\sqrt{\lambda_1\lambda_n}}{\lambda_1 + \lambda_n}, \qquad \sin\phi(A) = \frac{\lambda_n - \lambda_1}{\lambda_n + \lambda_1}. \tag{5}$$

The first antieigenvector (pair) is

$$x_\pm^1 = \pm\left(\frac{\lambda_n}{\lambda_1 + \lambda_n}\right)^{1/2} x_1 + \left(\frac{\lambda_1}{\lambda_1 + \lambda_n}\right)^{1/2} x_n \tag{6}$$

In (5) and (6) λ_1 and λ_n denote the smallest and largest eigenvalues of A. In (6) x_1 and x_n denote corresponding eigenvectors, with all of x_1, x_n, and $x_\pm^1$ normalized to norm 1. The first antieigenvectors $x_\pm^1$ are those which are turned the most by A. They always occur in pairs.

One more important result from the early days is the following Euler equation of the variational functional (2).

Theorem 7. *For A a strongly accretive bounded operator on a Hilbert space, the functional in (2) has Euler equation*

$$(7)\quad 2\|Ax\|^2\|x\|^2(Re\, A)x - \|x\|^2 Re\,\langle Ax,x\rangle A^*Ax - \|Ax\|^2 Re\,\langle Ax,x\rangle x = 0.$$

All first antieigenvectors, and all eigenvectors if A is a normal operator, satisfy (7). The functional (2) and the Euler equation (7) thus generalize the Rayleigh-Ritz characterization of eigenvectors to also include antieigenvectors.

Remark 2.1. All of the above became known in the period 1966–1968, 30 years ago. As my main interest had been to understand the semigroup generator BA dissipative product condition of Theorem 3, and as I was becoming interested in the problems of quantum scattering theory, I moved on. For example, I presented the Euler equation of Theorem 7 at the Third Symposium on Inequalities in Los Angeles in 1969, but in the Proceedings I just mentioned it is passing. Although the theory of semigroups entered my research again in the work on Kolmogorov-Prigogine theory in the 1980's, e.g. see [1, Part II, Chapter 1], except for a few momemtary exceptions I did not return seriously to the operator trigonometry until 1990. That return will be described in the next section.

Remark 2.2. Although the antigenvectors $x_\pm^1$ in (6) occur in pairs, their linear span does not form an anti-eigenspace. However, if the λ_1 or λ_n has multiplicity greater than one, then the respective x_1 or x_n in (6) may be taken arbitrarily of norm one from the respective eigenspace. This fact is easily verified but has not been pointed out heretofore. In this sense the first antieigenvectors take on the multiplicities of the eigenspaces.

3. Operator Trigonometry and Computational Linear Algebra

In the 1980's my research emphasis had turned to computational fluid dynamics. There one quickly learns that a major percentage (e.g., 70% of the computer time for applications in 2 space dimensions, 90% for those in three space dimensions) of the computational effort is consumed in the linear solvers for the large linear systems $Ax = b$ that result from discretizing the partial differential operators (e.g., the Laplacian) describing the fluid flows. Linear solvers may be roughly broken down into direct methods (e.g., Gauss, LU, or FFT-based schemes) or indirect methods (e.g., iterative Gradient or Splitting schemes). Being invited to speak at the 1990 conference [4], I decided to take that opportunity to examine the possible connections of the operator trigonometry to computational linear algebra. I had suspected such connections 30 years ago, I even mentioned it in one of the 1968 papers. Very quickly the following result was obtained [4].

Theorem 8. *In quadratic steepest descent solution of the symmetric positive definite linear matrix system $Ax = b$, the fundamental Kantorovich error bound*

$$E_a(x_{k+1}) \leqq \left(1 - \frac{4\lambda_1\lambda_n}{(\lambda_1 + \lambda_n)^2}\right) E_A(x_k) \tag{8}$$

is in fact trigonometric:

$$E_a(x_{k+1}) \leqq \sin^2 \phi(A) E_A(x_k). \tag{9}$$

In (8) and (9) E_A denotes the energy error inner product $E_A(x) = \langle (x - x^*), A(x - x^*)\rangle$, where x^* is the true solution of the system. Steepest descent is asymptotically a very slowly converging scheme but its analysis opens the way to that of better gradient schemes such as Conjugate Gradient, GMRES, others. See [5,6,7] for further information and bibliography.

When invited to speak at the 1996 conference [5], I decided I should also look at possible connections of the operator trigonometry to general iterative operator splitting schemes in computational linear algebra. The "grand-daddy" of all such schemes is the basic Richardson method $x_{k+1} = x_k + \alpha(b - Ax_k)$ with iteration matrix $G_\alpha = I - \alpha A$ and where one chooses the parameter α to produce an optimal convergence rate. The following result was obtained [5].

Theorem 9. *In Richardson iterative solution of $Ax = b$ for strictly accretive A, the optimal parameter α is*

$$\alpha = \epsilon_m = \frac{Re\langle Ax_\pm^1, x_\pm^1\rangle}{\|Ax_\pm^1\|^2} \tag{10}$$

where ϵ_m is the minimizing parameter for ν in (3). The optimal convergence rate of the Richardson scheme is $\sin\phi(A)$.

The operator trigonometry-computational linear algebra connection has recently been extended to Preconditioned Conjugate Gradient, Jacobi, Gauss-eidel, SOR, SSOR, Uzawa, AMLI, ADE, Multigrid, Domain Decomposition, and related iterative solution methods for $Ax = b$. More details may be found in [5,6,7] and work in progress.

Remark 3.1. In hindsight, I had independently derived my own version of the Kantorovich inequality, 30 years ago, when I obtained the expressions (5). Actually, I had obtained (5) in the form

$$\cos\phi(A) = \frac{2\sqrt{mM}}{m + M}, \qquad \sin\phi(A) = \frac{M - m}{M + m} \tag{11}$$

for any selfadjoint operator A with norm $\|A\| = M$ and $\|A^{-1}\|^{-1} = m$, using convexity properties of norms. In any case, the conclusion (9) from (8) was immediate once I looked it, given Theorems 5 and 6 from the operator trigonometry. But until I did that, no one had seen the natural geometry of the longstanding 1948 Kantorovich bound. Theorem 8 first appeared in the Proceedings [4], which were published in spite of the war breaking out in Yugoslavia only a few months after the June 1990 conference in Dubrovnik.

Remark 3.2. Originally, 30 years ago I defined higher antieigenvalues in terms of infima of functional (2) restricted to subspaces orthogonal to the preceding antieigenvectors. In the 1990's my computer experiments with the iterative linear solvers for $Ax = b$ indicated that it was better to define higher antieigenvalues combinatorially, e.g., replacing λ_n and λ_1 in (6) by λ_j and λ_i, $1 > i, j > n$, and the corresponding eigenvector components accordingly, thereby "nesting inward" through a sequence of decreasing critical turning angles. Let us observe here that these two philosophies are not necessarily so different. Consider any two vectors $x_+ = cx_1 + dx_2$ and $x_- = -cx_1 + dx_2$ where c and c are nonzero possibly complex coefficients and where x_1 and x_2 are linearly independent. Then x_+ and x_- have the same span as x_1 and x_2. Thus minimizing the functional (2) on the subspace orthogonal to the span of the first antieigenvector pair is the same as finding the next critical turning angle of A in the subspace orthogonal to the span of the first and last eigenvectors of A, at least in the cases (e.g., A symmetric or normal) where A's extreme eigenvectors may be expressed in terms of A's extreme eigenvectors as in (6). Pursuing this analysis a bit further, now also assume x_1 and $x_2 \equiv x_n$ to have norm one. Then (using (6) therein) one obtains the interesting expression

$$\begin{aligned} \langle x_+, x_- \rangle &= -|c|^2 + |d|^2 + 2i\mathrm{Im}\,(c\bar{d}\langle x_1, x_n\rangle) \\ &= \frac{\lambda_1 - \lambda_n}{\lambda_1 + \lambda_n} + i\,\frac{2\lambda_1^{1/2}\lambda_n^{1/2}}{\lambda_1 + \lambda_n}\,\mathrm{Im}\,\langle x_1, x_n\rangle \qquad (12) \\ &= -\sin\phi(A) + i\cos\phi(A)\mathrm{Im}\,\langle x_1, x_2\rangle. \qquad (13) \end{aligned}$$

In particular for A symmetric positive definite we have from (12) that $\langle x_+^1, x_-^1\rangle = -\sin\phi(A)$, i.e., the angle $\phi_\pm$ of the antieigenvector pair (6) is always $\phi(A) + \pi/2$.

As this result, and that of Remark 2.2, do not appear elsewhere, let us state them as follows.

Theorem 10. *Let A be a symmetric positive definite $n \times n$ matrix, $\lambda_1 \neq \lambda_n$. Then if the multiplicity of either λ_1 or λ_n is greater than one, antieigenvector pairs may be taken from (6) with x_1 and x_n any normalized eigenvectors corresponding to λ_1 and λ_n, respectively. The angle $\phi_\pm$ of the resulting antieigenvector pair $x_\pm^1$ is always $\phi(A) + \pi/2$, where $\phi(A)$ is A's operator angle. The same statements apply to higher antieigenvector pairs formed combinatorially* *as described above, with $\phi(A)$ replaced by $\phi_{ij}(A)$ defined by*

$$\sin \phi_{ij}(A) = \frac{\lambda_j - \lambda_i}{\lambda_j + \lambda_i}, \tag{14}$$

$i < j$, $\lambda_i < \lambda_j$.

4. Recent Developments

Here I will briefly mention two very recent developments of the operator trigonometry which I believe will have considerable future importance. The first is an intrinsic connection to Bell's fundamental inequality in quantum hidden variable theories. The second is a needed extension of the operator trigonometry to general operators A on a Hilbert space. Finally, a few other recent results not in the books [1,2] or the references [3–8] will be mentioned below.

Quantum Probabilities

The EPR paradox (Einstein,Podolski,Rosen,1935), Schrödinger's half dead-half alive cat, etc., all point to a desirability of additional localized variables in quantum theory. The ensuing 60 years has seen hugh debates and a large literature and finally a number of sophisticated experiments which, generally speaking, have refuted the possibility of thus far imagined suitable localized quantum hidden variables. Bell's inequality (1964) provided a very useful necessary condition for the existence of any classical probability model that any such local hidden variable theory might have. Important physical experiments were subsequently carried out to prove physical violation of Bell's inequality in a number of physical situations.

Recently I showed [8] that the Bell-Wigner-Accardi theories of quantum versus classical probabilities for quantum spin systems, are a special case of my operator trigonometry. Moreover the operator trigonometry should permit, in principle, the consideration of those quantum considerations in spaces of arbitrary dimension. Although I cannot give full details and background references here, we may state the result as follows [8,9].

Theorem 11. *The Bell (1964) Wigner (1970) Accardi (1982) necessary conditions for three dimensional quantum spin probabilities are special cases of the operator trigonometry. In particular, the essential triangle inequality*

$$\phi_{xz} \leqq \phi_{xy} + \phi_{yz} \tag{15}$$

for real angles between arbitrary unit vectors in a complex Hilbert space holds for arbitrary dimensions. The statistical meaning of the complex numbers in quantum mechanics derives fundmentally from the natural geometry of the operator trigonometry.

Extended Operator Trigonometry

As concerns the second recent development mentioned above, I have been generally negligent in extending the details of the operator trigonometry beyond the cases of A selfadjoint or A normal. Of course one should note that originally the theory was stated in a very general context, A an unbounded operator in a Banach space, and also the immediately apparent vagaries about real, imaginary, and complex operator angles were taken care of simply by defining real, imaginary, and total cosines. Because the original semigroup theory motivation dealt with accretive and dissipative operators, therefore I emphasized the real operator trigonometry. Also, one may easily rotate any desired half-plane theory to the real theory.

However in computational linear algebra important applications and hence interest has recently been turning to the case of A a general, nonsymmetric, perhaps at best sparse, perhaps very large, matrix, often $n \times n$, invertible, and with only real entries. Having these applications in the back of my mind, for the last couple of years I have thought about the most natural way to extend the operator trigonometry to arbitrary matrices, and my opinion now [10] is that one should use polar form. There are two strong contributing reasons for this, and I would like to expose them here. First, I have never been motivated in the operator trigonometry to think of uniformly turning operators A, e.g., those which rotate all vectors by a fixed angle. What we are interested in the operator trigonometry is the relative turning of vectors, just as in the classical eigenvalue theory we are interested in the relative stretching of vectors. Thus polar form $A = U|A|$ efficiently removes the "uniform turning", e.g., in U, and we already have the operator angle theory for $|A|$. Second, for invertible operators A, polar form is better than singular value decomposition for our purposes of an extended operator trigonometry, because we can show that the essential min-max Theorems 5 and 6 extend without incident. The key move [10] is to change the key definition (3) to

$$\sin \phi(A) = \inf_{\epsilon} \|\epsilon A - U\|. \tag{16}$$

Then, considering for example A to be arbitrary $n \times n$ nonsingular matrix with singular values $\sigma_1 \geqq \sigma_2 \geqq \cdots \geqq \sigma_n > 0$, we obtain from (14) that

$$\sin \phi(A) = \min_{\epsilon > 0} \|\epsilon |A| - I\| = \frac{\sigma_1(A) - \sigma_n(A)}{\sigma_1(A) + \sigma_n(A)}. \tag{17}$$

One may check [10] that the key min-max identity (4) is then satisfied if one modifies (2) to

$$\cos \phi(A) = \min_{x \neq 0} \frac{\langle |A|x, x\rangle}{\||A|x\|\|x\|}. \tag{18}$$

Cos$\phi(A)$ is then given as in (5) with λ_1 and λ_n replaced by σ_n and σ_1, respectively.

These considerations result in (see [10])

Theorem 12. *Let A be an arbitrary invertible operator on a complex Hilbert space X. From $A = U|A|$ polar form define the angle $\phi(A)$ according to (15) and $\cos\phi(A)$ according to (17). Then the min-max identity*

$$\sin^2\phi(A) + \cos^2\phi(A) = 1 \tag{19}$$

holds, and a full operator trigonometry of relative turning angles obtains for A from that of $|A|$. In this extended operator trigonometry, the Euler equation (7) is replaced by the simpler expression, $\|x\| = 1$ for simplicity,

$$\frac{|A|^2 x}{\langle |A|^2 x, x\rangle} - \frac{2|A|x}{\langle |A|x, x\rangle} + x = 0. \tag{20}$$

Other Recent Developments

The operator trigonometry of wavelet frames has been established in [11]. There the frame bounds A and B play the role of λ_1 and λ_n in the usual theory. A key notion of frame operator angle is introduced. The Wavelet Reconstruction Algorithm is shown to be Richardson iteration, with optimal parameters as in Theorem 9 above

Operator trigonometry is connected to control theory in [12]. The connection was apparent 30 years ago but not pursued. For example, given a first order finite dimensional ordinary differential equation system $\dot{x}(t) = Ax(t)$, $t > 0$, $x(0) = x_0$ given, the Lyapunov stability condition is that there exist some SPD matrix B such that

$$\langle (A^T B + BA)x, x\rangle < 0, \qquad x \neq 0 \tag{21}$$

However $A^T B + BA = 2\,\mathrm{Re}\,BA$ so the Lyapunov stability criterion is a special case of the multiplicative perturbation Theorem 3 above. The control theory has advanced substantially in the last 30 years, e.g. to algebraic Riccati equations and Schur complement formulations. Therefore further useful connections to the operator trigonometry remain to be developed.

The full operator trigonometry of the basic Dirichlet model problem of computational partial differential equations is worked out in [13]. In particular it is shown that if A_h denotes the usual centered 5-point stencil discretization of the Laplacian $-\Delta$, then the operator angle $\phi(A_h) = \pi/2 - \pi h$ where h is the grid size. All antieigenvectors are available and

$$\sin\phi(A_h) = \cos(\pi h), \qquad \cos\phi(A_h) = \sin(\pi h). \tag{22}$$

As the mesh becomes finer, $\phi(A_h)$ approaches 90°. This corresponds geometrically to the inverse discrete operator A_h^{-1} becoming more and more unbounded as $h \to 0$. The remarkable semi-iterative convergence improvements of the

ADI (Alternating-Direction-Implicit) scheme are seen in terms of interior operator angles. These semi-iterative scheme characteristics undoubtedly have interesting parallels in semi-group theory.

The Kantorovich-Wielandt inequalities of matrix theory are shown to be related to the operator trigonometry in [14]. The operator trigonometry is shown to provide a much simpler and more natural geoemtrical meaning to the Kantorovich-Wielandt condition number angle.

References

[1] K. Gustafson, *Lectures on Computational Fluid Dynamics, Mathematical Physics, and Linear Algebra*, Kaigai, Tokyo (1996), World Scientific, Singapore, (1997).

[2] K. Gustafson, *An extended operator trigonometry*, (to appear).

[3] K. Gustafson *Operator angles (Gustafson), Matrix singular angles (Wielandt), Operator deviations (Krein)*, Collected Works of Helmut Wielandt (B. Huppert and H. Schneider, eds.), De Gruyters, Berlin, (1996), 356–367.

[4] K. Gustafson, *Antieigenvalues in analysis*, Proceedings Fourth International Workshop in Analysis and its Applications, Dubrovnik, 1990, (C. Stanojevic, O. Hadzic, eds.), Novi Sad, Yugoslavia, (1991), 57–69.

[5] K. Gustafson, *Trigonometric interpretation of iterative methods*, Proceedings Conf. Algebraic Multilevel Iteration Methods with Applications (O. Axelsson and B. Polman, eds.), Nijmegen, Netherlands (1996), 23–29.

[6] K. Gustafson, *Operator trigonometry of iterative methods*, Num. Lin. Alg. with Applic. **4** (1997), 333-347.

[7] K. Gustafson, *Domain decomposition, operator trigonometry, Robin condition*, Contemporary Mathematics **218** (1998), 455–460.

[8] K. Gustafson, *The geometry of quantum probabilities*, On Quanta, Mind and Matter: Hans Primas in Context (A. Amann, H. Atmanspacher, U. Müller-Herold, eds.), Kluwer, Dordrecht, (1999), 151–164.

[9] K. Gustafson, *The trigonometry of quantum probabilities*, (to appear).

[10] K. Gustafson, *An extended operator trigonometry*, (to appear).

[11] K. Gustafson, *Operator trigonometry of wavelet frames*, Iterative Methods in Scientific Computation (J. Wang, M. Allen, B. Chen, T. Mathews, eds.), IMACS Series in Computational and Applied Mathematics **4**, New Brunswick, NJ (1998), 161–166.

[12] K. Gustafson, *Operator trigonometry of linear systems*, Proc. IFAC Symposium on Large Scale Systems (N. Koussoulas, P. Groumpos, eds.), Patras, Greece, July 15–17, 1998, Pergamon Press (1999), (to appear).

[13] K. Gustafson, *Operator Trigonometry of the model problem*, Num. Lin. Alg. with Applic. **5** (1999), (to appear).

[14] K. Gustafson, *The geometrical meaning of the Kantorovich–Wielandt inequalities*, Lin. Alg. & Applic., (to appear).

Karl Gustafson, Department of Mathematics, University of Colorado, Boulder, Colorado 80309, U.S.A.; International Solvay Institute, University of Brussels, CP 231, B-1050 Brussels, BELGIUM

E-mail address: gustafs@euclid.colorado.edu

Progress in Nonlinear Differential Equations
and Their Applications, Vol. 42

UNBOUNDED REPRESENTATIONS OF DISCRETE ABELIAN SEMIGROUPS

LÁSZLÓ KÉRCHY
UNIVERSITY OF SZEGED

This survey article contains the lecture delivered on the conference; it is based on the papers [5]–[9], where complete proofs can be found. In the first section we present results concerning single operators, and the second section is devoted to the study of representations of general discrete abelian semigroups. The main idea in both sections is to apply an appropriate normalizing gauge function in order to get connection with an isometry, or isometric representation.

I. Operators with regular norm-sequences

Let $\mathcal{H}$ be a (complex) Hilbert space, and let $\mathcal{L}(\mathcal{H})$ denote the set of all bounded, linear operators acting on $\mathcal{H}$. The operator $T \in \mathcal{L}(\mathcal{H})$ is called *power bounded*, if its powers of positive exponents form a bounded sequence: $\sup\{\|T^n\| : n \in \mathbf{N}\} < \infty$. Power bounded operators were investigated by many authors. Here we cite two results, which establish connection with a unitary operator.

Theorem A. (B. Sz.-Nagy, 1947) *Let $T \in \mathcal{L}(\mathcal{H})$ be a surjective power bounded operator, and let us assume that there exists a positive δ such that*

$$\|T^n x\| \geq \delta \|x\| \qquad \textit{holds, for every } x \in \mathcal{H}, n \in \mathbf{N}.$$

Then T is similar to a unitary operator U, acting on a Hilbert space $\mathcal{K}$.

Equivalently, the assumption above can be read that T is invertible, and T and T^{-1} are both power bounded operators. Since $\|T^{-n}\| = \|T^{*-n}\|$, it follows that $\|T^{*n}x\| \geq \delta\|x\|$ is also true. Relaxing the condition on the existence of a δ, independent of the vector x, quasisimilarity to a unitary operator can be proved. We recall that the operators $T \in \mathcal{L}(\mathcal{H})$ and $U \in \mathcal{L}(\mathcal{K})$ are quasisimilar, if there exist injective transformations $X \in \mathcal{L}(\mathcal{H}, \mathcal{K}), Y \in \mathcal{L}(\mathcal{K}, \mathcal{H})$ with dense ranges, satisfying the conditions $XT = UX$ and $YU = TY$. Let $\mathrm{Hlat}T$ stand for the lattice of hyperinvariant subspaces of T, that is the (closed) subspace $\mathcal{M}$ belongs to $\mathrm{Hlat}T$ if it is invariant for every operator in the commutant $\{T\}'$ of T. Since spectral subspaces are hyperinvariant for a unitary operator, the following theorem yields that the operator T has many hyperinvariant subspaces. We remind the reader that the existence of a nontrivial hyperinvariant (or invariant) subspace for a general Hilbert space operator is still open.

Theorem B. (B. Sz.-Nagy, C. Foias, 1966) *Let $T \in \mathcal{L}(\mathcal{H})$ be a power bounded operator such that*

$$\inf\{\|T^n x\| : n \in \mathbf{N}\} > 0 \quad and \quad \inf\{\|T^{*n} x\| : n \in \mathbf{N}\} > 0$$

hold, for every nonzero vector $x \in \mathcal{H}$.

Then T is quasisimilar to a unitary operator $U \in \mathcal{L}(\mathcal{K})$; furthermore, there exists an order-preserving, injective mapping $\varphi : \mathrm{Hlat}U \to \mathrm{Hlat}T$.

The technique applied in the proofs of the preceding theorems is the following. Let $\mathbf{Z}_+ := \mathbf{N} \cup \{0\}$ denote the set of nonnegative integers. Let us consider a Banach limit $L : \ell^\infty = \ell^\infty(\mathbf{Z}_+) \to \mathbf{C}$, that is L is a continuous, linear functional such that $L(\xi) = \lim \xi$ holds for every convergent sequence ξ, and $L(B\xi) = L(\xi)$ is true for every $\xi \in \ell^\infty$, where $B : (\xi_0, \xi_1, \dots) \mapsto (\xi_1, \xi_2, \dots)$. Given a power bounded operator $T \in \mathcal{L}(\mathcal{H})$, let us introduce a new semi-inner product $[x, y] := L(\{\langle T^n x, T^n y\rangle\}_{n=0}^\infty)$ on $\mathcal{H}$. Forming quotient space and after completion we obtain a new Hilbert space $\mathcal{K}$, where the following equations hold

$$[Tx, Ty] = L(\{\langle T^{n+1}x, T^{n+1}y\rangle\}_{n=0}^\infty) = L(\{\langle T^n x, T^n y\rangle\}_{n=0}^\infty) = [x, y].$$

It is natural to ask how this method can be extended for non-power bounded operators.

Let $T \in \mathcal{L}(\mathcal{H})$ be an arbitrary operator, and let us consider a sequence $p : \mathbf{Z}_+ \to (0, \infty)$ such that $\|T^n\| \le p(n)$ holds, for every $n \in \mathbf{Z}_+$. Let us introduce the functional

$$[x, y] := L(\{\langle T^n x, T^n y\rangle p(n)^{-2}\}_{n=0}^\infty), \quad x, y \in \mathcal{H}.$$

We are going to work with a sequence p such that the following equations hold:

$$\begin{aligned}[Tx, Ty] &= L(\{\langle T^{n+1}x, T^{n+1}y\rangle p(n+1)^{-2} p(n+1)^2 p(n)^{-2}\}_{n=0}^\infty) \\ &= cL(\{\langle T^{n+1}x, T^{n+1}y\rangle p(n+1)^{-2}\}_{n=0}^\infty) = c[x, y]\end{aligned}$$

with a positive constant c. This problem raises the question that under what conditions on a sequence $\eta \in \ell^\infty$ it is true that

$$L(\xi\eta) = cL(\xi) \quad \text{for all } \xi \in \ell^\infty \text{ and for all Banach limits } L.$$

Replacing the constant sequence $\mathbf{1}$ for ξ we are led to the particular question: when $L(\eta) = c$ holds for every Banach limit L. The answer is provided by the following classical result.

Theorem C. (G. G. Lorentz, 1948) *Let us be given a sequence $\eta \in \ell^\infty$ and a complex number $c \in \mathbf{C}$. Then for all Banach limits L we have $L(\eta) = c$ if and only if*

$$\lim_{N\to\infty} \sup_{t \in \mathbf{Z}_+} \left| \frac{1}{N} \sum_{n=t}^{t+N-1} \eta(n) - c \right| = 0.$$

If the condition of this theorem holds, then the sequence η is said to be *almost convergent* to c, in notation: a-lim$\eta = c$.

Now, it is not difficult to verify that $L(\xi\eta) = cL(\xi)$ is true, for every $\xi \in \ell^\infty$ and for all Banach limits L, exactly when a-lim$|\eta - c\mathbf{1}| = 0$.

Thus, it is reasonable to work with a normalizing sequence p satisfying the following conditions:

$$\text{a-lim}\left|\frac{p(n+1)}{p(n)} - c\right| = 0, \qquad \text{where } c > 0,$$

$$\|T^n\| \le p(n), \qquad \text{for all } n \in \mathbf{Z}_+, \text{ are true, and}$$

$$\text{a-lim}\frac{\|T^n\|}{p(n)} = 0 \quad \text{fails.}$$

If these conditions are fulfilled then p is said to be a *gauge function*, in notation: $p \in \mathcal{P}$, and the operator T is called to belong to the class $\mathcal{L}(p, \mathcal{H})$. We say that T has a *regular norm-sequence*, in notation: $T \in \mathcal{L}^{(r)}(\mathcal{H})$, if $T \in \mathcal{L}(p, \mathcal{H})$ holds with some gauge function p.

The limit c above turns out to be exactly the spectral radius of the operator T.

Proposition 1. *If $T \in \mathcal{L}(p, \mathcal{H})$ with a gauge function p, then* a-lim$p(n+1)/p(n) = r(T)$, *and so* $r(T) > 0$.

In the following theorem we get connection with an isometry, extending former results by Q. P. Vũ [15] and the author [4].

Theorem 2. *Let us assume that the Hilbert space $\mathcal{H}$ is separable, and let $T \in \mathcal{L}(p, \mathcal{H})$, where $p \in \mathcal{P}$.*

Then, there exist an isometry $V \in \mathcal{L}(\mathcal{K})$ and a transformation $X \in \mathcal{L}(\mathcal{H}, \mathcal{K})$ such that:

(a) $\ker X = \{h \in \mathcal{H} : \text{a-lim}\|T^n h\|/p(n) = 0\}$, $\quad (\text{ran}X)^- = \mathcal{K}$,

(b) $XT = r(T)VX$,

(c) *for every $C \in \{T\}'$ there exists a unique operator $D \in \{V\}'$, such that $XC = DX$.*

Furthermore, the mapping $\gamma : \{T\}' \to \{V\}'$, $C \mapsto D$ is a contractive algebra- homomorphism, and

$$\sigma(C) \supset \sigma(\gamma(C)), \quad \sigma_p(C^*) \supset \sigma_p(\gamma(C)^*) \quad \textit{for all } C \in \{T\}'.$$

In the nonseparable case, we are able to prove the statement (c) in a slightly modified form.

We can observe that the isometry V acts on a nonzero space if and only if the operator is asymptotically nonvanishing with respect to the gauge function p.

Exploiting the connection with V, existence of proper hyperinvariant subspaces can be shown. The theorems, what follow, extend former results due to Sz.-Nagy–Foias and B. Beauzamy.

Theorem 3. *Let us assume that $T \in \mathcal{L}(p, \mathcal{H})$, where $p \in \mathcal{P}$, and T is not scalar multiple of the identity. If there exist vectors $x, y \in \mathcal{H}$ such that*

$$\text{a-lim}\|T^n x\|/p(n) \neq 0 \quad \textit{and } \text{a-lim}\|T^{*n} y\|/p(n) \neq 0$$

hold, then T has a non-trivial hyperinvariant subspace.

Theorem 4. *Given $T \in \mathcal{L}(p, \mathcal{H})$, $p \in \mathcal{P}$, let us assume that* a-lim$\|T^n x\|/p(n) \neq 0$ *is true for every non-zero vector $x \in \mathcal{H}$. Let us assume, moreover, that there exists a sequence of vectors $\{x_n\}_{n \in \mathbf{Z}}$, indexed by the set of integers $\mathbf{Z}$, such that $T x_n = x_{n+1}$ holds for every $n \in \mathbf{Z}$, the vectors $\{x_n\}_{n \in \mathbf{N}}$ span an infinite dimensional subspace, and*

$$\sum_{n \in \mathbf{Z}} \frac{\log^* \left(\|x_n\| r(T)^{-n}\right)}{1 + n^2} < \infty,$$

where $\log^* t := 0$ if $0 \leq t \leq 1$, and $\log^* t := \log t$ if $t \geq 1$.

Then, there exists a sequence $\{\mathcal{H}_n\}_{n \in \mathbf{N}}$ of non-zero hyperinvariant subspaces, such that

$$\mathcal{H}_n \cap (\vee_{j \neq n} \mathcal{H}_j) = \{0\} \quad \textit{for every } \; n \in \mathbf{N}.$$

Furthermore, if $\sigma_p(T) \cap r(T)\mathbf{T} = \emptyset$, where $\mathbf{T} := \{z \in \mathbf{C} : |z| = 1\}$, then

$$\cap_{n \in \mathbf{N}} (\vee_{j \geq n} \mathcal{H}_j) = \{0\}.$$

Now, we proceed with statements on stability, that is we give conditions for the case when T is asymptotically vanishing with respect to the gauge function p, in the sense of almost convergence. The following theorems extend former results on power bounded operators by Ansari–Bourdon, Arendt–Batty–Lyubich–Vu and Katznelson–Tzafriri.

Theorem 5. *Let us assume that the operator $T \in \mathcal{L}(p, \mathcal{H})$ $(p \in \mathcal{P})$ is supercyclic, that is there exists a vector $x_0 \in \mathcal{H}$ such that the set $\{c_n T^n x_0 : n \in \mathbf{Z}_+, c_n \in \mathbf{C}\}$ is dense in $\mathcal{H}$.*

Then $\mathcal{H}_0(T, p) = \{x \in \mathcal{H} :$ a-lim$\|T^n x\|/p(n) = 0\}$ is a subspace of codimension at most 1, and — in contrast with the power bounded case — it can happen that $\mathcal{H}_0(T, p)$ is exactly of codimension 1.

Theorem 6. *If $T \in \mathcal{L}(p, \mathcal{H})$ $(p \in \mathcal{P})$, the peripheral spectrum $\sigma(T) \cap r(T)\mathbf{T}$ is countable and $\sigma_p(T^*) \cap r(T)\mathbf{T} = \emptyset$, then* a-lim$\|T^n x\|/p(n) = 0$ *is true for every $x \in \mathcal{H}$.*

Theorem 7. *If $T \in \mathcal{L}(p, \mathcal{H})$ $(p \in \mathcal{P})$ and $\sigma(T) \cap r(T)\mathbf{T} = \{\alpha_i\}_{i=1}^k$, then*

$$\text{a-lim}_n \frac{\|T^n \prod_{i=1}^k (T - \alpha_i I)\|}{p(n)} = 0.$$

Analogues of the previous theorems are true in general Banach spaces.

It is natural to ask that which operators are of regular norm-sequence. We have seen that positivity of the spectral radius is a necessary condition. In [7] several sufficient conditions were given in terms of the existence of arbitrarily long monotone (decreasing or increasing) sections in the derived sequence $\{\|T^n\|^{1/n}\}_{n=1}^\infty$. In the paper [8], joint with V. Müller, a necessary and sufficient condition was provided. In order to formulate this result, we need some notations.

Let $T \in \mathcal{L}(\mathcal{H})$ be an operator with positive spectral radius $r(T)$. For any $m \in \mathbf{Z}_+$ and $k \in \mathbf{N}$, let

$$M(m,k) := \max\{\|T^n\| r(T)^{-n} : m \le n < m+k\},$$

$$A(m,k) := \frac{1}{k} \sum_{n=m}^{m+k-1} \|T^n\| r(T)^{-n}.$$

Then the quantity $c(T)$ is defined by

$$c(T) := \inf_{k \in \mathbf{N}} \sup_{m \in \mathbf{Z}_+} A(m,k)/M(m,k).$$

Theorem 8. *The operator $T \in \mathcal{L}(\mathcal{H})$ has regular norm-sequence if and only if $r(T) > 0$ and $c(T) > 0$.*

We are able to show that there exists an operator T such that $r(T) > 0$ and $c(T) = 0$; thus positivity of the spectral radius solely is not sufficient for regularity of the norm-sequence.

To examine a power bounded operator T is the same as, from another point of view, to consider a bounded representation $\rho : \mathbf{Z}_+ \to \mathcal{L}(\mathcal{H})$ of the semigroup $\mathbf{Z}_+$. Bounded representations of arbitrary abelian semigroups were related to isometric representations by H. Bercovici [2] and Q. P. Vũ [15]. In the second section of this work we address the question that what kind of representations correspond to operators with regular norm-sequence.

II. Representations of abelian semigroups

Let S: be an arbitrary abelian, additive semigroup, with zero element. The functional $M : \ell^\infty(S) \to \mathbf{C}$ is an *invariant mean*, if M is linear, $\|M\| = 1 = M(\mathbf{1})$, and $M(\xi_t) = M(\xi)$ holds for every $\xi \in \ell^\infty(S)$ and $t \in S$, where $\xi_t(s) := \xi(s+t)$. Let $\mathcal{M}(S)$ stand for the set of all invariant means on $\ell^\infty(S)$. By a classical result of M. M. Day we know that $\mathcal{M}(S) \neq \emptyset$.

The sequence $\xi \in \ell^\infty(S)$ is called *almost convergent* to the complex number c, in notation: a-lim$\xi = c$, if $M(\xi) = c$ is true for every $M \in \mathcal{M}(S)$. We say that the sequence ξ *almost converges* to c *in the strong sense*, in notation: as-lim$\xi = c$, if a-lim$|\xi - c\mathbf{1}| = 0$.

Let us assume in the sequel that the semigroup S is cancellative, and that 0 is the only invertible element in S. The additive semigroup $\mathbf{Z}_+^n$ of n-tuples of nonnegative integers is an important example for the kind of semigroups we consider. The previous properties enable us to introduce a partial ordering on S by writing $s_1 \leq s_2$ if there exists $s_3 \in S$ such that $s_1 + s_3 = s_2$. Then S becomes a directed set, which makes a limiting process available.

Now, it can be verified that a-lim$\xi = c$ is true if and only if

$$q'(\mathrm{Re}\xi) = q(\mathrm{Re}\xi) = \mathrm{Re}c \quad \text{and} \quad q'(\mathrm{Im}\xi) = q(\mathrm{Im}\xi) = \mathrm{Im}c$$

are valid, where for any real sequence $\eta \in \ell^\infty(S)$:

$$q(\eta) := \inf\left\{\limsup \frac{1}{N}\sum_{i=1}^{N} \eta_{t_i} : N \in \mathbf{N}, t_1, \dots, t_N \in S\right\},$$

$$q'(\eta) := \sup\left\{\liminf \frac{1}{N}\sum_{i=1}^{N} \eta_{t_i} : N \in \mathbf{N}, t_1, \dots, t_N \in S\right\}.$$

The sequence $p : S \to (0, \infty)$ is called a *gauge function*, if for every $t \in S$ there exists a positive number $c_p(t)$ such that

$$\text{as-lim}\frac{p_t}{p} = c_p(t).$$

Let $\mathcal{P}(S)$ denote the set of all gauge functions on S.

Let $S^\#$ be the set of characters (nonzero complex homomorphisms) on S, and let $S_+^\# := \{\chi \in S^\# : \chi(s) > 0 \ \forall s \in S\}$ be the set of positive characters. It can be easily verified that, for any gauge function $p \in \mathcal{P}(S)$, the limit functional is a positive character: $c_p \in S_+^\#$.

Let $\mathcal{X}$ be a (complex) Banach space, and let $\rho : S \to \mathcal{L}(\mathcal{X})$ be a (unital) representation, that is $\rho(s+t) = \rho(s)\rho(t)$ holds for every $s, t \in S$ and $\rho(0) = I$. The norm- and spectral radius functions of ρ are defined by $|\rho|(s) := \|\rho(s)\|$ and $r_\rho(s) := r(\rho(s))$ $(s \in S)$.

We say that the representation ρ is *of regular norm-behaviour with respect to the gauge function* $p \in \mathcal{P}(S)$, in notation: $\rho \in \mathcal{R}(p, S, \mathcal{X})$, if $|\rho| \leq p$ and the quotient $|\rho|/p$ does not almost converge to zero. The representation ρ is called *of regular norm-behaviour*, in notation: $\rho \in \mathcal{R}(S, \mathcal{X})$, if $\rho \in \mathcal{R}(p, S, \mathcal{X})$ with some $p \in \mathcal{P}(S)$.

We can see that $\rho \in \mathcal{R}(p, \mathbf{Z}_+, \mathcal{X})$ if and only if $\rho(1) \in \mathcal{L}(p, \mathcal{X})$. On the other hand, if ρ is a bounded representation and $r_\rho \geq 1$, then $\rho \in \mathcal{R}(K\mathbf{1}, S, \mathcal{X})$ with $K \in [1, \infty)$.

It can be shown that if $\rho \in \mathcal{R}(p_1, S, \mathcal{X}) \cap \mathcal{R}(p_2, S, \mathcal{X})$ is true with two gauge functions $p_1, p_2 \in \mathcal{P}(S)$, then the corresponding limit functionals coincide: $c_{p_1} = c_{p_2}$. Therefore, for any $\rho \in \mathcal{R}(S, \mathcal{X})$ we can define the limit functional c_ρ of the representation ρ itself by $c_\rho := c_p$, where $p \in \mathcal{P}(S)$ is an arbitrary gauge function satisfying the condition $\rho \in \mathcal{R}(p, S, \mathcal{X})$.

Comparing with Proposition 1, a much looser connection can be stated between the limit functional and the spectral radius function in the general setting.

Proposition 9. *If* $\rho \in \mathcal{R}(S, \mathcal{X})$, *then* $c_\rho \leq r_\rho$.

The subset $S_r(\rho) := \{s \in S : c_\rho(s) = r_\rho(s)\}$ is called the *radial section of* S *with respect to* ρ. It is clear that $S_r(\rho)$ is a subsemigroup of S, containing the zero element. Under some conditions the radial section is a 'large' subset of the semigroup.

Proposition 10. (i) *If the semigroup* S *is generated by the elements* $\{s_1, \ldots, s_n\}$ *and* $r_\rho > 0$, *then*

$$S_r(\rho) \supset (s_1 + \cdots + s_n) + S.$$

(ii) *If* $\rho(s)$ *is invertible, for every* $s \in S$, *then* $S_r(\rho) = S$.

However, in contrast with the case $S = \mathbf{Z}_+$, the limit functional c_ρ usually does not coincide with the spectral radius function r_ρ, as it turns out from the following example.

Example 11. Let $S = \mathbf{Z}_+^2$ be the set of pairs of nonnegative integers, and let us consider the weight sequence $w : \mathbf{Z}_+^2 \to (0, \infty)$ defined by

$$w(i,j) = \begin{cases} 2^i, & \text{if } j = 0 \text{ and } i \in \mathbf{Z}_+, \\ 2^j, & \text{if } i = 0 \text{ and } j \in \mathbf{Z}_+, \\ 1 & \text{otherwise.} \end{cases}$$

We introduce the sequence space

$$\ell^2(\mathbf{Z}_+^2, w) := \left\{ f : \mathbf{Z}_+^2 \to \mathbf{C} : \|f\|^2 := \sum_{(i,j) \in \mathbf{Z}_+^2} |f(i,j)|^2 w(i,j)^2 < \infty \right\}.$$

The representation $\rho : \mathbf{Z}_+^2 \to \mathcal{L}(\ell^2(\mathbf{Z}_+^2, w))$ is the forward translation:

$$(\rho(m,n)f)(i,j) := \begin{cases} f(i-m, j-n), & \text{if } i \geq m \text{ and } j \geq n, \\ 0, & \text{otherwise.} \end{cases}$$

It can be easily seen that $\rho \in \mathcal{R}(p, \mathbf{Z}_+^2, \ell^2(\mathbf{Z}_+^2, w))$, where $p = w \in \mathcal{P}(\mathbf{Z}_+^2)$ with $c_p = \mathbf{1}$. Since $r_\rho = w$ and $c_\rho = \mathbf{1}$, we obtain that $S_r(\rho) = \{(0,0)\} \cup ((1,1) + \mathbf{Z}_+^2) \neq S$.

If S is not finitely generated then even $S_r(\rho) = \{0\}$ can occur.

Let us recall the definition of different kinds of spectra of a representation $\rho : S \to \mathcal{L}(\mathcal{X})$. In the definition of point spectrum and approximate point spectrum existence of common eigenvectors and approximate eigenvectors, respectively, is required. Namely, the *point spectrum* $\sigma_p(\rho)$ of ρ consists of those characters $\chi \in S^\#$ for which there exists a nonzero vector $x \in \mathcal{X}$, such that $\rho(s)x = \chi(s)x$ holds, for every $s \in S$. The character $\chi \in S^\#$ belongs to the *approximate point spectrum* $\sigma_{ap}(\rho)$ of ρ, if there exists a net $\{x_\nu\}_\nu$ of unit vectors in $\mathcal{X}$ such that $\lim_\nu \|\rho(s)x_\nu - \chi(s)x_\nu\| = 0$ is true, for every $s \in S$.

Let $\mathcal{A}_\rho$ denote the abelian (unital) Banach algebra generated by $\rho(S) := \{\rho(s) : s \in S\}$, and let $\Sigma(\rho)$ be the maximal ideal space of $\mathcal{A}_\rho$. The *algebraic spectrum* $\sigma_a(\rho)$ of ρ (called δ-spectrum in [11]) is defined as $\sigma_a(\rho) := \{h \circ \rho : h \in \Sigma(\rho)\} \subset S^\#$.

Now, we define the *peripheral spectrum* of a representation $\rho \in \mathcal{R}(S, \mathcal{X})$ by

$$\sigma_{per}(\rho) := \{\chi \in \sigma_a(\rho) : |\chi| = c_\rho\}.$$

The following proposition relates this spectrum to the approximate point spectrum. We say that a subset S' of the semigroup S is *absorbing*, if for every $s \in S$ there exists $s' \in S'$ such that $s + s' \in S'$.

Proposition 12. *Let us assume that $\rho \in \mathcal{R}(S, \mathcal{X})$. If the radial section $S_r(\rho)$ is absorbing in S — what happens if S is finitely generated or if ρ is a representation with invertible operators — then*

$$\sigma_{per}(\rho) \subset \sigma_{ap}(\rho).$$

The following theorem relates representations with regular norm-behaviour to isometric representations; it can be contrasted with [2, Theorem 2], [1, Proposition 3.1] or [14, Lemma 3.1].

Theorem 13. *For any representation $\rho \in \mathcal{R}(p, S, \mathcal{X}), p \in \mathcal{P}(S)$, there exist an isometric representation ψ of S on a Banach space $\mathcal{Y}$ and a transformation $X \in \mathcal{L}(\mathcal{X}, \mathcal{Y})$ such that:*

(i) $\ker X = \{x \in \mathcal{X} : \text{a-lim}|\rho x|/p = 0\}$ *and* $(\operatorname{ran} X)^- = \mathcal{Y}$;

(ii) $X\rho(s) = c_\rho(s)\psi(s)X$ *holds for every* $s \in S$;

(iii) *for every operator* $C \in \{\rho(S)\}'$, *there exists uniquely an operator* $D \in \{\psi(S)\}'$ *such that* $XC = DX$; *furthermore, the mapping* $\gamma : \{\rho(S)\}' \to \{\psi(S)\}'$, $C \mapsto D$ *is a contractive algebra-homomorphism;*

(iv) *the following relations are true for the spectra:*

$$\sigma_a(\rho) \supset \sigma_a(c_\rho\psi) = c_\rho\sigma_a(\psi),$$

$$\sigma_{per}(\rho) \supset \sigma_{per}(c_\rho\psi) = c_\rho\sigma_{per}(\psi) = c_\rho\sigma_{ap}(\psi),$$

$$\sigma_p(\rho^\#) \supset \sigma_p(c_\rho\psi^\#) = c_\rho\sigma_p(\psi^\#).$$

As an immediate consequence, we obtain that if there exists a vector $x \in \mathcal{X}$ such that a-lim$|\rho x|/p \neq 0$, then $\sigma_{per}(\rho) \neq \emptyset$.

If ρ is a representation on a separable Hilbert space $\mathcal{H}$, then the associated isometric representation ψ can be defined on a Hilbert space $\mathcal{K}$, as well. If ρ is a representation on a nonseparable Hilbert space $\mathcal{H}$ and the semigroup S is countably generated, then we are still able to define ψ on a Hilbert space $\mathcal{K}$, possessing all the stated properties of Theorem 13, with the only restriction that the operators $C \in \{\rho(S)\}'$ considered in (iii) should be particularly chosen.

Theorem 6 extends the classical Arendt–Batty–Lyubich–Vũ Theorem to individual, not necessarily power bounded operators. We are able to prove a stability statement of this type in our general setting, as well; this is the content of the following theorem.

Theorem 14. *Let* $\rho : S \to \mathcal{L}(\mathcal{X})$ *be a representation with regular norm-behaviour. If* $\sigma_{per}(\rho)$ *is countable and* $\sigma_p(\rho^\#) \cap \sigma_{per}(\rho) = \emptyset$, *then*

$$\text{a-lim}\frac{|\rho x|}{p} = 0$$

holds, for every vector $x \in \mathcal{X}$, *where* $p \in \mathcal{P}(S)$ *is any gauge function with the property* $\rho \in \mathcal{R}(p, S, \mathcal{X})$.

Acknowledgements. Research partially supported by Hungarian NFS Research grant no. T 022920.

References

[1] **C. J. K. Batty** and **Q. P. Vũ**, Stability of strongly continuous representations of abelian semigroups, *Math. Z.*, **209** (1992), 75–88.

[2] **H. Bercovici**, Commuting power bounded operators, *Acta Sci. Math. (Szeged)*, **57** (1993), 55–64.

[3] **M. M. Day**, Ergodic theorems for abelian semigroups, *Trans. Amer. Math. Soc.*, **51** (1942), 399–412.

[4] **L. Kérchy**, Isometric asymptotes of power bounded operators, *Indiana Univ. Math. J.*, **38** (1989), 173–188.
[5] **L. Kérchy**, Operators with regular norm-sequences, *Acta Sci. Math. (Szeged)*, **63** (1997), 571–605.
[6] **L. Kérchy**, Hyperinvariant subspaces of operators with non-vanishing orbits, *Proc. Amer. Math. Soc.*, **127** (1999), 1363–1370.
[7] **L. Kérchy**, Criteria of regularity for norm-sequences, *Integral Equations Operator Theory*, to appear.
[8] **L. Kérchy** and **V. Müller**, Criteria of regularity for norm-sequences. II, *Acta Sci.Math. (Szeged)*, to appear.
[9] **L. Kérchy**, Representations with regular norm-behaviour of discrete abelian semigroups, submitted to *Acta Sci. Math. (Szeged)*.
[10] **G. G. Lorentz**, A contribution to the theory of divergent sequences, *Acta Math.*, **80** (1948), 167–190.
[11] **Y. I. Lyubich**, *Introduction to the theory of Banach representations of groups*, OT 30, Birkhäuser, Basel–Boston–Berlin, 1988.
[12] **B. Sz.-Nagy**, On uniformly bounded linear transformations in Hilbert space, *Acta Sci. Math. (Szeged)*, **11** (1947), 152–157.
[13] **B. Sz.-Nagy** and **C. Foias**, *Harmonic Analysis of Operators on Hilbert Space*, North Holland – Akadémiai Kiadó, Amsterdam – Budapest, 1970.
[14] **Q. P. Vũ**, Theorems of Katznelson–Tzafriri type for semigroups of operators, *J. Funct. Anal.*, **103** (1992), 74–84.
[15] **Q. P. Vũ**, Almost periodic and strongly stable semigroups of operators, *Banach Center Publ.* **38**, Warsaw, 1994, 401–426.

L. KÉRCHY, BOLYAI INSTITUTE, UNIVERSITY OF SZEGED, ARADI VÉRTANÚK TERE 1, H-6720 SZEGED, HUNGARY
E-mail address: kerchy@math.u-szeged.hu

Progress in Nonlinear Differential Equations
and Their Applications, Vol. 42

SOME ESTIMATES OF SOLUTIONS FOR THE EQUATIONS OF MOTION OF COMPRESSIBLE VISCOUS FLUID IN AN 3D EXTERIOR DOMAIN

TAKAYUKI KOBAYASHI
KYUSHU INSTITUTE OF TECHNOLOGY

1. Introduction

In this talk, we consider the optimal decay rate of the solutions to the exterior initial boundary value problem of the equation which describes the motion of compressible viscous fluid. The equation is given by the following system of four equations for the density ρ and the velocity $\boldsymbol{v} = {}^T(v_1, v_2, v_3)$:

$$\begin{cases} \rho_t + \nabla\cdot(\rho\boldsymbol{v}) = 0, \\ \rho(\boldsymbol{v}_t + (\boldsymbol{v}\cdot\nabla)\boldsymbol{v}) + \nabla P(\rho) = \mu\Delta\boldsymbol{v} + \nu\nabla(\nabla\cdot\boldsymbol{v}), \end{cases} \tag{1.1}$$

where ${}^T(v_1, v_2, v_3)$ is the transposed (v_1, v_2, v_3), $P = P(\rho)$ the pressure, μ and ν the viscosity coefficients.

We consider the initial boundary value problem (IBVP) of (1.1) in the region $t \geq 0, x \in \Omega$ where Ω is an exterior domain in $\mathbb{R}^3$ with compact smooth boundary $\partial\Omega$. The boundary condition is supposed by

$$\begin{aligned} \boldsymbol{v}\,|_{\partial\Omega} &= 0, \\ \lim_{|x|\to\infty} \boldsymbol{v}(t,x) &= 0, \quad t > 0, \end{aligned} \tag{1.2}$$

and the initial condition is given by

$$(\rho, \boldsymbol{v})(0,x) = (\rho_0, \boldsymbol{v}_0)(x) \quad \text{in } \Omega. \tag{1.3}$$

The unique existence of smooth solutions of IBVP: (1.1), (1.2) and (1.3) globally in time near the constant state $(\bar\rho_0, 0, 0, 0)$ where $\bar\rho_0$ is a positive constant is proved by employing the same argument as in Matsumura and Nishida [9] in the L_2 framework. In this talk, we will give the rate of convergence of solutions as $t \to \infty$ under the assumption that $(\rho_0 - \bar\rho_0, \boldsymbol{v}_0)$ belongs to L_1, which is corresponding to the result for the Cauchy problem obtained by Hoff and Zumbrun [3]. In particular, this rate is almost sharp because Hoff and Zumbrun [4] gave the lower bounds of the Green's functions for the effective artificial linearized equations. Also, in [3], they showed that, asymptotically, the solution decomposes into the sum of two terms, one of which dominates in L_p for $p > 2$, the other for $p < 2$. The dominant term for $p > 2$ decays at the rate of a heat kernel. When $p < 2$, the dominant term decays more slowly than a heat kernel; in fact, the solution may grow without bound in L_p for p near 1. Roughly speaking, this implies the correct notion of diffusion wave in the context of compressible Navier-Stokes flows. In this talk, we shall state that the same assertion holds in the exterior domain case.

The basic assumption of this talk is the following.

(1) μ is a positive constant, ν is a non-negative constant

(2) P is a known function of ρ, smooth in a neighborhood of $\overline{\rho}_0$, where $\frac{\partial P}{\partial \rho} > 0$.

Notation. Before stating our main results precisely, at this point we shall explain our notation. Three dimensional row vector valued functions are denoted by bold-face letter which is corresponding to the velocity field, that is for example $\boldsymbol{v} = {}^T(v_1, v_2, v_3)$. Four dimensional row vector valued functions are denoted by black-boad bold letters, that is for example $\mathbb{U} = {}^T(u_1, u_2, u_3, u_4)$, where u_1 is corresponding to the density, ${}^T(u_2, u_3, u_4)$ the velocity field. If we write $(\rho, \boldsymbol{v})$, $\boldsymbol{v} = {}^T(v_1, v_2, v_3)$, then this also stands for the four row vector ${}^T(\rho, v_1, v_2, v_3)$. As usual, the $\cdot$ stands for the innerproduct in $\mathbb{R}^3$ and set

$$\partial_t = \partial/\partial t,\ u_t = \partial_t u,\ \partial_j = \partial/\partial x_j, \Delta = \partial_1^2 + \partial_2^2 + \partial_3^2, \nabla u = {}^T(\partial_1 u, \partial_2 u, \partial_3 u),$$

$$\partial_x^\alpha = \partial_1^{\alpha_1}\partial_2^{\alpha_2}\partial_3^{\alpha_3},\ \alpha = (\alpha_1, \alpha_2, \alpha_3),\ |\alpha| = \alpha_1 + \alpha_2 + \alpha_3,$$

$$\Delta \boldsymbol{u} = {}^T(\Delta u_1, \Delta u_2, \Delta u_3),\ \nabla \cdot \boldsymbol{v} = \textstyle\sum_{j=1}^3 \partial_j v_j,$$

$$(\boldsymbol{u} \cdot \nabla)\boldsymbol{v} = {}^T(\boldsymbol{u} \cdot \nabla v_1, \boldsymbol{u} \cdot \nabla v_2, \boldsymbol{u} \cdot \nabla v_3),$$

$$\partial_t^j \partial_x^\alpha \boldsymbol{v} = {}^T(\partial_t^j \partial_x^\alpha v_1, \partial_t^j \partial_x^\alpha v_2, \partial_t^j \partial_x^\alpha v_3),\ \partial_t^j \partial_x^\alpha \mathbb{U} = (\partial_t^j \partial_x^\alpha \rho, \partial_t^j \partial_x^\alpha \boldsymbol{v}),$$

$$(\mathbb{U} = (\rho, \boldsymbol{v})),$$

$$\partial_x^k u = (\partial_x^\alpha u \mid |\alpha| = k),\ \partial_x^k \boldsymbol{v} = (\partial_x^\alpha \boldsymbol{v} \mid |\alpha| = k),\ \partial_x^k \mathbb{U} = (\partial_x^\alpha \mathbb{U} \mid |\alpha| = k),$$

$$\partial_x^1 u = \partial_x u,\ \partial_x^1 \boldsymbol{v} = \partial_x \boldsymbol{v},\ \partial_x^1 \mathbb{U} = \partial_x \mathbb{U}.$$

Let D be any domain in $\mathbb{R}^3$. $L_p(D)$ denotes the usual L_p space on D with norm $\|\cdot\|_{p,D}$. Put

$$W_p^m(D) = \{u \in L_p(D) \mid \|u\|_{m,p,D} < \infty\},\ \ \|u\|_{m,p,D} = \textstyle\sum_{|\alpha|\le m} \|\partial_x^\alpha u\|_{p,D},$$

$$H^m(D) = W_2^m(D),\ \ W_p^0(D) = L_p(D),\ \ H^0(D) = L_2(D).$$

Sobolev spaces of vector valued functions are used as well as of scalar valued functions. For the function spaces of three and four dimensional row vector valued functions we use bold-face letter and the black boad bold letter, respectively. For example,

$$\mathbf{L}_p(D) = \{\boldsymbol{v} = (v_1, v_2, v_3) \mid v_j \in L_p(D)\},$$

$$\mathbb{L}_p(D) = \{\mathbb{U} = (\rho, \boldsymbol{v}) \mid \rho \in L_p(D), \boldsymbol{v} \in \mathbf{L}_p(D)\}.$$

Likewise for $\mathbf{W}_p^m(D)$, $\mathbb{W}_p^m(D)$, $\mathbf{H}^m(D)$ and $\mathbb{H}^m(D)$. Set

$$\|\boldsymbol{v}\|_{m,p,D} = \sum_{j=1}^{3} \|v_j\|_{m,p,D},\ \ \|\mathbb{U}\|_{m,p,D} = \sum_{j=1}^{4} \|u_j\|_{m,p,D}.$$

Set $\mathbb{P}\mathbb{U} = {}^T(0, \boldsymbol{v})$ and $(\mathbb{I} - \mathbb{P})\mathbb{U} = {}^T(\rho, 0, 0, 0)$ for $\mathbb{U} = (\rho, \boldsymbol{v})$. Set

$$\mathbb{W}_p^{k,m}(D) = \{\mathbb{U} \mid (\mathbb{I} - \mathbb{P})\mathbb{U} \in \mathbb{W}_p^k(D),\ \mathbb{P}\mathbb{U} \in \mathbb{W}_p^m(D)\},$$

$$\|\mathbb{U}\|_{\mathbb{W}_p^{k,m}(D)} = \|(\mathbb{I} - \mathbb{P})\mathbb{U}\|_{k,p,D} + \|\mathbb{P}\mathbb{U}\|_{m,p,D}.$$

When $D = \Omega$, we omit the index Ω. Namely,

$$\|\cdot\|_{p,\Omega} = \|\cdot\|_p, \quad \|\cdot\|_{k,p,\Omega} = \|\cdot\|_{k,p}, \quad \|\cdot\|_{\mathbb{B}W_p^{k,m}(\Omega)} = \|\cdot\|_{\mathbb{B}W_p^{k,m}}.$$

Likewise for L_p, $\mathbf{L}_p$, $\mathbb{B}L_p$, W_p^m, $\mathbf{W}_p^m$, $\mathbb{B}W_p^m$, H^m, $\mathbf{H}^m$, $\mathbb{B}H^m$ and $\mathbb{B}W_p^{k,m}$. Set

$$\begin{aligned} C^\ell([t_1,t_2];B) &= \{u(t) \mid \ \ell\text{-times continuously differentiable function of } \\ &\qquad t \in [t_1,t_2] \text{ with value in a Banach space } B\}, \\ L_2((t_1,t_2);B) &= \{u(t) \mid \ L_2\text{-function of } t \in [t_1,t_2] \text{ with value in } B\}. \end{aligned}$$

To denote various constants we use the same letter C and $C(A,B,\ldots)$ denotes the constant depending on the quantities $A, B, \ldots$. $O(\cdot)$ means the large order.

In order to state the existence of solutions according to Matsumura and Nishida [9], first we introduce the class of solutions $X^k(0,\infty)$ and some norms of solutions $N_k(0,\infty)$, $k = 1,2$:

$$\begin{aligned} X^k(0,\infty) &= \{\mathbb{B}U = (\rho, \boldsymbol{v}) \mid \rho - \bar{\rho}_0 \in \bigcap_{j=0}^{k} C^j([0,\infty); H^{k+2-j}), \\ &\qquad \partial_x \rho \in L_2((0,\infty); H^{k+1}), \partial_t^j \rho \in L_2((0,\infty); H^{k+2-j}), \ j = 1,k, \\ &\qquad \boldsymbol{v} \in \bigcap_{j=0}^{k} C^j([0,\infty); \mathbf{H}^{k+2-2j}), \partial_x \boldsymbol{v} \in L_2((0,\infty); \mathbf{H}^{k+2}), \\ &\qquad \partial_t^j \boldsymbol{v} \in L_2((0,\infty); \mathbf{H}^{k+3-2j}), \ j = 1,k\}, \\ N_1(0,\infty)^2 &= \sup_{0 \le t < \infty} \left(\|\mathbb{B}U(t) - \mathbb{B}\bar{U}_0\|_{3,2}^2 + \|\partial_t \mathbb{B}U(t)\|_{\mathbb{B}W_2^{2,1}}^2 \right) \\ &\quad + \int_0^\infty \left(\|\partial_x \mathbb{B}U(s)\|_{\mathbb{B}W_2^{2,3}}^2 + \|\partial_s \mathbb{B}U(s)\|_{2,2}^2 \right) ds, \\ N_2(0,\infty)^2 &= \sup_{0 \le t < \infty} \left(\|\mathbb{B}U(t) - \mathbb{B}\bar{U}_0\|_{4,2}^2 + \|\partial_t \mathbb{B}U(t)\|_{\mathbb{B}W_2^{3,2}}^2 \right. \\ &\qquad \left. + \|\partial_t^2 \mathbb{B}U(t)\|_{\mathbb{B}W_2^{2,0}}^2 \right) \\ &\quad + \int_0^\infty \left(\|\partial_x \mathbb{B}U(s)\|_{\mathbb{B}W_2^{3,4}}^2 + \|\partial_s \mathbb{B}U(s)\|_{3,2}^2 \right. \\ &\qquad \left. + \|\partial_s^2 \mathbb{B}U(s)\|_{\mathbb{B}W_2^{2,1}}^2 \right) ds, \end{aligned}$$

where $\mathbb{B}\bar{U}_0 = (\bar{\rho}_0, 0)$.

Now, let us explain some necessary condition on the initial data $(\rho_0, \boldsymbol{v}_0)$. To do this, for a while we assume that $(\rho, \boldsymbol{v}) \in X^k(0,\infty)$ is a solution of IBVP: (1.1), (1.2) and (1.3). Then, $\partial_t^j(\rho,\boldsymbol{v})|_{t=0}$, $j \ge 1$, are determined successively by the initial data $(\rho_0, \boldsymbol{v}_0)$ through the equation (1.1). Namely, if we put $(\rho_j, \boldsymbol{v}_j) = \partial_t^j(\rho,\boldsymbol{v})|_{t=0}$ and then $(\rho_j, \boldsymbol{v}_j)$ is generated inductively from $(\rho_0, \boldsymbol{v}_0)$ by means of the equation (1.1). For example,

$$\begin{aligned} \rho_1 &= -\nabla \cdot (\rho_0 \boldsymbol{v}_0), \\ \boldsymbol{v}_1 &= -(\boldsymbol{v}_0 \cdot \nabla)\boldsymbol{v}_0 + \tfrac{\mu}{\rho_0}\Delta \boldsymbol{v}_0 + \tfrac{\nu}{\rho_0}\nabla(\nabla \cdot \boldsymbol{v}_0) - \tfrac{\nabla P(\rho_0)}{\rho_0}. \end{aligned}$$

From the definition of $X^k(0,\infty)$, the solution $(\rho, \boldsymbol{v})$ belonging to $X^k(0,\infty)$ should satisfy the following condition at $t = 0$:

$$\rho_0 - \bar{\rho}_0 \in H^{k+2}, \quad \rho_j \in H^{k+2-j}, \quad (j = 1, k),$$
$$\boldsymbol{v}_j \in \mathbf{H}^{k+2-2j} \ (j = 0, 1, k), \quad \boldsymbol{v}_j|_{\partial\Omega} = 0 \ (j = 0, 1).$$

This is a requirement on the initial data $(\rho_0, \boldsymbol{v}_0)$ in order to obtain solutions belonging to $X^k(0,\infty)$. If $(\rho_0, \boldsymbol{v}_0)$ satisfies (1.4), we say that $(\rho_0, \boldsymbol{v}_0)$ satisfies the **k th order compatibility condition** and **regularity**.

Theorem 1.1. *Let $k = 1$ or 2. Assume that the assumptions (1) and (2) hold. Then, there exists an $\epsilon_0 > 0$ such that if $(\rho_0, \boldsymbol{v}_0)$ satisfies the k th order compatibility condition and regularity and $\|(\rho_0 - \bar{\rho}_0, \boldsymbol{v}_0)\|_{3,2} \leq \epsilon_0$, then (IBVP): (1.1), (1.2) and (1.3) admits a unique solution $(\rho, \boldsymbol{v}) \in X^k(0,\infty)$.*

Moreover, there exists a constant C such that

$$N_k(0,\infty) \leq C\|(\rho_0 - \bar{\rho}_0, \boldsymbol{v}_0)\|_{k+2,2}.$$

Matsumura and Nishida [9], they proved Theorem 1 with $k = 1$ for the equation which describes the motion of compressible viscous and heat conductive gases. Employing the same argument as in [9], we can prove Theorem 1 for the system (1.1), (1.2) and (1.3), and also in the case $k = 2$.

The main results of this talk is the following theorem concerning the rate of convergence of solutions to IBVP: (1.1), (1.2) and (1.3).

Theorem 1.2. *Assume that the assumptions (1) and (2) hold. Assume that $(\rho_0, \boldsymbol{v}_0)$ satisfies the 2nd order compatibility condition and regularity and $(\rho_0 - \bar{\rho}_0, \boldsymbol{v}_0) \in \mathbb{B}L_1$. Then, there exists an $\epsilon > 0$ such that if $\|(\rho_0 - \bar{\rho}_0, \boldsymbol{v}_0)\|_{4,2} \leq \epsilon$, then the solution $(\rho, \boldsymbol{v})$ of (IBVP): (1.1), (1.2) and (1.3) has the following asymptotic behavior as $t \to \infty$:*

$$\|(\rho - \bar{\rho}_0, \boldsymbol{v})(t)\|_2 = O(t^{-\frac{3}{4}});$$
$$\|\partial_x(\rho, \boldsymbol{v})(t)\|_{\mathbb{B}W_2^{1,2}} + \|\partial_t(\rho, \boldsymbol{v})(t)\|_{1,2} = O(t^{-\frac{5}{4}});$$
$$\|(\rho - \bar{\rho}_0, \boldsymbol{v})(t)\|_\infty = O(t^{-\frac{3}{2}});$$
$$\|\partial_x(\rho, \boldsymbol{v})(t)\|_p = O(t^{-\frac{3}{2}}), \ \ 3 < p < \infty;$$
$$\|(\rho - \bar{\rho}_0, \boldsymbol{v})(t)\|_1 = O(t^{\frac{1}{2}}).$$

Here ϵ depends on p. Moreover, if $\rho_0 - \bar{\rho}_0 \in \mathbb{B}W_1^1(\Omega)$, then

$$\|\partial_x \boldsymbol{v}(t)\|_1 = O(1) \quad as \quad t \to \infty.$$

Concerning the equation of motion of compressible viscous and heat-conductive gases, Deckelnick [1, 2] gave the decay rates of the solutions. But this rate is weaker compared with the decay rate obtained by Matsumura and Nishida [9] and Ponce [10] in Cauchy problem case, because the initial data are assumed to be in $\mathbb{B}H^3$ only. Kobayashi and Shibata [7] gave the optimal decay rate for

$p \geq 2$ in the case that the initial data belong to $\mathbb{B}L_1$, which is corresponding to the rate in the Cauchy problem case which was obtained by Matsumura and Nishida [9] and Ponce [10].

Our above results is corresponding to the rate in the Caucy problem case which was obtained by Hoff and Zumbrun [3].

2. Decay property of solutions to the linearized problem

In order to prove Theorem 1.2, we shall use the decay property of solutions to the corresponding linearized problem. If we linearize the equation (1.1) at the constant state $(\bar{\rho}_0, 0, 0, 0)$ and we make some linear transformation of the unknown function, then we have the following initial boundary value problem of the linear operators:

$$\begin{aligned} \rho_t + \gamma \nabla \cdot \boldsymbol{v} &= 0 \quad \text{in} \quad [0,\infty) \times \Omega, \\ \boldsymbol{v}_t - \alpha \Delta \boldsymbol{v} - \beta \nabla(\nabla \cdot \boldsymbol{v}) + \gamma \nabla \rho &= 0 \quad \text{in} \quad [0,\infty) \times \Omega, \\ \boldsymbol{v}|_{\partial\Omega} &= 0 \quad \text{on} \quad [0,\infty) \times \partial\Omega, \\ (\rho, \boldsymbol{v})|_{t=0} &= (\rho_0, \boldsymbol{v}_0) \quad \text{in} \quad \Omega, \end{aligned} \tag{2.1}$$

where α and γ are positive constants and β is a nonnegative constant. Let $\mathbb{B}A$ be the 4×4 matrix of the differential operators of the form:

$$\mathbb{B}A = \begin{pmatrix} 0 & \gamma\nabla \\ \gamma\nabla & -\alpha\Delta - \beta\nabla\nabla\cdot \end{pmatrix}$$

with the domain:

$$D_p(\mathbb{B}A) = \{\mathbb{B}U = (\rho, \boldsymbol{v}) \in \mathbb{B}W_p^{1,2} \mid \mathbb{B}P\mathbb{B}U|_{\partial\Omega} = 0\}$$

for $1 < p < \infty$. Then, (2.1) is written in the form:

$$\mathbb{B}U_t + \mathbb{B}A\mathbb{B}U = 0 \quad \text{for } t > 0, \quad \mathbb{B}U|_{t=0} = \mathbb{B}U_0,$$

where $\mathbb{B}U_0 = (\rho_0, \boldsymbol{v}_0)$ and $\mathbb{B}U = (\rho, \boldsymbol{v})$.

Moreover, if we apply some linear transformation to $(\rho - \bar{\rho}_0, \boldsymbol{v})$ (the resulting vector of functions being denoted by $\tilde{\mathbb{B}U} = (\tilde{\rho}, \tilde{\boldsymbol{v}})$, then we can reduce IBVP: (1.1), (1.2) and (1.3) to the problem:

$$\tilde{\mathbb{B}U}_t + \mathbb{B}A\tilde{\mathbb{B}U} = \mathbb{B}F(\tilde{\mathbb{B}U}) \quad \text{for } t > 0, \quad \tilde{\mathbb{B}U}|_{t=0} = \tilde{\mathbb{B}U}_0$$

with suitable nonlinear term $\mathbb{B}F(\tilde{\mathbb{B}U})$. Therefore, in order to prove Theorem 1.2, we have to obtain the suitable decay property of solutions to (2.1). To do this, we shall investigate the spectral analysis for the operator $\mathbb{B}A$.

Let $\rho(-\mathbb{B}A)$ be the resolvent set of the operator $-\mathbb{B}A$. Then, we have the following lemma.

Lemma 2.1. *(Resolvent estimates) Let $1 < p < \infty$. Then $-\mathbb{B}A$ is a closed linear operator in $\mathbb{B}W_p^{1,0}$ and*

$$\rho(-\mathbb{B}A) \supset \Sigma = \{\lambda \in \mathbb{B}C \mid CRe\lambda + (Im\lambda)^2 > 0\},$$

where C is a constant depending only on α, β and γ. Moreover, the following properties are valid : There exist positive constants λ_0 and $\delta < \frac{\pi}{2}$ such that for any $\lambda - \lambda_0 \in \Sigma_\delta = \{\lambda \in \mathbb{C} \mid | arg\lambda | \leq \pi - \delta\}$ and any $\mathbb{F} \in \mathbb{W}_p^{1,0}$,

$$|\lambda| \|(\lambda+\mathbb{A})^{-1}\mathbb{F}\|_{\mathbb{W}_p^{1,0}} + \|\mathbb{P}(\lambda+\mathbb{A})^{-1}\mathbb{F}\|_{2,p} \leq C(\lambda_0, \delta)\|\mathbb{F}\|_{\mathbb{W}_p^{1,0}}.$$

Moreover, if $\mathbb{F} \in \mathbb{W}_p^{2,1}$, then

$$|\lambda|^{1/2}\|(\mathbb{I} - \mathbb{P})(\lambda + \mathbb{A})^{-1}\mathbb{F}\|_{2,p} + |\lambda|^{-1/2}\|\mathbb{P}(\lambda + \mathbb{A})^{-1}\mathbb{F}\|_{3,p} \leq C(\lambda_0, \delta)\|\mathbb{F}\|_{\mathbb{W}_p^{2,1}}.$$

By Lemma 2.1 we see that $-\mathbb{A}$ generates an analytic semigroup $\{e^{-t\mathbb{A}}\}_{t\geq 0}$ on $\mathbb{W}_p^{1,0}$. Then we have

Theorem 2.1. *($L_p - L_q$ estimates) Let $\sigma = \frac{3}{2}\left(\frac{1}{q} - \frac{1}{p}\right)$ and $\eta = \frac{1}{p} - \frac{1}{q} + \frac{1}{2}$ where $1 \leq p, q \leq \infty$. Put $\mathbb{U}(t) = e^{-t\mathbb{A}}\mathbb{U}_0$. All the constant C in the theorem depends on the numbers p and q essentially.*
(I) Let $1 \leq q \leq 2 \leq p < \infty$.
(A) For $\mathbb{U}_0 \in \mathbb{W}_p^{1,0} \cap \mathbb{L}_q$ and $t \geq 1$, we have the following estimates:

$$\|\mathbb{U}(t)\|_p \leq Ct^{-\sigma}(\|\mathbb{U}_0\|_q + \|\mathbb{U}_0\|_{\mathbb{W}_p^{1,0}}), \quad 2 \leq p < \infty;$$

$$\|\partial_x\mathbb{U}(t)\|_p + \|\partial_t\mathbb{U}(t)\|_p \leq Ct^{-\sigma-\frac{1}{2}}(\|\mathbb{U}_0\|_q + \|\mathbb{U}_0\|_{\mathbb{W}_p^{1,0}}), \quad 2 \leq p \leq 3;$$

$$\|\partial_x\mathbb{U}(t)\|_p + \|\partial_t\mathbb{U}(t)\|_p \leq Ct^{-\frac{3}{2q}}(\|\mathbb{U}_0\|_q + \|\mathbb{U}_0\|_{\mathbb{W}_p^{1,0}}), \quad 3 < p < \infty;$$

$$\|\partial_x^2\mathbb{P}\mathbb{U}(t)\|_p \leq Ct^{-\frac{3}{2q}}(\|\mathbb{U}_0\|_q + \|\mathbb{U}_0\|_{\mathbb{W}_p^{1,0}}), \quad 2 \leq p < \infty;$$

$$\|\mathbb{U}(t)\|_{\mathbb{W}_\infty^{0,1}} \leq Ct^{-\frac{3}{2q}}(\|\mathbb{U}_0\|_q + \|\mathbb{U}_0\|_{\mathbb{W}_p^{1,0}}), \quad 3 < p < \infty.$$

(B) If $\mathbb{U}_0 \in \mathbb{W}_p^2 \cap \mathbb{L}_q$ and $\mathbb{P}\mathbb{U}_0 = 0$ on $\partial\Omega$, then we have the following estimates for $t \geq 1$:

$$\|\partial_x^2(\mathbb{I} - \mathbb{P})\mathbb{U}(t)\|_p + \|\partial_x^3\mathbb{P}\mathbb{U}(t)\|_p + \|\partial_x\partial_t\mathbb{U}(t)\|_p$$

$$hspace1.5in \leq Ct^{-\frac{3}{2q}}(\|\mathbb{U}_0\|_q + \|\mathbb{U}_0\|_{2,p}), \quad 2 \leq p < \infty;$$

$$\|\mathbb{U}(t)\|_{\mathbb{W}_\infty^{1,2}} \leq Ct^{-\frac{3}{2q}}(\|\mathbb{U}_0\|_q + \|\mathbb{U}_0\|_{2,p}), 3 < p < \infty.$$

(C) If we assume that $q > 1$ additionally, then for $\mathbb{U}_0 \in \mathbb{W}_p^{2,1} \cap \mathbb{W}_q^{1,0}$ and $t \geq 1$ we have the following estimates:

$$\|\partial_x^2(\mathbb{I} - \mathbb{P})\mathbb{U}(t)\|_p + \|\partial_x^3\mathbb{P}\mathbb{U}(t)\|_p + \|\partial_x\partial_t\mathbb{U}(t)\|_p \leq Ct^{-\frac{3}{2q}}(\|\mathbb{U}_0\|_{\mathbb{W}_q^{1,0}} + \|\mathbb{U}_0\|_{\mathbb{W}_p^{2,1}}), 2 \leq p < \infty;$$

$$\|\mathbb{U}(t)\|_{\mathbb{W}_\infty^{1,2}} \leq Ct^{-\frac{3}{2q}}(\|\mathbb{U}_0\|_{\mathbb{W}_q^{1,0}} + \|\mathbb{U}_0\|_{\mathbb{W}_p^{2,1}}), 3 < p < \infty.$$

(II) Let $1 \le q \le p < 2$.
(A) For $\mathbb{B}U_0 \in \mathbb{B}W_p^{1,0} \cap \mathbb{B}L_q$ *and* $t \ge 1$, *we have the following estimates:*

$$\begin{aligned}
\|\partial_x^k \mathbb{B}U(t)\|_p &\le Ct^{-\sigma+\eta-\frac{k}{2}}(\|\mathbb{B}U_0\|_q + \|\mathbb{B}U_0\|_{\mathbb{B}W_p^{1,0}}), \\
&\qquad 1 < p < 2,\ 0 \le k \le 1 \ \textit{integer}; \\
\|\partial_x^2 \mathbb{B}P\mathbb{B}U(t)\|_p &\le C max\{t^{-\frac{3}{2q}}, t^{-\sigma+\eta-1}\}(\|\mathbb{B}U_0\|_q + \|\mathbb{B}U_0\|_{\mathbb{B}W_p^{1,0}}), \\
&\qquad 1 < p < 2; \\
\|\mathbb{B}U(t)\|_1 &\le Ct^{\frac{1}{2}}(\|\mathbb{B}U_0\|_1 + \|\mathbb{B}U_0\|_{\mathbb{B}W_r^{1,0}}),\ 1 < r < \infty.
\end{aligned}$$

Moreover, if $\mathbb{B}U_0 \in \mathbb{B}W_1^{1,0}$, *then*

$$\|\partial_x \mathbb{B}P\mathbb{B}U(t)\|_1 \le C(\|\mathbb{B}U_0\|_{\mathbb{B}W_1^{1,0}} + \|\mathbb{B}U_0\|_{\mathbb{B}W_r^{1,0}}), \quad 1 < r < \infty.$$

(B) If $\mathbb{B}U_0 \in \mathbb{B}W_p^2 \cap \mathbb{B}L_q$ *and* $\mathbb{B}P\mathbb{B}U_0 = 0$ *on* $\partial\Omega$, *then we have the following estimates for* $t \ge 1$:

$$\begin{aligned}
\|\partial_x^2(\mathbb{B}I - \mathbb{B}P)\mathbb{B}U(t)\|_p &\le C max\{t^{-\frac{3}{2q}}, t^{-\sigma+\eta-1}\}(\|\mathbb{B}U_0\|_q + \|\mathbb{B}U_0\|_{2,p}), \\
&\qquad 1 < p < 2; \\
\|\partial_x^3 \mathbb{B}P\mathbb{B}U(t)\|_p &\le C max\{t^{-\frac{3}{2q}}, t^{-\sigma+\eta-\frac{3}{2}}\}(\|\mathbb{B}U_0\|_q + \|\mathbb{B}U_0\|_{\mathbb{B}W_p^{1,0}}), \\
&\qquad 1 < p < 2.
\end{aligned}$$

(C) If we assume that $q > 1$ *additionally, then for* $\mathbb{B}U_0 \in \mathbb{B}W_p^{2,1} \cap \mathbb{B}W_q^{1,0}$ *and* $t \ge 1$ *we have the following estimates:*

$$\|\partial_x^2(\mathbb{B}I - \mathbb{B}P)\mathbb{B}U(t)\|_p \le C max\{t^{-\frac{3}{2q}}, t^{-\sigma+\eta-1}\}(\|\mathbb{B}U_0\|_{\mathbb{B}W_q^{1,0}} + \|\mathbb{B}U_0\|_{\mathbb{B}W_p^{2,1}});$$

$$\|\partial_x^3 \mathbb{B}P\mathbb{B}U(t)\|_p \le C max\{t^{-\frac{3}{2q}}, t^{-\sigma+\eta-\frac{3}{2}}\}(\|\mathbb{B}U_0\|_{\mathbb{B}W_q^{1,0}} + \|\mathbb{B}U_0\|_{\mathbb{B}W_p^{2,1}}).$$

To prove Theorem 1.2, we reduce IBVP: (1.1), (1.2) and (1.3) to the following integral equation:

$$\mathbb{B}\tilde{U}(t) = e^{-t\mathbb{B}A}\mathbb{B}\tilde{U}_0 - \int_0^t e^{-(t-s)\mathbb{B}A}\mathbb{B}F(\mathbb{B}\tilde{U}(s))\,ds$$

(cf. (2.1), below). Applying Theorem 2.1 and using the fact that $N_2(0,\infty) \le C\|(\rho_0 - \bar{\rho}_0, \boldsymbol{v}_0)\|_{4,2}$ (cf. Theorem 1.1), we will have Theorem 1.2.

References

[1] Deckelnick, K. *Decay estimates for the compressible Navier-Stokes equations in unbounded domain,* Indiana Math. Z. 209, 2. pp. 115–130 (1992)

[2] Deckelnick, K. L^2*-decay for the compressible Navier-Stokes equations in unbounded domains,* Indiana Commun. in Partial Differential Equations. 18, pp. 1445-1476 (1993)

[3] Hoff, D and Zumbrun, K. *Multi-dimensional diffusion waves for the Navier-Stokes equations of compressible flow,* Indiana Univ. Math. J. 44, 2 pp. 604–676 (1995)

[4] Hoff, D and Zumbrun, K. *Pointwise decay estimates for multidimensional Navier-Stokes diffusion waves,* Z. angew. Math. Phys. 48 pp. 597–614 (1997)

[5] Kobayashi, T. *On a local energy decay of solutions for the equations of motion of viscous and heat-conductive gases in an exterior domain in* $\mathbf{R}^3$, Tsukuba. J. Math. 21 pp. 629–670 (1997)

[6] Kobayashi, T. *On the local energy decay of higher derivatives of solutions for the equations of motion of compressible viscous and heat-conductive gases in an exterior domain in* $\mathbf{R}^3$, Proc. Japan Acad. Ser. A 73 (7) pp. 126–129 (1997)

[7] Kobayashi, T and Shibata, Y. *Decay estimates of solutions for the equations of motion of compressible viscous and heat-conductive gases in an exterior domain in* $\mathbf{R}^3$, Commun. Math. Phys. 200 pp. 621–659 (1999)

[8] Matsumura, A. and Nishida, T. *The initial value problem for the equations of motion of compressible viscous and heat-conductive fluids,* Proc. Japan Acad. Ser. A. 55 pp. 337–342 (1979)

[9] Matsumura, A. and Nishida, T. *Initial boundary value problems for the equations of motion of compressible viscous and heat-conductive fluids,* Commun. Math. Phys. 89 pp. 445–464 (1983)

[10] Ponce, G. *Global existence of small solutions to a class of nonlinear evolution equations,* Nonlinear. Anal. TMA. 9 pp. 339–418 (1985)

TAKAYUKI KOBAYASHI, DEPARTMENT OF MATHEMATICS, KYUSHU INSTITUTE OF TECHNOLOGY, 1-1 SENSUI-CHO TOBATA-KU KITAKYUSYU 804-8550, JAPAN

E-mail address: `kybayasi@mns.kyutech.ac.jp`

Progress in Nonlinear Differential Equations
and Their Applications, Vol. 42

AN APPROXIMATION THEOREM OF LAX TYPE FOR SEMIGROUPS OF LIPSCHITZ OPERATORS

YOSHIKAZU KOBAYASHI
NIIGATA UNIVERSITY

SHINNOSUKE OHARU
HIROSHIMA UNIVERSITY

NAOKI TANAKA
OKAYAMA UNIVERSITY

Let X be a Banach space with norm $\|\cdot\|$ and D a subset of X. A one-parameter family $\{T(t)\}_{t\in[0,\infty)}$ of Lipschitz operators from D into itself is called a *semigroup of Lipschitz operators* on D if it satisfies the following conditions:

(S1) For $x \in D$ and $t, s \geq 0$,

$$T(t)T(s)x = T(t+s)x, \quad T(0)x = x.$$

(S2) For $x \in D$ and $t \geq 0$,

$$\lim_{s\to t}\|T(t)x - T(s)x\| = 0.$$

(S3) For $\tau > 0$, there exists $M_\tau \geq 1$ such that

$$\|T(t)x - T(t)y\| \leq M_\tau\|x - y\|$$

for x, $y \in D$ and $t \in [0,\tau]$.

Let $\{T(t)\}_{t\in[0,\infty)}$ be a semigroup of Lipschitz operators on D. For each $h \in (0, h_0]$, we introduce a Lipschitz operator C_h from D_h into itself, where D_h is a subset of X such that $D \subset D_h$. On the family $\{C_h\}_{h\in(0,h_0]}$, we impose two conditions, the *stability condition* (S) and the *consistency condition* (C) formulated as below:

(S) For $\tau > 0$, there exists $M_\tau \geq 1$ such that

$$\|C_h^n x - C_h^n y\| \leq M_\tau\|x - y\|$$

for x, $y \in D_h$, $h \in (0, h_0]$ and $n = 1,2,\ldots$ with $nh \in (0,\tau]$.

(C) There exists a dense subset $D_0 \subset D$ such that

$$\lim_{h\downarrow 0} \int_0^\tau \frac{1}{h}\|T(t+h)x - C_h T(t)x\|\, dt = 0$$

for $x \in D_0$ and $\tau > 0$.

Our objective here is to present an approximation theorem for semigroups of operators which asserts that a stable family of nonlinear operators consistent with a semigroup of Lipschitz operators converges necessarily.

1.

Our main result is stated as below, and this may be regarded as a nonlinear analogue of Lax's theorem [4].

Theorem 1. *Suppose that* $\{C_h\}_{h\in(0,h_0]}$ *satisfies conditions* (S) *and* (C). *Then*

$$\lim_{h\downarrow 0}\left(\sup_{t\in[0,\tau]} \|T(t)x - C_h^{[t/h]}x\|\right) = 0$$

for $x \in D$ *and* $\tau > 0$. *Here* $[a]$ *denotes the greatest integer not greater than* a.

Remark 1. This theorem was obtained under stronger assumptions by Chorin *et al.* In [3], (C) is replaced by the following condition:

(C′) There exits a dense subset $D_0 \subset D$ such that, for $x \in D_0$ and $\tau > 0$,

$$\lim_{h\downarrow 0}\left(\sup_{t\in[0,\tau]}\left(\frac{1}{h}\|T(t+h)x - C_h T(t)x\|\right)\right) = 0.$$

The theorem above follows directly from the following lemma.

Lemma 1. *Suppose that* $\{C_h\}_{h\in(0,h_0]}$ *satisfies condition* (S). *Then*

$$\|T(t)x - C_h^{[t/h]}x\| \le \frac{M_\tau}{h}\int_0^\tau \|T(s+h)x - C_h T(s)x\|\, ds$$
$$+(1+M_\tau)\sup\left\{\|T(s)x - T(\hat{s})x\|;\ s,\ \hat{s} \in [0, \tau + h_0],\ |s-\hat{s}| \le h\right\}$$

for $t \in [0,\tau]$, $h \in (0, h_0]$ *and* $x \in D$.

Proof. Let $x \in D$ and $u(t) = T(t)x$ for $t \in [0,\infty)$. It holds for $h \in (0, h_0]$ and $t \in [0,\tau]$ that

$$\|C_h^{[t/h]}u(0) - u(t)\| = \frac{1}{h}\int_0^h \|C_h^{[t/h]}u(0) - u(t)\|\, ds$$
$$\le \frac{1}{h}\int_0^h \|C_h^{[t/h]}u(0) - C_h^{[t/h]}u(s)\|\, ds + \frac{1}{h}\int_0^h \|C_h^{[t/h]}u(s) - u(s+[t/h]h)\|\, ds$$
$$+ \frac{1}{h}\int_0^h \|u(s+[t/h]h) - u(t)\|\, ds$$
$$= \text{(I)} + \text{(II)} + \text{(III)}.$$

Using (S), we have

$$(\mathrm{I}) + (\mathrm{III}) \leq (M_\tau + 1) \sup\Big\{ \|u(s) - u(\hat{s})\|;\ s,\ \hat{s} \in [0, \tau + h_0],\ |s - \hat{s}| \leq h \Big\}.$$

Let $s \in [0, h]$. Using (S) again, we also have

$$\begin{aligned}
&\|C_h^{[t/h]} u(s) - u(s + [t/h]h)\| \\
&\leq \sum_{j=0}^{[t/h]-1} \|C_h^{[t/h]-j} u(s + jh) - C_h^{[t/h]-j-1} u(s + (j+1)h)\| \\
&\leq M_\tau \sum_{j=0}^{[t/h]-1} \|C_h u(s + jh) - u(s + (j+1)h)\|,
\end{aligned}$$

which implies

$$\begin{aligned}
(\mathrm{II}) &\leq \frac{M_\tau}{h} \sum_{j=0}^{[t/h]-1} \int_0^h \|C_h u(s + jh) - u(s + (j+1)h)\|\, ds \\
&= \frac{M_\tau}{h} \sum_{j=0}^{[t/h]-1} \int_{jh}^{(j+1)h} \|C_h u(s) - u(s + h)\|\, ds \\
&\leq \frac{M_\tau}{h} \int_0^\tau \|C_h u(s) - u(s + h)\|\, ds.
\end{aligned}$$

Q.E.D.

The following proposition is useful to show that a family of nonlinear operators is stable.

Proposition 1. *The family $\{C_h\}_{h \in (0, h_0]}$ satisfies the stability condition (S) if and only if there exist $M \geq 1$, $\omega \geq 0$ and functionals $V_h(\cdot, \cdot) : D_h \times D_h \to [0, \infty)$, $h \in (0, h_0]$, such that*

$$V_h(C_h x, C_h y) \leq e^{\omega h} V_h(x, y)$$

and

$$\|x - y\| \leq V_h(x, y) \leq M \|x - y\|$$

for $x, y \in D_h$ and $h \in (0, h_0]$.

Proof. Since the sufficiency of the condition is trivial, we only prove the necessity. Suppose that $\{C_h\}_{h \in (0, h_0]}$ satisfies (S). Let $\tau > h_0$. Then there exists $M_\tau \geq 1$ such that $\|C_h^n x - C_h^n y\| \leq M_\tau \|x - y\|$ for $x, y \in D_h$, $n = 1, 2, \ldots$ and $h \in (0, h_0]$ with $nh \in (0, \tau]$. The assertion follows if we choose $\omega = \log M_\tau / (\tau - h_0)$, $M = M_\tau$ and

$$V_h(x, y) = \sup\Big\{ e^{-nh\omega} \|C_h^n x - C_h^n y\|;\ n = 0, 1, 2, \ldots \Big\}$$

for $x,\ y \in D_h$. In fact, for $x, y \in D_h$, $h \in (0, h_0]$, $n = 0, 1, 2, \ldots$, and nonnegative integers p and q such that $n = p[\tau/h] + q$ and $0 \le q < [\tau/h]$, we have

$$\begin{aligned}\|C_h^n x - C_h^n y\| \le & M_\tau^{p+1}\|x - y\| = M_\tau \exp(p\omega(\tau - h_0))\|x - y\| \\ \le & M \exp(p\omega(([\tau/h] + 1)h - h_0))\|x - y\| \\ \le & M \exp(p\omega[\tau/h]h)\|x - y\| \le M \exp(nh\omega)\|x - y\|.\end{aligned}$$

Q.E.D.

Let A_0 be the infinitesimal generator of the semigroup $\{T(t)\}_{t\in[0,\infty)}$ defined by

$$D(A_0) = \left\{x \in D;\ \lim_{h\downarrow 0} \frac{1}{h}(T(h)x - x) \text{ exists}\right\}$$

and

$$A_0 x = \lim_{h\downarrow 0} \frac{1}{h}(T(h)x - x) \quad \text{for } x \in D(A_0).$$

We assume that there exists a dense subset D_0 of D such that, for any $x \in D_0$, $u(t) = T(t)x$ is a (unique) strong solution of the initial-value problem:

$$u'(t) = A_0 u(t),\ t > 0; \quad u(0) = x.$$

Note that this assumption implies that $D(A_0)$ is dense in D. We define an operator $A_h : D_h \to X$ by

$$A_h x = \frac{1}{h}(C_h x - x)$$

for $x \in D_h$ and $h \in (0, h_0]$. By means of the infinitesimal generator A_0, one can formulate a criterion for the consistency condition (C).

Proposition 2. *Condition* (C″) *below implies the consistency condition* (C).

(C″) *For* $x \in D_0$ *and* $\tau > 0$,

$$\lim_{h\downarrow 0}\left(\int_0^\tau \|A_h T(t)x - A_0 T(t)x\|\, dt\right) = 0.$$

Proof. Let $x \in D_0$, $\tau > 0$ and $u(t) = T(t)x$ for $t \in [0, \infty)$. Then,

$$\begin{aligned}&\int_0^\tau \frac{1}{h}\|T(t + h)x - C_h T(t)x\|\, dt = \int_0^\tau \frac{1}{h}\|u(t + h) - C_h u(t)\|\, dt \\ &= \int_0^\tau \|\frac{1}{h}(u(t + h) - u(t)) - A_h u(t)\|\, dt \\ &\le \int_0^\tau \|\frac{1}{h}(u(t + h) - u(t)) - u'(t)\|\, dt + \int_0^\tau \|A_0 u(t) - A_h u(t)\|\, dt.\end{aligned}$$

Letting $h \downarrow 0$, we obtain the desired assertion, since $u(t)$ is supposed to be absolutely continuous on $[0, \tau]$ and strongly differentiable at almost all $t \in (0, \tau)$.

Q.E.D.

2.

We here consider a simple application of our results to the initial value problem for the following quasilinear wave equation of Kirchhoff type with damping term:

$$
\text{(IVP)} \qquad \begin{cases} \partial_t u(t,x) = \partial_x v(t,x), \quad t>0, \ -\infty < x < \infty, \\ \partial_t v(t,x) = \beta'(\int_{-\infty}^{\infty} |u(t,x)|^2\,dx)\partial_x u(t,x) - \nu v(t,x), \\ \qquad\qquad\qquad\qquad\qquad\qquad t>0, \ -\infty < x < \infty, \\ u(0,x) = u_0(x),\ v(0,x) = v_0(x), \quad -\infty < x < \infty. \end{cases}
$$

Here $\nu > 0$ is a constant, $\beta \in C^2[0,\infty)$ with $\beta(0) = 0$, and it is assumed that there exist constants $\delta_0 > 0$ and $M_0 > 0$ such that $\beta'(r) \geq \delta_0$ and $0 \leq \beta''(r) \leq M_0$ for $r \in [0,\infty)$.

To this problem (IVP), we formulate the following difference scheme of Friedrichs-Lax type:

$$
\begin{cases} \frac{1}{h}\left(u_h^n(x) - \frac{1}{2}(u_h^{n-1}(x+k) + u_h^{n-1}(x-k))\right) \\ \quad = \frac{1}{2k}\left(v_h^{n-1}(x+k) - v_h^{n-1}(x-k)\right), \quad n = 1,2,\ldots, \ -\infty < x < \infty \\ \frac{1}{h}\left(v_h^n(x) - \frac{1}{2}(v_h^{n-1}(x+k) + v_h^{n-1}(x-k))\right) \\ \quad = \beta'(\int_{-\infty}^{\infty} |u_h^n(x)|^2\,dx)\frac{1}{2k}\left(u_h^{n-1}(x+k) - u_h^{n-1}(x-k)\right) \\ \qquad -\frac{\nu}{2}\left(v_h^{n-1}(x+k) + v_h^{n-1}(x-k)\right), \quad n = 1,2,\ldots, \ -\infty < x < \infty, \\ u_h^0(x) = u_0(x),\ v_h^0(x) = v_0(x), \quad -\infty < x < \infty, \end{cases}
$$

where $h > 0$ and $k > 0$ represent differencing in time and differencing in space, respectively.

Theorem 2. *Let $E_0 > 0$. Assume that the mesh ratio $r = h/k$ is kept as a constant as $h \downarrow 0$, and that $r\sqrt{\beta'(2E_0/\delta_0)} < 1$. If*

$$
\beta\left(\int_{-\infty}^{\infty} |u_0(x)|^2\,dx\right) + \int_{-\infty}^{\infty} |v_0(x)|^2\,dx \leq E_0
$$

and

$$
\int_{-\infty}^{\infty} \left(|u_0|^2 + |v_0|^2 + |\partial_x u_0|^2 + |\partial_x v_0|^2\right)\,dx
$$

is sufficiently small, then

$$
\sup_{t\in[0,\tau]} \left(\int_{-\infty}^{\infty} \left(|u_h^{[t/h]}(x) - u(t,x)|^2 + |v_h^{[t/h]}(x) - v(t,x)|^2\right)\,dx\right) \to 0
$$

as $h \downarrow 0$, where

$$
\begin{aligned}
&(u(t,\cdot), v(t,\cdot)) \\
\in & W^{1,1}_{loc}([0,\infty); L^2(\boldsymbol{R}) \times L^2(\boldsymbol{R})) \cap L^\infty((0,\infty); H^1(\boldsymbol{R}) \times H^1(\boldsymbol{R}))
\end{aligned}
$$

is the solution of (IVP) *with initial data* (u_0, v_0).

Proof. We only make a sketch of the proof. To this end, we choose $X = L^2(\boldsymbol{R}) \times L^2(\boldsymbol{R})$ with the norm $\|(u,v)\|_X = (\int_{-\infty}^{\infty} |u|^2 + |v|^2\,dx)^{1/2}$ for $(u,v) \in X$. We define two functionals $V : X \times X \to [0,\infty)$ and $E : X \to [0,\infty)$ by

$$
\begin{aligned}
&V((u,v),(\hat{u},\hat{v})) \\
&= \left(\beta'(\int_{-\infty}^{\infty} |u(x)|^2\,dx) \int_{-\infty}^{\infty} |u(x) - \hat{u}(x)|^2\,dx + \int_{-\infty}^{\infty} |v(x) - \hat{v}(x)|^2\,dx\right)^{1/2}
\end{aligned}
$$

and

$$
E(u,v) = \beta\left(\int_{-\infty}^{\infty} |u(x)|^2\,dx\right) + \int_{-\infty}^{\infty} |v(x)|^2\,dx
$$

for (u,v), $(\hat{u},\hat{v}) \in X$, respectively.

Let $h_0 > 0$ and $r_h > 0$ for $h \in (0,h_0]$. We then set

$$
\begin{aligned}
D_h = \Big\{ &(u,v) \in L^2(\boldsymbol{R}) \times L^2(\boldsymbol{R});\ E(u,v) \le E_0, \\
&E(u,v) + \beta'(\int_{-\infty}^{\infty} |u(x)|^2\,dx) \int_{-\infty}^{\infty} |\frac{1}{2k}(u(x+k) - u(x-k))|^2\,dx \\
&+ \int_{-\infty}^{\infty} |\frac{1}{2k}(v(x+k) - v(x-k)) + \frac{\nu}{2}(u(x+k) + u(x-k))|^2\,dx \le r_h \Big\}
\end{aligned}
$$

and define an operator $C_h : D_h \to X$ by

$$
C_h(u,v) = (u_h, v_h) \quad \text{for} \quad (u,v) \in D_h \text{ and } h \in (0,h_0],
$$

where

$$
\left\{
\begin{aligned}
u_h(x) &= \frac{1}{2}(u(x+k) + u(x-k)) + \frac{h}{2k}(v(x+k) - v(x-k)) \\
v_h(x) &= \frac{1}{2}(v(x+k) + v(x-k)) \\
&\quad + \beta'(\int_{-\infty}^{\infty} |u_h(x)|^2\,dx) \frac{h}{2k}(u(x+k) - u(x-k)) \\
&\qquad - \frac{\nu h}{2}(v(x+k) + v(x-k))
\end{aligned}
\right.
$$

for $h \in (0,h_0]$. We may choose $h_0 > 0$ and $r_h > 0$ for $h \in (0,h_0]$ so that $\bigcap_{h \in (0,h_0]} D_h \ne \emptyset$ and, for each $h \in (0,h_0]$, C_h may map D_h into itself. Then, there exists $\omega > 0$ such that

$$
V(C_h(u,v), C_h(\hat{u},\hat{v})) \le e^{h\omega} V((u,v),(\hat{u},\hat{v}))
$$

for $(u,v), (\hat{u},\hat{v}) \in D_h$ and $h \in (0,h_0]$. Proposition 1 then implies that $\{C_h\}_{h\in(0,h_0]}$ satisfies the stability condition (S).

We next choose an $r_0 > 0$ sufficiently small and define an operator $A : D(\subset \bigcap_{h\in(0,h_0]} D_h) \to X$ by

$$D = \Big\{ (u,v) \in H^1(\boldsymbol{R}) \times H^1(\boldsymbol{R});\ E(u,v) \le E_0,\\ E(u,v) + \beta'\Big(\int_{-\infty}^{\infty} |u(x)|^2\,dx\Big) \int_{-\infty}^{\infty} |\partial_x u(x)|^2\,dx \\ + \int_{-\infty}^{\infty} |\partial_x v(x) + \nu u(x)|^2\,dx \le r_0 \Big\}$$

and

$$A(u,v) = (\partial_x v, \beta'\Big(\int_{-\infty}^{\infty} |u|^2\,dx\Big)\partial_x u - \nu v) \quad \text{for } (u,v) \in D(A) = D.$$

Then, it is shown that there exists a semigroup $\{T(t)\}_{t\in[0,\infty)}$ on D with the following properties:

(1) The operator A is the infinitesimal generator of the semigroup $\{T(t)\}_{t\in[0,\infty)}$.

(2) For any $(u_0,v_0) \in D$, $(u(t,\cdot),v(t,\cdot)) = T(t)(u_0,v_0)$ is the strong solution of the initial-value problem:

$$\frac{d}{dt}(u(t,\cdot),v(t,\cdot)) = A(u(t,\cdot),v(t,\cdot)),\ t>0; \quad (u(0,\cdot),v(0,\cdot)) = (u_0,v_0).$$

(3) There exists $\omega > 0$ such that, for $(u,v), (\hat{u},\hat{v}) \in D$ and $t \ge 0$,

$$V(T(t)(u,v), T(t)(\hat{u},\hat{v})) \le e^{\omega t} V((u,v),(\hat{u},\hat{v})).$$

(4) For $(u_0,v_0) \in D$ and $\tau > 0$,

$$\lim_{h\downarrow 0} \int_0^{\tau} \|A_h T(t)(u_0,v_0) - AT(t)(u_0,v_0)\|_X\,dt = 0$$

where

$$A_h(u,v) = \frac{1}{h}(C_h(u,v) - (u,v)) \quad \text{for } (u,v) \in D_h \text{ and } h \in (0,h_0].$$

From these properties and Proposition 2, it follows that $\{C_h\}_{h\in(0,h_0]}$ satisfies the consistency condition (C). Theorem 1 then implies that

$$\lim_{h\downarrow 0} \left(\sup_{t\in[0,\tau]} \|T(t)(u_0,v_0) - C_h^{[t/h]}(u_0,v_0)\|_X \right) = 0$$

for $(u_0,v_0) \in D$ and $\tau > 0$. Q.E.D.

References

[1] H. Brézis and A. Pazy, Convergence and approximation of semigroups of nonlinear operators in Banach spaces, *J. Funct. Anal.*, **7**(1972), 63–74.

[2] P. R. Chernoff and J. E. Marsden, Properties of Infinite Dimensional Hamiltonian Systems, *Lecture Note in Math.*, **425**, Springer-Verlag, New York, 1974.

[3] A. J. Chorin, T. J. R. Hughes, M. F. McCracken and J. E. Marsden, Product formulas and numerical algorithms, *Comm. Pure Appl. Math.*, **31**(1978), 205–256.

[4] P. D. Lax and R. D. Richtmyer, Survey of the stability of linear finite difference equations, *Comm. Pure Appl. Math.*, **9**(1956), 267–293.

[5] J. E. Marseden, On product formulas for nonlinear semigroups, *J. Funct. Anal.*, **13**(1973), 51–72.

[6] I. Miyadera and Y. Kobayashi, Convergence and approximation of nonlinear semigroups, *Proceedings of Japan–France Seminar on Functional Analysis and Numerical Analysis*, Japan Society for the Promotion of Science, Tokyo, 277-295, 1978.

[7] I. Miyadera and S. Oharu, Approximation of semi-groups of nonlinear operators, *Tôhoku Math. J.*, **22**(1970), 24–47.

[8] R. D. Richtmyer and K. W. Morton, Difference Methods for Initial Value Problems, 2nd Edition, Wiley-Interscience, New York, 1967.

[9] Y. Yamada, On some quasilinear wave equations with dissipative terms, *Nagoya Math. J.*, **87**(1982), 17–39.

Yoshikazu Kobayashi, Applied Mathematics Department, Faculty of Engineering, Niigata University, Niigata 950-2181, JAPAN
E-mail address: kobayasi@info.eng.niigata-u.ac.jp

Shinnosuke Oharu, Department of Mathematics, Faculty of Science, Hiroshima University, Hiroshima 739-8526, JAPAN
E-mail address: oharu@math.sci.hiroshima-u.ac.jp

Naoki Tanaka, Department of Mathematics, Faculty of Science, Okayama University, Okayama 700-8530, JAPAN
E-mail address: tanaka@math.okayama-u.ac.jp

Progress in Nonlinear Differential Equations
and Their Applications, Vol. 42

INVARIANT SUBSPACES AND QUASIAFFINE TRANSFORMS OF UNITARY OPERATORS

CARLOS S. KUBRUSLY
CATHOLIC UNIVERSITY (PUC-RIO)

Abstract. A classical conjecture on nontrivial invariant subspaces for Hilbert-space contractions reads as follows. "a $\mathcal{C}_{1\cdot}$-contraction has a nontrivial invariant subspace". This turns out to be equivalent to a second conjecture, namely, "if a contraction is a quasiaffine transform of a unitary operator, then it has a nontrivial invariant subspace". Although these are still unsolved problems, it can be proved that *if a $\mathcal{C}_{1\cdot}$-contraction has no nontrivial invariant subspace, then it is a quasiaffine transform of its own unitary extension, which is reductive and has an invariant dense and totally cyclic linear manifold.* This paper presents a brief review, based on [7] and [9], on the equivalence between the above conjectures.

1. Preliminaries

Throughout the paper $\mathcal{H}$ and $\mathcal{K}$ stand for infinite-dimensional complex separable Hilbert spaces, and $\mathcal{B}[\mathcal{H},\mathcal{K}]$ stands for the Banach space of all bounded linear transformations of $\mathcal{H}$ into $\mathcal{K}$. Let $\mathcal{N}(X) \subseteq \mathcal{H}$ and $\mathcal{R}(X) \subseteq \mathcal{K}$ denote the null space (i.e. the kernel) and range of $X \in \mathcal{B}[\mathcal{H},\mathcal{K}]$, respectively. Set $\mathcal{B}[\mathcal{H}] = \mathcal{B}[\mathcal{H},\mathcal{H}]$ for short. If T lies in $\mathcal{B}[\mathcal{H}]$, then we say that T is an operator on $\mathcal{H}$, and T^* in $\mathcal{B}[\mathcal{H}]$ denotes the adjoint of T. By a subspace of $\mathcal{H}$ we mean a closed linear manifold of $\mathcal{H}$, so that the closure $\mathcal{R}^-$ of a linear manifold $\mathcal{R}$ of $\mathcal{H}$ is a subspace of $\mathcal{H}$. A set $\mathcal{S} \subseteq \mathcal{H}$ is invariant for an operator T on $\mathcal{H}$ if $T(\mathcal{S}) \subseteq \mathcal{S}$. A subspace $\mathcal{M}$ of $\mathcal{H}$ is nontrivial if $\{0\} \neq \mathcal{M} \neq \mathcal{H}$. $\mathcal{M}$ is a reducing subspace for $T \in \mathcal{B}[\mathcal{H}]$ if it is invariant for both T and T^* (equivalently, if both $\mathcal{M}$ and its orthogonal complement, $\mathcal{M}^\perp$, are invariant for T). An operator is reducible if it has a nontrivial reducing subspace, and reductive if all its invariant subspaces are reducing.

An operator T on $\mathcal{H}$ is strongly stable (notation: $T^n \xrightarrow{s} O$) if the power sequence $\{T^n\}_{n\geq 0}$ converges strongly to the null operator (i.e. if $T^n x \to 0$ as $n \to \infty$ for every $x \in \mathcal{H}$). By a contraction we mean an operator T such that $\|T\| \leq 1$. As usual (cf. [11, p.72]), a contraction T is of class $\mathcal{C}_{0\cdot}$ if it is strongly stable, and of class $\mathcal{C}_{\cdot 0}$ if its adjoint T^* is strongly stable. Let $\mathcal{C}_{1\cdot}$ and $\mathcal{C}_{\cdot 1}$ be the classes of all contractions for which $T^n x \not\to 0$ and $T^{*n} x \not\to 0$, respectively, for every nonzero x in $\mathcal{H}$.

If T is a contraction on $\mathcal{H}$, then $\{T^{*n}T^n\}_{n\geq 0}$ is a monotone bounded sequence of self-adjoint operators (in fact a nonincreasing sequence of nonnegative contractions) so that it converges strongly. Since T^* is a contraction whenever T is, the sequence $\{T^n T^{*n}\}_{n\geq 0}$ also converges strongly. Hence, associated with each contraction T on $\mathcal{H}$, there exist operators A and A_* on $\mathcal{H}$ which are the strong limits of $\{T^{*n}T^n\}_{n\geq 0}$ and $\{T^nT^{*n}\}_{n\geq 0}$, respectively. That is,

$$T^{*n}T^n \xrightarrow{\;s\;} A \quad \text{and} \quad T^nT^{*n} \xrightarrow{\;s\;} A_*.$$

A few well-known properties of these strong limits, that will be required in the sequel, are displayed below (see e.g. [5] and [8, Ch.3]).

$$(1) \qquad O \leq A \leq I \quad \text{and} \quad \|A\| = 1 \text{ whenever } A \neq O,$$

$$(2) \qquad T^*AT = A \quad (\text{i.e. } \|A^{\frac{1}{2}}Tx\| = \|A^{\frac{1}{2}}x\| \text{ for every } x \in \mathcal{H}),$$

$$(3) \qquad \mathcal{N}(A) = \{x \in \mathcal{H}:\; T^nx \to 0 \text{ as } n \to \infty\},$$

$$(4) \qquad \mathcal{N}(I - A) = \{x \in \mathcal{H}:\; \|T^nx\| = \|x\| \text{ for all } n \geq 0\}.$$

Clearly, replacing T with T^* (and vice versa) we get similar properties for A_*. Property (3) says that $T \in \mathcal{C}_{0\cdot}$ if and only if $A = O$ (i.e. if and only if $\mathcal{N}(A) = \mathcal{H}$), and $T \in \mathcal{C}_{1\cdot}$ if and only if $\mathcal{N}(A) = \{0\}$. Therefore,

$$\begin{aligned}
T \in \mathcal{C}_{00} &\iff A = A_* = O,\\
T \in \mathcal{C}_{01} &\iff A = O \text{ and } \mathcal{N}(A_*) = \{0\},\\
T \in \mathcal{C}_{10} &\iff \mathcal{N}(A) = \{0\} \text{ and } A_* = O,\\
T \in \mathcal{C}_{11} &\iff \mathcal{N}(A) = \mathcal{N}(A_*) = \{0\}.
\end{aligned}$$

Now consider a linear transformation $V_0\colon \mathcal{R}(A^{\frac{1}{2}}) \to \mathcal{R}(A^{\frac{1}{2}})$ defined by $V_0A^{\frac{1}{2}}x = A^{\frac{1}{2}}Tx$ for every $x \in \mathcal{H}$. The definition of V_0 makes sense because A is self-adjoint (property (1)) so that $A|_{\mathcal{R}(A)}\colon \mathcal{R}(A) \to \mathcal{R}(A)$ is injective. Observe that V_0 acts isometrically on $\mathcal{R}(A^{\frac{1}{2}})$. This is ensured by property (2). Indeed, $\|V_0A^{\frac{1}{2}}x\| = \|A^{\frac{1}{2}}Tx\| = \|A^{\frac{1}{2}}x\|$ for every $x \in \mathcal{H}$. Extend V_0 over $\mathcal{R}(A^{\frac{1}{2}})^- = \mathcal{R}(A)^-$ and get the isometry

$$V\colon \mathcal{R}(A)^- \to \mathcal{R}(A)^-.$$

Thus each contraction T on $\mathcal{H}$ also induces an isometry V on $\mathcal{R}(A)^-$ such that (see e.g. [5] and [8, Ch.3]).

$$(5) \qquad A^{\frac{1}{2}}T = VA^{\frac{1}{2}} \quad (\text{i.e. if } A \neq O, \text{ then } A^{\frac{1}{2}} \text{ intertwines } T \text{ to } V).$$

Recall that an operator $T \in \mathcal{B}[\mathcal{H}]$ is intertwined to an operator $L \in \mathcal{B}[\mathcal{K}]$ if there exists a nonzero transformation $X \in \mathcal{B}[\mathcal{H}, \mathcal{K}]$ such that $XT = LX$. In such a case we say that X intertwines T to L. If $\mathcal{R}(X)^- = \mathcal{K}$, then T is said to be densely intertwined to L. A transformation $X \in \mathcal{B}[\mathcal{H}, \mathcal{K}]$ is quasiinvertible if it is injective and has a dense range (i.e. if $\mathcal{N}(X) = \{0\}$ and $\mathcal{R}(X)^- = \mathcal{K}$). $T \in \mathcal{B}[\mathcal{H}]$ is a quasiaffine transform of $L \in \mathcal{B}[\mathcal{K}]$ if there exists a quasiinvertible $X \in \mathcal{B}[\mathcal{H}, \mathcal{K}]$ intertwining T to L.

Suppose T is a $\mathcal{C}_{1\cdot}$-contraction so that $\mathcal{N}(A) = \{0\}$ or, equivalently, $\mathcal{R}(A)^- = \mathcal{H}$ (reason: A is self-adjoint). In this case we can define a new inner product on $\mathcal{H}$, say $\langle\ ;\ \rangle_A : \mathcal{H} \times \mathcal{H} \to \mathbb{C}$, given by

$$\langle x; y\rangle_A \;=\; \langle Ax; y\rangle$$

for every $x, y \in \mathcal{H}$, which induces the norm $\|\ \|_A$ on $\mathcal{H}$ such that

$$0 < \|A^{\frac{1}{2}}x\| = \|x\|_A \leq \|x\|$$

for every nonzero vector x in $\mathcal{H}$. Consider the linear transformation $X_0 : (\mathcal{H}, \langle\ ;\ \rangle_A) \to (\mathcal{R}(A^{\frac{1}{2}}), \langle\ ;\ \rangle)$ defined by $X_0 x = A^{\frac{1}{2}}x$ for every $x \in \mathcal{H}$. Let $\mathcal{H}_A$ be the completion of $(\mathcal{H}, \langle\ ;\ \rangle_A)$ and extend X_0 over the Hilbert space $\mathcal{H}_A$ to get the surjective isometry (i.e. the unitary operator)

$$X_A : \mathcal{H}_A \to \mathcal{H}.$$

In fact, $\mathcal{R}(X_A) = \mathcal{R}(A^{\frac{1}{2}})^- = \mathcal{R}(A)^- = \mathcal{H}$ and $\|X_A x\| = \|A^{\frac{1}{2}}x\| = \|x\|_A$ for every $x \in \mathcal{H}$. Note that T acts isometrically on $(\mathcal{H}, \langle\ ;\ \rangle_A)$. Actually, by property (2),

$$\|Tx\|_A \;=\; \|A^{\frac{1}{2}}Tx\| \;=\; \|A^{\frac{1}{2}}x\| \;=\; \|x\|_A$$

for every $x \in \mathcal{H}$. Extend T over $\mathcal{H}_A$ and get the isometry

$$T_A : \mathcal{H}_A \to \mathcal{H}_A$$

such that $T_A x = Tx$ for each $x \in \mathcal{H}$. The unitary operator $X_A : \mathcal{H}_A \to \mathcal{H}$ intertwines the isometry $T_A : \mathcal{H}_A \to \mathcal{H}_A$ to the isometry $V : \mathcal{H} \to \mathcal{H}$, so that T_A and V are unitarily equivalent. Indeed, property (5) leads to

$$X_A T_A x \;=\; A^{\frac{1}{2}}Tx \;= VA^{\frac{1}{2}}x \;=\; VX_A x$$

for every $x \in \mathcal{H}$. Let U be the minimal unitary extension of the isometry V. In light of the above unitary equivalence between the isometries T_A and V it is usual to refer to U as *the unitary extension of* T.

2. Introduction

Does every operator on a (separable) Hilbert space of dimension greater than one have a nontrivial invariant subspace? This is perhaps the most celebrated open question in operator theory. Its relevance is linked to the Spectral Theorem (normal operators do have nontrivial invariant subspaces) and also to the canonical Jordan form (every operator on a finite-dimensional space has a nontrivial invariant subspace). This makes the search for nontrivial invariant subspaces a natural, although challenging and recalcitrant one. The collection of all invariant subspaces for a given operator T is invariant under scalar multiplication. Therefore, if $T \neq O$ and $0 < \alpha \leq \|T\|^{-1}$, then T and αT (which is a contraction) share exactly the same lattice of invariant subspaces. Thus the invariant subspace problem is reduced to the class of all contractions: does every contraction have a nontrivial invariant subspace?

Isometries comprise a rather special class of $\mathcal{C}_{1\cdot}$-contractions. They are those contractions T on $\mathcal{H}$ for which $\|Tx\| = \|x\|$ for every $x \in \mathcal{H}$. Equivalently, those contractions T for which $A = I$ (see property (4), or simply recall that T is an isometry if and only if $T^{*n}T^n = I$ for all $n \geq 0$). It is easy to show that every isometry has a nontrivial invariant subspace. Actually, the next proposition says much more than this.

PROPOSITION 1. *If a contraction T on $\mathcal{H}$ has no nontrivial invariant subspace, then it is either a $\mathcal{C}_{00}$-contraction, a $\mathcal{C}_{01}$-contraction such that $\|A_* x\| < \|x\|$ for every nonzero x in $\mathcal{H}$, or a $\mathcal{C}_{10}$-contraction such that $\|Ax\| < \|x\|$ for every nonzero x in $\mathcal{H}$.*

PROOF. See [8, Ch.5]. □

Since a unitary operator is precisely a normal isometry, it follows that the unitaries comprise a set of particularly well-known operators. There are, however, some interesting open questions even for this special set of Hilbert-space operators. Sample: is a unitary operator weakly stable if and only if it is a bilateral shift or a direct summand of a bilateral shift? (An operator T is weakly stable if the power sequence $\{T^n\}_{n\geq 0}$ converges weakly to the null operator).

Some invariant subspace problems for contractions are equivalent to open questions on unitary operators. A class of such equivalent problems was considered in [7] and [9]. The purpose of this paper is to present a brief and unified survey on the results of [7] and [9].

3. Equivalent Invariant Subspace Problems

A classical open question in operator theory reads as follows.

(Q_0) *Does a contraction not in $\mathcal{C}_{00}$ have a nontrivial invariant subspace?*

Observe that this question asks whether the conclusion in Proposition 1 can be sharpened to $T \in \mathcal{C}_{00}$. That is, whether a contraction without a nontrivial invariant subspace is of class $\mathcal{C}_{00}$. In other words, whether a contraction without a nontrivial invariant subspace is strongly stable with a strongly stable adjoint. Recall that T has a nontrivial invariant subspace if and only if T^* has. Thus Proposition 1 leads to the following reformulation of question Q_0.

(Q_1) *Does a $\mathcal{C}_{1\cdot}$-contraction have a nontrivial invariant subspace?*

Another classical open question in operator theory is: does a quasiaffine transform of a normal operator have a nontrivial invariant subspace? (See e.g. [10, p.194].) A particular case of it, referring to contractions, reads as follows.

(Q_2) *Does a contraction, which is a quasiaffine transform of a unitary operator, have a nontrivial invariant subspace?*

Since a unitary operator is a normal isometry, and an isometry is a $C_{1\cdot}$-contraction, questions (Q_1) and (Q_2) can be generalized as follows.

(Q_3) *Does a contraction, which is intertwined to a $C_{1\cdot}$-contraction, have a nontrivial invariant subspace?*

It was shown in [7] that these questions are all pairwise equivalent, and hence the invariant subspace problems stated on them are reduced to the open question (Q_2) on unitary operators.

THEOREM 1. *Every $C_{1\cdot}$-contraction has a nontrivial invariant subspace if and only if every contraction which is a quasiaffine transform of a unitary operator has a nontrivial invariant subspace.*

4. Invariant Subspaces for $C_{1\cdot}$-contractions

Every $C_{1\cdot}$-contraction T is a quasiaffine transform of the isometry V, which turns out to be its unitary extension when T has no nontrivial invariant subspace. To prove this, we first state a couple of propositions whose proofs are based on standard results of single operator theory.

PROPOSITION 2. *If T is a quasiaffine transform of a hyponormal operator L, then the spectrum of T includes the spectrum of L.*

PROOF. See [3] — also see [4, p.94]. □

Recall that an operator L is hyponormal if $LL^* \leq L^*L$. Clearly, every isometry is a hyponormal contraction.

PROPOSITION 3. *If an operator is densely intertwined to a nonunitary isometry, then it has a nontrivial invariant subspace.*

PROOF. See e.g. [9] or [12]. □

The next lemma is a deep result whose proof is not elementary at all.

LEMMA 1. *A contraction whose spectrum includes the unit circle has a nontrivial invariant subspace.*

PROOF. See [2] — also see [1]. □

For a $C_{1\cdot}$-contraction, the assumption on its spectrum in Lemma 1 can be replaced by the same assumption on the spectrum of its unitary extension.

THEOREM 2. *Let T be a $C_{1\cdot}$-contraction. The spectrum of T includes the spectrum of the isometry V. Moreover, if the spectrum of the unitary extension of T is the unit circle, then T has a nontrivial invariant subspace.*

PROOF. Let T be a $C_{1\cdot}$-contraction on $\mathcal{H}$. Since $\mathcal{N}(A^{\frac{1}{2}}) = \mathcal{N}(A) = \{0\}$ and

$\mathcal{R}(A^{\frac{1}{2}})^- = \mathcal{R}(A)^- = \mathcal{H}$, it follows that $A^{\frac{1}{2}}$ is a quasiinvertible operator which, according to (5), intertwines T to the isometry V. Hence T is a quasiaffine transform of V. Thus, by Proposition 2, the spectrum of the isometry V is included in the spectrum of T: $\sigma(V) \subseteq \sigma(T)$. From now on suppose T has no nontrivial invariant subspace. In this case V is unitary (Proposition 3) so that V is the unitary extension of T. Moreover, Lemma 1 ensures that the unit circle Γ is not included in $\sigma(T)$. Therefore $\sigma(V) \neq \Gamma$, so that the spectrum of the unitary extension of T is not the whole unit circle. □

Recall: if a unitary operator U is nonreductive, then $\sigma(U) = \Gamma$. This shows that the next result, which is quite important in its own right and whose original proof in [6] precedes Lemma 1, may be obtained as an immediate corollary of Theorem 2.

COROLLARY 1. *If the unitary extension of a $C_{1\cdot}$-contraction T is nonreductive, then T has a nontrivial invariant subspace.*

Let $\mathrm{Lat}(T)$ denote the lattice of all invariant subspaces for an arbitrary operator T on $\mathcal{H}$. The orbit of $x \in \mathcal{H}$ under T is the set $\{T^n x\}_{n\geq 0}$, whose (linear) span is the linear manifold $\mathrm{span}\{T^n x\}_{n\geq 0} = \{p(T)x\colon p \text{ is a polynomial}\}$ of $\mathcal{H}$. Its closure, $\bigvee\{T^n x\}_{n\geq 0}$, clearly lies in $\mathrm{Lat}(T)$. These are the cyclic subspaces in $\mathrm{Lat}(T)$: $\mathcal{M} \in \mathrm{Lat}(T)$ is cyclic if $\mathcal{M} = \bigvee\{T^n x\}_{n\geq 0}$ for some $x \in \mathcal{H}$. If $\bigvee\{T^n x\}_{n\geq 0} = \mathcal{H}$, then x is said to be a cyclic vector for T. We shall say that a linear manifold $\mathcal{R}$ of $\mathcal{H}$ is *totally cyclic* for T if every nonzero vector in $\mathcal{R}$ is cyclic for T. Recall that T has no nontrivial invariant subspace (i.e. $\mathrm{Lat}(T) = \{\{0\}, \mathcal{H}\}$) if and only if $\mathcal{H}$ is totally cyclic for T.

PROPOSITION 4. *Suppose $T \in \mathcal{B}[\mathcal{H}]$ is densely intertwined to $L \in \mathcal{B}[\mathcal{K}]$. Let $X \in \mathcal{B}[\mathcal{H}, \mathcal{K}]$ be a transformation with dense range intertwining T to L. If $x \in \mathcal{H}$ is cyclic for T, then $Xx \in \mathcal{K}$ is cyclic for L. If a linear manifold $\mathcal{R}$ of $\mathcal{H}$ is totally cyclic for T, then $X(\mathcal{R}) \subseteq \mathcal{K}$ is totally cyclic for L.*

PROOF. See [9]. □

Corollary 1 points the investigation to $C_{1\cdot}$-contractions that have a reductive unitary extension, while Theorem 1 brings it back to question (Q_2).

THEOREM 3. *If a $C_{1\cdot}$-contraction has no nontrivial invariant subspace, then it is a quasiaffine transform of its unitary extension, which is reductive and has an invariant dense and totally cyclic linear manifold.*

PROOF. Let T be a $C_{1\cdot}$-contraction on $\mathcal{H}$. Observe by property (5) that $\mathcal{R}(A^{\frac{1}{2}})$ is invariant for V and that T is a quasiaffine transform of the isometry V (for $\mathcal{N}(A^{\frac{1}{2}}) = \mathcal{N}(A) = \{0\}$ and $\mathcal{R}(A^{\frac{1}{2}})^- = \mathcal{R}(A)^- = \mathcal{H}$ — in particular, T is densely intertwined to V). From now on suppose T has no nontrivial invariant subspace. Thus Proposition 3 ensures that V is unitary, and hence it is the unitary extension of T (which is reductive according to Corollary 1). Moreover,

$\mathcal{H}$ is totally cyclic for T so that $\mathcal{R}(A^{\frac{1}{2}})$, which is dense in $\mathcal{H}$, is totally cyclic for V by Proposition 4. □

Acknowledgements. Partially supported by CNPq and FAPERJ.

References

[1] H. Bercovici, Notes on invariant subspaces, *Bull. Amer. Math. Soc.* **23** (1990) 1–36.

[2] S.W. Brown, B. Chevreau and C. Pearcy, On the structure of contractions operators II, *J. Functional Anal.* **76** (1988) 30–55.

[3] W.S. Clary, Equality of spectra of quasi-similar hyponormal operators, *Proc. Amer. Math. Soc.* **53** (1975) 88–90.

[4] J.B. Conway, *The Theory of Subnormal Operators* (Mathematical Surveys and Monographs Vol. 36, Amer. Math. Soc., Providence, 1991).

[5] E. Durszt, Contractions as restricted shifts, *Acta Sci. Math.* (*Szeged*) **48** (1985) 129–134.

[6] L. Kérchy, Invariant subspaces of $C_{1\cdot}$-contractions with non-reductive unitary extensions, *Bull. London Math. Soc.* **19** (1987) 161–166.

[7] C.S. Kubrusly, Equivalent invariant subspace problems, *J. Operator Theory* **37** (1997) 1–6.

[8] C.S. Kubrusly, *An Introduction to Models and Decompositions in Operator Theory* (Birkhäuser, Boston, 1997).

[9] C.S. Kubrusly, Invariant subspaces for a class of $C_{1\cdot}$-contractions, *Adv. Math. Sci. Appl.* **9** (1999) 129–135.

[10] H. Radjavi and P. Rosenthal, *Invariant Subspaces* (Springer, New York, 1973).

[11] B. Sz.-Nagy and C. Foiaş, *Harmonic Analysis of Operators on Hilbert Spaces* (North-Holland, Amsterdam, 1970).

[12] K. Takahashi, The reflexivity of contractions with nonreductive *-residual parts, *Michigan Math. J.* **34** (1987) 153–159.

Carlos J. Kubrusly, Catholic University (PUC-Rio) and CNPq, 22543-900 Rio de Janeiro, BRAZIL

E-mail address: carlos@ele.puc-rio.br

Progress in Nonlinear Differential Equations
and Their Applications, Vol. 42

INTEGRATED C-SEMIGROUPS AND C-COSINE FUNCTIONS OF HERMITIAN AND POSITIVE OPERATORS

YUAN-CHUAN LI
SEN-YEN SHAW
NATIONAL CENTRAL UNIVERSITY, TAIWAN

Abstract. Some recent results on n-times integrated C-semigroups and C-cosine functions of hermitian and positive operators on Banach spaces are discussed. The following are some interesting properties: 1) A hermitian n-times integrated C-semigroup $T(\cdot)$ (resp. C-cosine function $C(\cdot)$) is $(n-1)$-th continuously differentiable in operator norm on $[0,\infty)$ and $T^{(n-1)}(\cdot)$ (resp. $C^{(n-1)}(\cdot)$) is a norm continuous hermitian once integrated C-semigroup (resp. C-cosine function) if $n \geq 2$; 2) A hermitian C-semigroup is infinitely differentiable in operator norm on $(0,\infty)$; 3) A hermitian C-cosine function is norm continuous either at all points or at none point of $[0,\infty)$; 4) A positive C-semigroup (resp. C-cosine function) which dominates C is infinitely differentiable in operator norm on $[0,\infty)$; moreover, if it is nondegenerate, then its generator A must be bounded and $T(t) = \sum_{n=0}^{\infty} \frac{t^n}{n!} A^n C$ (resp. $C(t) = \sum_{n=0}^{\infty} \frac{t^{2n}}{(2n)!} A^n C$); 5) Hermitian n-times integrated semigroups and hermitian n-times integrated cosine functions are exponentially bounded.

1. Introduction

Let X be a (complex) Banach space, and denote by $B(X)$ the space of all bounded linear operators on X. Let $C \in B(X)$, and let $T(\cdot) \equiv \{T(t); t \geq 0\}$ be a strongly continuous family in $B(X)$. For $n \geq 1$, $T(\cdot)$ is called an *n-times integrated C-semigroup* on X ([11], [13]) if it satisfies: $T(t)C = CT(t)$, $T(0) = 0$, and

$$T(s)T(t)x = \frac{1}{(n-1)!}\left(\int_t^{s+t} - \int_0^s\right)(s+t-r)^{n-1}T(r)Cx\,dr.1 \tag{1.1}$$

for $x \in X$, s, $t \geq 0$ (see also [1], [18], [23] for the case $C = I$). $T(\cdot)$ is called a (0-times integrated) *C-semigroup* (see [3], [4], [17], [24]) on X if it satisfies: $T(0) = C$, and

$$T(s)T(t) = T(s+t)C \text{ for } s,\ t \geq 0..2 \tag{1.2}$$

The classical C_0-semigroups are C-semigroups with C equal to the identity operator I.

Let $C(\cdot) \equiv \{C(t); t \geq 0\}$ be a strongly continuous family in $B(X)$. For $n \geq 1$, $C(\cdot)$ is called an *n-times integrated C-cosine function* on X (see [10], [11], [20]) if it satisfies: $C(t)C = CC(t)$, $C(0) = 0$, and

$$\begin{aligned} 2C(t)C(s)x &= \frac{1}{(n-1)!}\Big\{(-1)^n \int_0^{|s-t|} (|s-t|-r)^{n-1}C(r)Cx\,dr \\ &+ \Big[\int_0^{s+t} - \int_0^t - \int_0^s\Big](t+s-r)^{n-1}C(r)Cx\,dr \qquad (1.3) \\ &+ \int_0^t (s-t+r)^{n-1}C(r)Cx\,dr + \int_0^s (t-s+r)^{n-1}C(r)Cx\,dr\Big\} \end{aligned}$$

for all $x \in X$ and $s,t \geq 0$. It is called a (0-times integrated) C-*cosine function* if $C(0) = C$ and

$$2C(t)C(s) = [C(t+s) + C(|t-s|)]C \text{ for } s,t \geq 0..4 \qquad (1.4)$$

See [8], [9], [27], [28] (also [26] for applications to partial differential equations). The classical cosine operator functions ([6], [21]) are C-cosine functions with C equal to the identity operator I.

$T(\cdot)$ (resp. $C(\cdot)$) is said to be *nondegenerate* if $T(t)x = 0$ (resp. $C(t)x = 0$) for all $t > 0$ implies $x = 0$. In order that $T(\cdot)$ (resp. $C(\cdot)$) be nondegenerate it is necessary (and sufficient in case $n = 0$) that C is injective. The *generator* A of a nondegenerate n-times integrated C-semigroup $T(\cdot)$ is the closed operator A defined as:

$$x \in D(A) \text{ and } Ax = y \Leftrightarrow T(t)x - t^nCx/n! = \textstyle\int_0^t T(s)y\,ds \text{ for } t \geq 0.$$

We know that $R(\int_0^t T(s)ds) \subset D(A)$ and

$$\int_0^t T(s)ds A \subset A\int_0^t T(s)ds = T(t) - \frac{t^n}{n!}C \text{ for } t \geq 0..5 \qquad (1.5)$$

The *generator* A of a nondegenerate n-times integrated C-cosine function $C(\cdot)$ is defined as:

$$x \in D(A) \text{ and } Ax = y \Leftrightarrow C(t)x - t^nCx/n! = \textstyle\int_0^t (t-s)C(s)y\,ds \text{ for } t \geq 0.$$

We have $R\Big(\int_0^t (t-s)C(s)ds\Big) \subset D(A)$ and

$$\int_0^t (t-s)C(s)ds A \subset A\int_0^t (t-s)C(s)ds = C(t) - \frac{t^n}{n!}C \text{ for } t \geq 0..6 \qquad (1.6)$$

An operator T is *hermitian* if its (algebra) *numerical range* $V(T) := \{F(T); F \in B(X)', \|F\| = F(I) = 1\}$ is contained in the real line $\mathbb{R}$, or equivalently, if $\|\exp(itT)\| = 1$ for all $t \in \mathbb{R}$. The set of all hermitian (resp. positive) operators is closed with respect to the weak operator topology. If T is hermitian, then $V(T)$ is equal to the closed convex hull of the spectrum $\sigma(T)$, and the norm and spectral radius of T are equal (see [2], §26). An n-times integrated C-semigroup $T(\cdot)$ (or C-cosine function $C(\cdot)$) is said to be *hermitian* if $T(t)$ (or $C(t)$) is hermitian for all $t \geq 0$.

In papers [12], [14], and [15], we investigated hermitian and positive integrated C-semigroups and C-cosine functions. The purpose of this note is: 1) to survey some interesting properties concerning differentiability and exponential boundedness; 2) to present some supplementary results. N-times integrated C-semigroups will be discussed in Section 2 and n-times integrated C-cosine functions will be discussed in Section 3.

2. N-times Integrated C-semigroups

Theorem 2.1. *Let $T(\cdot)$ be a hermitian n-times integrated C-semigroup on a Banach space X.*

(i) *If $n \geq 2$, then $T(\cdot)$ is $(n-1)$-th continuously differentiable in operator norm on $[0, \infty)$ and $T^{(n-1)}(\cdot)$ is a norm continuous hermitian once integrated C-semigroup.*

(ii) *If $n = 1$, then $T(\cdot)$ is norm continuous on $[0, \infty)$ and continuously differentiable in operator norm on $(0, \infty)$, $T'(\cdot)$ is hermitian, locally bounded on $[0, \infty)$, and norm continuous on $(0, \infty)$, and $T'(t + s)C = T'(t)T'(s)$ for all $t, s \geq 0$, where we define $T'(0) = C$. Moreover, if $T(\cdot)$ is nondegenerate and has generator A, then $T(t)x$ is differentiable at 0 for all x in $X_1 := \overline{D(A)}$; and $T'(\cdot)|X_1$ is a hermitian $C|X_1$-semigroup on X_1 which is norm continuous on $(0, \infty)$.*

(iii) *If $n = 0$, then $T(\cdot)$ is infinitely differentiable in operator norm on $(0, \infty)$.*

(i) and (ii) are proved in [12, Theorem 2.3 and Corollary 2.4], and (iii) is proved in [15, Theorem 2.4].

Although a hermitian C-semigroup is infinitely differentiable in operator norm on $(0, \infty)$, it is not necessarily norm continuous at 0.

Example 2.2. The positive C_0-semigroup $T_1(\cdot)$, defined by $T_1(t)x := (e^{-nt}x_n)$ $(x = (x_n) \in \ell_2,\ t \geq 0)$ is such an example because its generator A, defined as $Ax := (-nx_n)$ with natural domain, is unbounded.

Although a hermitian integrated C-semigroup $T(\cdot)$ is norm differentiable on $(0, \infty)$ and $T'(\cdot)$ is norm continuous and satisfies $T'(t)T'(s) = T'(t + s)C$ on $(0, \infty)$, $T(\cdot)$ may not be strongly differentiable at 0 and $T'(t)$ may not be strongly convergent to C as $t \to 0^+$, i.e., $T'(\cdot)$ is not necessarily a C-semigroup. For example, the left multiplications $L_{S(t)}, t \geq 0$, on the space $B(X)$ by the integrals $S(t)$ of the above C_0-semigroup $T_1(\cdot)$ form such a positive integrated semigroup on $B(X)$, because $\|L'_{S(t)} - L_I\| = \|T_1(t) - I\| \not\to 0$ as $t \to 0^+$. From this fact and Theorem 2.1(ii) we infer that the generator of $L_{S(\cdot)}$ is not densely defined in $B(X)$.

The following corollary is deduced from Theorem 2.1 for those $T(\cdot)$ which have generators with dense domain.

Corollary 2.3. *If the generator A of a hermitian nondegenerate n-times integrated C-semigroup $T(\cdot)$ on X is densely defined, then $T(\cdot)$ is infinitely differentiable in operator norm on $(0,\infty)$ and is n-th strongly differentiable at 0, and $T^{(n)}(\cdot)$ is a hermitian C-semigroup with generator A. When $n \geq 1$, $C(\cdot)$ is $(n-1)$-th continuously differentiable in operator norm at 0.*

While $R(C) \subset \overline{D(A)}$ holds for all nondegenerate C-semigroups [13, Corollary 5.7], it is not necessarily true for nondegenerate n-times integrated C-semigroups when $n \geq 1$; there are integrated semigroups with generators not densely defined (cf. [1, Example 6.4]). But, as is shown in [12, p. 634], this inclusion relation is true for hermitian nondegenerate integrated C-semigroups on a reflexive space X. On the other hand, as an injective hermitian operator, C has dense range in X (an argument is to be given in the proof of Corollary 3.3). Therefore, the generator of a hermitian nondegenerate n-times integrated C-semigroup is densely defined if the space X is reflexive, and in this case, the conclusion of Corollary 2.3 holds.

In Corollary 2.3, while $T(\cdot)$ is infinitely differentiable in operator norm on $(0,\infty)$, it is not necessarily n-th differentiable in operator norm at 0 and $T^{(n)}(\cdot)$ is not necessarily norm continuous at 0. The C_0-semigroup $T_1(\cdot)$ in Example 2.2 is an example for the case $n=0$. Note that $T_1(\cdot)$ is a positive C_0-semigroup satisfying $0 \leq T_1(t) \leq I$ for all $t \geq 0$. A sharp contrast between this semigroup and a positive C-semigroup which dominates C is demonstrated by the following theorem from [15, Theorem 2.5].

Theorem 2.4. *If $T(\cdot)$ is a C-semigroup such that $T(t) \geq C \geq 0$, then $T(\cdot)$ is infinitely differentiable in operator norm on $[0,\infty)$ and $T^{(n)}(t)$ is positive for all $t \geq 0$ and $n \geq 1$.*

Example 2.5. Consider the operator function $T(\cdot)$ on ℓ_1, defined by $T(t)x := (\alpha_n e^{nt-n^2} x_n)$ for $x = (x_n) \in \ell_1$ and $t \geq 0$, where $\alpha_n \geq 0$. This is a C-semigroup with $Cx := (\alpha_n e^{-n^2} x_n)$. Clearly, $T(t) \geq C \geq 0$ for all $t \geq 0$. Hence it follows from Theorem 2.4 that $T(\cdot)$ is infinitely differentiable in operator norm on $[0,\infty)$. Since $\|T(2n)\| \geq \|T(2n)e_n\| = \alpha_n e^{n^2}$, $n \geq 1$, $T(\cdot)$ is not exponentially bounded. Clearly, $T(\cdot)$ is nondegenerate if and only if $\alpha_n > 0$ for all $n \geq 1$. In this case, the generator A, defined by $Ax := (nx_n)$, is unbounded.

On some particular kinds of spaces, when a C-semigroup $T(\cdot)$ is nondegenerate and satisfies $T(t) \geq C \geq 0$ for all $t \geq 0$, the exponential boundedness of $T(\cdot)$ is equivalent to the boundedness of its generator.

Theorem 2.6. ([12, Theorem 3.3 and Corollary 3.4]) *Let A be the generator of a C-semigroup $T(\cdot)$ that satisfies $T(t) \geq C \geq 0$ for all $t \geq 0$. Then*

(i) $\|T(t)\| \leq Me^{wt}$ *for all $t \geq 0$ if and only if A satisfies:*
$R(C) \subset D(A^n)$, $A^nC \in B(X)$, $A^nC \geq 0$, *and* $\|A^nC\| \leq Mw^n$ *for $n \geq 0$.*
In this case, we have $T(t) = \sum_{n=0}^{\infty} \frac{t^n}{n!} A^n C$.

(ii) *In each of the cases:* (1) $X = L^p(\mu)\,(1 < p < \infty)$; (2) $X = L^1(S,\mu)$ *or* $C_0(\Omega)$ *(with* Ω *a locally compact space) and* $\overline{R(C)} = X$, $T(\cdot)$ *is exponentially bounded if and only if* $A \in B(X)$, $A \geq 0$, *and* $AC = CA$.

Note that the condition that $T(t) \geq C \geq 0$ for all $t \geq 0$ is essential for (ii) of Theorem 2.6 to hold. The following example justifies this fact.

Example 2.7. The positive C-semigroup $T(\cdot)$, defined by $T(t)x := (e^{-nt-n^2}x_n)$ $(x = (x_n) \in \ell_2,\, t \geq 0)$, satisfies $0 \leq T(t) \leq C$ and hence $\|T(t)\| \leq \|C\|$ for all $t \geq 0$, but its generator is unbounded.

When $C = I$, we even have the following stronger phenomenon. See [12, Theorem 2.3(g)] for its proof.

Theorem 2.8. *Every hermitian n-times integrated semigroup* $T(\cdot)$ *is exponentially bounded, i.e.,* $\|T(t)\| \leq Me^{wt}$ *for some* $M > 0$ *and all* $t \geq 0$*; in case* $n = 0$*, one can take* $M = 1$.

Without the assumption of "hermitian", the exponential boundedness is not guaranteed. For examples of integrated semigroups which are not exponentially bounded, see [7] and [5, p.110].

3. N-times Integrated C-cosine functions

Theorem 3.1. *Let* $C(\cdot)$ *be a hermitian n-times integrated C-cosine function on a Banach space X.*

(i) *If* $n \geq 2$*, then* $C(\cdot)$ *is* $(n-1)$*-th continuously differentiable in operator norm on* $[0,\infty)$ *and* $C^{(n-1)}(\cdot)$ *is a norm continuous hermitian once integrated C-cosine function.*

(ii) *If* $n = 1$*, then* $C(\cdot)$ *is norm continuous on* $[0,\infty)$*. Moreover, if* $C(\cdot)$ *is nondegenerate and has generator A, then* $C(\cdot)x$ *is continuously differentiable on* $[0,\infty)$ *for all x in* $X_1 := \overline{D(A)}$*, and* $C'(\cdot)|X_1$ *is a hermitian* $C|X_1$*-cosine function on* X_1.

(iii) *If* $n = 0$*, then* $C(\cdot)$ *is either norm continuous on* $[0,\infty)$*, or not norm continuous at any point of* $[0,\infty)$.

Proof. (i) and the first part of (ii) are proved in [14, Theorem 2.4]. (iii) is proved in [15, Theorem 3.3]. We now prove the second part of (ii).

Define $A_h(t) := h^{-1}[C(t+h) - C(t)]$ for $t \geq 0$ and $h \neq 0$ with $t + h \geq 0$. Let $m_{C(\cdot)}$ be the state space of the Banach algebra $\mathbf{A}_{C(\cdot)}$ generated by $C(\cdot)$, C and I, and let $\phi \in m'_{C(\cdot)} := \{\phi \in m_{C(\cdot)}; \phi(C(\cdot)) \not\equiv 0\}$. By the argument in the proof of Theorem 2.4 in [14], there is an $\alpha_\phi \in \mathbb{R} \cup i\mathbb{R}$ such that $\phi(C(t)) = \int_0^t \cosh(\alpha_\phi s)ds\phi(C)$. Hence for all $t \geq 0$ and $h \neq 0$ with $t + h \geq 0$, we have

$$\begin{aligned}|\phi(A_h(t))| &= |h^{-1}\phi[C(t+h)-C(t)]| \\ &= |h^{-1}\int_t^{t+h}\cosh(\alpha_\phi s)\phi(C)ds| \\ &\leq \beta_{t+|h|} := \sup\{|\cosh(\alpha_\phi s)\phi(C)|;\phi \in m'_{C(\cdot)},\ 0\leq s\leq t+h\}.\end{aligned}$$

It was shown in the proof of Theorem 2.4 of [14] that $\beta_t < \infty$ for all $t \geq 0$.

Since $A_h(t)$ is hermitian, we must have $\|A_h(t)\| = \sup\{|\phi(A_h(t)|;\ \phi \in m'_{C(\cdot)}\} \leq \beta_{t+|h|}$. This implies the local uniform boundedness of $A_h(\cdot)$ on $[0,\infty)$. Since $C(t)x - tCx = \int_0^t\int_0^s C(r)Axdrds$ for $x \in D(A)$, we have for $x \in D(A)$

$$\lim_{h\to 0} A_h(t)x = C'(t)x = Cx + \int_0^t C(r)Axdr,$$

which is strongly continuous in t on $[0,\infty)$. Then, the local uniform boundedness of $A_h(\cdot)$ on $[0,\infty)$ implies that for each $x \in X_1$ $C'(t)x = \lim_{h\to 0} A_h(t)x$ exists uniformly for t in bounded subsets of $[0,\infty)$ and hence is continuous on $[0,\infty)$. Straightforward

differentiation shows that $C'(\cdot)|_{X_1}$ is a $C|_{X_1}$-cosine function. Since $A_h(t)|_{X_1}$ is hermitian, so is $C'(t)|_{X_1}$. The proof is complete.

As is asserted by Theorem 2.1(ii), a hermitian once integrated C-semigroup is continuously differentiable in operator norm on $(0,\infty)$. But it is not clear whether a hermitian once integrated C-cosine function $C(\cdot)$ is even only strongly differentiable on $(0,\infty)$. When $C(\cdot)$ has a densely defined generator, $C(\cdot)$ is strongly differentiable on $[0,\infty)$ (by (ii) of Theorem 3.1), but $C'(\cdot)$ may be not norm continuous.

Example 3.2. The hermitian once-integrated cosine function $C(\cdot)$, defined by $C(t)x := (\frac{1}{n}\sin(nt)x_n)$ $(x := (x_n) \in \ell_1)$ for $t \geq 0$, is strongly differentiable and $C'(\cdot)$ is a cosine function, because its generator A, defined by $Ax := (-n^2x_n)$, has dense domain. Since A is unbounded, $C'(\cdot)$ is not continuous in operator norm on $[0,\infty)$.

Corollary 3.3. *If the generator A of a hermitian nondegenerate n-times integrated C-cosine function $C(\cdot)$ on X is densely defined, then $C(\cdot)$ is $(n-1)$-th continuously differentiable in norm on $[0,\infty)$ in case $n \geq 2$, n-th strongly differentiable on $[0,\infty)$ in case $n \geq 1$, and $T^{(n)}(\cdot)$ is a hermitian C-cosine function with generator A. This holds in particular when the space is reflexive.*

Proof. The conclusion except the last sentence follows from Theorem 3.1. To prove the last sentence we first show that the inclusion $R(C) \subset \overline{D(A)}$ holds for all hermitian nondegenerate integrated C-cosine functions on a reflexive.

Let $A_h(t) = h^{-1}[C(t+h) - C(t)]$ for $t \geq 0$ and $h \neq 0$ with $t+h \geq 0$. We have for an arbitrary $x \in X$ and $t \geq 0$

$$\|A_h(t)x\| \leq \beta_{t+|h|}\|x\| \text{ for } h \neq 0 \text{ with } t+h \geq 0.$$

Let y be an arbitrary weak cluster element of the net $A_h(t)x(h \to 0)$. Then $y = w\text{-}\lim_{n\to\infty} A_{h_n}(t)x$ exists for some sequence $\{h_n\} \to 0$. Then the closeness of A implies that $\int_0^t C(r)xdr \in D(A)$ and

$$\begin{aligned} A_{h_n}(t)x &= h_n^{-1}[C(t+h_n)x - C(t)x] \\ &= h_n^{-1}A\Big\{\int_0^{t+h_n}\int_0^s C(r)xdrds - \int_0^t\int_0^s C(r)xdrds\Big\} + Cx \end{aligned}$$

converges weakly to $A\int_0^t C(r)xdr + Cx$ as $n \to \infty$. This means that the vector $A\int_0^t C(r)xdr + Cx$ is the unique weak cluster point of the net $\{A_h(t)x\}$ $(h \to 0)$. Hence $C(\cdot)$ is weakly differentiable on $[0,\infty)$ and

$$A\int_0^t C(r)xdr = C'(t)x - Cx \text{ for all } x \in X \text{ and } t \geq 0.$$

This implies that $C'(0)x = Cx$ and $\int_0^t C(s)xds \in D(A)$ for all $t \geq 0$ and $x \in X$. Taking weak derivatives twice we obtain that $C'(t)x \in w\text{-cl}(D(A)) = \overline{D(A)}$ for all $t \geq 0$ and $x \in X$. Hence $R(C) = R(C'(0)) \subset \overline{D(A)}$.

Since C is hermitian, $\{\exp(itC); t \in \mathbb{R}\}$ is a unitary group. Therefore, it follows from the mean ergodic theorem for C_0-semigroups on reflexive spaces (see e.g. [10]) that $X = N(C) \oplus \overline{R(C)}$. The non degeneracy of $C(\cdot)$ implies that C is injective, and so we have $X = \overline{R(C)} = \overline{D(A)}$. The proof is complete.

The following theorem (taken from [14, Lemma 3.1] and [15, Theorem 3.4]) shows that positive C-cosine functions possess the best regularity.

Theorem 3.4. *A C-cosine function $C(\cdot)$ satisfies $C(t) \geq 0$ for all $t \geq 0$ if and only if $C(t) \geq C \geq 0$ for all $t \geq 0$. In this case, $C(\cdot)$ is infinitely differentiable in operator norm on $[0,\infty)$ and $C^{(n)}(t)$ is positive for all $t \geq 0$ and $n \geq 1$.*

Example 3.5. The C-cosine function $C(\cdot)$ on ℓ_2, defined by $C(t)x := ((n-1)e^{-n^2}\cosh(nt)x_n)$ for $x = (x_n) \in \ell_2$ and $t \geq 0$, is a degenerate positive C-cosine function with $Cx := ((n-1)e^{-n^2}x_n)$. $C(\cdot)$ is infinitely differentiable in operator norm on $[0,\infty)$, by Theorem 3.4. Since $\|C(2n)\| \geq \|C(2n)e_n\| > \frac{(n-1)}{2}e^{n^2}$ for $n \geq 1$, this C-cosine function is not exponentially bounded.

When $C(\cdot)$ is nondegenerate and positive on some particular spaces, the exponential boundedness of $C(\cdot)$ is characterized by the boundedness of the generator.

Theorem 3.6. *Let A be the generator of a positive C-cosine function $C(\cdot)$. Then*

(i) $\|C(t)\| \leq Me^{wt}$ *for all* $t \geq 0$ *if and only if* A *satisfies:*
$R(C) \subset D(A^n)$, $A^nC \in B(X)$, $A^nC \geq 0$, $\|A^nC\| \leq Mw^{2n}$ *for all* $n \geq 0$.
In this case, we have $C(t) = \sum_{n=0}^{\infty} \frac{t^{2n}}{(2n)!}A^nC$.

(ii) *For either of the cases:* (1) $X = L^p(\mu)\,(1 < p < \infty)$; (2) $X = L^1(S,\mu)$ *or* $C_0(\Omega)$ *(with* Ω *a locally compact space) and* $\overline{R(C)} = X$, $C(\cdot)$ *is exponentially bounded if and only if* $A \in B(X)$, $A \geq 0$, *and* $AC = CA$.

Proof. (i) is proved in [14, Theorem 3.6].

To prove the sufficiency part of (ii), we first note that each of the spaces $C_0(\Omega)$ and $L^p(S,\mu), 1 \leq p \neq 2 < \infty$, has the property that a bounded operator on X is hermitian (resp. positive) if and only if it is a multiplication operator by a bounded, real (resp. positive) valued function (see [16], [22], and [25]). Thus, on

these spaces, a product of hermitian (resp. positive) operators is still hermitian (resp, positive). Note also that the product of two commuting positive operators on a Hilbert space is still positive. Then the sufficiency part of (ii) follows from the sufficiency part of (i).

To show the necessity part of (ii), we first note that, as multiplication operators, positive operators on spaces $X = C_0(\Omega)$, $L^p(\mu), 1 \leq p < \infty, p \neq 2$, have the property:

$$(*) \qquad 0 \leq A \leq B \text{ implies } \|Ax\| \leq \|Bx\| \text{ for all } x \in X.$$

Commuting operators on a Hilbert space H also share this property (see [12, Corollary 3.4]).

From the proof of Theorem 3.5 of [14] we can see that $0 \leq AC \leq w^2C$, so that by (*) we have $\|ACx\| \leq w^2\|Cx\|$ for all $x \in X$. Since, as seen in the proof of Corollary 3.3, injective hermitian operators on reflexive spaces have dense ranges, we have $\|A\| \leq w^2$ for the case $X = L^p(\mu)(1 < p < \infty)$. This also holds for$L^1(\mu)$ and $C_0(\Omega)$ under the assumption $\overline{R(C)} = X$.

Next, we show that $A \geq 0$. Since both C and $CA(= C(C^{-1}AC) = AC)$ are positive operators, for the case $p = 2$, $C^{1/2}$ is an injective positive operator with dense range, and $\langle AC^{1/2}x, C^{1/2}x\rangle = \langle ACx, x\rangle \geq 0$ for all $x \in L^2$. Since A is bounded, this implies that A is positive on L^2. For other cases, there are positive functions h_1 and h_2 such that $Cf = h_1 f$ and $CAf = h_2 f$ for all $f \in X$. The injectivity of C implies that $h_1(s) > 0$ for all $s \in \Omega$ in case $X = C_0(\Omega)$ (resp. a.e. $[\mu]$ in case $X = L^p(\mu)$). It follows that $Af = h_1^{-1}h_2 f$ for all $f \in X$, and so A is positive. This proves the necessity part of (ii).

Example 3.7. The positive C-cosine function $C(\cdot)$ on ℓ_1, defined by $C(t)x := (e^{-n^2}\cosh(nt)x_n)$ for $x = (x_n) \in \ell_1$ and $t \geq 0$, is a nondegenerate C-cosine function with $Cx := (e^{-n^2}x_n)$. Clearly, the generator A, defined by $Ax := (n^2x_n)$ with natural domain, is unbounded. Hence, by (ii) of Theorem 3.6, this C-cosine function is not exponentially bounded. In fact, $\|C(2n)\| \geq \|C(2n)e_n\| > e^{n^2}/2$ for all $n \geq 1$.

That the positivity is essential for (ii) of Theorem 3.6 to hold is illustrated by the next example.

Example 3.8. The C-cosine function $C(\cdot)$, defined by

$$C(t)x := (e^{-n^2}\cos(nt)x_n)\ (x = (x_n) \in \ell_1,\ t \geq 0),$$

is hermitian, non-positive, and contractive;

$$\|C(t)x\|_1 = \sum_{n=1}^{\infty} e^{-n^2}|\cos(nt)||x_n| \leq \|x\|_1$$

for all $x \in \ell_1$ and $t \geq 0$. But its generator A, defined by $Ax := (-n^2 x_n)$, is unbounded.

When $C = I$, we even have the following stronger result about exponential boundedness. See [14, Theorem 2.4(c)] for its proof.

Theorem 3.9. *Every hermitian n-times integrated cosine function $C(\cdot)$ is exponentially bounded, i.e., $\|C(t)\| \leq Me^{wt}$ for some $M > 0$ and all $t \geq 0$; in case $n = 0$, one can take $M = 1$.*

Acknowledgement. Research supported in part by the National Science Council of Taiwan.

References

[1] W. Arendt, Vector valued Laplace transforms and Cauchy problems, *Israel J. Math.* **59** (1987), 327-352.

[2] F. F. Bonsall and J. Duncan, Numerical Ranges II, London Math. Soc. Lecture Series No. 10, Cambridge, 1973.

[3] E. B. Davies and M. M. Pang, The Cauchy problem and a generalization of the Hille-Yosida theorem, *Proc. London Math. Soc.* **55** (1987), 181-208.

[4] R. deLaubenfels, C-semigroups and the Cauchy problem, *J. Funct. Anal.* **111** (1993), 44-61.

[5] R. deLaubenfels, Existence Families, Functional Calculi and Evolution Equations, Lect. Notes Math., Vol. 1570, Springer, 1994.

[6] H. O. Fattorini, Ordinary differential equations in linear topological spaces. I, *J. Differential Equations* **5** (1968), 72-105.

[7] H. Kellermann and M. Hieber, Integrated semigroups, *J. Funct. Anal.* **84** (1989), 321-349.

[8] C.-C. Kuo and S.-Y. Shaw, C-cosine functions and the abstract Cauchy problem, I, *J. Math. Anal. Appl.* **210** (1997), 632-646.

[9] C.-C. Kuo and S.-Y. Shaw, C-cosine functions and the abstract Cauchy problem, II, *J. Math. Anal. Appl.* **210** (1997), 647-666.

[10] Y.-C. Li, Integrated C-semigroups and C-cosine Functions of Operators on Locally Convex Spaces, Ph.D. dissertation, National Central University, 1991.

[11] Y.-C. Li and S.-Y. Shaw, On generators of integrated C-semigroups and C-cosine functions, *Semigroup Forum* **47** (1993), 29-35.

[12] Y.-C. Li and S.-Y. Shaw, Hermitian and positive C-semigroups on Banach spaces, *Publ. RIMS Kyoto Univ.* **31** (1995), 625-644.

[13] Y.-C. Li and S.-Y. Shaw, N-times integrated C-semigroups and the abstract Cauchy problem, *Taiwanese J. Math.* **1** (1997), 75-102.

[14] Y.-C. Li and S.-Y. Shaw, Hermitian and positive integrated C-cosine functions on Banach spaces, *Positivity* **2** (1998), 281-299.

[15] Y.-C. Li and S.-Y. Shaw, Infinite differentiability of hermitian and positive C-semigroups and C-cosine functions, *Publ. RIMS Kyoto Univ.* **34** (1998), 579-590.

[16] G. Lumer, Isometries of reflexive Orlicz spaces, *Ann. Inst. Fourier (Grenoble)* **13** (1963), 99-109.

[17] I. Miyadera, On the generators of exponentially bounded C-semigroups, *Proc, Japan Acad, Ser. A Math. Sci.* **62** (1986), 239-242.

[18] F. Neubrander, Integrated semigroups and their application to the abstract Cauchy problem, *Pacific J. Math.* **135** (1988), 111-155.

[19] S.-Y. Shaw, Mean ergodic theorems and linear functional equations, *J. Funct. Anal.* **87** (1989), 428-441.

[20] S.-Y. Shaw and Y.-C. Li, On n-times integrated C-cosine functions, in Evolution Equations, Marcel Dekkar, 1995, pp. 393-406.

[21] M. Sova, Cosine operator functions, *Rozprawy Math.* **49** (1966), 1-47.

[22] K. W. Tam, Isometries of certain function spaces, *Pacific J. Math.* **31** (1969), 233-246.

[23] N. Tanaka and I. Miyadera, Exponentially bounded C-semigroups and integrated semigroups, *Tokyo J. Math.* **12** (1989), 99-115.

[24] N. Tanaka and I. Miyadera, C-semigroups and the abstract Cauchy problem, *J. Math. Anal. Appl.* **170** (1992), 196-206.

[25] E.Torrance, Adjoints of Operators on Banach Spaces, Ph.D. thesis, Illinois, 1968.

[26] Q. Zheng, Abstract Differential Operators and Evolution Equations, Doctoral dissertation, Tübingen University, 1997.

[27] Q. Zheng and Y. Lei, Exponentially bounded C-cosine functions of operators, *J. Systems Sci. Math, Sci.* **16** (1996), 242-252. (in Chinese).

[28] Q. Zheng and L. Liu, Almost periodic regularized group, semigroups, and cosine functions, *J. Math. Anal. Appl.* **197** (1996), 90-112.

YUAN-CHUAN LI, SEN-YEN SHAW, DEPARTMENT OF MATHEMATICS, NATIONAL CENTRAL UNIVERSITY, CHUNG-LI, TAIWAN

E-mail address: `shaw@math.ncu.edu.tw`

Progress in Nonlinear Differential Equations
and Their Applications, Vol. 42

ON L^p-THEORY OF C_0-SEMIGROUPS GENERATED BY ELLIPTIC SECOND ORDER DIFFERENTIAL EXPRESSIONS

VITALI LISKEVICH
UNIVERSITY OF BRISTOL

Abstract. We study well-posedness in L^p of the Cauchy problem for second order parabolic equations with time-independent measurable coefficients by means of constructing corresponding C_0-semigroups. Lower order terms are considered as form-bounded perturbations of the generator of the symmetric Markov semigroup associated with the Dirichlet form. It is shown that the C_0-semigroup corresponding to the Cauchy problem exists in a certain interval in the scale of L^p-spaces which depends only on form-bounds of perturbations. Examples show that this interval cannot be extended. A new method is developed for studying the L^p-independence of the spectrum. The L^p-independence of the spectrum is proved for non-symmetric generators of semigroups, without assumptions of Gaussian-type estimates.

This note is a short survey of the results on L^p-properties of C_0-semigroups generated by second order elliptic operators with measurable coefficients. This theory is being extensively developed nowadays mainly because of applications to partial differential equations. We start with some properties of symmetric Markov semigroups and their generators for a variety of examples of which we refer to [2]. In particular, selfadjoint second order elliptic operators in divergence form are Markov generators. Then we develop an abstract perturbation theory of such operators, one of possible application of which is perturbation by zero order terms (potentials). The last (and the main) section deals with second order operators containing first and zero order terms. We construct semigroups on L^p-spaces and derive some of their important properties.

Markov semigroups

Let $(M, \mathcal{M}, \mu)$ be a measure space with σ-finite measure μ and A be a non-negative selfadjoint operator in L^2. Let us consider a C_0-semigroup of contractions $(e^{-tA}, t \geq 0)$ acting on $L^2(M, \mu) \equiv L^2$. It is said to be *Markov semigroup* and its generator A is called *symmetric Markov generator* if for $t \geq 0$ the following two conditions are satisfied:

$$\begin{aligned} e^{-tA}[L^2_+] \quad &\subset \quad L^2_+, \\ (f \in L^2,\ |f| \leq 1) \quad &\Longrightarrow \quad (|e^{-tA} f| \leq 1 \ \ \mu-\text{a.e.}) \end{aligned}$$

It is a standard fact that this semigroup defines a family of C_0-semigroups of contractions in L^p, $\ p \in [1, \infty)$: $e^{-tA_p} := (e^{-tA} \upharpoonright [L^2 \cap L^p])^{\sim}_{L^p \to L^p}$.

Since e^{-tA_p} is a contraction the operator A_p is accretive with respect to the semi-inner product $[u,v] = \langle u, |v|^{p-1} sgn\ v\rangle \|v\|_p^{2-p}$ in L^p:

$Re\ [A_p u, u] \geq 0$ for all $u \in \mathcal{D}(A_p)$ and $[A_p u, u] \geq 0$ for all $u \in \mathcal{D}(A_p) \cap Re\ L^p$.

The following theorem makes this property more precise for symmetric Markov generators.

Theorem 1. *Let $f \in \mathcal{D}(A_p)$ for some $p \in (1,\infty)$. Then $f_p := f|f|^{\frac{p-2}{2}} \in \mathcal{D}(A^{1/2})$, and*

$$\begin{aligned} 4\frac{p-1}{p^2}\|A^{1/2} f_p\|_2^2 &\leq Re\ \langle A_p f, |f|^{p-1} sgn\ f\rangle \leq \varkappa(p)\|A^{1/2} f_p\|_2^2, \\ \varkappa(p) &= \sup\{(1+s^{\frac{1}{p}})(1+s^{\frac{1}{p'}})(1+s^{\frac{1}{2}})^{-2};\ s \in (0,1)\}, \\ |Im\ \langle A_p f, |f|^{p-1} sgn\ f\rangle| &\leq \frac{|p-2|}{2\sqrt{p-1}} Re\ \langle A_p f, |f|^{p-1} sgn\ f\rangle. \end{aligned}$$

$-A_p$ generates a bounded holomorphic semigroup of angle $\theta = \arccos\left|1 - \frac{2}{p}\right|$.

For the proof we refer to [10], [7] (see also [9]). The sector of holomorphy in Theorem 1 is the best possible (see [16]).

Perturbations

Let A and B be symmetric Markov generators. If $\mathcal{D}(A^{1/2}) \cap \mathcal{D}(B^{1/2})$ is dense in L^2 one can define the form-sum $A \dot{+} B$ which is the non-negative self-adjoint operator in L^2 associated with the sesquilinear form $t_{A+B}[u,v] = t_A[u,v] + t_B[u,v]$, $u, v \in \mathcal{D}(A^{1/2}) \cap \mathcal{D}(B^{1/2})$. It is not difficult to show that $A \dot{+} B$ is a symmetric Markov generator and $(A \dot{+} B)_p \supset A_p + B_p$, $\forall p \in (1,\infty)$.

We say that B is form bounded with respect to A with bound $\beta > 0$ if

$$\mathcal{D}(B^{1/2}) \supset \mathcal{D}(A^{1/2}) \text{ and } B \leq \beta A + c(\beta) \text{ in the sense of quadratic forms}$$

for some $c(\beta) \geq 0$. The next theorem allows one to construct an extension of the difference $A_p - B_p$ in a certain interval of the L^p-scale. This is an L^p-version of the well-known KLMN-theorem. We refer to [10] for the proof of this and related results.

Theorem 2. *Let A and B be symmetric Markov generators. Suppose that $B \in PK_\beta(A)$ for some $\beta \in (0,1)$. Let $t(\beta) = t_1$ and $t'(\beta) = t_2$ be the corresponding roots of the equation $\beta\varkappa(q) = 4\frac{q-1}{q^2}$, $1 < q < \infty$. Then for every $p \in [t(\beta), t'(\beta)]$ and $t \geq 0$*

$$(i) \qquad \|e^{-t(A \dot{-} B)} f\|_p \leq e^{\varkappa(p)c(\beta)t}\|f\|_p, \quad f \in L^2 \cap L^p$$

and consequently this defines a family of operators $\{(A \dot{-} B)_p\}_{t(\beta)\leq p \leq t'(\beta)}$ which generate quasi-contraction C_0-semigroups.

$$(ii) \qquad (A \dot{-} B)_p \supset A_p - B_p$$

(iii)
$\forall p \in (t_1, t_2)$ *the semigroup* $e^{-t(A\dot{-}B)_p}$ *is holomorphic with respect to* t *of angle*

$$\theta_1 = \frac{\pi}{2} - \arctan \frac{|p-2|}{2\sqrt{p-1}} \frac{4\frac{p-1}{p^2} + \beta\varkappa(p)}{4\frac{p-1}{p^2} - \beta\varkappa(p)},$$

(iv) $\forall p \in (t_1, t_2)\ \forall f \in \mathcal{D}((A\dot{-}B)_p)\ :\ f_p := f|f|^{\frac{p-2}{2}} \in Q(A)$ *and*

$$\begin{aligned} \mathit{Re}\ \langle (A\dot{-}B)_p f, \mathit{sgn}\ f \cdot |f|^{p-1}\rangle &\geq 4\frac{p-1}{p^2}\|A^{\frac{1}{2}} f_p\|_2^2 - \varkappa(p)\|B^{\frac{1}{2}} f_p\|_2^2 \\ &\geq (4\frac{p-1}{p^2} - \beta\varkappa(p))\|A^{\frac{1}{2}} f_p\|_2^2 - \varkappa(p)c(\beta)\|f\|_p^p. \end{aligned}$$

The main tool in proving Theorem 2 is the first inequality in the statement of Theorem 1. Using the full power of Theorem 1 one can treat non-symmetric operators of the type $A - B \pm iC$ with A, B, C being symmetric Markov generators, and B, C form-bounded with respect to A (see [10] for details). The corresponding quadratic form is sectorial. In case of perturbations by a potential one can avoid this restriction. For perturbation theory by a complex singular potential in a much more general context see [6].

Uniqueness

So far we have considered abstract generators of Markov semigroups. In this section we treat the uniqueness problem for the Markov generator perturbed by a potential. By uniqueness we mean here uniqueness of the operator extension of $A+V \restriction \mathcal{E}$ in L^p which is a generator of a C_0-semigroup. Here $\mathcal{E}$ is some "nice" domain which is a common core for A and V in L^p. In general Markov generator A may not possess a core consisting of functions with compact support. The natural domain of uniqueness to be treated is $\mathcal{E} = (A+1)^{-1}[L^1 \cap L^\infty] \subset L^\infty$. So in the following abstract result the condition $V \in L^p$ is a necessary one.

Theorem 3. *Let* $V \in PK_\beta(A) \cap L^{k_p(\beta)} \cap L^p$ *for some* $\beta \in (0,1)$, $k_p(\beta) = \frac{1+\sqrt{1-\beta}}{1+\sqrt{1-\beta}-2/p'}$ *and* $p \in (t(\beta), t'(\beta))$. *Then* $(A_p - V_p \restriction \mathcal{E})^{\sim}_{L^p \to L^p} = (A\dot{-}V)_p$.

See [10] for the proof. The main difference between Theorem 3 and traditional known results of this sort is that the negative part of the potential V is not assumed to be strongly relatively bounded with respect to the unperturbed generator A in any L^p, but only form-bounded. The next theorem complements Theorem 3 in case of form-bounded potential with zero form bound.

Theorem 4. *Let* A *be a symmetric Markov generator satisfying the Sobolev type inequality* $\|f\|_{2j}^2 \leq c_j\|A^{1/2} f\|_2^2$, $j > 1$ *and* $e^{-tA_1^*}1 = 1$. *Let* $p > 1$ *and* $1/p + 1/j \geq 1$. *Suppose that* $0 \leq V \in L^p$, $V \in \cap_{\beta>0} PK_\beta(A)$, $V \ln_+ V \in PK_\beta(A) \cap L^p$ *and* $\beta < 4(j-1)$. *Then* $(A_p - V_p \restriction \mathcal{E})^{\sim}_{L^p \to L^p} = (A\dot{-}V)_p$.

Second order elliptic operators in L^p

In this section we come to our main topic, L^p-theory of generators of C_0-semigroups associated with the second order elliptic differential expression

$$\hat{T} = -\nabla \cdot a \cdot \nabla + b \cdot \nabla + V := -\sum_{j,k=1}^{d} \partial_j a_{jk} \partial_k + \sum_{j=1}^{d} b_j \partial_j + V.$$

Let $a : \Omega \longrightarrow \mathbb{R}^d \otimes \mathbb{R}^d$ be a matrix-valued measurable function, $b : \Omega \longrightarrow \mathbb{R}^d$ be a vector-valued measurable function, $V : \Omega \longrightarrow \mathbb{R}$ be a measurable function. We assume that the following conditions on the coefficients are satisfied:

A1. $a_{jk} \in L^1_{loc}(\Omega)$ and $a_{jk} = a_{kj}$ for $1 \leqslant j, k \leqslant d$.

A2. a is strictly elliptic, i.e. $a(x) \geqslant \sigma I$ in matrix sense, for some $\sigma > 0$ and almost all $x \in \Omega$.

Let $\mathbf{a}$ be the following sesquilinear form in L^2:

$$\mathbf{a}[u, v] := \langle \nabla u, a \cdot \overline{\nabla v} \rangle := \int_\Omega \sum_{j,k=1}^{d} a_{jk}(x) \frac{\partial u(x)}{\partial x_k} \frac{\overline{\partial v(x)}}{\partial x_j} dx, \; D(\mathbf{a}) := C_c^\infty(\Omega).$$

It is well-known that the form $\mathbf{a}$ is closable in L^2 and, without additional assumptions on Ω and a_{jk}, may have infinitely many closed extensions. We confine ourselves to ones which are *Dirichlet forms*. In particular, we define the following extensions of the form $\mathbf{a}$:

$$\tau_D := \bar{\mathbf{a}}, \quad \text{the closure of } \mathbf{a},$$
$$\tau_N[u, v] := \langle \nabla u, a \cdot \nabla v \rangle, \; D(\tau_N) := \{u \in W^{1,2}(\Omega); \langle \nabla u \cdot a \cdot \overline{\nabla u} \rangle < \infty\}.$$

These forms are Dirichlet forms (cf. [2]), so the associated selfadjoint operators A_D, A_N, respectively, generate Markov semigroups. Below we denote by τ one of the above forms and by A the associated operator.

In order to formulate our assumptions on the first order term we need to introduce the function $b \cdot a^{-1} \cdot b := \sum_{j,k=1}^{d} b_j b_k a^{jk}$ where a^{jk} are the entries of the inverse matrix a^{-1}.

B. $b \cdot a^{-1} \cdot b \in L^1_{loc}(\Omega)$ and there exist $\beta, c(\beta) \geqslant 0$ such that $b \cdot a^{-1} \cdot b \leqslant \beta A + c(\beta)$ in the form sense.

V. $V = V^+ - V^-$, $V^\pm \geqslant 0$, V^+ *admissible* with respect to e^{-tA} (i.e. $D(\tau) \cap Q(V^+)$ is dense in L^2;
$V^- \leqslant \gamma A + c(\gamma)$ in the form sense for some $\gamma, c(\gamma) \geqslant 0$.

FS. $\sqrt{\beta} + \gamma < 1$.

The last condition ensures form smallness of the lower order terms with respect to the main part of the operator. This in turn guarantees that the form

$$\mathbf{t}[u,v] := \tau[u,v] + \langle b\cdot\nabla u, v\rangle + \langle Vu, v\rangle,\quad D(\mathbf{t}) = D(\tau)\cap Q(V^+)$$

is a densely defined closed sectorial form in L^2. By the representation theorem (cf. [4], Ch. VI, Thm. 2.1) $\mathbf{t}$ is associated with an m-sectorial operator T which generates a holomorphic semigroup e^{-tT} on L^2. The following theorem was proved in [5].

Theorem 5. *Let the conditions A1,A2, B, V, FS be satisfied. Then for* $p \in [p_-, p_+] := [\frac{4}{2-\sqrt{\beta}+\sqrt{(2-\sqrt{\beta})^2-4\gamma}}, \frac{4}{2-\sqrt{\beta}-\sqrt{(2-\sqrt{\beta})^2-4\gamma}}]$ *(and* $p \in [\frac{2}{2-\sqrt{\beta}}, \infty)$ *in the case* $\gamma = 0$*, i.e.* V^- *bounded)*

$$\|e^{-tT}f\|_p \le e^{(\frac{c(\beta)}{2(p-1)}+c(\gamma))t}\|f\|_p \quad \forall f\in L^2\cap L^p$$

and $e^{-tT}\restriction_{L^\infty_c}$ *extends to a* C_0*-semigroup* e^{-tT_p} *on* L^p*. It is holomorphic, positivity preserving, and also* L^∞*-contractive if* $V^- = 0$*.*

The proof is based on accretivity estimates and some properties of Dirichlet forms. For the Schrödinger operators this result was obtained in [1].

Remarks. *1. For* $V = 0$ *one can also treat the case* $\beta \in [1, 4)$*. Of course in this case the operator* T *cannot be constructed by means of the quadratic forms method. However, the semigroup can be constructed as a limit of semigroups with bounded drifts. See [5] for the details.*
2. In [5] one can find also an application of Theorem 5 to non-divergence type operators.

Recently it was revealed that for the case of bounded a_{jk} the interval $[p_-, p_+]$ can be extended. The following theorem holds (joint result with Z.Sobol and H.Vogt, the paper is in preparation).

Theorem 6. *In addition to the conditions of Theorem 5 suppose that* $d \ge 3$ *and* $a_{jk} \in L^\infty$*. Then for* $p \in \left(\frac{p_- d}{2p_-+d-2}, \frac{p_+ d}{d-2}\right)$ *the semigroup* $e^{-tT}\restriction_{L^\infty_c}$ *extends to a* C_0*-semigroup* e^{-tT_p} *on* L^p*. It is holomorphic and positivity preserving.*

Using operator $-\Delta + c_1\frac{x}{|x|^2}\cdot\nabla - c_2\frac{1}{|x|^2}$ one can show that the interval obtained in Theorem 6 is the maximal interval of well-posedness of the Cauchy problem in L^p.

Next we discuss the problem of L^p-independence of the spectrum for T_p. The problem of L^p-independence of the spectrum for elliptic operators goes back to [15] where the question was posed for Schrödinger operators. The main breakthrough was made in [3] where the question was answered in the affirmative for the case that the negative part of the potential is from the Kato class. In [13] L^p-independence was established for the Schrödinger operator with form bounded negative part of the potential. In this case the operator exists only in L^p for p from a certain interval around $p = 2$. The ideas from [13] were put in a more general context in [12]. Further progress was made in [14]. Below we follow [11]. To formulate our result on L^p independence of the spectrum we need the following notion (due to Yu. Semenov [14]). We call a function $\psi : \mathbb{Z}^d \longrightarrow \mathbb{R}^d$ *L^1-regular*, if

$$\sup_{k\in\mathbb{Z}^d} \sum_{j\in\mathbb{Z}^d} e^{-|\psi(k)-\psi(j)|} =: M_\psi < \infty.$$

A function $\psi : \mathbb{R}^d \longrightarrow \mathbb{R}^d$ is called L^1-regular, if it is Lipschitz continuous and $\psi\restriction_{\mathbb{Z}^d}$ is L^1-regular.

Theorem 7. *Let A be one of the operators A_D, A_i, A_N defined above. In the case $A = A_N$ assume that Ω satisfies the cone property. Let $\psi : \mathbb{R}^d \longrightarrow \mathbb{R}^d$ be L^1-regular. Assume that there exist constants c_0, c_1 such that the inequality*

$$\nabla(\xi\psi)\cdot a\cdot\nabla(\xi\psi) \leqslant c_0(A\dot{+}V^+) + c_1 \tag{1}$$

is fulfilled in the form sense for all $\xi\in\mathbb{R}^d$, $|\xi|\leqslant 1$. Then $\sigma(T_p) = \sigma(T)$ for all $p\in[p_-,p_+]$ (and for $p\in[\frac{2}{2-\sqrt{\beta}},\infty)$ in the case $\gamma = 0$).

Remarks. *1. For the uniformly elliptic case (a_{jk} bounded), condition (1) is obviously satisfied with $\psi(x) = x$. The a_{jk} are also allowed to grow to infinity as V^+ does. Since ψ can be constant on compact sets, the a_{jk} may have local L^1-singularities on these sets.*
2. The result of Theorem 7 carries over to the case of divergence form elliptic operators containing singular magnetic vector potentials and an additional imaginary scalar potential (see [6] for the corresponding constructions).

The proof of Theorem 7 is based upon the following abstract result which is of independent interest.

Theorem 8. *Given $1\leqslant p<q<\infty$ let T_p and T_q be generators of consistent C_0-semigroups ($e^{-T_p}\restriction_{L^p\cap L^q} = e^{-T_q}\restriction_{L^p\cap L^q}$). If there exist $\varepsilon>0$, $C<\infty$, $\lambda_0\in\rho(T_p)\cap\rho(T_q)$ and a L^1-regular function $\psi : \mathbb{R}^d \longrightarrow \mathbb{R}^d$ such that*

$$\|e^{\xi\psi}(\lambda_0 - T_p)^{-1}e^{-\xi\psi}\restriction_{L_c^\infty}\|_{p\to q} \leqslant C \text{ for all } \xi\in\mathbb{R}^d,\ |\xi|\leqslant\varepsilon,$$

then $\sigma(T_p) = \sigma(T_q)$, so the spectra are p-independent. ($\|A\|_{p\to q}$ means the norm of the operator A as a mapping from L^p to L^q.)

The main tool needed for the proof of Theorem 8 is the following result

Proposition 9. *Let $1\leqslant p\leqslant r\leqslant s\leqslant q\leqslant\infty$, $\varepsilon>0$ and $\psi : \mathbb{R}^d \longrightarrow \mathbb{R}^d$ be L^1-regular. For a linear operator $A : L_c^\infty(\Omega) \longrightarrow L^1_{loc}(\Omega)$ define*

$$\|A\|_{p\to q,\varepsilon} := \sup\{\|e^{\xi\psi}Ae^{-\xi\psi}\|_{p\to q}; \xi\in\mathbb{R}^d, |\xi|\leqslant\varepsilon\} \quad (\in[0,\infty]).$$

(i) *There exists* $C = C(\varepsilon, \psi) < \infty$ *such that*

$$\|A\|_{r\to s} \leqslant C \cdot \|A\|_{p\to q,\varepsilon}.$$

(ii) *For* $\varepsilon_0 < \varepsilon$ *we have* $\|A\|_{r\to s,\varepsilon_0} \leqslant C(\varepsilon - \varepsilon_0, \psi) \cdot \|A\|_{p\to q,\varepsilon}$.

(iii) *There exists* $c(\xi) = c(\xi; \varepsilon, \psi)$ *with* $c(\xi) \to 0$ $(\xi \to 0)$ *such that*

$$\|e^{\xi\psi} A e^{-\xi\psi} - A\|_{r\to s} \leqslant c(\xi) \cdot \|A\|_{p\to q,\varepsilon} \quad \text{for all } |\xi| < \varepsilon.$$

Remark. *For* $\psi(x) = x$ *assertions (i) and (ii) are due to G. Schreieck and J. Voigt [13], (iii) is proved in [12]. For the case of* L^1*-regular* ψ *assertion (i) can be found in [14].*

Acknowledgement. The author would like to acknowledge financial support of EPSRC and the Royal Society.

References

[1] A.G. BELYI AND YU.A.SEMENOV, On the L^p-theory of Schrödinger semigroups.II (Russ), *Sibirsk. Math. J.* **31**(1990), 16–26 (English transl. in *Siberian Math.J.* **31** (1991), 540–549).

[2] M. FUKUSHIMA, Y. OSHIMA AND M. TAKEDA, *Dirichlet Forms and Symmetric Markov Processes*, Walter de Gruyter, Berlin-New York, 1994.

[3] R. HEMPEL AND J. VOIGT, The spectrum of a Schrödinger operator on $L_p(\mathbb{R}^\nu)$ is p-independent, *Comm. Math. Phys.* **104** (1986), 243–250.

[4] T. KATO, *Perturbation Theory of Linear Operators*, 2nd edition, Springer-Verlag, Berlin-Heidelberg-New York, 1980.

[5] V. LISKEVICH, On C_0-semigroups generated by elliptic second order differential expressions on L^p-spaces, *Differential Integral Equations* **9** (1996), no.4, 811–826.

[6] V.A.LISKEVICH AND A.MANAVI, Dominated semigroups with singular complex potentials, *Journal of Functional Analysis*, **151** (1997), 281–305.

[7] V.A.LISKEVICH AND M.A.PERELMUTER, Analyticity of submarkovian semigroups, *Proceedings of AMS* **123** (1995), 1097–1104.

[8] V.A.LISKEVICH, M.A.PERELMUTER YU.A. AND SEMENOV, Form-bounded perturbations of generators of submarkovian semigroups, *Acta Applicandae Mathematicae*, **44** (1996), 353–377.

[9] V.A.LISKEVICH AND YU.A.SEMENOV, Some inequalities for submarkovian generators and their applications to the perturbation theory, *Proc. AMS* **119** (1993), 1171–1177.

[10] V.A.LISKEVICH AND YU.A.SEMENOV, Some problems on Markov semigroups, in *Advances in Partial Differential Equations*, Vol. 11, Akademie Verlag Berlin, 1996, 163–217.

[11] V. LISKEVICH AND H. VOGT, On L^p-spectra and essential spectra of second order elliptic operators, *Proceedings of the London Math. Soc.*, to appear.

[12] G. SCHREIECK, *L_p-Eigenschaften der Wärmeleitungshalbgruppe mit singulärer Absorption*, doctoral dissertation, Shaker Verlag, Aachen, 1996.

[13] G. SCHREIECK AND J. VOIGT, Stability of the L_p-spectrum of Schrödinger operators with form small negative part of the potential, in "*Functional Analysis*" Proc. Essen 1991, Bierstedt, Pietsch, Ruess, Vogt eds., Marcel-Dekker, New York, 1994.

[14] Yu. A. SEMENOV, Stability of L^p-spectrum of generalized Schrödinger operators and equivalence of Green's functions, *Internat. Math. Res. Notices* **12** (1997), 574–593.

[15] B. SIMON, Schrödinger semigroups, *Bull. Amer. Math. Soc. (N.S.)* **7** (1982), 447–526.

[16] J. VOIGT, The sector of holomorphy for symmetric submarkovian semigroups, in Dierolf, S. et all (eds.), Functional Analysis, de Gruyter, 1996, pp.449–453.

VITALI LISKEVICH, SCHOOL OF MATHEMATICS, UNIVERSITY OF BRISTOL, BRISTOL BS8 1TW, UNITED KINGDOM
E-mail address: v.liskevich@bristol.ac.uk

Progress in Nonlinear Differential Equations
and Their Applications, Vol. 42

RATES OF LOCAL ERGODIC LIMITS OF N-TIMES INTEGRATED SOLUTION FAMILIES

HSIANG LIU
SEN-YEN SHAW
NATIONAL CENTRAL UNIVERSITY, TAIWAN

Abstract. We consider local ergodic limits of n-times integrated solution families for the linear Volterra equation $u(t) = f(t) + \int_0^t a(t-s)Au(s)ds$. Rates of optimal convergence and non-optimal convergence, and sharpness of non-optimal rate are discussed. Specialization of the result to n-times integrated semigroups and cosine functions are observed.

1. Introduction

Let X be a Banach space and $B(X)$ the space of all bounded linear operators on X. Let A be a (not necessarily densely defined) closed linear operater in X and $a \in L^1_{loc}(\mathbb{R}^+)$ be a non-negative kernel $\not\equiv 0$. Consider the Volterra equation:

$$VE, A, a, f \qquad u(t) = f(t) + \int_0^t a(t-s)Au(s)ds,\ t \geq 0,$$

for $f \in C([0,\infty); X)$.

A family $\{S(t); t \geq 0\}$ in $B(X)$ is called an *n-times integrated solution family* for (VE, A, a, f) (see [2], [13]) if

(S1) $S(\cdot)$ is strongly continuous on $[0,\infty)$ and $S(0)=I$ if $n=0$ and 0 if $n \geq 1$;

(S2) $S(t) \in D(A)$ and $AS(t)x = S(t)Ax$ for all $x \in D(A)$ and $t \geq 0$;

(S3) $a * S(t)x \in D(A)$ and $S(t)x = \frac{t^n}{n!}x + A\int_0^t a(t-s)S(s)xds$ for all $x \in X$ and $t \geq 0$.

A 0-times integrated solution family is also called a *solution family* or *resolvent family* ([9], [12], [14]). Since A is closed, (S2) and (S3) imply

(S4) $S(t) = \frac{t^n}{n!}x + \int_0^t a(t-s)S(s)Axds$ for all $x \in D(A)$ *and* $t \geq 0$.

The notion of an n-times integrated solution family is an extension of the concepts of *n-times integrated semigroups* ([1], [11]) and *n-times integrated cosine functions* [15] (corresponding to the cases $a \equiv 1$ and $a(t) = t$, respectively). The existence of a n-times integrated solution family enables one to find the solution for the equation (VE, A, a, f) (see [13]).

The rate of strong convergence of a resolvent family $S(t)$ as $t \to 0^+$ has been studied in [6]. In this paper, we shall prove similar result for the more general case where $S(t)$ is an n-times integrated solution family. What we are concerned with now is the convergence rates of $Q_0(t)x := \frac{n!}{t^n}S(t)x\,(x \in X)$

and of some related approximation processes $Q_m(t)$, $m \geq 1$, which are certain means of $Q_0(t)$. For the existence of the limits as $t \to 0^+$, the fulfilment of the condition $S(t) = O(t^n)\,(t \to 0^+)$ is required. Thus, throughout the paper we assume that

$$\|S(t)\| \leq Mt^n \text{ for all } t \in [0,1].$$

Just like [6], we shall characterize the rates of *optimal* and *non-optimal* convergence, and verify the sharpness of non-optimal convergence rates. We first prepare with some preliminaries in Section 2, then we discuss optimal convergence in Section 3 and non-optimal rates and their sharpness in Section 4. Finally, in Section 5 we discuss specializations to integrated semigroups and integrated cosine functions.

2. Preliminaries

As usual the rates of covergence will be characterized by means of *K-functional* and *relative completion*, which we recall now.

Let X be a Banach space with norm $\|\cdot\|_X$, and Y a submanifold with seminorm $\|\cdot\|_Y$. The *K-functional* is defined by

$$K(t,x) := K(t,x,X,Y,\|\cdot\|_Y) = \inf_{y\in Y}\{\|x-y\|_X + t\|y\|_Y\}.$$

If Y is also a Banach space with norm $\|\cdot\|_Y$, then the *completion of* Y *relative to* X is defined as

$$\tilde{Y}^X := \{x \in X : \exists\{x_m\} \subset Y \text{ s.t. } \lim_{m\to\infty} \|x_m - x\|_X = 0 \text{ and } \sup \|x_m\|_Y < \infty\}.$$

It is known [3] that $K(t,x)$ is a bounded, continuous, monotone increasing and subadditive function of t for each $x \in X$, and $K(t,x,X,Y,\|\cdot\|_Y) = O(t)\,(t \to 0^+)$ if and only if $x \in \tilde{Y}^X$.

For convenience, we define the following notations. Put $j_n(t) = \frac{t^n}{n!}$ for $t \geq 0$ and $n \geq 0$ and denote by a_0 the *Dirac measure* δ_0 at 0. For each $m \geq 0$, let $a_{m+1}(t) = a * a_m(t)$ for $t \geq 0$, let

$$\begin{cases} k_m(0) &= 0; \\ k_m(t) &= \dfrac{a_{m+1} * j_n(t)}{a_m * j_n(t)} \quad \text{for } t > 0 \end{cases}$$

and define

$$Q_m(t)x = \frac{a_m * S(t)x}{a_m * j_n(t)} \quad \text{for } x \in X \text{ and } t > 0.$$

Note that $a_m(t)$ and $a_m * j_n(t)$ are nondecreasing positive functions of t. Therefore

$$k_m(t) = \frac{1}{a_m * j_n(t)} \int_0^t a(t-s)(a_m * j_n)(s)ds \leq \int_0^t a(s)ds \to 0 \tag{2.1}$$

as $t \to 0^+$. Note further that

$$\begin{aligned}\|Q_m(t)x\| &\leq \frac{1}{a_m * j_n(t)} \int_0^t a_m(t-s)\|S(s)x\|ds \\ &\leq \frac{M\|x\|}{a_m * j_n(t)} \int_0^t a_m(t-s)s^n ds = Mn!\|x\|\end{aligned}$$

and we have

$$\|Q_m(t)\| \leq Mn! \qquad (m \geq 0,\ 0 < t \leq 1). \tag{2.2}$$

The next lemma relates A, $Q_m(t)$, and $Q_{m+1}(t)$, and shows that $\{Q_m(t)\}_{t>0}$ is an approximation process on the invariant subspace $X_1 := \overline{D(A)}$.

Lemma 2.1. *Suppose that $\|S(t)\| \leq Mt^n$ for all $t \in [0,1]$. Then*

(a) *$Q_m(t)x \in D(A)$ and $AQ_m(t)x = Q_m(t)Ax$ for all $x \in D(A)$, $m \geq 0$ and $t > 0$;*

(b) *$R(Q_{m+1}(t)) \subset D(A)$, and $Q_{m+1}(t)A \subset AQ_{m+1}(t) = \frac{1}{k_m(t)}[Q_m(t) - I]$ for all $m \geq 0$ and $t > 0$;*

(c) *$Q_0(t)x = \frac{n!}{t^n}S(t)x \to x$ as $t \to 0^+$ for all $x \in \overline{D(A)}$;*

(d) *For $m \geq 0$, $Q_{m+1}(t)x \to x$ as $t \to 0^+$ if and only if $x \in \overline{D(A)}$.*

Proof. Part (a) and the first assertion of part (b) follow easily by (S2), (S3), Theorem 3.7.11 of [10], and induction. For $m \geq 0$, $t > 0$ *and* $x \in X$, we have, by (S3),

$$\begin{aligned}AQ_{m+1}(t)x &= \frac{1}{a_{m+1} * j_n(t)}[a_m * A(a * S(t)x)] \\ &= \frac{1}{a_{m+1} * j_n(t)}[(a_m * S(t)x) - a_m * j_n(t)x] \\ &= \frac{1}{k_m(t)}[Q_m(t)x - x].\end{aligned}$$

This proves (b). For (c), since $\|S(t)\| \leq Mt^n$ for $t \in [0,1]$, we have

$$\begin{aligned}\|Q_0(t)x - x\| &= \|\frac{n!}{t^n}S(t)x - x\| = \|\frac{n!}{t^n}\int_0^t a(t-s)S(s)Axds\| \\ &\leq n!M\|Ax\| \int_0^t a(t-s)ds \to 0\end{aligned}$$

as $t \to 0^+$ for all $x \in D(A)$ and hence for all $x \in \overline{D(A)}$. Now suppose $Q_m(t)x \to x$ for some $m \geq 0$ and all $x \in \overline{D(A)}$. Then we have

$$\begin{aligned}
&\|Q_{m+1}(t)x - x\| \\
&= \frac{1}{a_{m+1} * j_n(t)} \left\| \int_0^t a(t-s)(a_m * S(s)x ds - \int_0^t a(t-s)(a_m * j_n)(s)x ds \right\| \\
&\le \frac{1}{a_{m+1} * j_n(t)} \int_0^t a(t-s)(a_m * j_n)(s)\|Q_m(s)x - x\| ds \to 0,
\end{aligned}$$

as $t \to 0^+$. This proves the sufficiency part of (d) by induction. The necessity follows from (b). Note that $\overline{D(A)} = X$ if $n = 0$. In fact, if $n = 0$, then for all $x \in$ we have $Q_0(t)x = S(t)x \to x$ and so $Q_1(t)x \to x$ as $t \to 0^+$, which implies $x \in \overline{D(A)}$, by (d). When $n > 0$, A is not necessarily densely defined in X and so X_1 is often only a proper subspace of X. But the part of A in X_1 has domain dense in X_1.

Proposition 2.2. *Let $S(\cdot)$ be an n-times integrated solution family for (VE, A, a, f) such that $\|S(t)\| \le Mt^n$ for all $t \in [0,1]$, and let B be the part of A in $X_1 := \overline{D(A)}$. Then*

(a) *$Q_{m+1}(t)X_1 \subset D(B)$ and $Q_{m+1}(t)B \subset BQ_{m+1}(t)|_{X_1} = \frac{1}{k_m(t)}(Q_m(t) - I)|_{X_1}$ for all $m \ge 0$ and $t > 0$;*

(b) *B is densely defined in X_1;*

(c) *for $x \in D(B)$ and $m \ge 0$, we have $\lim_{t\to 0^+} \frac{1}{k_m(t)}[Q_m(t)x - x] = Bx$.*

Proof. Note that Lemma 2.1(a) implies that X_1 is invariant under $Q_m(t)$. To show (a), let $x \in X_1$. Then there is a sequence $\{x_n\} \subset D(A)$ converging to x. Since $AQ_{m+1}(t)x_n = Q_{m+1}(t)Ax_n$ lies in $D(A)$ and converges to $AQ_{m+1}(t)x \in X_1$, we have $Q_{m+1}(t)x \in D(B)$ and $BQ_{m+1}(t)x = AQ_{m+1}(t)x$ for all $x \in X_1$. When $x \in D(B$, we have $x \in D(A)$ and $Ax \in X_1$ so that $Q_{m+1}(t)Bx = Q_{m+1}(t)Ax = AQ_{m+1}(t)x = BQ_{m+1}(t)x$. These facts and part (b) of Lemma 2.1 imply assertion (a).

Since by (a), $Q_m(t)x \in D(B)$ for all $x \in X_1$ and $m \ge 1$, (b) follows by part (c) of Lemma 2.1. For (c), let $x \in D(B)$. Then $x \in D(A)$ and $Bx = Ax \in X_1$, so that assertion (c) follows from (a) and Lemma 2.1(c).

We also need the following proposition.

Proposition 2.3. *Under the assumption in Proposition 2.2, we have that A is bounded if and only if $\|(Q_m(t) - I)|_{X_1}\| \to 0^+$ as $t \to 0^+$ for some (all) $m \ge 0$. In this case, we have $\|(Q_m(t) - I)|_{X_1}\| = O(k_m(t))(t \to 0^+)$ for all $m \ge 0$.*

Proof. If A is bounded, then B is a bounded, closed operator with domain $D(B)$ dense in $\overline{D(A)}$(Proposition 2.2(b)). Hence the closed graph theorem implies $D(B) = \overline{D(B)} = \overline{D(A)} = D(A)$, so that $A = B$. Moreover, the boundedness of A implies that the graph norm $\|x\|_A := \|x\| + \|Ax\|$ is equivalent to $\|x\|$ for $x \in D(A)$. These facts together with Proposition 2.2(c) and the uniform boundedness principle imply that $\|(Q_m(t) - I)|_{X_1}\| = O(k_m(t))(t \to 0^+)$ for all $m \ge 0$. Since $k_m(t) \to 0$ as $t \to 0^+$ for all $m \ge 0$, this proves the necessity.

Conversely, if $\|(Q_m(t)-I)|_{X_1}\| \to 0$ as $t \to 0^+$ for some $m \geq 0$, then for small t, $Q_m(t)|_{X_1}$ is invertible so that, by part (a) of Proposition 2.2, $D(A) \supset R(Q_m(t)|_{X_1}) = X_1$, and so $D(A)$ is closed and hence, by the closedness, A must be bounded.

3. Saturation

The following optimal convergence theorem shows that $\{Q_m(t)\}_{t>0}$ has the saturation class $\widetilde{D(B)}^{X_1}$.

Theorem 3.1. *Let $S(\cdot)$ be an n-times integrated solution family for (VE, A, a, f) such that $\|S(t)\| \leq Mt^n$ for all $t \in [0,1]$. For all $m \geq 0$, and $x \in X_1 = \overline{D(A)}$ we have:*

(i) *$\|Q_m(t)x - x\| = o(k_m(t))(t \to 0^+)$ for some(all) $m \geq 0$ if and only if $x \in N(B) = N(A)$, the null space of B and A.*

(ii) *The following are equivalent:*
(a) *$\|Q_m(t)x - x\| = O(k_m(t))(t \to 0^+)$.*
(b) *$x \in \widetilde{D(B)}^{X_1}$, where $X_1 = \overline{D(A)}$;*
(c) *$x \in D(B)$, if X is a reflexive Banach space.*

Proof. The sufficiency part of (i) follows by part (c) of Proposition 2.2, and the necessity follows by part (b) of Proposition 2.2.

(a)⇒(b). Suppose that $\|Q_m(t)x - x\| = O(k_m(t))\,(t \to 0^+)$ and define $x_\nu = Q_{m+1}(\frac{1}{\nu})x$ for $\nu \geq 1$. Then $x_\nu \to x$ by Lemma 2.1(c), and Proposition 2.2(a) implies that that $\{x_\nu\} \subset D(B)$ and

$$\|Bx_\nu\| = \|\frac{1}{k_m(\frac{1}{\nu})}[Q_m(\frac{1}{\nu})x - x]\| = O(1)\,(\nu \to \infty),$$

whence $x \in \widetilde{D(B)}^{X_1}$.

(b)⇒(a). By proposition 2.2(c) and the uniform boundedness principle, there is an $M_1 > 0$ such that

$$\frac{1}{k_m(t)}\|Q_m(t)x - x\| \leq M_1\|x\|_{D(B)} \qquad (x \in D(B), t > 0)$$

If $x \in \widetilde{D(B)}^{X_1}$, there is a sequence $\{x_\nu\} \subset D(B) \subset D(A)$ such that $x_\nu \to x$ and $\|x_\nu\|_{D(B)} \leq M_2$, for some $M_2 > 0$. Therefore

$$\frac{1}{k_m(t)}\|Q_m(t)x_\nu - x_\nu\| \leq M_1\|x_\nu\|_{D(B)} \leq M_1M_2$$

for all $t > 0$, $\nu \in N$. Let $\nu \to \infty$, and we have (a).

Corollary 3.2. *If X is reflexive and (VE, A, a, f) admits an n-times integrated solution family $S(\cdot)$ satisfying $\|S(t)\| = O(t^n)\,(t \to 0^+)$ and $\|S(t)\| = O(e^{\omega t})\,(t \to \infty)$, then A is densely defined.*

Proof. From the estimate in the proof of Theorem 3.1, we see that if $x \in D(A)$, then (a) (with $m = 0$) holds, and hence (b) holds, i.e., $D(A) \subset \widetilde{D(B)}^{X_1}$. If X is reflexive, then we have $\widetilde{D(B)}^{X_1} = D(B)$. Thus $D(A) = D(B)$, which implies that the range of $\lambda - A$ is contained in $\overline{D(A)} \neq X$, provided that A is nondensely defined. This means that the resolvent set of A is empty. But, by taking Laplace transform of the equations in (S3) and (S4), it is easy to see that the resolvent set of A is non-empty.

4. Non-optimal Convergence

Lemma 4.1. *Let $\|S(t)\| \le Mt^n$ for all $t \in [0,1]$. Then for $x \in \overline{D(A)}$ we have*

$$\begin{aligned} K(t,x,X,D(A),\|\cdot\|_{D(A)}) &\le K(t,x,\overline{D(A)},D(B),\|\cdot\|_{D(B)}) \\ &\le Mn!K(t,x,X,D(A),\|\cdot\|_{D(A)}). \end{aligned}$$

Proof. The first inequality is obvious from the definition of the K-functional. To show the second inequality, fix $\epsilon > 0$, $x \in \overline{D(A)}$, and $m \in N$ arbitrarily. By part (c) of Lemma 2.1, there is a sufficiently small $\tau > 0$ such that $\|x - Q_m(\tau)x\| < \epsilon$. Choose a $y_1 \in D(B)$ such that

$$\|Q_m(\tau)x - y_1\| + t\|y_1\|_{D(B)} \le K(t, Q_m(\tau)x, \overline{D(A)}, D(B), \|\cdot\|_{D(B)}) + \epsilon.$$

We obtain

$$\begin{aligned} K(t,x,\overline{D(A)},D(B),\|\cdot\|_{D(B)}) &\le \|x-y_1\| + t\|y_1\|_{D(B)} \\ &\le \|x - Q_m(\tau)x\| + \|Q_m(\tau)x - y_1\| + t\|y_1\|_{D(B)} \\ &\le K(t,Q_m(\tau)x,\overline{D(A)},D(B),\|\cdot\|_{D(B)}) + 2\epsilon. \end{aligned}$$

Next choose $y_2 \in D(A)$ such that

$$\|x - y_2\| + t\|y_2\|_{D(A)} \le K(t,x,X,D(A),\|\cdot\|_{D(A)}) + \frac{\epsilon}{Mn!}.$$

Since $Q_m(\tau)y_2 \in D(B)$, by Proposition 2.2(a), we have

$$\begin{aligned} &K(t,Q_m(\tau)x,\overline{D(A)},D(B),\|\cdot\|_{D(B)}) \\ &\quad\le \|Q_m(\tau)x - Q_m(\tau)y_2\| + t\|Q_m(\tau)y_2\|_{D(B)} \\ &\quad\le Mn!(\|x-y_2\| + t\|y_2\|_{D(A)}) \\ &\quad\le Mn!K(t,x,X,D(A),\|\cdot\|_{D(A)}) + \epsilon. \end{aligned}$$

Combining this and the previous inequality we obtain

$$K(t,x,\overline{D(A)},D(B),\|\cdot\|_{D(B)}) \le Mn!K(t,x,X,D(A),\|\cdot\|_{D(A)}) + 3\epsilon.$$

Since $\epsilon > 0$ is arbitrary, we complete the proof.

Theorem 4.2. *Let $\|S(t)\| \le Mt^n$ for all $t \in [0,1]$, $m \ge 0$, $0 < \beta \le 1$, and $x \in \overline{D(A)}$. If $K(k_m(t), x, X, D(A), \|\cdot\|_{D(A)}) = O((k_m(t))^\beta)$, then $\|Q_m(t)x - x\| = O((k_m(t))^\beta)$. The converse is also true if $k_m(t)$ is nondecreasing for t near 0.*

Proof. Part (c) of Proposition 2.2 and the principle of uniform boundedness imply that there exists an $M_3 > 0$ such that $\frac{1}{k_m(t)}\|Q_m(t)y - y\| \le M_3\|y\|_{D(B)}$ for all $y \in D(B)$ and $0 < t \le 1$. Thus, for $y \in D(B)$ and $0 < t \le 1$, we have

$$\begin{aligned} \|Q_m(t)x - x\| &\le \|(Q_m(t) - I)(x - y)\| + \|Q_m(t)y - y\| \\ &\le (Mn! + 1)\|x - y\| + k_m(t)M_3\|y\|_{D(B)} \\ &\le \max(Mn! + 1, M_3)(\|x - y\| + k_m(t)\|y\|_{D(B)}). \end{aligned}$$

Hence

$$\begin{aligned} \|Q_m(t)x - x\| &\le \max(Mn! + 1, M_3)K(k_m(t), x, D(A), D(B), \|\cdot\|_{D(B)}) \\ &\le Mn! \max(Mn! + 1, M_3)K(k_m(t), x, X, D(A), \|\cdot\|_{D(A)}) \\ &= O((k_m(t))^\beta), \end{aligned}$$

by Lemma 4.1. If $k_m(t)$ is nondecreasing in t, the converse implication holds because, by Lemma 2.1(b) and (2.2),

$$\begin{aligned} &K(k_m(t), x, X, D(A), \|\cdot\|_{D(A)}) \\ &\quad\le \|x - Q_{m+1}(t)x\| + k_m(t)[\|AQ_{m+1}(t)x\| + \|Q_{m+1}(t)x\|] \\ &\quad\le \sup_{0<s\le t} \|Q_m(s)x - x\| + \|Q_m(t)x - x\| + k_m(t)Mn!\|x\| \\ &\quad= \sup_{0<s\le t} O((k_m(s))^\beta) + O((k_m(t))^\beta) = O((k_m(t))^\beta). \end{aligned}$$

To consider the sharpness of approximation, we need the following result of Davydov [8].

Theorem A. *Let X be a Banach space and X^+ be the set of all nonnegative sublinear, real-valued function S on X for which the norm $\|S\|_{X^+} := \sup\{Sx : x \in X,\ \|x\| \le 1\}$ is bounded. Furthermore, let H be an unbounded set of continuous seminorms. If the set $\{x \in X : \lim_{\|h\|_{X^+} \to \infty} h(x) = 0,\ h \in H\}$ is dense in X, then there exists an element $x^* \in X$ such that $\sup_{h \in H} h(x^*) = 1$ and $\overline{\lim}_{\|h\|_{X^+} \to \infty} h(x^*) = 1$.*

Theorem 4.3. *Let $\|S(t)\| \le Mt^n$ for all $t \in [0,1]$. Then A is unbounded if and only if for some(each) $0 < \beta < 1$ and $m \ge 0$ there exists an $x^*_{m,\beta} \in X_1 = \overline{D(A)}$ such that*

$$\|Q_m(t)x^*_{m,\beta} - x^*_{m,\beta}\| \left\{ \begin{array}{l} = O((k_m(t))^\beta) \\ \ne o((k_m(t))^\beta) \end{array} \right. \quad (t \to 0^+).$$

If $\beta = 1$ and $A \ne 0$, $x^*_{m,1}$ can be any element in $\widetilde{D(B)}^{X_1} \backslash N(B)$.

Proof. If A is bounded, then by Proposition 2.3 we have $\|(Q_m(t) - I)|_{X_1}\| = O(k_m(t))$ so that $\|(Q_0(t) - I)|_{X_1}\| = o((k_m(t))^\beta)(t \to 0^+)$ for $0 < \beta < 1$. This shows the sufficiency.

For the necessity, define $H := \{h_t : h_t(x) = (k_m(t))^{-\beta}\|Q_m(t)x - x\|,\ x \in X_1 = \overline{D(A)}\}$. If H satisfies the hypothesis of Theorem A, we complete the proof. By (2.2), we see that $\|h_t\|_{X_1^+} \le (Mn! + 1)(k_m(t))^{-\beta}$ and this implies that if $\|h_t\|_{X_1^+} \to \infty$, then $t \to 0^+$. By part (c) of Theorem 2.3, the set $\{x \in X_1 : \lim_{\|h\|_{X_1^+} \to \infty} h(x) = 0,\ h \in H\}$ contains $D(B)$ and hence is dense in $\overline{D(A)}$. If A is unbounded, then by Proposition 2.3, we have $\overline{\lim}_{t\to 0^+}\|(Q_m(t) - I)|_{X_1}\| > 0$ so that $\overline{\lim}_{t\to 0^+}\|h_t\|_{X_1} = \overline{\lim}_{t\to 0^+}(k_m(t))^{-\beta}\|(Q_m(t) - I)|_{X_1}\| = \infty$. Thus H really satisfies the hypothesis of Theorem B with X replaced by X_1. Hence there exists an $x^*_{m,\beta} \in X_1$ such that $\sup_t h_t(x_{m,\beta}) \le 1$ and $\overline{\lim}_{t\to 0^+} h_t(x_{m,\beta}) = 1$, i.e., $x^*_{m,\beta}$ satisfies $\|Q_m(t)x^*_{m,\beta} - x^*_{m,\beta}\| = O((k_m(t))^\beta)(t \to 0^+)$ and $\|Q_m(t)x_{m,\beta} - x^*_{m,\beta}\| \neq o((k(t))^\beta)(t \to 0^+)$.

The final assertion follows immediately from Theorems 3.2.

5. Specialization

In this section we discuss applications to n-times integrated semigroups and n-times integrated cosine functions.

5.I. n-times integrated semigroups

A strongly continuous family $\{T(t); t \ge 0\}$ in $B(X)$ is called an *n-times integrated semigroup* on X [1] if $T(0) = 0$ and

$$T(t)T(s)x = \frac{1}{(n-1)!}\Big(\int_t^{t+s} (s+t-r)^{n-1}T(r)x\,dr - \int_0^s (t+s-r)^{n-1}T(r)x\,dr\Big)$$

for $x \in X$ and $t, s \ge 0$. $T(\cdot)$ is *nondegenerate* if $T(t)x = 0$ for all $t > 0$ implies $x = 0$. The *generator* A of a nondegenerate n-times integrated semigroup $T(\cdot)$ is defined as follows:

$$x \in D(A) \text{ and } Ax = y \text{ if and only if } T(t)x = \int_0^t T(r)y\,dr + \frac{t^n}{n!}x \text{ for } t \ge 0.$$

It is known that $T(\cdot)$ is an n-times integrated solution family for (VE, A, a, f) with $a \equiv 1$. Clearly, we have $k_m(t) = \frac{t}{n+m+1}$ and

$$Q_m(t) = \frac{(n+m)!}{t^{n+m}} \int_0^t \frac{(t-s)^{m-1}}{(m-1)!} T(s)x\,ds.$$

Thus we deduce from Theorems 3.1 and 4.2 the following theorem.

Theorem 5.1. *Let $T(\cdot)$ be an n-times integrated semigroup satisfying $\|T(t)\| = O(t^n)\,(t \to 0^+)$, let B denote the part of the generator A in $X_1 = \overline{D(A)}$, $m \geq 1$, $x \in X_1$.*

(i) $\|\frac{n!}{t^n}T(t)x - x\| = o(t)(t \to 0^+)$ *if and only if* $\|\frac{(n+m)!}{t^{n+m}}\int_0^t \frac{(t-s)^{m-1}}{(m-1)!}T(s)xds - x\| = o(t)(t \to 0^+)$, *if and only if* $x \in N(B)$, *the null space of* B.

(ii) *The following are equivalent for* $0 < \beta \leq 1$:

(a) $\|\frac{n!}{t^n}T(t)x - x\| = O(t^\beta)(t \to 0^+)$;

(b) $\|\frac{(n+m)!}{t^{n+m}}\int_0^t \frac{(t-s)^{m-1}}{(m-1)!}T(s)xds - x\| = O(t^\beta)(t \to 0^+)$;

(c) $K(t, x, X, D(A), \|\cdot\|_{D(A)}) = O(t^\beta)(t \to 0^+)$.

If $\beta = 1$ *the assertions* (a), (b), *and* (c) *are also equivalent to*

(d) $x \in \widetilde{D(B)}^{X_1}$, *where* $X_1 = \overline{D(A)}$;

(e) $x \in D(B)$ *if* X *is additionally a reflexive Banach space.*

From Theorem 4.3 we deduce the next theorem.

Theorem 5.2. *Under the hypothesis of Theorem 5.1, the generator A is unbounded if and only if for each $0 < \beta < 1$ there exists $x^*_{\beta 1}$ and $x^*_{\beta 2}$ in X_1 such that*

$$\|\frac{n!}{t^n}T(t)x^*_{\beta 1} - x^*_{\beta 1}\| \begin{cases} = O(t^\beta) \\ \neq o(t^\beta) \end{cases} \qquad (t \to 0^+);$$

$$\|\frac{(n+m)!}{t^{n+m}}\int_0^t \frac{(t-s)^{m-1}}{(m-1)!}T(s)x^*_{\beta 2}ds - x^*_{\beta 2}\| \begin{cases} = O(t^\beta) \\ \neq o(t^\beta) \end{cases} \qquad (t \to 0^+).$$

Remark. The above results for $m = 0, 1$ were proved in [7]. When $n = 0$, Theorems 5.1 reduces to Butzer and Berens' results in [4, Chapter II] and [3]; Theorem 5.2 becomes the result of Butzer, Dickmeis and Davydov [8, p. 441].

5.II. n-times integrated cosine functions

A strongly continuous family $\{C(t); t \geq 0\}$ of bounded linear operators on X is called an *n-times integrated cosine function* on X [15] if $C(0) = 0$ and

$$\begin{aligned} 2C(t)C(s)x \;=\; & \frac{1}{(n-1)!}\Big\{(-1)^n \int_0^{|s-t|} (|s-t| - u)^{n-1}C(u)xdu \\ & + \Big[\int_0^{s+t} - \int_0^t - \int_0^s\Big](t+s-u)^{n-1}C(u)xdu \\ & + \int_0^t (s-t+u)^{n-1}C(u)xdu + \int_0^s (t-s+u)^{n-1}C(u)xdu\Big\} \end{aligned}$$

for all $x \in X$ and $s,\ t > 0$. It is called a (0-times integrated) cosine function if

$$C(0) = I \text{ and } 2C(t)C(s) = C(t+s) + C(t-s) \text{ for } t \geq s \geq 0.$$

$C(\cdot)$ is said to be *nondegenerate* if $C(t)x = 0$ for all $t > 0$ implies $x = 0$. The *generator* A of a nondegenerate n-times integrated cosine function $C(\cdot)$ is defined as:

$$x \in D(A) \quad \text{and} \quad Ax = y \Leftrightarrow C(t)x - \frac{t^n}{n!}x = \int_0^t (t-u)C(u)y du \ \text{ for } \ t \geq 0.$$

For properties of n-times integrated cosine functions see [15].

It is known that $C(\cdot)$ is an n-times integrated solution family for (VE, A, a, f) with $a(t) = t$. Clearly, we have $k_m(t) = \frac{t^2}{(n+2m+2)(n+2m+1)}$ and $Q_m(t) = \frac{(n+2m)!}{t^{n+2m}} \int_0^t \frac{(t-s)^{2m-1}}{(2m-1)!} C(s)xds$. Thus we deduce from Theorems 3.1 and 4.2 the following theorem.

Theorem 5.3. *Let $C(\cdot)$ be an n-times integrated cosine function satisfying $\|C(t)\| = O(t^n)\,(t \to 0^+)$, let B denote the part of the generator A in $X_1 = \overline{D(A)}$, $m \geq 1$, $x \in X_1$.*

(i) $\|\frac{n!}{t^n}C(t)x - x\| = o(t^2)$ $(t \to 0^+)$ *if and only if* $\|\frac{(n+2m)!}{t^{n+2m}} \int_0^t (t-s)^{2m-1} C(s)xds - x\| = o(t^2)$ $(t \to 0^+)$, *if and only if* $x \in N(B)$.

(ii) *The following are equivalent for* $0 < \beta \leq 2$:

(a) $\|\frac{n!}{t^n}C(t)x - x\| = O(t^\beta) \quad (t \to 0^+)$;

(b) $\|\frac{(n+2m)!}{t^{n+2m}} \int_0^t \frac{(t-s)^{2m-1}}{(2m-1)!} C(s)xds - x\| = O(t^\beta) \quad (t \to 0^+)$;

(c) $K(t^2, x, X, D(A), \|\cdot\|_{D(A)}) = O(t^\beta) \quad (t \to 0^+)$.

If $\beta = 2$ *the assertions* (a), (b), *and* (c) *are also equivalent to*

(d) $x \in \widetilde{D(B)}^{X_1}$, *where* $X_1 = \overline{D(A)}$;

(e) $x \in D(B)$ *if* X *is additionally a reflexive Banach space.*

From Theorem 4.3 we deduce the next theorem.

Theorem 5.4. *Under the hypothesis of Theorem 5.3, the generator A is unbounded if and only if for each $0 < \beta < 1$ there exists $x^*_{\beta 3}$ and $x^*_{\beta 4}$ in X_1 such that*

$$\|\frac{n!}{t^n}C(t)x^*_{\beta 3} - x^*_{\beta 3}\| \begin{cases} = O(t^\beta) \\ \neq o(t^\beta) \end{cases} \quad (t \to 0^+);$$

$$\|\frac{(n+2m)!}{t^{n+2m}} \int_0^t \frac{(t-s)^{2m-1}}{(2m-1)!} C(s)x^*_{\beta 4}ds - x^*_{\beta 4}\| \begin{cases} = O(t^\beta) \\ \neq o(t^\beta) \end{cases} \quad (t \to 0^+).$$

Acknowledgement. Research supported in part by the National Science Council of Taiwan

References

[1] W. Arendt, *Vector-valued Laplace transforms and Cauchy problems*, Israel J. Math **59** (1987), 327–352.

[2] W. Arendt and H. Kellerman, *Integrated solutions of Volterra integrodifferential equations and applications*, Volterra Integrodifferential Equations in Banach Spaces, Pitman Res. Notes in Math. vol. 190, 1989, pp. 21–51.

[3] H. Berens, *Interpolationsmethoden sur Behandlung von Approximationsprozessen auf Banachraumen*, Lect. Notes Math. 64. Springer-Verlag, Berlin-Heidelberg-New York, 1968

[4] P.L. Butzer and H. Berens, *Semi-Groups of Operators and Approximation*, Springer-Verlag, New York, 1967

[5] P. L. Butzer and A. Gessinger, *Ergodic theorems for semigroups and cosine operator functions at zero and infinity with rates; applications to partial differential equation: A survey*, Contemp. Math. **190** (1995), 67–94.

[6] J.-C. Chang and S.-Y. Shaw, *Rates of approximation and ergodic limits of resolvent families*, Arch. Math. **66** (1996), 320–330.

[7] J.-C. Chang and S.-Y. Shaw, *Optimal and Non-optimal Rates of Approximation for Integrated Semigroups and Cosine Functions*, J. Approx. Theory **90** (1997), 200–223.

[8] O. V. Davydov, *Some condensation theorem for semigroup operators*, Manuscripta Math. **79** (1993), 435–446.

[9] W. Desch and J. Prüss, *Counterexamples for abstract linear Volterra equations*, J. Integral Equations Appl. **5** (1993), 29–45.

[10] E. Hille and R.S. Phillips, *Functional Analysis and Semigroups*, The American Mathmatical Society, 1965.

[11] Y.-C. Li and S.-Y. Shaw, *N-times integrated C-semigroups and the abstract Cauchy problem*, Taiwanese J. Math. **1** (1997), 75–102.

[12] C. Lizama, *On Volterra equations associated with a linear operator*, Proc. Amer. Math. Soc. **118** (1993), 1159–1166.

[13] H. Oka, *Linear Volterra equations and integrated solution families*, Semigroup Forum **53** (1996), 278–297.

[14] J. Prüss, *Evolutionary Integral Equations and Applications*, Birkhauser, Basel, 1993.

[15] S.-Y. Shaw and Y.-C. Li, *On n-times integrated C-cosine functions*, Evolution Equations, Marcel Dekkar, 1995, pp. 393–406.

Hsiang Liu, Sen-Yen Shaw, Department of Mathematics, National Central University, Chung-Li, TAWIAN

E-mail address: `shaw@math.ncu.edu.tw`

Progress in Nonlinear Differential Equations
and Their Applications, Vol. 42

ABSTRACT WELL-POSED AND ILL-POSED CAUCHY PROBLEMS FOR INCLUSIONS

I.V. MELNIKOVA
URAL STATE UNIVERSITY

Introduction

A wide range of mathematical models in physics, engineering, finance, biology etc., leads to the degnerate Cauchy problems

$$Mu'(t) = LU(t), t \in [0,T), u(0) = x; \quad \frac{d}{dt}Mv(t) = Lv(t), t \in [0,T), Mv(0) = x; \tag{1}$$

with linear operators $M, L : X \to Y$ in Banach spaces X, Y ($\ker M \neq \{0\}$) and to the Cauchy problem with a multivalued linear operator $\mathcal{A} : X \to X$

$$(ICP) \qquad u'(t) \in \mathcal{A}u(t),\ t \in [0,T),\ T \leq \infty,\ u(0) = x.$$

In [14] for (ICP) the well-posedness criterion extending Miydera-Feller-Phillips-Hille-Yosida (MFPHY) criterion is proved.

Theorem 1. *Let $\mathcal{A}$ be a multivalued linear operator in a Banach space X and $\rho(\mathcal{A}) \neq \emptyset$. Let $X_1 := \overline{D(\mathcal{A})}$, $Au := \mathcal{A}u \cap X_1$, $D(A) := \{u \in X : Au \neq \emptyset\}$. Then the following statements are equivalent*

- **(I)** *(ICP) is uniformly well-posed on $D(\mathcal{A})$ (that is there exists a unique solution $u(t)$ such that* $\sup_{t\leq\tau<T} \|u(t)\| \leq C\|x\|$*).*
- **(II)** *$D(A) = D(\mathcal{A})$, operator A is the simple restriction of $\mathcal{A}$ and the generator of a C_0 semigroup on X_1.*
- **(III)** *The (MFPHY) condition for $R_A(\lambda) = (\lambda - A)^{-1}$*

$$\exists K > 0, \omega \in \mathbf{R} : \left\| \frac{d^k}{d\lambda^k} (\lambda - A)^{-1} \right\| \leq \frac{Kk!}{(\Re\lambda - \omega)^{k+1}},\ \Re\lambda > \omega,\ k = 0,1,2,\dots$$

and the decomposition

$$X = X_1 \oplus \mathcal{A}0 \tag{2}$$

hold.

A wide range of papers is devoted to investigation of decision, uniqueness and smoothnes questions for the degenerate Cauchy problems, the Cauchy problem (ICP) and the integrated Cauchy problem:

$$v(t) \in \frac{t^n}{n!}x + \mathcal{A}\int_0^t v(s)\,ds, \quad x \in X, \tag{3}$$

[6, 20, 16, 2, 8], nevertheless a well-posedness criterion was not obtained in the general case. The paper is devoted to the Cauchy problem (ICP) which is not uniformly well-posed on $D(\mathcal{A})$ or equivalently to (ICP) with unfilfilled (MFPHY) condition. For such problems we consider 1) semigroup approach, 2) abstract distributions approach, 3) regularization approach.

The strategic idea of the first two approaches is the relaxation of the notion "well-posedness" so that a Cauchy problem which is not well-posed in the classical sense becomes well-posed in some other sense. Using degenerate integrated and C-regularized semigroups we construct a solution for x from some subset of $D(A)$ stable with respect to some norm of x stronger than X norm. Using degenerate distribution semigroups we construct a solution for any $x \in X$ stable in the space of abstract distributions. In the regularization approach assuming that for some x there exists a solution $u(t)$ and x_δ ($\|x - x_\delta\| \le \delta$) is given, we construct a solution $u_\alpha(t)$ of a well-posed (regularized) problem that depends on a regularization parameter α. In this connection we chose $\alpha = \alpha(\delta)$ such that $\alpha(\delta) \to 0$ and $u_{\alpha(\delta)}(t) \to u(t)$ as $\delta \to 0$.

We demonstrate the profound connections between these at first sight absolutely different methods. These connections in turn demonstrate that all these methods are properly regularization methods for solving the problem in different statements.

1. Semigroup Approach. Degenerate semigroups

Definition 1. *The Cauchy problem (ICP) is called (n,ω)-well-posed on $\mathcal{X} \subset X$ if for any $x \in \mathcal{X}$, there exists a unique solution $u(t)$, $t \ge 0$, such that*

$$\|u(t)\| \le Ce^{\omega t}\|x\|_{\mathcal{A},n}, \quad \|x\|_{\mathcal{A},n} := \sum_{k=0}^{k=n} \inf_{y \in \mathcal{A}^k x} \|y\|, \quad C > 0. \tag{4}$$

For investigation of (n,ω)-well-posedness we introduce a definition of degenerate integrated semigroups based on ideas of the integrated and K-convoluted semigroups definition via relations to their generator [18, 3, 4].

Definition 2. *Let $k \in \mathbf{N}$ and A be a closed linear operator on X. A one-parameter family $\{V(t), t \in [0,T)\}$ of bounded linear strongly continuous in t operators satisfying*

$$V(t)x - \frac{t^k}{k!}x = \int_0^t V(s)\mathcal{A}x\,ds, \quad x \in D(\mathcal{A}) \tag{5}$$

$$V(t)x - \frac{t^k}{k!}x \in \mathcal{A}\int_0^t V(s)x\,ds, \quad x \in X \tag{6}$$

is said to be degenerate k-times integrated semigroup generated by $\mathcal{A}$, and $\mathcal{A}$ the generator of $V(t)$.

For the introduced $V(t)$ the "semigroup" relation holds:

$$\textbf{(V1)}\quad (k-1)!\,V(t)V(s) = \int_0^s [(s-r)^{k-1}V(t+r) - (t+s-r)^{k-1}V(r)]\,dr,\qquad s,t,t+s \in [0,T),$$

and for $\mathcal{A}$ generated $V(t)$

$$D(\mathcal{A}) = \left\{x \in X \mid \exists y : V(t)x = \frac{t^n}{n!}x + \int_0^t V(s)y\,ds,\ t \in [0,T)\right\},\quad \{y\} = \mathcal{A}x.$$

For an exponentially bounded semigroup Definition 2 is equivalent to the definition via (V1) and the equality

$$(\lambda - \mathcal{A})^{-1}x = \int_0^\infty \lambda_k e^{-\lambda t}V(t)x\,dt,\quad \Re\lambda > \omega. \tag{7}$$

(Applying $(\lambda - \mathcal{A})$ to (7) from the left and right we have (5), (6)). The equality (7) means that any degenerate semigroup has the multivalued generator and conversely, if a multivalued operator is the generator of a semigroup, then the semigroup is degenerate.

Theorem 2. *Let $\mathcal{A}$ be a closed linear multivalued operator on X and $\rho(\mathcal{A}) \neq \emptyset$. If the Cauchy problem (ICP) is (n,ω)-well-posed on $D_{n+1} := D(\mathcal{A}^{n+1})$, then $\mathcal{A}$ is the generator of $(n+1)$-times integrated exponentially bounded semigroup and (MFPHY)-type condition:*

$$\left\| \frac{d^k}{d\lambda^k}\frac{(\lambda - \mathcal{A})^{-1}}{\lambda^{n+1}} \right\| \le \frac{Mk!}{(\lambda-\omega)^{k+1}},\quad \lambda > \omega,\ k = 0,1,2,\dots \tag{8}$$

holds for some $M > 0$.

Proof. Let $x \in D_{n+1}$ and $u(t)$, $t \ge 0$ be the unique solution of (ICP) (with $T = \infty$) corresponding to x. We introduce the solution operator $U(t)$ on D_{n+1} by $U(t)x := u(t)$. Due to (4) we can extend $U(t)$ to the space $[D_{n+1}]_n$, the closure of D_{n+1} in the norm $\|\cdot\|_n$. Due to uniqueness for any $x \in D_{n+1}$, $R(\lambda)U(t)x = U(t)R(\lambda)x$, and

$$\|U(t)R(\lambda)x\| \le Ce^{\omega t}\|R(\lambda)x\|_n \le C_1e^{\omega t}\|x\|_{n-1}.$$

Next, integrating the equality $R(\lambda)U'(t)x = R(\lambda)\mathcal{A}U(t)x = \lambda R(\lambda)U(t)x - U(t)x$, we have

$$\int_0^t U(s)x\,ds = -U(t)R(\lambda)x + R(\lambda)x + \lambda\int_0^t U(s)R(\lambda)x\,ds,\ x \in D_{n+1},$$

and on D_n we define

$$U_1(t)x := -U(t)R(\lambda)x + R(\lambda)x + \lambda\int_0^t U(s)R(\lambda)x\,ds. \tag{9}$$

The defined $U_1(t)$ commutes with $R(\lambda)$ on D_n, is exponentially bounded with respect to $\|x\|_{n-1}$, and can be continued to $[D_n]_{n-1}$. In the general case,

$$(10)\ U_k(t)x := -U_{k-1}(t)R(\lambda)x + \frac{t^{k-1}}{(k-1)!}R(\lambda)x + \lambda\int_0^t U_{k-1}(s)R(\lambda)x\,ds.$$

$U_k(t)$ is defined on D_{n+1-k} and may be continued to $[D_{n+1-k}]_{n-k}$. In particular, $U_n(t)$ is defined and commutes with $R(\lambda)$ on $D(\mathcal{A})$ and may be continued to $\overline{D(\mathcal{A})}$; $V(t) := U_{n+1}(t)$ is defined, exponentially bounded and strongly continuous in $t \geq 0$ on the whole space X. To show $V(t)$ is a $(n+1)$-times integrated semigroup with the generator $\mathcal{A}$ it is sufficient to prove (5), (6) with $k = n+1$. We can prove the equation and inclusion for $k = 1$ and then by induction for $k > 1$. Due to the equation and inclusion for $U(t)$:

$$U(t)x - x = \int_0^t U(\tau)\mathcal{A}x\,d\tau, \quad x \in D_{n+1},$$

$$U(t)x - x \in \mathcal{A}\int_0^t U(\tau)x\,d\tau, \quad x \in [D_{n+1}]_n,$$

$U_1(t)$ fits (5) with $k = 1$ for $x \in D_n$ and (6) for $x \in [D_n]_{n-1}, \ldots, V(t)$ fits (5), (6) with $k = n+1$. From (5), (6) for strongly continuous exponentially bounded function $V(t)$ the equality (7) follows. Hence, for $V(t)$ we have (7) with $k = n+1$ and the estimates (8) hold true. □

Theorem 3. *Let $\mathcal{A}$ be a multivalued linear operator on X. Suppose that for some $M > 0$ and $\omega \in \mathbf{R}$*

$$(11)\qquad \left\| \frac{d^k}{d\lambda^k}\frac{(\lambda - \mathcal{A})^{-1}}{\lambda^n} \right\| \leq \frac{Mk!}{(\lambda-\omega)^{k+1}}, \quad k = 0, 1, 2, \ldots, \ \lambda > \omega.$$

Then the Cauchy problem (ICP) is (n,ω)-well-posed on $R^{n+1}(\lambda)\overline{D(\mathcal{A})}$.

Proof. Given (11), operator $\mathcal{A}$ is the generator of an $(n+1)$-times integrated exponentially bounded semigroup $\{V(t),\ t \geq 0\}$ (defined via (V1), (7)) with the property

$$(12)\qquad \|V(t+h) - V(t)\| \leq Me^{\omega t}h, \quad t \geq 0,\ h \geq 0.$$

[1]. As it was mentioned above, for the $(n+1)$-times integrated exponentially bounded semigroup $V(t)$ the equation (5) and the inclusion (6) with $k = n+1$ hold. Hence

$$(13)\quad V'(t)x = \frac{t^n}{n!}x + V(t)\mathcal{A}x, \quad V'(t)x \in \frac{t^n}{n!}x + \mathcal{A}V(t)x, \quad x \in D(\mathcal{A}).$$

Using property (12) and closedness of $\mathcal{A}$ we can extend the inclusion in (13) to $\overline{D(\mathcal{A})}$. Now take $x \in D(\mathcal{A}^2)$ (that is $\mathcal{A}x \cap D(\mathcal{A}) \neq \emptyset$), then from (13) we have

$$\mathcal{A}V'(t)x = \frac{t^n}{n!}\mathcal{A}x + \mathcal{A}V(t)\mathcal{A}x,$$

$$V'(t)\mathcal{A}x \in \frac{t^n}{n!}\mathcal{A}x + V(t)\mathcal{A}^2x \subset \frac{t^n}{n!}\mathcal{A}x + \mathcal{A}V(t)\mathcal{A}x, \quad x \in D(\mathcal{A}^2),$$

therefore $V'(t)\mathcal{A}x \in \mathcal{A}V'(t)x$ and

$$V''(t)x = \frac{t^{n-1}}{(n-1)!}x + V'(t)\mathcal{A}x, \quad V''(t)x \in \frac{t^{n-1}}{(n-1)!}x + \mathcal{A}V'(t)x, \quad x \in D(\mathcal{A}^2). \tag{14}$$

Again using the property (12) and closedness of $\mathcal{A}$ we obtain that the inclusion from (14) holds for all $x \in R(\lambda)\overline{D(\mathcal{A})}$, and the inclusion for $V^{(3)}(t)x$ holds on $R^2(\lambda)\overline{D(\mathcal{A})}$. We continue the process and obtain

$$V^{(n+2)}(t)x \in \mathcal{A}V^{(n+1)}(t)x, \quad x \in R^{n+1}(\lambda)\overline{D(\mathcal{A})}, \quad V^{(n+1)}(0)x = x.$$

Therefore, $V^{(n+1)}(t)x$, $x \in R^{n+1}(\lambda)\overline{D(\mathcal{A})}$, is a solution of the Cauchy problem (ICP). For any other exponentially bounded solution $y(t)$ we have

$$\int_0^\infty e^{-\lambda t}y(t)\,dt = \frac{1}{\lambda}\left\{x + \int_0^\infty e^{-\lambda t}y'(t)\,dt\right\} \in \frac{1}{\lambda}\left\{x + \mathcal{A}\int_0^\infty e^{-\lambda t}y(t)\,dt,\right\}$$

$$x \in (\lambda - \mathcal{A})\int_0^\infty e^{-\lambda t}y(t)\,dt,$$

hence

$$R(\lambda)x = \int_0^\infty \lambda^{n+1}e^{-\lambda t}V(t)x\,dt = \int_0^\infty e^{-\lambda t}V^{(n+1)}(t)x\,dt = \int_0^\infty e^{-\lambda t}y(t)\,dt.$$

Therefore, $V^{(n+1)}(t)x = y(t)$. Considering that $V^{(n+1)}(t)x$ is expressed in terms of $V'(t)$ on elements $\mathcal{A}^kx$, $k \le n$; for $x \in D_{n+1}$, $\mathcal{A}^kx \subset D(\mathcal{A})$, and $V'(t)$ is bounded on $\overline{D(\mathcal{A})}$, we have stability of $V^{(n+1)}(t)x$ with respect to $\|x\|_{\mathcal{A},n}$ and hence with respect to $\|x\|_n$. □

Note that in contrast to Theorem 1, considering (n,ω)-well-posedness of (ICP) we have not got any decomposition of X, and by this reason we obtain in Theorem 2 necessary condition (8) and in Theorem 3 sufficient condition (11). Now we consider the decomposition

$$X = X_{n+1} \oplus \ker R^{n+1}(\lambda) = X_{n+1} \oplus \mathcal{A}^{n+1}0, \tag{15}$$

the extension of (2) to the case of a degenerate n-times integrated semigroup. Given (15) we can prove the necessary and sufficient condition for the (n,ω)-well-posedness of (ICP).

Theorem 4. *Let $\mathcal{A}$ be a multivalued linear operator on X, $\rho(\mathcal{A}) \neq \emptyset$, and the decomposition (15) hold. Then the Cauchy problem (ICP) is (n,ω)-well-posed on D_{n+1} if and only if (11) is satisfied.*

Remark. Ascertaining similar to the nondegenerate case the connection between integrated semigroups and C-semigroups [12] we can prove that under (15) the Cauchy problem (ICP) is (n,ω)-well-posed on D_{n+1} if and only if $\mathcal{A}$ is the generator of C-semigroup with $C = R^n_{\mathcal{A}}(\lambda)$.

Now we consider well-posedness conditions for (ICP) in the local case: $T<\infty$.

Theorem 5. *Let $\mathcal{A}$ be a linear multivalued operator on X and $\rho(\mathcal{A}) \neq \emptyset$. The Cauchy problem (ICP) is n-well-posed on D_{n+1} for some n if and only if the following condition holds:*

(R) *there exists a parameter $m \in \mathbf{R}$ and a set*

$$\Lambda := \left\{ \lambda \in \mathbf{C} : \Re\lambda > \frac{m}{\tau} \ln(1+|\lambda|) + \frac{C}{\tau}, \ \tau \in (0,T), \, C > 0 \right\}$$

such that $\|R_{\mathcal{A}}(\lambda)\| \leq M|\lambda|^m$ for some $M > 0$ and any $\lambda \in \Lambda$.

Proof. Let (ICP) be n-well-posed on D_{n+1}, that is for any $x \in D_{n+1}$ there exists a unique solution $u(t)$ such that

$$\sup_{t\leq\tau<T} \|u(t)\| \leq C_\tau \|x\|_n, \quad \|x\|_n := \inf_{y:R^n(\lambda)y=x} \|y\|.$$

As it is proved in Theorem 2, the constructed family $\{V(t), \ t \in [0,T)\}$ satisfies (5) and (6) with $k = n+1$; hence by Definition 2, $V(t)$ is the local degenerate $(n+1)$-times integrated semigroup with generator $\mathcal{A}$. Equivalence between existence of a $(n+1)$-times integrated semigroup and condition **(R)** may be proved by Tanaka-Okazawa scheme [17, 12]: from (5), (6) with $k = n+1$ it follows

$$\begin{aligned} R(\lambda,\tau)(\lambda - \mathcal{A})x &= (I - G(\lambda))x, \quad x \in D(A), \\ (\lambda - \mathcal{A})R(\lambda,\tau)x &\ni (I - G(\lambda))x, \quad x \in X, \end{aligned}$$

where

$$R(\lambda,\tau)x := \int_0^\tau \lambda^{n+1} \exp(-\lambda t) V(s)x \, ds,$$

$$G(\lambda)x := \lambda^n e^{-\lambda\tau} V(\tau)x + \sum_{k=0}^{n-1} \frac{(\lambda\tau)^k}{k!} e^{-\lambda t} x, \quad x \in X,$$

and $\|G(\lambda)\| < \gamma < 1$, $\|(1 - G(\lambda))^{-1}\| < 1/(1-\gamma)$ for $\lambda \in \Lambda$. Therefore, for $\lambda \in \Lambda$ there exists the bounded operator $(\lambda - \mathcal{A})^{-1} = R_{\mathcal{A}}(\lambda)$, and condition **(R)** is fulfilled with parameter m equal to $(n+1)$. The converse. If **(R)** holds, then for any $p \geq 2m+2$, the function

$$V_p(t) := \frac{1}{2\pi i} \int_\Gamma \lambda^{-p} \exp(\lambda t) R_{\mathcal{A}}(\lambda) \, d\lambda \tag{16}$$

is defined on X and satisfies (5), (6) with $k = p$. Then for any $x \in D(\mathcal{A}^{p+2})$, $u(t) = V_p^{(p)}x$ is proved to be the unique solution of (ICP) stable with respect to $\|x\|_n$. $\square$

Note that in contrast to condition (11), in the local case even with given decomposition (15) condition **(R)** does not bound parameter n and m rigidly.

2. Distribution methods

Using semigroup methods for any $x \in D(\mathcal{A}^{n+1})$ we constructed a unique solution of (ICP) stable with respect to $\|x\|_n$. Now we consider the Cauchy problem (ICP) for any $x \in X$ in the space of abstract distributions $\mathcal{D}'(X) := \mathcal{L}(\mathcal{D}, X)$. If $u(t)$ is a classical solution of (ICP), then the distribution $U \in \mathcal{D}_0'(X)$:

$$U(\phi) := \langle u(t)H(t), \phi \rangle = \int_0^\infty \phi(t)u(t)\,dt, \quad \phi \in \mathcal{D}$$

is a solution of the inclusion

$$\text{(DCP)} \qquad \mathcal{P} * U \ni \delta \otimes x, \quad \mathcal{P} := \delta' \otimes I - \delta \otimes \mathcal{A} \in \mathcal{D}_0'(\mathcal{L}([D(\mathcal{A})], X)).$$

Here $H(t)$ is the Heaviside function, $\mathcal{D}$ is the Schwartz space of infinitely differentiable functions with compact supports, $\mathcal{D}_0'(X) \subset \mathcal{D}'(X)$ is the subspace of distributions equal to zero on $(-\infty, 0)$, $\langle \delta \otimes \mathcal{A}, \phi \rangle := \langle \delta, \phi \rangle \mathcal{A}$, $[D(\mathcal{A})] := \{D(\mathcal{A}), \|x\|_{\mathcal{A}} := \|x\| + \|\mathcal{A}x\|\}$, $\|\mathcal{A}x\|$ is the norm of the element $\{\mathcal{A}x\}$ in the factor-space $X/\mathcal{A}0$.

Definition 3. *The Cauchy problem (ICP) is called well-posed in the sense of distributions (or (DCP) is well-posed) if for any $x \in X$, there exists an unique solution of (DCP) $U \in \mathcal{D}_0'([D(\mathcal{A})])$ (degenerate on some subspace Z) and for any $x_n \to 0$, $U_n \to 0$ in $\mathcal{D}_0'([D(\mathcal{A})])$.*

Theorem 6. *Let $\mathcal{A} : X \to X$ be a linear closed multivalued operator. Then the following statements are equivalent.*

(W) *(DCP) is well-posed.*

(S) *There exists a solution operator $S \in \mathcal{D}_0'(\mathcal{L}(X, [D(\mathcal{A})]))$:*

$$(17) \qquad (\mathcal{P} * S)x \ni \delta \otimes x, \quad x \in X, \quad (S * \mathcal{P})x = \delta \otimes x, \quad x \in D(\mathcal{A}),$$

dengerate on $\mathcal{A}^n 0$ for some n.

(I) *For any $T > 0$, $\mathcal{A}$ is the generator of a local $k(T)$-times integrated degenerate on $\mathcal{A}0$ semigroup $\{V(t),\ t \in [0, T)\}$.*

Proof. **(W)** $\Leftrightarrow$ **(S)**. Similarly to the singlevalued case [5, 13], this equivalence is based on the equality

$$\langle U, \phi \rangle = \langle Sx, \phi \rangle := \langle S, \phi \rangle x, \quad \phi \in \mathcal{D},\ x \in X.$$

(S) $\Rightarrow$ **(I)**. Let $S \in \mathcal{D}'_0(\mathcal{L}(X, [D(\mathcal{A})])$ be a solution of (17). The distribution S can be extended to $\mathcal{D}^p[-1,T]$, where p is the local order of S on $(-1,T)$. We construct a $k = (p+2)$-times integrated semigroup just as the primitive for a distribution in the distribution theory. Let $t \in [0,T]$, consider the function $\psi_{t,p}(s) = \chi(s)\eta_p(t-s) \in \mathcal{D}^p[-1,T]$, where

$$\chi(s) = \begin{cases} 0, & s \le -1, \\ 1, & s \ge 0, \end{cases} \quad \chi(s) \in C^\infty(\mathbf{R}), \qquad \eta_p(t) = \begin{cases} \frac{t^{p+1}}{(p+1)!}, & t \ge 0, \\ 0, & t < 0, \end{cases}$$

and define $V(t) = \langle S, \psi_{t,p} \rangle$. We can prove similar to the nondegenerate case that $V(t)$ satisfies (5), (6).

(I)$\Rightarrow$**(S)**. We construct $V_m(t)$ by formula (16) for $m \ge 2k+2$. This integral defines a continuous function on $(-\infty, \tau_m)$, where $\tau_m = \frac{\tau}{k}(m-k-1)$, $\tau < T$, and $S := V_m^{(m)}$ on $(-\infty, \tau_m)$. Now we consider the structure of Sx on the degeneration set Z. Let $x \in Z$. Since for every $\phi \in \mathcal{D}_0$, $S(\phi)x = 0$, we have $Sx = 0$ on $(0,\infty)$. Owing to $S \in \mathcal{D}'_0(\mathcal{L}(X, [D(\mathcal{A})]))$, we have $Sx = 0$ on $(-\infty, 0)$. Hence $\operatorname{supp} Sx = \{0\}$ and $Sx = \sum_{i=0}^{p} \delta^{(i)} z_i$. Since Sx is a solution of (17), we have

$$\delta^{(p+1)} \otimes z_p + \delta^{(p)} \otimes (Z_{p-1} - \mathcal{A}z_p) + \cdots + \delta' \otimes (z_0 - \mathcal{A}z_1) - \delta \otimes \mathcal{A}z_0 \ni \delta \otimes x,$$

then $-x \in \mathcal{A}z_0$, $z_0 \in \mathcal{A}z_1$, ..., $z_{p-1} \in \mathcal{A}z_p$, $z_p = 0$ or $z_p = 0$, $z_{p-1} \in \mathcal{A}0$, ..., $z_0 \in \mathcal{A}^{p-1}0$, $-x \in \mathcal{A}^p 0$.

3. Regularization methods

In the previous sections by the semigroup and distribution methods we studied the Cauchy problem (ICP) with $\mathcal{A}$ having regular points in some right half-plane or in the domain Λ. For the case more narrow domains of regular points we can use K-convoluted semigroup and ultradistribution methods [4, 9, 15]. The limit case, when $\mathcal{A}$ has no regular points in a right half-plane, is the case of the back heat type. Ill-posed Cauchy problems for such equations used to be investigated by regularization methods.

We consider the ill-posed Cauchy problem (ICP) with $\mathcal{A}$ such that $-\mathcal{A}$ generates the uniform well-posed Cauchy problem. Then by Theorem 1, the simple restriction of $-\mathcal{A}$ operator $-A$ is the generator of a semigroup of class C_0. We construct a regularizing operator for (ICP) as the superposition of the projection on $X_1 = \overline{D(\mathcal{A})}$ and a regularizing operator for the ill-posed Cauchy problem with singlevalued operator A:

$$u'(t) = AU(t), \quad t \ge 0, \ u(0) = x_\delta. \tag{18}$$

That is we construct a regularizing operator for (18) with $x_\delta = P_{X_1}\tilde{x}_\delta$, where $\|x - \tilde{x}_\delta\| \le \delta$ and for x there exists a solution of (ICP).

Definition 4. *The operator $R_{\varepsilon,t} : X \to C\{[0,T], X\}$ is called the regularizing operator for the Cauchy problem (18) if*

(1) *Operator $R_{\varepsilon,t}$ is defined for all $x_\delta \in X$. (Usually $R_{\varepsilon,t}$ supposed to be bounded.)*

(2) *There exists a dependence $\varepsilon = \varepsilon(\delta)$ $(\varepsilon(\delta) \to_{\delta\to 0} 0)$ such that $\|R_{\varepsilon(\delta),t}x_\delta - u(t)\| \to_{\delta\to 0} 0$.*

Main "differential" regularization methods for ill-posed Cauchy problem (18) are the quasireversion method:

$$u'_{\varepsilon,\delta}(t) = (A - \varepsilon A^2)u_{\varepsilon,\delta}(t) =: A_\varepsilon u_{\varepsilon,\delta}(t), \quad 0 < t \le T, \ u_{\varepsilon,\delta}(0) = x_\delta;$$

the auxiliary boundary conditions (ABC) method:

$$\tilde{u}'_{\varepsilon,\delta}(t) = A\tilde{u}_{\varepsilon,\delta}(t), \quad 0 < t < T, \quad \tilde{u}'_{\varepsilon,\delta}(0) + \varepsilon\tilde{u}'_{\varepsilon,\delta}(T) = x_\delta, \quad \varepsilon > 0,$$

and the method of reducing to a Dirichlet problem (Carasso's method). The regularizing operators obtained by the quasireversion and ABC method are the following [10, 7]

$$R_{\varepsilon,t}x_\delta := V_{A_\varepsilon}(t)x_\delta = -\frac{1}{2\pi i}\int_\gamma e^{(\mu-\varepsilon\mu^2)t}R(\mu, A)x_\delta\, d\mu.$$

$$\tilde{R}_{\varepsilon,t}x_\delta := -\frac{1}{2\pi i}\int_\gamma \frac{e^{\lambda t}}{1+\varepsilon e^{\lambda T}}R(\lambda, A)x_\delta\, d\lambda.$$

The following theorem demonstrates relation of regularization methods to semigroup methods.

Theorem 7. *[11] Suppose $-A$ generates a strongly continuous semigroup on X. Then the following statements are equivalent: (I) A is the generator of a local C_ε-semigroup $\{S(t), 0 \le t < T\}$ with operator C_ε strongly convergent to the identity operator as $\varepsilon \to 0$; (II) there exists a linear bounded regularizing operator $R_{\varepsilon,t}$ for (18). It is invertible and commutes with A.*

Acknowledgements. The paper is partially supported by grants MGPE-98 and RFBR No 99-01-00142.

References

[1] Arendt, W. *Vector valued Laplace transforms and Cauchy problems*, Israel J. Math., **59** (1987), 327–352.

[2] Arendt, W. and Favini, A. *Integrated solutions to implicit differential equations.* Rend. Sem. Mat. Univ. Pol. Torino, **51**, No 4 (1993), Partial Diff. Eqs., 315–329.

[3] Arendt, W., El-Mennaoui, J. and Keyantuo, V. *Local integrated semigrpus: evolution with jumps of regularity.* J. of Math. Analysis and Appl., **186**, No 2 (1994), 572–595.

[4] Cioranescu, I. and Lumer, G. *Regularizations of evolution equations via kernels $K(t)$ K-evolution operators and convoluted semigroups, generation theorems. Seminar notes in funct. analysis and PDE, Louisiana State Univer., Baton Rouge.* 1994. 45–52.

[5] Fattorini, H.O. *The Cauchy problem. Encyclop. Math. Appl.*, **18**. Reading, Mass.: Addison-Wesley, 1983.

[6] Favini, A. and Yagi. *Degenerate differential equations in Banach spaces.* New York, Basel, Hong Kong: Marcel Dekker, Inc., 1999.

[7] Ivanov, V.K., Melnikova, I.V. and Filinkov, A.I. *Differential-operator equations and ill-posed problems* Moscow: Nauka, 1995.

[8] Knuckles, C. and Neubrander F. *Remarks on the Cauchy problem for multi-valued linear operators. Math. Res.*, **82**. Berlin: Academie-Verlag, 1994. 174–187.

[9] Komatsu H. *Ultradistributions I: Structure theorems and characterization. J. Fac. Sci. Univ. Tokyo*, **20**, No 1 (1973), 25–105.

[10] Melnikova I.V. *Regularization of ill-posed differential problems. Siberian Math. J.*, **33**, No 2 (1992), 125–134.

[11] Melnikova, I.V. and Filinkov, A.I. *Integrated semigroups and C-semigroups. Well-posedness and regularization of differential-operator problems. Russian Math. Surveys*, **49**, No 6 (1994), 115–155.

[12] Melnikova, I.V. and Alshansky. *Well-posedness of the Cauchy problem in a Banach space: regular and degenerate cases. Journal of Math. Sci.*, **87**, No 4 (1997), 3732–3777.

[13] Melnikova, I.V. *Properties of Lion's d-semigroups and generalized well-posedness of the Cauchy problem. Functional Analysis and its Applications*, **31**, No 3 (1997), 23–37.

[14] Melnikova, I.V. and Gladchenko, A.V. *Well-posedness of the Cauchy problem for inclusions in Banach spaces. Soviet Math. Dokl.*, **361**, No 6 (1998).

[15] Melnikova, I.V. Anufrieva, U.A. and Filinkov, A.I. *Lacplace tranform of K-semigroups and well-posedness of the Cauchy problem, J. Int. Trans. and Spec. Funct.* **8**, No 1-2 (1999), 1–20.

[16] Neubrander, F. and Abdelasis, N.Y. *Dengerate abstract Cauchy problem, Seminar notes in funct. analysis and PDE, Louisiana State Univ., Baton Rouge.* 1991-1992. 1–12.

[17] Tanaka, N. and Okazawa, N. *Local C-semigroups and local integrated semigroups. Proc. London Math. Soc.*, **61**, No 3 (1990), 63–90.

[18] Thieme, H.R. *Integrated semigroups and integrated solutions to abstract Cauchy problems. J. of Math. Anal. and Appl.*, **152** (1990), 416–447.

[19] Yagi, A. *Generation theorem of semigroup for multivalued linear operators. Osaka J. Math.*, **28** (1991), 385–410.

[20] Zaidman, S. *Well-posed Cauchy problem and related semigroups of operators for the equation $Bu'(t) = Au(t)$, $t \geq 0$ in Banach spaces. Libertas Mathematica*, **12** (1992), 147–159.

I.V. Melnikova, Department of Mathematics, Ural State University, Lenina St. 51, Ekaterinburg 620151, RUSSIA
E-mail address: `irina.melnikova@usu.ru`

Progress in Nonlinear Differential Equations
and Their Applications, Vol. 42

A BOUNDED COMPACT SEMIGROUP ON HILBERT SPACE NOT SIMILAR TO A CONTRACTION ONE

CHRISTIAN LE MERDY
UNIVERSITÉ DE FRANCHE-COMTE

Let H be a Hilbert space and let $B(H)$ be the Banach space of all bounded operators on H. By definition we say that $T \in B(H)$ is similar to a contraction if there exists an invertible operator $S \in B(H)$ such that $\|S^{-1}TS\| \leq 1$. Likewise, we say that a c_0-semigroup $(T_t)_{t\geq 0}$ on H is similar to a contraction semigroup if there exists an invertible operator $S \in B(H)$ such that $\|S^{-1}T_tS\| \leq 1$ for any $t \geq 0$. Clearly any $T \in B(H)$ similar to a contraction is power bounded, i.e., $\sup\{\|T^n\| : n \in \mathbb{N}\} < \infty$, and any c_0-semigroup $(T_t)_{t\geq 0}$ similar to a contraction one is bounded, i.e., $\sup\{\|T_t\| : t \geq 0\} < \infty$. However it has been known for a long time that these conditions are not sufficient to ensure similarity to contraction ([F], [P]). Besides some deep research made to characterize power bounded operators similar to contraction (see [Pa, Bo, Pi1, Pi2]), a recent attempt was made by Q.-P. Vu and F. Yao ([VY]) and by the author ([L]) to classify bounded c_0-semigroups similar to contraction ones. The situations for single operators and for semigroups are quite different. For example, it was noticed both in [VY] and [L] that the property that $\lim_{t\to\infty} \|T_t\| = 0$ does not imply that $(T_t)_{t\geq 0}$ is similar to a contraction semigroup, whereas any $T \in B(H)$ such that $\lim_{n\to\infty} \|T^n\| = 0$ is similar to a contraction (Rota's Theorem).

A classical theorem due to Nagy asserts that if $T \in B(H)$ is power bounded and compact, then it is similar to a contraction ([N]). In [VY] it was shown that if $(T_t)_{t\geq 0}$ is a bounded, compact, and uniformly continuous semigroup on H, then it is similar to a contraction one. However the question of whether Nagy's Theorem is true for general (non-uniformly continuous) c_0-semigroups was left open in that paper. The purpose of this note is to provide the following negative answer.

Theorem. There exists a Hilbert space H and a bounded c_0-semigroup $(T_t)_{t\geq 0}$ on H which satisfies the following properties.

(i) For any $t > 0$, the operator T_t is compact.

(ii) There is no invertible $S \in B(H)$ such that $\|S^{-1}T_tS\| \leq 1$ for any $t \geq 0$.

Proof of Theorem. The operators T_t will be defined as diagonal operators with respect to a conditional Schauder basis. The idea of using Schauder bases to produce exotic examples of semigroups and sectorial operators seems to go back to [BC]. Let H be a separable infinite dimensional Hilbert space. We recall that by definition, a Schauder basis on H is a sequence $(e_j)_{j\geq 1} \subset H$ with the following property. for any x in H, there is a unique sequence $(\alpha_j)_{j\geq 1} \subset \mathbb{C}$ such that $x = \sum_{j=1}^{\infty} \alpha_j e_j$. As a consequence of that definition, there exists a constant $C > 0$ such that for any finite sequence $(\alpha_j)_{j\geq 1} \subset \mathbb{C}$ for any $n \geq 1$,

$$\left\| \sum_{j=1}^{n} \alpha_j e_j \right\| \leq C \left\| \sum_{j=1}^{\infty} \alpha_j e_j \right\| . \tag{1}$$

Let us denote by $\mathcal{V} = \mathrm{Span}\{e_j : j \geq 1\} \subset H$ the dense subspace spanned by the e_j's. Given any complex sequence $a = (a_j)_{j\geq 1} \subset \mathbb{C}$, we let $M_a \colon \mathcal{V} \to \mathcal{V}$ be the diagonal linear mapping defined by $M_a(e_j) = a_j e_j$ for any $j \geq 1$. We say that the basis $(e_j)_{j\geq 1}$ is conditional if:

(2) There exists a bounded sequence $a = (a_j)_{j\geq 1}$ such that M_a is unbounded.

From now on we assume that $(e_j)_{j\geq 1}$ is a conditional basis on H. Although M_a may be unbounded when a is a bounded sequence, we have the following simple criteria at our disposal. Let

$$|||a||| = \limsup_{j\to\infty} |a_j| + \sum_{j=1}^{\infty} |a_{j+1} - a_j|.$$

Then M_a is bounded if $|||a|||$ is finite and in that case,

$$\|M_a\| \leq C|||a|||, \tag{3}$$

where C is the constant appearing in (1). Indeed this follows from a simple Abel transformation.

For any $t \geq 0$, let $a^t = (a^t_j)_{j\geq 1}$ be the sequence defined by $a^t_j = e^{-2^j t}$. If $t > 0$, a^t is a nonincreasing sequence converging to 0 as j goes to ∞ hence $|||a^t||| \leq 1$. It therefore follows from (3) that each diagonal mapping M_{a^t} extends to a bounded operator $T_t \colon H \to H$, with

$$\forall t \geq 0, \quad \|T_t\| \leq C. \tag{4}$$

For any $x \in \mathcal{V}$, we obviously have the semigroup property $T_t T_s(x) = T_{t+s}(x)$ and the continuity property $\lim_{t\to 0} T_t(x) = x$. Using (4) and a standard equicontinuity argument, we deduce that $(T_t)_{t\geq 0}$ is a bounded c_0-semigroup on H.

Let us now check that T_t is compact if $t > 0$. For any $n \geq 1$, let $T^n_t : H \to H$ be the (unique) bounded operator satisfying $T^n_t(e_j) = T_t(e_j)$ if $1 \leq j \leq n$ and $T^n_t(e_j) = 0$ if $j \geq n+1$. Then the restriction of $T_t - T^n_t$ to $\mathcal{V}$ is the diagonal operator associated to the sequence $(0, \cdots, 0, a^t_{n+1}, a^t_{n+2}, \cdots)$. It therefore follows from (3) that

$$\forall t > 0, \quad \forall n \geq 1, \quad \|T_t - T^n_t\| \leq C e^{-2^{n+1} t}.$$

Since $\lim_{n\to\infty} e^{-2^{n+1}t} = 0$ and each T^n_t has a finite rank we deduce that T_t is compact.

We now turn to the proof of (ii). We shall make use of the congenerator for semigroups on Hilbert space and of von Neumann's inequality for contractions, for which we refer to [NF]. We assume that there exists an invertible operator $S \in B(H)$ such that $(S^{-1}T_tS)_{t\geq 0}$ is a contraction semigroup. Let A be the infinitesimal generator of $(T_t)_{t\geq 0}$ and let $T = (A+I)(A-I)^{-1} \in B(H)$ be the cogenerator of $(T_t)_{t\geq 0}$. Then $S^{-1}TS$ is the cogenerator of $(S^{-1}T_tS)_{t\geq 0}$ hence $S^{-1}TS$ is a contraction.

Letting $D = \{z \in \mathbf{C}\colon |z| < 1\}$ be the open unit disc of $\mathbf{C}$, we denote by $H^\infty(D)$ the Banach algebra of all bounded holomorphic functions $f : D \to \mathbf{C}$, equipped with the norm

$$\|f\|_\infty = \sup\{|f(z)| : z \in D\}.$$

Given any $f \in H^\infty(D)$ and any $r \in (0,1)$, we let $f_r \in H^\infty(D)$ be the Poisson approximation of f defined by $f_r(z) = f(rz)$. Then it follows from von Neumann's inequality that:

$$\forall f \in H^\infty(D),\ \ \forall r \in (0,1),\ \ \|f_r(T)\| \leq \|S\|\ \|S^{-1}\|\ \|f\|_\infty. \tag{5}$$

Let $\tau = (\tau_j)_{j\geq 1}$ be the sequence defined by:

$$\tau_j = \frac{2^j - 1}{2^j + 1}\,. \tag{6}$$

Using the obvious fact that $\mathcal{V}$ is included in the domain of A, we see that T is the bounded extension of M_τ. Likewise, for any $f \in H^\infty(D)$ and any $r \in (0,1)$, the operator $f_r(T)$ is the bounded extension of $M_{f_r(t)}$. Using (5) and letting r tend to 1, we obtain that:

$$\forall f \in H^\infty(D), \quad M_{f(\tau)} \ \text{ is bounded.} \tag{7}$$

However it follows from Carleson's theorem (see, e.g., [G, Section VII.1]) that the sequence τ defined by (6) is an interpolating sequence for D. That is, given any bounded sequence $a = (a_j)_{j\geq 1}$, there exists a function $f \in H^\infty(D)$ such that $f(\tau_j) = a_j$ for any $j \geq 1$. Comparing (2) and (7), we obtain a contradiction. □

Remark. Let $S_1(H)$ be the space of all trace class operators T on H, with norm $\|T\|_1 = \operatorname{tr}|T|$. Recall that any element T of $S_1(H)$ is compact and that $\|T\| \leq \|T\|_1$. It can be deduced from the above proof that condition (i) in our Theorem can be strengthened as:

(i)′ For any $t > 0$, the operator T_t belongs to $S_1(H)$ and $\lim_{t\to\infty} \|T_t\|_1 = 0$.

This is left to the reader.

References

[BC] J.-B. Baillon and P. Clément. Examples of unbounded imaginary powers of operators, J. Funct. Anal. 100 (1991), 419–434.

[Bo] J. Bourgain. On the similarity problem for polynomially bounded operators on Hilbert space, Israel. J. Math. 54 (1986), 227–241

[F] S.R. Foguel. A counterexample to a problem of Sz.-Nagy, Proc. Amer. Math. Soc. 15 (1964), 788–790.

[G] J.B. Garnett. Bounded analytic functions. Academic Press, New York, 1981.

[L] C. Le Merdy. The similarity problem for bounded analytic semigroups on Hilbert space, Semigroup Forum 56 (1998), 205–224.

[N] B. Sz.-Nagy. Completely continuous operators with uniformly bounded iterates, Publ. Math. Inst. Hungarian Acad. Sci 4 (1959), 89–92.

[NG] B. Sz.-Nagy and C. Foias. Harmonic analysis of operators on Hilbert space. Akademiai Kiado, Budapest, 1970.

[P] E.W. Packel. A semigroup analogue of Foguel's counterexample, Proc. Amer. Math. Soc. 21 (1969), 240–244.

[Pa] V.I. Paulsen. Every completely polynomially bounded operator is similar to a contraction, J. Func. Anal. 55 (1984), 1–17.

[Pi1] G. Pisier. Similarity problems and completely bounded maps, Lecture Notes in Math. 1618. Springer, 1996.

[Pi2] G. Pisier. A polynomially bounded operator which is not similar to a contraction. J. Amer. Math. Soc. 10 (1997), 351–369.

[VY] Q.-P. Vu and F. Yao. On similarity to contraction semigroups in Hilbert space. Semigroup Forum 56 (1998), 197–204.

Christian Le Merdy, Equipe de Mathematiques, Université de Franche-Comte, Route de Gray, 25030 Besancon Cedex, FRANCE

E-mail address: `lemerdy@math.uh.edu`

Progress in Nonlinear Differential Equations
and Their Applications, Vol. 42

PROPERTIES OF STABILIZATION FUNCTIONAL FOR PARABOLIC CAUCHY PROBLEM

A.B. MURAVNIK
10TH COMMUNAL MEDICAL UNIT OF THE VORONEZH CITY

Abstract. This paper is devoted to properties of sets of initial-value functions providing the stabilization of Cauchy problem solutions for parabolic equations. These problems arise in modelling of heat processes and diffusion processes. The main result is: the set of the inititial-value functions which satisfy the necessary and sufficient condition of stabilization for the classical solution (or the solution almost everywhere) of Cauchy problem for a parabolic equation is an unseparable subspace of the space of continuous bounded functions (or of the space of measurable essentially bounded functions).

1. Introduction

We will use so-called "operator view" on the stabilization problem (see [4]): there is an operator, translating the initial-value function of Cauchy problem for a parabolic equation into the limit (as $t \to \infty$) of the solution of this problem. In these terms the necessary and sufficient condition of stabilization (see for instance [1]) is determined by the domain of such operator. If we consider the case of a bounded initial-value function (and a bounded solution), then this operator is a functional and its domain is a subset of the set of bounded functions. We determine features of this subset.

The reasoning for the investigation of these features is the following.

Suppose that the initial-value function does not provide the stabilization of the solution. Then an important for optimum control applications question about the best approximation of this initial-value function by elements of the domain of the stabilization functional arises. Since the space of bounded continuous functions (as well as $L_\infty(\mathbf{R}^m)$) is not Hilbert space, this question is not trivial and so we need an information about the structure of the above-mentioned domain.

The following notations are used:

$K(r)$ is the ball in $\mathbf{R}^m$ with the radius r and with the centre in the origin.

$S(r)$ is the sphere in $\mathbf{R}^m$ with the radius r and with the centre in the origin; ds is an element of its area.

$B(D)$ is the space of continuous and bounded on D functions with the norm $\sup\limits_{x\in D} |f(x)|$.

Consider the following problem:

$$Lu = 0, \ x \in \mathbf{R}^m, \ t > 0 \tag{1}$$

$$u_{|_{t=0}} = f(x), \ x \in \mathbf{R}^m \tag{2}$$

Here (1) is a parabolic equation satisfying to the assumptions of [1], f is a continuous and bounded function.

It is well known (see for instance [3]) that the classical bounded solution of (1)-(2) exists and is unique. There is (see [1]) a necessary and sufficient condition of (pointwise) stabilization of this solution:

$$\lim_{t\to\infty} u(x,t) \text{ exists for each } x \in \mathbf{R}^m \text{ if and only if } \lim_{r\to\infty} \frac{1}{r^m} \int\limits_{K(r)} f(x)dx \text{ exists.}$$

Let E is the set of such functions $f \in B(\mathbf{R}^m)$, which satisfy to this stabilization condition.

2. Properties of the domain of the stabilization functional

The following assertion is correct:

Theorem A. *E is an unseparable subspace of $B(\mathbf{R}^m)$.*

Proof. Obviously E is a linear manifold. E is a closed set; it was established in [4] by means of maximum principle, but there is a direct proof of this fact:

Let $\{f_n(x)\}_{n=1}^\infty \subset B(\mathbf{R}^m)$ and let $\{f_n(x)\}_{n=1}^\infty$ converges in the norm of $B(\mathbf{R}^m)$ i.e. uniformly. Then its limit belongs to $B(\mathbf{R}^m)$ too. For any $\varepsilon > 0$ there exists $N(\varepsilon)$ such that for any $n \geq N(\varepsilon)$ for any natural p for any $x \in \mathbf{R}^m$

$$|f_n(x) - f_{n+p}(x)| < \frac{\varepsilon}{|K(1)|}$$

($|A|$ is the volume of A).
Hence $-\frac{\varepsilon}{|K(1)|} < f_n(x) - f_{n+p}(x) < \frac{\varepsilon}{|K(1)|}$; therefore for any $r > 0$

$$-\frac{1}{r^m} \int\limits_{K(r)} \frac{\varepsilon}{|K(1)|} dx \leq \frac{1}{r^m} \int\limits_{K(r)} [f_n(x) - f_{n+p}(x)]dx \leq \frac{1}{r^m} \int\limits_{K(r)} \frac{\varepsilon}{|K(1)|} dx$$

$$\text{i.e.} \left| \frac{1}{r^m} \int\limits_{K(r)} f_n(x)dx - \frac{1}{r^m} \int\limits_{K(r)} f_{n+p}(x)dx \right| < \varepsilon.$$

In the last inequality we pass to the limit as $r \to \infty$; we obtain that $|l_n - l_{n+p}| < \varepsilon$, where $l_k = \lim\limits_{r\to\infty} \frac{1}{r^m} \int\limits_{K(r)} f_k(x)dx$ for each natural k .

Hence $\{l_n\}_{n=1}^{\infty}$ is fundamental, hence it converges. Denote its limit by l. Then for any positive r and for any natural n

$$\left|\frac{1}{r^m}\int\limits_{K(r)} f(x)dx - l\right| =$$

$$\left|\frac{1}{r^m}\int\limits_{K(r)} f(x)dx - \frac{1}{r^m}\int\limits_{K(r)} f_n(x)dx + \frac{1}{r^m}\int\limits_{K(r)} f_n(x)dx - l_n + l_n - l\right| \leq$$

$$\left|\frac{1}{r^m}\int\limits_{K(r)} f(x)dx - \frac{1}{r^m}\int\limits_{K(r)} f_n(x)dx\right| + \left|\frac{1}{r^m}\int\limits_{K(r)} f_n(x)dx - l_n\right| + |l_n - l|.$$

Fix positive ε and choose n so large that $|l_n - l| < \frac{\varepsilon}{3}$ and $|f_n(x) - f(x)| < \frac{\varepsilon}{3|K(1)|}$ for any $x \in \mathbf{R}^m$.

Now choose r so large that $\left|\frac{1}{r^m}\int\limits_{K(r)} f_n(x)dx - l_n\right| < \frac{\varepsilon}{3}.$

We obtain: $\left|\frac{1}{r^m}\int\limits_{K(r)} f(x)dx - l\right| < \varepsilon$. ε was chosen arbitrarily hence

$\lim\limits_{r\to\infty}\frac{1}{r^m}\int\limits_{K(r)} f(x)dx$ exists and this limit is equal to l. Hence, $f \in E$, i.e. E is a closed set.

Thus, E is a subspace of $B(\mathbf{R}^m)$.

Now we will prove the unseparability of E.

We will use a corollary of Wiener Tauberian theorem (see [2, pp. 1003-1004]), which we quote here for completeness:

Let $\psi_0(\tau) \in L_1(\mathbf{R}^1_+)$, $f(\tau) \in L_\infty(\mathbf{R}^1_+)$ and suppose that for each real x

$$\int\limits_0^\infty \psi_0(\tau)\tau^{ix}d\tau \neq 0.$$

Assume that there exists a real number A such that

$$\lim_{r\to\infty}\frac{1}{r}\int\limits_0^\infty \psi_0\left(\frac{\tau}{r}\right)f(\tau)d\tau = A\int\limits_0^\infty \psi_0(\tau)d\tau.$$

Then for each $\psi(\tau) \in L_1(\mathbf{R}^1_+)$

$$\lim_{r\to\infty}\frac{1}{r}\int\limits_0^\infty \psi\left(\frac{\tau}{r}\right)f(\tau)d\tau = A\int\limits_0^\infty \psi(\tau)d\tau.$$

Now let E_0 is the set of spherically symmetric functions $g(x)$ satisfying the following assumptions:

$g(x) = \varphi(|x|)$ ($\varphi(\rho)$ is a function of a scalar variable);

if $\rho \in [10^n + 1,\ 10^{n+1}]$ then $\varphi(\rho) = 0$;

if $\rho \in [10^n,\ 10^n + 1]$ then

either $\varphi(\rho) = 0$ or $\varphi(\rho) = \begin{cases} 2(\rho - 10^n) \text{ for } \rho \in [10^n,\ 10^n + \frac{1}{2}] \\ 2(10^n + 1 - \rho) \text{ for } \rho \in [10^n + \frac{1}{2},\ 10^n + 1] \end{cases}$.

Thus $0 \le g(x) \le 1$.

For any $r \in [10,\ 100]$ $\quad \dfrac{1}{r}\displaystyle\int_0^r \varphi(\rho)d\rho \le \dfrac{2}{10}$;

for any $r \in [100,\ 1000]$ $\quad \dfrac{1}{r}\displaystyle\int_0^r \varphi(\rho)d\rho \le \dfrac{3}{100}$;

for any $r \in [1000,\ 10000]$ $\quad \dfrac{1}{r}\displaystyle\int_0^r \varphi(\rho)d\rho \le \dfrac{4}{1000}$;

...

for any $r \in [10^n,\ 10^{n+1}]$ $\quad \dfrac{1}{r}\displaystyle\int_0^r \varphi(\rho)d\rho \le \dfrac{n+1}{10^n}$.

Since sequence $\{\dfrac{n+1}{10^n}\}$ monotonously tends to zero then

$$\lim_{r\to\infty} \frac{1}{r}\int_0^r \varphi(\rho)d\rho = 0.$$

$$\frac{1}{r^m}\int_{K(r)} g(x)dx = \frac{1}{r^m}\int_0^r\int_{S(\rho)} g(x)dsd\rho = \frac{1}{r^m}\int_0^r \varphi(\rho)\int_{S(\rho)} dsd\rho$$

$$= \frac{2\pi^{\frac{m}{2}}}{\Gamma(\frac{m}{2})r^m}\int_0^r \rho^{m-1}\varphi(\rho)d\rho.$$

Now we set:

$$\psi_0(\tau) \stackrel{\text{def}}{=} \begin{cases} 1 \text{ for } 0 < \tau < 1 \\ 0 \text{ otherwise} \end{cases},\quad \psi(\tau) \stackrel{\text{def}}{=} \begin{cases} \tau^{m-1} \text{ for } 0 < \tau < 1 \\ 0 \quad \text{otherwise} \end{cases},\quad f(\tau) \stackrel{\text{def}}{=} \varphi(\tau),$$

and we can apply the above-mentioned corollary of Wiener Tauberian theorem.

Thus we obtain that

$$\frac{1}{r^m}\int\limits_{K(r)} g(x)dx = 0$$

i.e. $E_0 \subset E$.

On the other hand there is one-to-one correspondence between E_0 and the set of the sequences which consist of zeroes and unities (if $\varphi(\rho) \equiv 0$ on $[10^n, 10^{n+1}]$ then zero is on the nth position otherwise unity is on the nth position) but the power of such set is the power of $\mathbf{R}^1$ because it is the set of the infinite decimal fractions.

After all $\|g_1 - g_2\| = 1$ for any different g_1, g_2 from E_0.

Now suppose to the contrary that E is a separable space. Let $\varepsilon_0 = \frac{1}{3}$. Then all elements of E are in a denumerable set of balls with radius $\frac{1}{3}$ (the centres of these balls constitute a denumerable dense set in E). Since E_0 is a set of the power of the continuum at least one of this balls contains two different elements of E_0, i.e the distance between these elements is less than 1.

We obtain a contradiction. □

3. Remark

Theorem A is also correct correct if we replace $B(\mathbf{R}^m)$ by $L_\infty(\mathbf{R}^m)$. In this case the coefficients of (1) are measurable functions and the solution of (1) is an almost everywhere differentiable function from $L_\infty(\mathbf{R}^m)$.

References

[1] V.N. Denisov and V.D. Repnikov, The stabilization of a solution of a Cauchy problem for parabolic equations, Differential Equations, **20**, No. 1 (1984), 16-33.

[2] N. Dunford and J.T. Schwartz, *Linear Operators, Part 2*, Interscience Publishers, New York 1963.

[3] A. Friedman, *Partial Differential Equations of Parabolic Type*, Prentice-Hall, Englewood Cliffs, New York 1964.

[4] V.V. Žikov, On the stabilization of solutions of parabolic equations, Math. USSR Sb., **33**, No. 4 (1977), 519-537.

A.B. Muravnik, Computer Centre, 10th Communal Medical Unit of the Voronezh City, Lizyukova 36A, Voronezh 394053, RUSSIA

E-mail address: muravnik@hotmail.com

Progress in Nonlinear Differential Equations
and Their Applications, Vol. 42

ON THE SPECTRUM OF SOME FUNCTIONAL DIFFERENTIAL EQUATIONS

TOSHIKI NAITO
NGUYEN VAN MINH
UNIVERSITY OF ELECTRO-COMMUNICATIONS, TOKYO

JONG SON SHIN
KOREA UNIVERSITY, TOKYO

1. Introduction

The main result(Theorem 2) is that a left half plane is contained in the residual spectrum of the generator of a solution semigroup of evolution equations with infinite delay on the phase space of some continuous functions. As a result, the half plane is contained in the essential spectrum of the generator, and the growth bound of the semigroup is bounded below by the constant which is independent of the equation but depends on the phase space.

For the functional differential equation with finite delay, the spectrum of the generator of the solution semigroup consists of only normal eigenvalues. They are given by the set of zeros of the determinant of the characteristic matrix of the equation, cf. [1]. The main reason for the appearance of only normal eigenvalues is that the solution operator becomes compact after the delay time.

In the case of infinite delay, the whole distribution of the spectrum of the generator of solution semigroup is not yet known even for the equation in the finite dimensional space. Long ago [6], the first author proved that the spectrum of the generator consists of only normal eigenvalues in some right half plane; but nothing is said about the spectrum in the rest set of that half plane; see [5], too.

Since the solution operator is not compact in general, it is reasonable to expect that the essential spectrum will appear in some left half plane. The present paper gives an answer to this left problem for the case of simple evolution equation with infinite delay on the special phase space. We are now studying some generalizations, and the results will soon appear in the subsequent papers.

2. The spectrum of solution semigroups

We will compute the spectrum of the solution semigroup of a functional differential equation with infinite delay of the form

$$x'(t) = Ax(t) + L(x_t),$$

where $x(t)$ is a function taking the value in a Banach space E, $x_t : (-\infty, 0] \to E$ is defined as $x_t(\theta) = x(t+\theta), \theta \in (-\infty, 0]$, and x_t is called the t-section of x.

The operator A is the infinitesimal generator of a C_0-semigroup $T(t), t \geq 0$, and L a bounded linear operator on the following function space $\mathcal{C}_\gamma$ to E. $\mathcal{C}_\gamma$ is a Banach space of continuous functions $\phi : (-\infty, 0] \to E$ such that the limit

$$\tilde{\phi}(-\infty) := \lim_{\theta \to -\infty} e^{-\gamma\theta}\phi(\theta)$$

exists, and the norm in $\mathcal{C}_\gamma$ is defined by

$$\|\phi\| := \sup\{|e^{-\gamma\theta}\phi(\theta)| : \theta \in (-\infty, 0]\}.$$

For any $x \in E$ and any complex number λ, define the function $\varepsilon_\lambda \otimes x$: $(-\infty, o] \to E$ by

$$(\varepsilon_\lambda \otimes x)(\theta) = e^{\lambda\theta}x, \quad \text{for} \quad \theta \in (-\infty, 0].$$

It belongs to $\mathcal{C}_\gamma$ provided $\Re\lambda > \gamma$.

Suppose that $u : (-\infty, \infty) \to E$ is a function such that $\phi := u_0 \in \mathcal{C}_\gamma$ ant it is continuous on $[0, \infty)$. We say that u is a mild solution of the above functional differential equation if it satisfies the integral equation

$$u(t) = T(t)u(0) + \int_0^t T(t-s)L(u_s)ds \quad \text{for} \quad t \geq 0,$$

Then there exists a unique mild solution $u(t) := u(t, \phi)$ for any $\phi \in \mathcal{C}_\gamma$, cf. [9, 10].

Denote by $u_t(\phi)$ the t-section of $u(t, \phi)$: $u_t(\phi)(\theta) = u(t+\theta, \phi), \theta \in (-\infty, 0]$. Define the solution operator $U_L(t)$ by $U_L(t)\phi = u_t(\phi)$ for $\phi \in \mathcal{C}_\gamma$ and $t \geq 0$. Then the family $U_L(t), t \geq 0$, forms a C_0- semigroup of bounded linear operators on $\mathcal{C}_\gamma$. Let A_L be its infinitesimal generator.

To describe the spectrum of A_L, we introduce the characteristic operator of the functional differential equation discussed now. Define the bounded linear operator $L_\lambda : E \to E$ for $\Re\lambda > \gamma$ by $L_\lambda x = L(\varepsilon_\lambda \otimes x)$ for $x \in E$. Then the characteristic operator $\Delta_L(\lambda)$ is defined by

$$\Delta_L(\lambda)x = (\lambda I - A - L_\lambda)x \quad \text{for} \quad x \in D(A).$$

It is a closed linear operator on E. Set

$$\mathbf{C}^\gamma = \{\lambda \in \mathbf{C} : \Re\lambda > \gamma\} \quad \text{and} \quad P(\Delta_L) = \{\lambda \in \mathbf{C}^\gamma : N(\Delta_L(\lambda)) \neq \{0\}\}.$$

If $R(\lambda, A) = (\lambda I - A)^{-1}$ exists, then $\Delta_L(\lambda) = (I - L_\lambda R(\lambda, A))(\lambda I - A)$.

For a linear operator T on a Banach space, let $\rho(T)$ be the resolvent set of T, $\sigma(T)$ the spectrum of T, $P_\sigma(T)$ the point spectrum of T, $R_\sigma(T)$ the residual spectrum of T, $E_\sigma(T)$ the essential spectrum of T, and set $N_\sigma(T) := \sigma(T) \setminus E_\sigma(T)$. The points in $N_\sigma(T)$ are called normal eigenvalues of T. Recall that, by definition, λ is a normal eigenvalue of T if it is an isolated point of $\sigma(T)$, the range $R(T - \lambda I)$ is closed and the generalized eigenspace $M_\lambda(T)$ is of finite dimension.

First of all the following result holds, see [7].

Lemma 1. *If $\Re\lambda > \gamma$, then $\lambda \in P_\sigma(A_L)$ if and only if $N(\Delta_L(\lambda)) \neq \{0\}$, and for such a λ the nontrivial eigenfunctions $\phi \in N(\Delta_L(\lambda))$ are given as $\phi = \varepsilon_\lambda \otimes x$ for $x \neq 0$ such that $\Delta_L(\lambda)x = 0$.*

The spectrum of the semigroup of the translations on $C[0,\infty]$ is computed in the book [2, pp.530-533]. The spectrum of its generator d/dt on $C[0,\infty]$ consists of left half plane $\Re\lambda \leq 0$; the points on the imaginary axis except for $\lambda = 0$ are in the continuous spectrum and the rest set of the half plane is in the point spectrum. Comparing this result, we have the following main theorem.

Theorem 2. $\{\lambda : \Re\lambda < \gamma\} \subset R_\sigma(A_L)$, *and hence,* $\{\lambda : \Re\lambda \leq \gamma\} \subset E_\sigma(A_L)$.

Proof The generator A_L is given as follows. $D(A_L)$ consists of functions $\phi \in \mathcal{C}_\gamma$ such that $\phi(0) \in D(A)$, the derivative $\phi' \in \mathcal{C}_\gamma$ and that $\phi'(0) = A\phi(0) + L(\phi)$; and $A_L\phi = \phi'$. We consider the range of $\lambda I - A_L$. The equation $(\lambda I - A_L)\phi = \psi$ is equivalent that

$$\lambda\phi(\theta) - \phi'(\theta) = \psi(\theta) \quad \text{for} \quad \theta \leq 0,$$

$$\phi(0) \in D(A), \quad \lambda\phi(0) - A\phi(0) - L(\phi) = \psi(0).$$

The solution of the first differential equation is given as

$$\phi(\theta) = e^{\lambda\theta}\left(\phi(0) + \int_\theta^0 e^{-\lambda s}\psi(s)ds\right).$$

Suppose that $\Re\lambda < \gamma$. This solution ϕ belongs to the space $\mathcal{C}_\gamma$ if and only if

$$\phi(0) + \int_{-\infty}^0 e^{-\lambda s}\psi(s)ds = 0.$$

Hence, for every $\psi \in \mathcal{C}_\gamma$, the differential equation has a unique solution $\phi \in \mathcal{C}_\gamma$ such that $\phi = -K_\lambda\psi$, where

$$K_\lambda\psi(\theta) = \int_{-\infty}^\theta e^{\lambda(\theta - s)}\psi(s)ds.$$

Furthermore, K_λ is a bounded linear operator on $\mathcal{C}_\gamma$ such that $\|K_\lambda\| \leq 1/(\gamma - \Re\lambda)$.

The condition $\phi(0) \in D(A)$ and $(\lambda I - A)\phi(0) - L(\phi) = \psi(0)$ becomes $K_\lambda\psi(0) \in D(A)$ and

$$D_\lambda\psi := \psi(0) + (\lambda I - A)((K_\lambda\psi)(0)) + L(-K_\lambda\psi) = 0.$$

Since

$$|(K_\lambda\psi)(0)| \leq (1/(\gamma - \Re\lambda))\|\psi\|,$$

$D_\lambda : \mathcal{C}_\gamma \to E$ is a linear operator on $\mathcal{C}_\gamma$ into E which is a sum of a bounded operator on $\mathcal{C}_\gamma$ and a closed operator whose domain is $\{\psi : (K_\lambda\psi)(0) \in D(A)\}$. Since the null space of such a linear operator is closed in general, $N(D_\lambda)$ is a closed subspace of $\mathcal{C}_\gamma$. Hence $R(\lambda I - A_L) = N(D_\lambda)$, and it is a closed subspace of $\mathcal{C}_\gamma$.

If $R(\lambda I - A_L)$ is dense, it follows that $N(D_\lambda) = \mathcal{C}_\gamma$. Let $a \in D(A)$, and set $\psi^n(\theta) = e^{(\gamma+n)\theta}a, n = 1, 2, \cdots$. Since $\psi^n \in \mathcal{C}_\gamma = N(D_\lambda)$, we have that $D_\lambda \psi^n = 0$. From the elementary computation, we have that

$$D_\lambda \psi^n = a + \frac{1}{\gamma + n - \lambda}((\lambda I - A)a - L(\varepsilon_{\gamma+n} \otimes a)).$$

If n is sufficiently large, the right side does not vanish. This is a contradiction.

Therefore $R(\lambda I - A)$ is not dense, that is, $\lambda \in R_\sigma(A)$ provided $\Re\lambda < \gamma$.

An accumulation point of the spectrum lies in the essential spectrum by definition. Hence $\{\lambda : \Re\lambda \le \gamma\} \subset E_\sigma(A_L)$.

To get the next theorem, we recall standard results on the spectral theory of the C_0-semigroup $T(t)$ on a Banach space; see [13, Proposition 4.13]. The growth bound $\omega_s(T)$, and the essential growth bound $\omega_e(T)$ of $T(t)$ are defined by

$$\omega_s(T) := \lim_{t\to\infty} \frac{\log \|T(t)\|}{t} = \inf_{t>0} \frac{\log \|T(t)\|}{t},$$

$$\omega_e(T) := \lim_{t\to\infty} \frac{\log \alpha(T(t))}{t} = \inf_{t>0} \frac{\log \alpha(T(t))}{t},$$

where $\|T(t)\|$ stands for the operator norm of $T(t)$ and $\alpha(T(t))$ is the measure of noncompactness of $T(t)$ which is described by the Kuratowskii measure of noncompactness of bounded sets in E(cf.[13]). Then the spectral radius $r_s(T(t))$ and the essential spectral radius $r_e(T(t))$ are given as $r_s(T(t)) = \exp(t\omega_s(T))$ and $r_e(T(t)) = \exp(t\omega_e(T))$. Let A be the generator of $T(t)$. Then

$$\sup\{\Re\lambda : \lambda \in \sigma(A)\} \le \omega_s(T), \quad \sup\{\Re\lambda : \lambda \in E_\sigma(A)\} \le \omega_e(T) \tag{1}$$

and

$$\omega_s(T) = \max\{\omega_e(T), s(A)\},$$

where

$$s(A) := \sup\{\Re\lambda : \lambda \in N_\sigma(A)\}.$$

We apply these results for the semigroup $U_L(t)$. As a fundamental result, if $T(t)$ is a compact semigroup, then we have $\omega_e(U_L) \le \gamma$; see Proposition 4.6 (1) in [11]. Then it follows at once $E_\sigma(A_L) = \{\lambda : \Re\lambda \le \gamma\}$.

We get the following theorem by combining these results and Theorem 2.

Theorem 3. *If $T(t)$ is a compact semigroup, then*

$$\{\lambda : \Re\lambda \le \gamma\} = E_\sigma(A_L), \quad \{\lambda : \Re\lambda > \gamma\} = N_\sigma(A_L) \cup \rho(A_L),$$

$$N_\sigma(A_L) = \{\lambda : N(\Delta_L(\lambda)) \ne \{0\}\}.$$

3. The stability of solutions

As an application to the stability, we have the following result from Theorem 3.

Proposition 4. *If $T(t)$ is a compact semigroup, then, $r_e(U_L(t)) = e^{\gamma t}$, and for any $\epsilon > 0$ there exists an $M_\epsilon > 0$ such that*

$$e^{\gamma t} \leq \|U_L(t)\| \leq M_\epsilon e^{(\omega+\epsilon)t} \quad \textit{for} \;\; t > 0,$$

where $\omega = \max\{\gamma, \sup\{\Re\lambda : \lambda \in P(\Delta_L)\}\}$.

Theorem 5. *Suppose that $\gamma < 0$, $T(t)$ is a C_0-compact semigroup and that $\omega_s(U_L) = 0$ for the operator L.*

(i) *If $\Delta_L(\lambda)^{-1}$ has a pole of order greater than 2 at some point on the imaginary axis, then $\|U_L(t)\phi\| \to \infty$ as $t \to \infty$ for some ϕ.*

(ii) *If $\Delta_L(\lambda)^{-1}$ has at most poles of order 1 at every point on the imaginary axis, then $1 \leq \|U_L(t)\| \leq M$ for $t \geq 0$ and for some constant $M > 0$.*

Proof. Sine $\omega_e(U_L) = \gamma < 0 = \omega_s(U_L)$, $\sigma(A_L)$ in $\mathbf{C}^\gamma$ consists of only normal eigenvalues with real part not greater than 0, and there exist normal eigenvalues with 0 real parts.

We can prove that, if $\Delta_L(\lambda)^{-1}$ has a pole of order k at some point $\mu \in \mathbf{C}^\gamma$, then $R(\lambda, A_L)$ has the pole of the same order at μ.

The assertions (i) and (ii) then are derived from the the results of the general C_0-simegroup $T(t)$ with the generator A: if x belongs to the null space $N((A-\lambda I)^m), I$ being the identity, A being the generator of C_0-simegroup $T(t)$, then

$$T(t)x = e^{\lambda t} \sum_{k=0}^{m-1} \frac{t^k}{k!} (A - \lambda I)^k x.$$

Finally, we give a simple example.

Example (cf. [7]) Let $E = L^2([0,\pi], \mathbf{C})$, the set of square integrable functions on $[0,\pi]$. Consider the equation

$$x'(t) = Ax(t) + b \int_{-\infty}^{t} e^{-c(t-s)} x(s) ds,$$

where A is defined as $Af = f''$ for $f \in E$ such that f is continuously differentiable, the derivative f' is absolutely continuous, $f'' \in E$, and that $f(0) = f(\pi) = 0$. Then

$$\sigma(A) = P_\sigma(A) = \{-n^2 : n = 1, 2, \ldots\},$$

and A generates a C_0-compact semigroup $T(t)$ on E.

If $c > -\gamma$, then, for $\phi \in \mathcal{C}_\gamma$,

$$|L(\phi)| := \left| b \int_{-\infty}^{0} e^{c\theta}\phi(\theta)d\theta \right| \le |b|\|\phi\|/(c+\gamma).$$

The characteristic operator $\Delta_L(\lambda)$ becomes

$$\Delta_L(\lambda)f = \lambda f - Af - b\int_{-\infty}^{0} e^{(c+\lambda)\theta}d\theta f = (h(\lambda)I - A)f, \quad f \in D(A), \ \Re\lambda > \gamma > -c,$$

where $h(\lambda) = \lambda - b/(c+\lambda)$. Therefore

$$P(\Delta_L) = \{\lambda : h(\lambda) = -n^2 \ \text{ for some } \ n \ge 1, \Re\lambda > \gamma\}.$$

Since $R(\lambda, A)$ has a pole of order 1 at $\lambda = -n^2, n \ge 1$, $\Delta_L(\lambda)$ has a pole of order 1 at every point of $P(\Delta_L)$. The roots of $h(\lambda) = -n^2$ are given as

$$\lambda_n = [-(c+n^2) + \sqrt{D_n}]/2 \quad \kappa_n = [-(c+n^2) - \sqrt{D_n}]/2,$$

where $D_n = (c+n^2)^2 - 4(cn^2 - b)$. If $b > 0$, then

$$\lambda_1 > \lambda_2 > \cdots > \lambda_n \to -c \quad (n \to \infty).$$

If $b = c > 0$, then $\lambda_1 = 0$, and moreover, if $-c < \gamma < 0$, then there exists an $M \ge 1$ such that $1 \le \|U_L(t)\| \le M, t \ge 0$.

Refer to [7, 8, 12] for other stability properties of solutions to the above equation.

Dedication. The first author dedicates this paper for the memory of his mother.

References

[1] Hale J.K. and Lunel S.M.V., *Introduction to Functional Differential Equations*, Springer-Verlag, 1993.

[2] Hille E. and Phillips R.S., *Functional Analysis and Semi-Groups,* American Math. Soc., 1957.

[3] Grabosch A. and Moustakas U., A semigroup approach to retarded differential equations, *One-Parameter Semigroups of Positive Operators, Lecture Notes in Math., 1184*, Springer-Verlag, 1986, 219-232.

[4] Grabosch A. and Moustakas U., A semigroup approach to retarded equations, ibid., 356-367.

[5] Kaashoek M.A. and Lunel S.M.V., Characteristic matrices and spectral properties off evolutionary systems, Trans. Amer. Math. Soc., 334(1992), 479-517.

[6] Naito T., On linear autonomous retarded equations with an abstract phase space for infinite delay, J. Differential Equations, 33(1979), 74-91.

[7] Naito T., Shin J.S., Murakami S., On solution semigroups of general functional differential equations, Nonlinear Anal., 30-7(1997), 4565-4576.

[8] Naito T. and Shin J.S., Evolutions equations with infinite delay, RIMS Kokyuroku, 984(1997), 147-160.

[9] 5 Naito T., Shin J.S., Murakami S., The generator of the solution semigroup for the general linear functional differential equations, Bull. of the University of Electro-Communications, 11-1,(1998), 29-38.

[10] Shin J.S., Comparison theorems and uniqueness of mild solutions to semilinear functional differential equations in a Banach space, Nonlinear Anal., 23(1994), 825-847.

[11] Shin J.S., and Naito T., Semi-Fredholm operators and periodic solutions for linear functional differential equations, J. Differential Equations, 153(1999), 407-441.

[12] Shin J.S., Naito T., and Nguyen V.M., On stability of solutions in linear autonomous functional differential equations, RIMS Kokyuroku, 1083(1999),231-242.

[13] Webb G.F., *Theory of Nonlinear Age-Dependent Population Dynamics,* Marcell Dekker, INC, 1985

TOSHIKI NAITO, THE UNIVERSITY OF ELECTRO-COMMUNICATIONS, CHOFU, TOKYO 182-8585, JAPAN
E-mail address: naito@e-one.uec.ac.jp

NGUYEN VAN MINH, THE UNIVERSITY OF ELECTRO-COMMUNICATIONS, CHOFU, TOKYO 182-8585, JAPAN
E-mail address: minh@im.uec.ac.jp

JONG SON SHIN, KOREA UNIVERSITY, KODAIRA, TOKYO 182-8560, JAPAN
E-mail address: shinjs@tech.korea-u.ac.jp

Progress in Nonlinear Differential Equations
and Their Applications, Vol. 42

LOGARITHMIC CHARACTERIZATION OF BOUNDED IMAGINARY POWERS

NOBORU OKAZAWA
SCIENCE UNIVERSITY OF TOKYO

Introduction

Let A be a closed linear operator with domain $D(A)$ and range $R(A)$ in a Banach space X. Our basic assumption consists of three conditions:

(i) $\rho(-A) \supset \mathbf{R}_+$ and $\exists M \geq 1$ such that $\|\xi(A+\xi)^{-1}\| \leq M \;\forall\, \xi > 0$.

(ii) $\overline{D(A)} = X$.

(iii) $\overline{R(A)} = X$.

It was forty years ago that the fractional powers A^α ($\Re\,\alpha > 0$) of a general closed linear operator A satisfying condition (i) was constructed by Balakrishnan [**1**]. Later the method of his construction was generalized by Komatsu [**6**] to cover any complex powers including the imaginary powers A^{it} ($t \in \mathbf{R}$). On the other hand, it was thirty years ago that the logarithm of a general closed linear operator A satisfying conditions (i)–(iii) was constructed by Nollau [**12**] (see also Yoshikawa [**19**]).

A simple relationship between $\{A^{it}; t \in \mathbf{R}\}$ and $i \log A$ has just been made clear by Okazawa [**13**, Addendum]:

Theorem 1. *Let A be a closed linear operator in X satisfying conditions* (i) – (iii). *Then the following are equivalent:*

(BIP) *The family $\{A^{it}; t \in \mathbf{R}\}$ forms a C_0-group on X.*

(LOG) *The operator $i \log A$ is the generator of a certain C_0-group on X.*

In this case, the C_0-group $\{A^{it}; t \in \mathbf{R}\}$ is generated by $i \log A$:

$$A^{it} = \exp(it \log A) := \operatorname*{s-lim}_{n\to\infty}\Big(1 - i\frac{t}{n}\log A\Big)^{-n}.$$

The proof in [**13**] is based on the semigroup theory and the integral representations for A^{it} and $\log A$: for $u \in D(A) \cap R(A)$,

(0.1)
$$A^{it}u = \frac{\sinh \pi t}{\pi t}u + \frac{\sinh \pi t}{\pi}\int_0^1 [s^{-it}(1+sA)^{-1}Au - s^{it}(1+sA^{-1})^{-1}A^{-1}u]\,ds,$$

$$(\log A)u = \int_0^1 [(1+sA)^{-1}Au - (1+sA^{-1})^{-1}A^{-1}u]\,ds. \tag{0.2}$$

In this note we try to present a proof of Theorem 1 based on the formulas which look simpler than (0.1) and (0.2):

$$A^{it}u = \frac{\sinh \pi t}{\pi t}\int_0^\infty s^{it}A(A+s)^{-2}u\,ds, \tag{0.3a}$$

$$(\log A)u = \int_0^\infty (\log s)A(A+s)^{-2}u\,ds, \tag{0.4a}$$

where $(A+s)^{-2}$ is used instead of $(A+s)^{-1}$. (0.3a) is known while (0.4a) seems to be new. Here $u \in D(A) \cap R(A)$. In fact, (0.3a) and (0.4a) can be rewritten as (0.1) and (0.2):

$$A^{it}u = \frac{\sinh \pi t}{\pi t}\int_0^1 [s^{-it}(1+sA)^{-2}Au + s^{it}(1+sA^{-1})^{-2}A^{-1}u]\,ds, \tag{0.3b}$$

$$(\log A)u = \int_0^1 \left(\log \frac{1}{s}\right)[(1+sA)^{-2}Au - (1+sA^{-1})^{-2}A^{-1}u]\,ds. \tag{0.4b}$$

These formulas have nice "symmetry" with respect to A and A^{-1} so that they immediately imply that for $u \in D(A) \cap R(A)$,

$$(A^{-1})^{it}u = A^{-it}u \quad \text{and} \quad (\log A^{-1})u = -(\log A)u. \tag{0.5}$$

For other characterizations of condition (BIP) see McIntosh [**9**] (in terms of H_∞ functional calculus in Hilbert spaces) and Monniaux [**11**] (in terms of analytic generators in *UMD*-Banach spaces); see also Le Merdy [**8**, Remark 2.6].

1. The logarithm

To prove the formula (0.4) for $\log A$ we start with the definition of logarithms employed in [**13**]. Let A be a closed linear operator in X satisfying conditions (i)–(iii). First we note that condition (i) yields that

$$N(A) \cap \overline{R(A)} = \{0\}$$

(see Yosida [**20**, Section VIII.4]). It then follows from condition (iii) that the null space of A is trivial: $N(A) = \{0\}$. Therefore, a closed linear operator A in X may be called *positive* if it satisfies conditions (i)–(iii). In fact, A is *non-negative* (see Komatsu [**7**]) and invertible. We then have

$$A \text{ is positive} \iff A^{-1} \text{ is positive.}$$

Now $\log A$ is defined as the closure of $L(A, A^{-1})$:

$$D(L(A,A^{-1})) := D(\log(1+A)) \cap D(\log(1+A^{-1})),$$

$$L(A,A^{-1})u := \log(1+A)u - \log(1+A^{-1})u, \quad u \in D(L(A,A^{-1})).$$

The closability of $L(A, A^{-1})$ is proved as follows. Let $B = A$ or A^{-1}. Then $\log(1+B)$ is defined as the negative (infinitesimal) generator of a C_0-semigroup $\{(1+B)^{-t}; t \geq 0\}$ of bounded linear operators on X (note that $D(B)$ is dense in X):

$$(1+B)^{-t} = \exp[-t\log(1+B)], \quad t \geq 0.$$

Let $0 < t < \ell$ ($\ell \in \mathbf{N}$). Then

$$(1+B)^{-t} := \frac{\Gamma(\ell)}{\Gamma(t)\Gamma(\ell-t)} \int_0^1 s^{t-1}(1-s)^{\ell-t-1}(1+sB)^{-\ell}\, ds.$$

For $u \in D(B)$ and $0 < \alpha < 1$ we have

$$\alpha^{-1}[(1+B)^{-\alpha}u - u] = -\frac{\sin \pi\alpha}{\pi\alpha} \int_0^1 \left(\frac{s}{1-s}\right)^{\alpha} (1+sB)^{-1}Bu\, ds.$$

Letting $\alpha \downarrow 0$, we are led to the integral representation of $\log(1+B)$:

$$\log(1+B)u = \int_0^1 (1+sB)^{-1}Bu\, ds, \quad u \in D(B)$$

(cf. [**13**, Theorem 2.4]), which implies that

$$\text{(1.1)} \qquad \|L(A, A^{-1})u\| \leq M\|Au\| + (1+M)\|A^{-1}u\|, \quad u \in D(A) \cap R(A).$$

Furthermore, $D(L(A, A^{-1}))$ is invariant under $C := (1+A)^{-1}(1+A^{-1})^{-1}$, with

$$\text{(1.2)} \qquad L(A, A^{-1})Cw = CL(A, A^{-1})w, \quad w \in D(L(A, A^{-1})).$$

The closability of $L(A, A^{-1})$ is a consequence of the identity (1.2) and the estimate (1.1). In fact, it follows from $w_n \to 0$ and $L(A, A^{-1})w_n \to z$ that $Cz = 0$ and hence $z = 0$; note that $L(A, A^{-1})C$ is a bounded linear operator on X. In this connection we have

Lemma 2 ([**13**, Theorem 4.6]). *Let A be as in* Theorem 1. *Then*

(a) $(\log A)u = \lim_{\alpha\downarrow 0} \alpha^{-1}(A^{\alpha}u - u), \quad u \in D(A) \cap R(A)$.

(b) $D(A) \cap R(A)$ *is a core for* $\log A$, *i.e.,* $u_n := (1+n^{-1}A)^{-1}(1+n^{-1}A^{-1})^{-1}u \to u$ *and* $(\log A)u_n \to (\log A)u$ *for* $u \in D(\log A)$ *as* $n \to \infty$.

We can now prove (0.4). The proof is based on Komatsu's formula (see [**7**]) for general complex powers of a "non-negative" operator in X:

$$\text{(1.3)}\; A^{\alpha}u = \frac{\sin \pi\alpha}{\pi\alpha} \int_0^{\infty} s^{\alpha}A(A+s)^{-2}u\, ds, \quad u \in D(A) \cap R(A), \quad |\Re\alpha| < 1;$$

note that (0.3a) is nothing but (1.3) with $\alpha = it$ ($t \in \mathbf{R}$). Setting $\alpha = 0$ in (1.3), we have

$$\text{(1.4)} \qquad u = \int_0^{\infty} A(A+s)^{-2}u\, ds, \quad u \in D(A) \cap R(A).$$

Thus we see from (1.3) and (1.4) that

$$\frac{1}{\alpha}(A^\alpha u - u) = \frac{\sin \pi\alpha}{\pi\alpha}\int_0^\infty \frac{1}{\alpha}(s^\alpha - 1)A(A+s)^{-2}u\,ds + \frac{1}{\alpha}\Big(\frac{\sin\pi\alpha}{\pi\alpha} - 1\Big)u.$$

Letting $\alpha \downarrow 0$, we can obtain (0.4a) by virtue of Lemma 2 (a). Incidentally, the proof of Lemma 2 (a) is based on well-known Balakrishnan's formula:

$$A^\alpha u = \frac{\sin\pi\alpha}{\pi}\int_0^\infty s^{\alpha-1}(A+s)^{-1}Au\,ds,\quad u \in D(A),\ \ 0 < \Re\alpha < 1$$

(see [**13**, Proposition 4.3]).

Remark 1. Let $B = A$ or A^{-1}. Then it follows that

$$\log(1+B)u = \int_0^1 \Big(\log\frac{1}{s}\Big)(1+sB)^{-2}Bu\,ds,\quad u \in D(B).$$

2. Imaginary powers and the C-regularized group

For completeness let us recall the definition of imaginary powers for a "positive" operator. Let A be a closed linear operator in X satisfying conditions (i)–(iii). Then A^{it} is defined as the closure of $I(A, A^{-1})$:

$$I(A,A^{-1})u := \frac{\sinh\pi t}{\pi t}\int_0^1 [s^{-it}(1+sA)^{-2}Au + s^{it}(1+sA^{-1})^{-2}A^{-1}u]\,ds,$$

$$u \in D(I(A,A^{-1})) := D(A)\cap R(A).$$

Let C be as in Section 1: $C = (1+A)^{-1}(1+A^{-1})^{-1}$. To prove the closability of $I(A,A^{-1})$ it suffices (as we have seen for $L(A,A^{-1})$ in Section 1) to note that

$$I(A,A^{-1})Cu = CI(A,A^{-1})u,\quad u \in D(A)\cap R(A), \tag{2.1}$$

$$\|I(A,A^{-1})C\| \le 2M^2(1+M)^2\,\frac{\sinh\pi t}{\pi t},\quad t \in \mathbf{R}. \tag{2.2}$$

Once (0.3) and (0.4) are established, we can easily prove

Lemma 3. *Let A be as in* Theorem 1. *Then for $u \in D(A)\cap R(A)$*

$$\lim_{t\to 0} t^{-1}(A^{it}u - u) = i(\log A)u. \tag{2.3}$$

In fact, we see from (0.3a) and (1.4) that

$$\frac{1}{t}(A^{it}u - u) = \frac{\sinh\pi t}{\pi t}\int_0^\infty \frac{1}{t}(s^{it} - 1)A(A+s)^{-2}u\,ds + \frac{1}{t}\Big(\frac{\sinh\pi t}{\pi t} - 1\Big)u.$$

Letting $t \to 0$, we can obtain (2.3) by virtue of (0.4a).

To prove Theorem 1 we need one more lemma:

Lemma 4. *Let A be as in* Theorem 1. *Then for $u \in D(A) \cap R(A)$*

$$A^{it}u = \lim_{n\to\infty} \left(1 - i\frac{t}{n} \log A\right)^{-n} u. \tag{2.4}$$

The proof of Lemma 4 is based on

Proposition 5. *Let A be as in* Theorem 1. *Set*

$$S(t) := A^{it}C = A^{it}(1+A)^{-1}(1+A^{-1})^{-1}, \quad t \in \mathbf{R}.$$

Then $\{S(t); t \in \mathbf{R}\}$ forms an exponentially bounded C-regularized group on X, with

$$\|S(t)\| \le 2M^2(1+M)^2 e^{\pi|t|}$$

and complete infinitesimal generator $i \log A$.

The notion of exponentially bounded C-regularized semigroups was introduced independently by Da Prato [**2**] and Davies-Pang [**4**]. The term "C-regularized semigroup" is a mixed name proposed by Hieber, Holderrieth and Neubrander [**5**]. Let C be a bounded linear operator on X, with $N(C) = \{0\}$ and $\overline{R(C)} = X$. Then a one-parameter family $\{S(t); t \ge 0\}$ of bounded linear operators on X is called an *exponentially bounded C-regularized semigroup* on X if it satisfies

(R1) $S(t+s)C = S(t)S(s)\ \forall\, t, s \ge 0$, and $S(0) = C$.
(R2) $S(\cdot)$ is strongly continuous on $\overline{\mathbf{R}_+} := [0, \infty)$.
(R3) $\exists K \ge \|C\|$, $\exists \omega \ge 0$ such that $\|S(t)\| \le Ke^{\omega t}\ \forall\, t \ge 0$. It is worth

noticing that a C_0-semigroup on X is a 1-regularized semigroup on X.

Let $\{S(t)\}$ and C be as in Proposition 5. Then $N(C) = \{0\}$ and $\overline{R(C)} = X$. Roughly speaking, (R1) and (R2) follow from the well-known additivity and analyticity of $A^\alpha u$, $u \in D(A) \cap R(A)$, with respect to α ($|\Re\alpha| < 1$). On the other hand, (R.3) is nothing but (2.2) which follows from (0.3b). Thus $\{S(t)\}$ forms an exponentially bounded C-regularized group on X.

Among others the *complete infinitesimal generator* is a useful notion in the theory of (exponentially bounded) C-regularized semigroups. As an extension of the Hille-Yosida generation theorem it was characterized by Tanaka [**16**].

The complete infinitesimal generator G of a C-regularized semigroup $\{S(t)\}$ is defined as the closure of the *infinitesimal generator* G_0:

$$D(G_0) := \{u \in R(C); \exists \lim_{t\downarrow 0} t^{-1}[C^{-1}S(t)u - u]\},$$

$$G_0 u := \lim_{t\downarrow 0} t^{-1}[C^{-1}S(t)u - u], \quad u \in D(G_0).$$

To prove the closability of G_0 it suffices to note that

$$S(t)u - Cu = \int_0^t S(s)G_0 u\, ds.$$

In fact, it follows from $u_n \to 0$ and $G_0 u_n \to v$ that $Cv = \lim_{t\downarrow 0} t^{-1} \int_0^t S(s)v\, ds = 0$ and hence $v = 0$.

Again, let $\{S(t)\}$ and C be as in Proposition 5. Then we see from (2.1) that

$$C^{-1}S(t)u = A^{it}u \quad \text{for } u \in R(C) = D(A) \cap R(A). \tag{2.5}$$

Therefore it follows from Lemma 3 that G_0 coinsides with the restriction of $i \log A$ to $R(C)$. In view of Lemma 2 (b) we obtain

$$G := \overline{G_0} = i \log A. \tag{2.6}$$

This completes the proof of Proposition 5.

Let $\{S(t)\}$ be a C_0-semigroup on X, with generator G. Then the exponential representation formula is well-known:

$$S(t) = \operatorname*{s-lim}_{n\to\infty} (1 - (t/n)G)^{-n} \tag{2.7}$$

(see e.g. Pazy [**14**]). If $\{S(t)\}$ is a C-regularized semigroup on X with complete infinitesimal generator G, then as an analogue of (2.7) we have

$$S(t)v = \lim_{n\to\infty} (1 - (t/n)G)^{-n}Cv, \quad v \in X;$$

hence we obtain

$$C^{-1}S(t)u = \lim_{n\to\infty} (1 - (t/n)G)^{-n}u, \quad u \in R(C) \tag{2.8}$$

(see Tanaka [**16**, Remark to Section 1]). In view of (2.5) and (2.6), (2.4) is nothing but (2.8). This proves Lemma 4.

Remark 2. Let $B = A$ or A^{-1}. Then, making the substitution $1 + s = r^{-1}$, we see from (0.3a) that

$$(1 + B)^{it}(1 + B)^{-1} = \frac{\sinh \pi t}{\pi t} \int_0^1 r^{-it}(1 - r)^{it}(1 + rB)^{-2}\, dr, \quad t \in \mathbf{R}. \tag{2.9}$$

This shows that $\{(1 + B)^{it}(1 + B)^{-1}; t \in \mathbf{R}\}$ forms a $(1 + B)^{-1}$-regularized group on X, with

$$\|(1 + B)^{it}(1 + B)^{-1}\| \le M^2 \frac{\sinh \pi t}{\pi t}$$

and complete infinitesimal generator $i \log(1 + B)$.

3. Proof of Theorem 1

The proof is divided into two parts.

Proof of "(BIP) ⇒ (LOG)". Suppose that $\{A^{it}\}$ is a C_0-group on X with generator G:

$$A^{it} = \exp(tG), \quad t \in \mathbf{R}. \tag{3.1}$$

Then (2.3) yields that $G = i \log A$ on $R(C) = D(A) \cap R(A)$. Since A^{it} is a bounded linear operator on X, we see from (2.1) that

$$\exp(tG)C = C \exp(tG), \quad t \in \mathbf{R}.$$

It then follows that $R(C)$ is a core for the generator G (see Davies [**3**, Theorem 1.9]). Therefore we can conclude by Lemma 2 (b) that $G = i \log A$ and hence (3.1) is nothing but $A^{it} = \exp(it \log A)$. □

Proof of "(LOG) ⇒ (BIP)". Suppose that $\{\exp(it \log A); t \in \mathbf{R}\}$ is a C_0-group on X. Then (2.4) is replaced with

$$A^{it}u = \operatorname*{s-lim}_{n \to \infty} (1 - i(t/n) \log A)^{-n}u = \exp(it \log A)u, \quad u \in R(C).$$

Since $\overline{R(C)} = X$, it follows from the closedness of A^{it} that $A^{it} = \exp(it \log A)$. Thus $\{A^{it}\}$ forms a C_0-group on X. □

This completes the proof of Theorem 1.

Remark 3. Let A be as in Theorem 1. Suppose that condition (BIP) or (LOG) is satisfied. Then

$$A^{it} = (1 + A)^{it}(1 + A^{-1})^{-it}, \quad t \in \mathbf{R},$$

where $(1 + B)^{it}$ is given by (2.9).

In fact, it follows from (BIP) that $(1 + A)^{it}$ is also bounded on X; of course the boundedness of $(1 + A^{-1})^{it}$ follows from that of $(A^{-1})^{it}$ [see (0.5)]. This implication is proved by Prüss and Sohr [**15**, Theorem 3] (see also Monniaux [**10**] and Uiterdijk [**18**]). Therefore we see from the definition of $\log A$ that

$$\begin{aligned} A^{it} &= \exp(it \log A) = \exp[it \log(1 + A)] \exp[-it \log(1 + A^{-1})] \\ &= (1 + A)^{it}(1 + A^{-1})^{-it}. \end{aligned}$$

Remark 4. Let A be a positive operator in X. Then $\log A$ is well-defined as we have seen above. In this connection Nollau [**12**] proved that the resolvent set of $\log A$ is non-empty. Moreover, one can show that $\pi \pm i \log A$ are nonnegative operators in X (see [**13**, Lemma 5.1]). Thus we may ask whether the family $\{A^{it}; t \in \mathbf{R}\}$ (of "unbounded" imaginary powers) forms an "integrated" group on X or not. In the present case integrated groups are the same as $(1 + \pi + i \log A)^{-1}$-regularized groups (see [**17**]). The author does not yet know the answer.

Acknowledgement. The author expresses his hearty thanks to Professor H. Komatsu for inviting him to the first International Conference on Semigroups of Operators: Theory and Applications, organized by Professor A.V. Balakrishnan, one of the pioneers in the theory of fractional powers of operators.

References

[1] A.V. Balakrishnan, *Fractional powers of closed operators and the semi-groups generated by them*, Pacific J. Math. **10** (1960), 419–437.

[2] G. Da Prato, *Semigruppi regolarizzabili*, Ricerche Mat. **15** (1966), 223–246.

[3] E.B. Davies, *One-Parameter Semigroups*, London Mathematical Society Monographs, No. 15, Academic Press, London and New York, 1980.

[4] E.B. Davies and M.M. Pang, *The Cauchy problem and a generalization of the Hille-Yosida theorem*, Proc. London Math. Soc. **55** (1987), 181–208.

[5] M. Hieber, A. Holderrieth and F. Neubrander, *Regularized semigroups and systems of linear partial differential equations*, Ann. Scuola Norm. Sup. Pisa Cl. Sci. (4) **19** (1992), 363–379.

[6] H. Komatsu, *Fractional powers of operators*, Pacific J. Math. **19** (1966), 285–346.

[7] H. Komatsu, *Fractional powers of operators,* VI: *Interpolation of non-negative operators and imbedding theorems*, J.Fac.Sci.Univ.Tokyo, Sec.IA **19** (1972), 1–63.

[8] C. Le Merdy, *The similarity problem for bounded analytic semigroups on Hilbert space*, Semigroup Forum **56** (1998), 205–224.

[9] A. McIntosh, *Operators which have an H_∞ functional calculus*, Miniconference on Operator Theory and Partial Differential Equations, 1986. Proceedings of the Centre for Mathematical Analysis, vol. 14, Australian National University, Canberra, 1986, pp. 210–231.

[10] S. Monniaux, *A perturbation result for bounded imaginary powers*, Arch. Math. **68** (1997), 407–417.

[11] S. Monniaux, *A new approach to the Dore-Venni theorem*, Math. Nachr. **204** (1999), 163–183.

[12] V. Nollau, *Über den Logarithmus abgeschlossener Operatoren in Banachschen Räumen*, Acta Sci. Math. **30** (1969), 161–174.

[13] N. Okazawa, *Logarithms and imaginary powers of closed linear operators*, Integral Equations Operator Theory, to appear.

[14] A. Pazy, *Semigroups of Linear Operators and Applications to Partial Differential Equations*, Applied Math. Sciences, vol. 44, Springer-Verlag, Berlin and New York, 1983.

[15] J. Prüss and H. Sohr, *On operators with bounded imaginary powers in Banach spaces*, Math. Z. **203** (1990), 429–452.

[16] N. Tanaka, *On the exponentially bounded C-semigroups*, Tokyo J. Math. **10** (1987), 107–117.

[17] N. Tanaka and I. Miyadera, *Some remarks on C-semigroups and integrated semigroups*, Proc. Japan Acad. **63A** (1987), 139–142.

[18] M. Uiterdijk, *A note on the functional calculus for sectorial operators*, Indag. Math. (N.S.) **10** (1999), 131–143.

[19] A. Yoshikawa, *On the logarithm of closed linear operators*, Proc. Japan Acad. **49** (1973), 169–173.

[20] K. Yosida, *Functional Analysis*, Grundlehren Math. Wissenschaften 123, Springer-Verlag, Berlin and New York, 1965; 6th ed., 1980.

Noboru Okazawa, Department of Mathematics, Science University of Tokyo, Wakamiya-cho 26, Shinjuku-ku, Tokyo 162-0827, JAPAN

E-mail address: okazawama.kagu.sut.ac.jp

Progress in Nonlinear Differential Equations
and Their Applications, Vol. 42

UNILATERAL SHIFTS IN WAVELET THEORY AND ALGORITHM

JOSEPH N.Q. PHAM
UNIVERSITY OF CALFORNIA, LOS ANGELES

Abstract. The construction of wavelets is usually derived through a multiresolution analysis and formulated as a pair of quadrature mirror filters $\{h(n), g(n)\}$. In this construction there is a shift structure of subspaces associated with the dilation operator, which is a bilateral shift of infinite multiplicity. We show that associated with each wavelet construction through multiresolution analysis are two unilateral shifts of infinite multiplicity on $l^2(\mathbb{Z})$. The generating wandering subspaces of these shifts, as well as the shifts themselves, are given in terms of the functions $h(n)$ and $g(n)$. The two shifts are duals of each other in a way that allows us to decompose $l^2(\mathbb{Z})$ into the direct sum of the two generating wandering subspaces. It is shown that the decomposition of functions via the first of these two unilateral shifts is equivalent to the algorithm used in the wavelet transform. Thus, we provide a Hilbert space setting for the wavelet algorithm, often referred to as the discrete wavelet transform.

1. Introduction

A function $\psi(t)$ of the function space $L^2(\mathbb{R})$ is called an orthogonal wavelet if the functions

$$\left(\sqrt{2}\right)^m \psi(2^m t - n), \quad m, n \in \mathbb{Z}, \ t \in \mathbb{R},$$

are orthonormal and span the space. The usual construction of wavelets is through multiresolution analysis [4], where the key characters in the construction are the filters $h(n)$ and $g(n)$, known in the signal processing literature as quadrature mirror filters. Obvious in the structure of a multiresolution analysis is the shift structure corresponding to the dilation bilateral shift. The purpose of this paper is to show that there are two other natural shifts corresponding to wavelets constructed through a multiresolution. These shifts are unilateral shifts of infinite multiplicity on $l^2(\mathbb{Z})$.

The outline of this paper is as follows. We start Section 2 by defining our notation. We then review the basic definitions regarding shift operators and recall the important results concerning wavelets and multiresolution analysis. Section 3 contains the main results of the paper. Here, we show that associated with each multiresolution of $L^2(\mathbb{R})$ are two unilateral shifts on $l^2(\mathbb{Z})$. These shifts, as well as their generating wandering subspace, are given in terms of the filters $h(n)$ and $g(n)$ characterizing the multiresolution. A key result is that the decomposition of $l^2(\mathbb{Z})$ into the m-wandering subspaces of the first of these shifts provides us the Hilbert space setting for the wavelet algorithm.

We conclude the paper in Section 4 by mentioning these shift operators in the context of wavelet packets.

2. Preliminaries

2.1. Notation

In this paper, let $\mathcal{H}$ in general denote a Hilbert space with inner product $[\cdot,\cdot]$ and norm $\|\cdot\|$. We will always assume our Hilbert spaces to be separable. The set of real numbers is denoted by $\mathbb{R}$, the set of complex numbers by $\mathbb{C}$, the set of integers is by $\mathbb{Z}$, the set of nonnegative integers by $\mathbb{Z}^+$, and the unit circle by $\mathbb{T} = \{z \in \mathbb{C} : |z| = 1\}$. Finally, we will only work with the spaces of square integrable functions $L^2(\mathbb{R})$ and $l^2(\mathbb{Z})$, as well as their frequency counterparts $L^2(i\mathbb{R})$ and $L^2(\mathbb{T})$.

2.2. Shift Operators

Let B be a linear bounded operator on $\mathcal{H}$. A closed subspace W of $\mathcal{H}$ is *wandering* for B, or is a wandering subspace of B, if $W \perp B^n W$, $\forall n > 0$. Clearly, if W is wandering for B, then for each $m > 0$, $B^m W$ is still wandering. We shall call $B^m W$ a *m-wandering* subspace and refer to the subspaces $\{B^m W\}_{m \geq 0}$ as the *m-wandering subspaces* of B. Note that the m-wandering subspaces are mutually orthogonal.

An isometry S_+ on $\mathcal{H}$ is a *unilateral shift* if it admits a generating wandering subspace W, i.e., if $\mathcal{H}$ admits the orthogonal decomposition $\mathcal{H} = \bigoplus_{n=-\infty}^{\infty} S_+^n W$. Similarly, a *bilateral shift* $S\colon \mathcal{H} \to \mathcal{H}$ is a unitary operator which admits a generating wandering subspace W, i.e., if $\mathcal{H} = \bigoplus_{n=-\infty}^{\infty} S^n W$. In both cases, the dimension of the generating wandering subspaace is called the *multiplicity* of the shift. Shift operators, as we will see, play a fundamental role in the construction of wavelets. For more information regarding shifts, we refer to [2], [6], and [7].

Define the operations of dilation D, translation T, and modulation E on $L^2(\mathbb{R})$ as follows:

$$Df(t) = \sqrt{2}\, f(2t), \quad Tf(t) = f(t-1), \quad Ef(t) = e^{it} f(t).$$

It is clear that all three are unitary operators. They are, in fact, all bilateral shifts of infinite multiplicity. The operators D, T, and E, when defined in the obvious manner $L^2(i\mathbb{R})$ are still unitary. We will not use a different notation for the operators acting on $L^2(i\mathbb{R})$ as opposed to $L^2(\mathbb{R})$, since it will be clear which space they are defined on. finally, it is also useful to define T_2 on $l^2(\mathbb{Z})$ and E_2 on $L^2(\mathbb{T})$ by

$$T_2 f(n) = f(n-2), \quad E_2 F(e^{i\omega}) = e^{i2\omega} F(e^{i\omega}).$$

2.3. Wavelets

We now recall some important results concerning the construction of wavelets through multiresolution analysis. More details may be found in [1] and [4].

Definition 2.1. A *multiresolution analysis* of $L^2(\mathbb{R})$ is a sequence of closed subspaces $\{V_n\}_{n\in\mathbb{Z}}$ of $L^2(\mathbb{R})$ with the following properties:

1. $V_n \subset V_{n+1}$

2. $\overline{\bigvee_{n=-\infty}^{\infty} V_n} = L^2(\mathbb{R}), \quad \bigwedge_{n=-\infty}^{\infty} V_n = \{0\}$

3. $f(t) \in V_n \Leftrightarrow f(2t) \in V_{n+1} \;\; \forall n \in \mathbb{Z}$

4. $f(t) \in V_n \Rightarrow f(t - 2^{-n}K) \in V_n \;\; \forall k \in \mathbb{Z}$

5. $\exists$ a *scaling* function $\phi(t)$ such that $\{T^n\phi(t)\}_{n\in\mathbb{Z}}$ is an orthonormal basis for V_0.

Define W_n to be the orthogonal complement of V_n in V_{n+1}. It is easy to show that conditions *1* and *3* of multiresolution implies that $W_n = D^n W_0$ for all integers n. Furthermore, the subspaces $\{W_n\}_{n\in\mathbb{Z}}$ are mutually orthogonal, i.e., they are n-wandering subspaces of D. The addition of property *2* implies that these subspaces are complete in $L^2(\mathbb{R})$. W_0 is then a generating wandering subspace of D on $L^2(\mathbb{R})$ and we have the decomposition $L^2(\mathbb{R}) = \bigoplus_{n=-\infty}^{\infty} D^n W_0$. If we can find a function $\psi(t)$ whos integer translates $\{T^n\psi(t)\}_{n\in\mathbb{Z}}$ are an orthonormal basis for W_0, then $\{D^m T^n \psi(t)\}_{n\in\mathbb{Z}}$ will be an orthonormal basis for $L^2(\mathbb{R})$, which would make $\psi(t)$ a wavelet. Thus a wavelet decomposition of $L^2(\mathbb{R})$ is a decomposition into the n-wandering subspaces of the bilateral shift D on $L^2(\mathbb{R})$.

It turns out that the wavelet is uniquely determined (up to an integer translation) by the scaling function $\phi(t)$. Knowledge of $\phi(t)$ and $\psi(t)$ is equivalent to knowledge of their Fourier coefficients $h(n)$ and $g(n)$ in the space V_1. The pair of filters $\{h(n), g(n)\}$ are known in the signal processing literature as Quadrature Mirror Filters (QMF's). They are defined from the equations

$$\phi(t) = \sum_{k=-\infty}^{\infty} h(k) D T^k \phi(t), \quad \psi(t) = \sum_{k=-\infty}^{\infty} g(k) D T^k \phi(t). \tag{2.1}$$

Lemma 2.1. *Let $\phi(t)$ be a scaling function and $\psi(t)$ be the corresponding wavelet in a given multiresolution. Then the orthogonality of $\{\phi(t-n)\}$ in V_0 and $\{\psi(t-n)\}$ in W_0 implies that*

$$\sum_{n=-\infty}^{\infty} |\Phi(i(\omega + 2\pi n))|^2 = 1, \quad \sum_{n=-\infty}^{\infty} |\Psi(i(\omega + 2\pi n))|^2 = 1. \tag{2.2}$$

Moreover, $V_0 \perp W_0$ implies that

$$\sum_{n=-\infty}^{\infty} \Phi(i(\omega + w\pi n))\, \overline{\Psi(i(\omega + 2\pi n))} = 0. \tag{2.3}$$

Lemma 2.2. *Let $h(n)$ and $g(n)$ be the quadrature mirror filters of a given multiresolution, as defined in* (2.1). *Then*

$$|H(e^{i\omega})|^2 + \left|H(e^{i(\omega+\pi)})\right|^2 = 2, \quad |G(e^{i\omega})|^2 + \left|G(e^{i(\omega+\pi)})\right|^2 = 2, \tag{2.4}$$

and

$$G(e^{i\omega})\,\overline{H(e^{i\omega})} + G(e^{i(\omega+\pi)})\,\overline{H(e^{i(\omega+\pi)})} = 0. \tag{2.5}$$

Wavelet theory is developed in $L^2(\mathbb{R})$, but in practice one is usually more interested in decomposing functions in $l^2(\mathbb{Z})$. The usual method is to associate the $l^2(\mathbb{Z})$ sequence as the Fourier coefficients of an $L^2(\mathbb{R})$ function in one of the subspaces V_m. The algorithm then proceeds by continually decomposing the coefficients in V_m to coefficients in V_{m-1} and W_{m-1}. Let $P_{V_m} : \mathcal{H} \to V_m$ be the orthogonal projection from $\mathcal{H}$ onto the closed subspace V_m. Similarly, let $P_{W_m} : \mathcal{H} \to W_m$ be the orthogonal projections from $\mathcal{H}$ onto the closed subspace W_m. We have

$$P_{V_m} f(t) = \sum_{n=-\infty}^{\infty} [f, D^m T^n \phi(t)] D^m T^n \phi(t),$$

$$P_{W_m} f(t) = \sum_{n=-\infty}^{\infty} [f, D^m T^n \psi(t)] D^m T^n \psi(t).$$

Define the $l^2(\mathbb{Z})$ sequences

$$c_m(n) = [f(t), D^m T^n \phi(t)], \quad d_m(n) = [f(t), D^m T^n \psi(t)]. \tag{2.6}$$

By the mechanics of orthogonal projections, it is easy to show the decomposition formulas

$$c_m(n) = \sum_{l=-\infty}^{\infty} \overline{h(l-2n)}\, c_{m+1}(l), \quad d_m(n) = \sum_{l=-\infty}^{\infty} \overline{g(l-2n)}\, c_{m+1}(l) \tag{2.7}$$

as well as the reconstruction formula

$$c_{m+1}(n) = \sum_{l=-\infty}^{\infty} h(n-2l) c_m(l) + \sum_{l=-\infty}^{\infty} g(n-2l) d_m(l). \tag{2.8}$$

3. Main Results

3.1. The Isometric Operators in the Wavelet Algorithm

Define the operator S_h on $l^2(\mathbb{Z})$ by

$$S_h x(n) = \sum_{k=-\infty}^{\infty} h(n-2k) x(k). \tag{3.1}$$

It is easy to show that S_h is an isometry, and that its adjoint is given by

$$S_h^* x(n) = \sum_{k=-\infty}^{\infty} \overline{h(k-2n)}\, x(k). \tag{3.2}$$

Most of our calculations of interest are "cleaner" in the frequency domain. This motivates the following.

Proposition 3.1. *The operator S_h is unitarily equivalent via the Fourier transform to the operator S_H, i.e., $\mathcal{F}S_h = S_H\mathcal{F}$, where $\mathcal{F}$ is the Fourier transform and*

$$S_H X(e^{i\omega}) = H(e^{i\omega})X(e^{i2\omega}). \tag{3.3}$$

Proof. The proof is straightforward by a direct calculation. □

It follows that the operator S_H is also an isometry. Moreover, $S_H^* = \mathcal{F}S_h^*\mathcal{F}^*$, which yields

$$S_H^* Y(e^{i\omega}) = \frac{1}{2}\left\{\overline{H(e^{i\frac{\omega}{2}})}\, Y(e^{i\frac{\omega}{2}}) + \overline{H\left(e^{i(\frac{\omega}{2}+\pi)}\right)}\, Y\left(e^{i(\frac{\omega}{2}+\pi)}\right)\right\}. \tag{3.4}$$

Define the operators S_g, S_g^*, S_G, and S_G^* accordingly and respectively from (3.1), (3.2), (3.3), and (3.4) with the role of h replaced by g. The decomposition formulas of (2.7) are then written as

$$c_m(n) = S_h^*\, c_{m+1}(n), \quad d_m(n) = S_g^*\, c_{m+1}(n),$$

and the reconstruction formula of (2.8) is

$$c_{m+1}(n) = S_h\, c_m(n) + S_g\, d_m(n).$$

Since S_h and S_g are isometries, $S_hS_h^*$ and $S_gS_g^*$ are projections from $l^2(\mathbb{Z})$ onto subspaces in $l^2(\mathbb{Z})$. Let these subspaces be Q_0 and P_0, respectively. We then have the decomposition

$$l^2(\mathbb{Z}) = Q_0 \oplus Q_0^\perp, \quad l^2(\mathbb{Z}) = P_0 \oplus P_0^\perp.$$

$I - S_hS_h^*$ is then the orthogonal projection from $l^2(\mathbb{Z})$ onto $Q_0^\perp$ and $I - S_gS_g^*$ is the projection from $l^2(\mathbb{Z})$ onto $P_0^\perp$. The next lemma shows that we actually have

$$l^2(\mathbb{Z}) = Q_0 \oplus P_0.$$

Lemma 3.2.

$$S_hS_h^* + S_gS_g^* = I. \tag{3.5}$$

Proof. It is of course equivalent to show the $L^2(\mathbb{T})$ version of (3.5) and that is the path that we shall follow. For all $X(e^{i\omega}) \in L^2(\mathbb{T})$, we have

$$\begin{aligned} S_H S_H^* X(e^{i\omega}) &= \frac{1}{2} S_H \left\{ \overline{H(e^{i\frac{\omega}{2}})}\, X(e^{i\frac{\omega}{2}}) + \overline{H\left(e^{i(\frac{\omega}{2}+\pi)}\right)}\, X\left(e^{i(\frac{\omega}{2}+\pi)}\right) \right\} \\ &= \frac{1}{2} \left\{ |H(e^{i\omega})|^2 X(e^{i\omega}) + H(e^{i\omega})\, \overline{H(e^{i(\omega+\pi)})}\, X(e^{i(\omega+\pi)}) \right\}. \end{aligned}$$

Writing down the expression for $(S_H S_H^* + S_G S_G^*)$ and the using (2.4) with the fact that $G(e^{i\omega}) = e^{in\omega}\, \overline{H(e^{i(\omega+\pi)})}$, for any n odd, will yield the desired result. □

We now derive two unilateral shifts on $L^2(\mathbb{T})$ corresponding to the multiresolution analysis on $L^2(\mathbb{R})$. The following simple result is useful in what is to follow.

Lemma 3.3. *Let $\Phi(i\omega)$ be the Fourier transform of a scaling function $\phi(t)$. Let $A(e^{i\omega})$ and $B(e^{i\omega})$ be arbitrary in $L^2(\mathbb{T})$. Then*

$$[A(e^{i\omega})\Phi(i\omega),\, B(e^{i\omega})\Phi(i\omega)]_{L^2(i\mathbb{R})} = [A(e^{i\omega}),\, B(e^{i\omega})]_{L^2(\mathbb{T})}. \tag{3.6}$$

Proof. First note that the inner product on the left hand side of (3.6) is an $L^2(i\mathbb{R})$ inner product while the one on the right hand side is an $L^2(\mathbb{T})$ inner product. We have

$$\begin{aligned} &[A(e^{i\omega})\Phi(i\omega),\, B(e^{i\omega})\Phi(i\omega)] \\ &\qquad = \frac{1}{2\pi} \int_{-\infty}^{\infty} A(e^{i\omega})\, \overline{B(e^{i\omega})}\, |\Phi(i\omega)|^2\, d\omega \\ &\qquad = \frac{1}{2\pi} \sum_{k=-\infty}^{\infty} \int_{-\pi}^{\pi} A(e^{i\omega})\, \overline{B(e^{i\omega})}\, |\Phi(i(\omega+2\pi k))|^2\, d\omega \\ &\qquad = \frac{1}{2\pi} \int_{-\pi}^{\pi} A(e^{i\omega})\, \overline{B(e^{i\omega})} \sum_{k=-\infty}^{\infty} |\phi(i(\omega+2\pi k))|^2\, d\omega. \end{aligned}$$

Substituting from (2.2) yields the desired result. □

Proposition 3.4. *The subspace $Q_0 \subset L^2(\mathbb{T})$ defined by*

$$Q_0 = \overline{span}\{E_2^{-k} G(e^{i\omega})\}_{k\in\mathbb{Z}}$$

is a wandering subspace of the isometry S_H.

Proof. We need to show that $Q_0 \perp S_H^p Q_0$, $\forall p > 0$. This turns out to be a direct result from the fact that W_n is orthogonal to W_{n+p}. Start with

$$0 = [T^k \psi(t),\, D^p T^l \psi(t)], \quad \forall p \in \mathbb{Z}.$$

After a fair amount of manipulation, one can show that this implies

$$0 = \left[E_2^{-2^p k} G(e^{i2^p\omega}) D^p \Phi(i\omega),\, E_2^{-l} G(e^{i\omega}) \Phi(i\omega) \right], \quad \forall p \in \mathbb{Z}. \tag{3.7}$$

It can be shown that $D^p\Phi(i\omega) = \left\{ \prod_{m=0}^{p-1} H(e^{i2^m\omega}) \right\} \Phi(i\omega)$, so that (3.7) reads

$$0 = \left[E_2^{-2^p k} G(e^{i2^p\omega}) \prod_{m=0}^{p-1} H(e^{i2^m\omega}) \Phi(i\omega),\, E_2^{-l} G(e^{i\omega}) \Phi(i\omega) \right].$$

The functions $E_2^{-2^p k} G(e^{i2^p\omega}) \prod_{m=0}^{p-1} H(e^{i2^m\omega})$ are in $L^2(\mathbb{T})$, and so from Lemma (3.3) we get

$$0 = \left[E_2^{-2^p k} G(e^{i2^p\omega}) \prod_{m=0}^{p-1} H(e^{i2^m\omega}),\, E_2^{-l} G(e^{i\omega}) \right], \quad k, l \in \mathbb{Z},\ p \in \mathbb{Z}^+.$$

Finally, it is straightforward to verify that $\left\{ E_2^{-2^p k} G(e^{i2^p\omega}) \prod_{m=0}^{p-1} H(e^{i2^m\omega}) \right\}_{k\in\mathbb{Z}}$ are precisely the orthonormal basis functions of $S_H^p Q_0$, thus completing the proof of the proposition. □

Let $Q_n = S_H^n Q_0$, $n \geq 0$. From the remarks in Section 2.2, these n-wandering subspaces are mutually orthogonal and it is straightforward to show that they are indeed just shifts of one another, i.e., $S_H Q_n = Q_{n+1}$. We may then define a subspace $\mathcal{M} \subseteq L^2(\mathbb{T})$ by $\mathcal{M} = \bigoplus_{n=0}^{\infty} S_H^n Q_0$. The subspace $\mathcal{M}$ is obviously invariant for S_H, hence, the restriction $S_H|_{\mathcal{M}}$ is a unilateral shift. This motivates the following theorem, where what remains to be shown is that $\mathcal{M}$ is actually all of $L^2(\mathbb{T})$.

Theorem 3.5. *Let $H(e^{i\omega})$ and $G(e^{i\omega})$ be the quadrature mirror filters in a given multiresolution analysis on $L^2(\mathbb{R})$. Then the operator S_H is a unilateral shift of infinite multiplicity on $L^2(\mathbb{T})$. Therefore, $L^2(\mathbb{T})$ admits the decomposition*

$$L^2(\mathbb{T}) = \bigoplus_{p=0}^{\infty} s_H^p Q_0, \tag{3.8}$$

where the generating wandering subspace Q_0 is given by

$$Q_0 = \overline{span}\{E_2^{-n} G(e^{i\omega})\}_{n\in\mathbb{Z}}.$$

Moreover, the functions $\{E_2^{-n} G(e^{i\omega})\}_{n\in\mathbb{Z}}$ are orthonormal in Q_0, so that the functions $\{S_H^m E_2^{-n} G(e^{i\omega})\}_{m\in\mathbb{Z}^+, n\in\mathbb{Z}}$, form an orthonormal basis for $L^2(\mathbb{T})$.

Proof. We have already shown that S_H is a unilateral shift on $\mathcal{M} \subseteq L^2(\mathbb{T})$, where

$$\mathcal{M} = \bigoplus_{n=0}^{\infty} S_H^n Q_0 = Q_0 \oplus \bigoplus_{n=1}^{\infty} S_H^n Q_0.$$

since S_H is a unilateral shift, $S_H S_H^*$ is the orthogonal projection from $L^2(\mathbb{T})$ onto $\bigoplus_{n=1}^{\infty} S_H^n Q_0$, and $I - S_H S_H^*$ is the projection operator onto Q_0. But from Lemma 3.2, $S_H S_H^* = I - S_G S_G^*$, which is the projection from $L^2(\mathbb{T})$

to $Q_0^\perp$, the orthogonal complement of Q_0 in $L^2(\mathbb{T})$. In other words, we have $L^2(\mathbb{T}) = Q_0 \oplus Q_0^\perp$ and $Q_0^\perp = \bigoplus_{n=1}^\infty S_H^n Q_0$, so that $\mathcal{M}$ is actually all of $L^2(\mathbb{T})$. The result that $\{S_H^m E_2^{-n} G(e^{i\omega})\}_{m\in\mathbb{Z}^+, n\in\mathbb{Z}}$ is an orthonormal basis for $L^2(\mathbb{T})$ is a trivial conclusion from the fact that $\{E_2^{-n} G(e^{i\omega})\}_{n\in\mathbb{Z}}$ is orthonormal in Q_0, a generating wandering subspace of the unilateral shift S_H on $L^2(\mathbb{T})$. □

Note that since S_H is a unilateral shift on $L^2(\mathbb{T})$, the generating wandering subspace Q_0 is actually $\mathcal{N}(S_H^*)$. Indeed, let $X(e^{i\omega}) \in \mathcal{N}(S_H^*)$. Then from (3.4), we get

$$S_H^* X(e^{i\omega}) = \frac{1}{2}\left\{\overline{H(e^{i\frac{\omega}{2}})}\, X(e^{i\frac{\omega}{2}}) + \overline{H\left(e^{i(\frac{\omega}{2}+\pi)}\right)}\, X\left(e^{i(\frac{\omega}{2}+\pi)}\right)\right\} = 0.$$

This is precisely (2.5) with $G(e^{i\omega})$ replaced by $X(e^{i\omega})$. We know that solutions to this equation are of the form $X(e^{i\omega}) = E_2^{-k} G(e^{i\omega})$ for some integer k. Therefore

$$\mathcal{N}(S_H^*) = \overline{\text{span}}\{E_2^{-k} G(e^{i\omega})\}_{k\in\mathbb{Z}},$$

which is Q_0.

It turns out that the duality between the quadrature mirror filters $H(e^{i\omega})$ and $G(e^{i\omega})$ give rise to another unilateral shift S_G. Reversing the roles of $H(e^{i\omega})$ and $G(e^{i\omega})$ in the above, we get:

Theorem 3.6. *Let $H(e^{i\omega})$ and $G(e^{i\omega})$ be the quadrature mirror filters in a given multiresolution analysis on $L^2(\mathbb{R})$. Then the operator S_G is a unilateral shift of infinite multiplicity on $L^2(\mathbb{T})$. Therefore, $L^2(\mathbb{T})$ admits the decomposition*

$$L^2(\mathbb{T}) = \bigoplus_{m=0}^{\infty} S_G^m P_0, \tag{3.9}$$

where the generating wandering subspace P_0 is given by

$$P_0 = \overline{\text{span}}\{E_2^{-n} H(e^{i\omega})\}_{n\in\mathbb{Z}}.$$

Moreover, the functions $\{E_2^{-n} H(e^{i\omega})\}_{n\in\mathbb{Z}}$ are orthonormal in P_0, so that the functions $\{S_G^m E_2^{-n} H(e^{i\omega})\}_{m\in\mathbb{Z}^+, n\in\mathbb{Z}}$ form an orthonormal basis for $L^2(\mathbb{T})$.

We now have two shifts on $L^2(\mathbb{T})$, with the representations of $L^2(\mathbb{T})$ given by (3.8) and (3.9). The following theorem gives one more decomposition of $L^2(\mathbb{T})$, namely, as a direct sum of the generating wandering subspaces of the unilateral shifts S_H and S_G.

Theorem 3.7. *Let $H(e^{i\omega})$ and $G(e^{i\omega})$ be the quadrature mirror filters in a given multiresolution analysis on $L^2(\mathbb{R})$. Let S_H and S_G be the unilateral shifts on $L^2(\mathbb{T})$ corresponding to this multiresolution. Then*

$$L^2(\mathbb{T}) = Q_0 \oplus P_0,$$

where Q_0 and P_0 are the generating wandering subspaces of the unilateral shifts S_H and S_G, respectively.

Proof. This proof follows from the relationship of S_H and S_G in Lemma 3.2. □

The results of the previous three theorems have their counterparts in $l^2(\mathbb{Z})$. We summarize this in the following.

Corollary 3.8. *Let $h(n)$ and $g(n)$ be the quadrature mirror filters in a given multiresolution analysis on $L^2(\mathbb{R})$. Then the operators S_h and S_g on $l^2(\mathbb{Z})$ defined by*

$$S_h x(n) = \sum_{k=-\infty}^{\infty} h(n-2k)x(k), \quad S_g(n) = \sum_{k=-\infty}^{\infty} g(n-2k)x(k).$$

are unilateral shifts of infinite multiplicity on $l^2(\mathbb{Z})$. Therefore $l^2(\mathbb{Z})$ may then be represented by

$$l^2(\mathbb{Z}) = \bigoplus_{m=0}^{\infty} S_h^m Q_0, \quad Q_0 = \overline{span}\{T_2^l g(n)\}_{l\in\mathbb{Z}}, \tag{3.10}$$

or

$$l^2(\mathbb{Z}) = \bigoplus_{m=0}^{\infty} S_g^m P_0, \quad P_0 = \overline{span}\{T_2^l h(n)\}_{l\in\mathbb{Z}}. \tag{3.11}$$

In addition,

$$l^2(\mathbb{Z}) = Q_0 \oplus P_0.$$

We now turn our attention to deriving an efficient algorithm for the decomposition of functions into our shifted basis functions. First, consider the representation of $l^2(\mathbb{Z})$ as in (3.10). Any $f(n) \in l^2(\mathbb{Z})$ can then be written as

$$f(n) = \sum_{m=0}^{\infty} \sum_{l=-\infty}^{\infty} [f(n),\, S_h^m T_2^l\, g(n)] S_h^m T_2^l\, g(n).$$

Define a sequence of $l^2(\mathbb{Z})$ functions $\alpha_m(l) = [f(n),\, S_h^m T_2^l g(n)]$, $m \in \mathbb{Z}^+$. We have

$$\alpha_0(l) = [f(n),\, T_2^l g(n)] = \sum_{n=-\infty}^{\infty} f(n)\,\overline{g(n-2l)} = S_g^* f(l),$$

where $f(l)$ can be viewed as the Fourier coefficients of $f(n)$ in the standard basis $e_l(n) = \delta_{n,l}$. Next, for each $m > 0$, we get

$$\alpha_m(l) = [f(n), S_h^m T_2^l g(n)] = \left[S_h^{*m} f(n),\, T_2^l g(n)\right].$$

Thus $\alpha_m(l) = S_g^* S_h^{*m} f(l)$. Define $\tilde{\alpha}_{m-1}(l) = S_h^{*m} f(l)$, so that $\alpha_m(l) = S_g^* \tilde{\alpha}_{m-1}(l)$. This gives us a recursive procedure in finding the Fourier coefficients in each of the m-wandering subspace Q_m. The reconstruction procedure is also straightforward. Given $\alpha_m(l)$ and $\tilde{\alpha}_m(l)$, we have

$$\begin{aligned} S_h\tilde{\alpha}_m(l) + S_g\alpha_m(l) &= S_h S_h^{*m+1} f(l) + S_g S_g^* S_h^{*m} f(l) \\ &= (S_h S_h^* + S_g S_g^*) S_h^{*m} f(l) \\ &= \tilde{\alpha}_{m-1}(l) \end{aligned}$$

since $S_h S_h^* + S_g S_g^* = I$ from Lemma 3.2. *This is exactly the same algorithm that is used in a wavelet decomposition!*

This shows that the wavelet algorithm, commonly referred to as the discrete wavelet transform, is precisely a decomposition into the m-wandering subspaces of the unilateral shift S_h. In other words, the representation

$$l^2(\mathbb{Z}) = \bigoplus_{m=0}^{\infty} S_h^m Q_0, \quad Q_0 = \overline{\text{span}}\{T_2^l g(n)\}_{l\in\mathbb{Z}}. \tag{3.12}$$

provides the Hilbert space setting for the wavelet decomposition and reconstruction algorithm.

The decomposition via the shift S_g is found in a similar manner and the decomposition into the generating wandering subspaces is simply

$$f(n) = \sum_{l=-\infty}^{\infty} \alpha_0(l) g(n-2l) + \sum_{l=-\infty}^{\infty} \beta_0(l) h(n-2l)$$

where $\alpha_0(l) = S_g^* f(l)$ and $\beta_0(l) = S_h^* f(l)$.

4. Concluding Remarks

We have shown that associated with each multiresolution decomposition are two unilateral shifts S_h and S_g of infinite multiplicity on $l^2(\mathbb{Z})$. The decomposition of $l^2(\mathbb{Z})$ into the m-wandering subspaces of S_h is precisely the wavelet decomposition. The basis functions from these two shifts are actually just two possible sets of basis vectors from the more general decompositions via *Wavelet packets* [8]. They are, however, the only decompositions with the shift structure. Further results regarding these shifts and other results concerning operator theoretic issues in wavelet theory can be found in [5].

References

[1] Daubechies, Ingrid. *Ten Lectures on Wavelets*, CBMS-NSF Regional Conf. Ser. in Appl. Math., vol. 61. Society for Industrial and Applied Mathematics, Philadelphia, 1992.

[2] Halmos, Paul. Shifts on Hilbert Spaces, *J. Reine Angew. Math* **208** (1961), 102–112.

[3] Lax, P. and Phillips, R.S. *Scattering theory*, Pure and Applied Mathematics, vol. 26. Academic Press, Inc., New York, 1967.

[4] Mallat, Stephane. Multiresolution approximations and wavelet orthonormal bases of $L^2(\mathbb{R})$, *Trans. Amer. Math. Soc.* **315** (1989), 69–87.

[5] Pham, Joseph. *An Operator Theoretic Approach to Wavelet Theory with Applications.* Ph.D. dissertation, University of California, Los Angeles, 1999.

[6] Rosenblum, M. and Rovnyak, J. *Hardy Classes and Operator Theory.* Oxford University Press, New York, 1985.

[7] Sz.-Nagy, B. and Foias, C. *Harmonic Analysis of Operators on Hilbert Space.* American Elsevier Publishing Company, Inc., New York, 1970.

[8] Wickerhauser, M.V. *Adapted Wavelet Analysis from Theory to Software.* A.K. Peters, Wellesley, Massachusetts, 1994.

Joseph N.Q. Pham, Department of Electrical Engineering, University of California, Los Angeles, 66-124 Engineering 4, Los Angeles, California 90095-1594, U.S.A.

E-mail address: pham@ee.ucla.edu

Progress in Nonlinear Differential Equations
and Their Applications, Vol. 42

ASYMPTOTIC BEHAVIOR OF SEMIGROUPS OF HOLOMORPHIC MAPPINGS

MARK ELIN
BRAUDE COLLEGE

SIMEON REICH
TECHNION-ISRAEL INSTITUTE OF TECHNOLOGY

DAVID SHOIKHET
BRAUDE COLLEGE
TECHNION-ISRAEL INSTITUTE OF TECHNOLOGY

Abstract. We present several new results on the asymptotic behavior of nonlinear semigroups of holomorphic mappings on the open unit balls of complex Banach and Hilbert spaces.

Let X be a complex Banach space and let $D \subset X$ be a domain (that is, an open connected subset of X). Recall that a mapping $f : D \to X$ is called holomorphic if it is Fréchet differentiable at each point of D [9]. The set of all holomorphic mappings from D into X will be denoted by $Hol(D, X)$.

Definition 1. *Let D be a domain in X and let $g \in Hol(D, X)$. The mapping g is said to be a semi-complete vector field on D if the Cauchy problem*

$$\left\{ \begin{array}{l} \frac{dv}{dt} + g(v) = 0 \\ v(0) = x \end{array} \right. \tag{1}$$

has a solution $v(\cdot, x) : \mathbb{R}^+ \to D$ which is well-defined on all of $\mathbb{R}^+$ for each initial datum $x \in D$.

Note that since any $g \in Hol(D, X)$ is locally bounded (hence locally Lipschitzian), this solution is unique and the family $\{S(t)\}_{t \geq 0}$, defined by $S(t) := v(t, \cdot)$, is a one-parameter semigroup (flow) of holomorphic self-mappings of D, i.e.,

$$\left\{ \begin{array}{l} S(t+s) = S(t) \circ S(s) \quad \text{for all} \quad t, s, \geq 0 \\ \\ S(0) = I, \end{array} \right. \tag{2}$$

where $\circ$ denotes the composition operation and I is the restriction of the identity operator on X to D. In the case where this flow consists of automorphisms of D it can be extended to a one parameter group and the Cauchy problem (1) has a unique solution $u(\cdot, x) : \mathbb{R} \to D$ defined on all of $\mathbb{R} = (-\infty, \infty)$ for each $x \in D$. The converse is also true. In such a situation the mapping g is said to be a complete vector field (see [4, 15]).

We observe that if the solution $S(t) = v(t, \cdot)$ of (1) is known, then $g \in Hol(D, X)$ can be recovered as the strong limit

$$g(x) = \lim_{t \to 0+} \frac{x - S(t)x}{t} \tag{3}$$

$(S(t)x = v(t, x))$, i.e., $-g$ is the right derivative of the semigroup $S(t)$ at zero. Therefore the mapping g is seen to be the infinitesimal generator of the semigroup.

As a matter of fact, for hyperbolic domains the converse is also true: If for a given semigroup $\{S(t) : t \geq 0\}$ of holomorphic self-mappings of D, which is continuous in $t \geq 0$, the strong limit $g(x)$ in (3) exists for all $x \in D$, then $v(t, x) = S(t)x$ is the solution of the Cauchy problem (1) for all $t \geq 0$ and $x \in D$. In other words, $g \in Hol(D, X)$ is a generator of a flow on D if and only if it is a semi-complete vector field (see, for example, [12]).

Let D be a domain in X and let g be a semi-complete vector field on D. Suppose that $S(t)$ is the semigroup of holomorphic mappings generated by g. The uniqueness of the solution of the Cauchy problem (1) implies that the null point set of g in D coincides with the common fixed point set of $S(t)$, i.e.,

$$Null_D g = \bigcap_{t \geq 0} Fix_D S(t). \tag{4}$$

In the theory of evolution equations this set is usually called the stationary point set of the semigroup.

Definition 2. *Let g be a semi-complete vector field on a domain D in X with $Null_D g \neq \phi$. A point $a \in Null_D g$ is said to be locally uniformly attractive if the semigroup $S(t)$ generated by g converges to a in the topology of local uniform convergence over D (see [5], [10]).*

Definition 3. *Let D be a domain in a Banach space X and let $\mathcal{G}(D)$ be the family of all semi-complete vector fields on D. A mapping $g \in \mathcal{G}(D)$ is said to be a strongly semi-complete vector field if it has a unique null point in D which is a locally uniformly attractive fixed point for the semigroup generated by g.*

Let $\sigma(A)$ denote the spectrum of a bounded linear operator $A : X \to X$. It is known [12] that if D is a bounded domain, then $g \in \mathcal{G}(D)$ with $g(\tau) = 0$, $\tau \in D$, is strongly semi-complete if and only if there is $\varepsilon > 0$ such that $Re\lambda \geq \varepsilon > 0$ for all $\lambda \in \sigma(g'(\tau))$. Such a point τ is sometimes said to be strictly regular.

In this paper we will give several sufficient conditions for $f \in Hol(D, X)$ to be strongly semi-complete on the open unit ball D of X and obtain rates of convergence for the semigroups generated by such mappings.

Let X^* be the dual of X. By $\langle x, x^* \rangle$ we denote the action of a linear functional x^* in X^* on an element x of X.

The mapping $J : X \to 2^{X^*}$ defined by

$$J(x) = \left\{ x^* \in X^* : \langle x, x^* \rangle = \|x\|^2 = \|x^*\|^2 \right\}, \quad x \in X,$$

is called the (normalized) duality mapping.

Theorem 4. *Let D be the open unit ball in X and let $g \in Hol(D, X)$ satisfy*

$$Re \langle g(x), x^* \rangle \geq \alpha(\|x\|)\|x\|, \quad x \in D, \quad x^* \in J(x), \tag{5}$$

where α is a real continuous function on $[0,1]$ such that

$$\alpha(1) = \omega > 0. \tag{6}$$

Then

(i) g is strongly semi-complete;

(ii) if $\{S(t)\}_{t \geq 0}$ is the semigroup generated by g, then for each pair of points x and y in D the following estimate holds:

$$\rho(S(t)x, S(t)y) \leq e^{-\frac{\omega}{2}t} \rho(x, y), \tag{7}$$

where ρ is the hyperbolic metric on D. In particular, if τ is the null point of g, then

$$\rho(S(t)x, \tau) \leq e^{-\frac{\omega}{2}t} \rho(x, \tau), \tag{8}$$

for all $x \in D$.

Proof. Consider for each $n = 1, 2, \ldots$ the mappings $g_n \in Hol(D, X)$ defined by

$$g_n(x) = x + \frac{t}{n} g(x) - y, \quad x \in D, \tag{9}$$

where $t \geq 0$ and $y \in D$. Let D_r be the open ball of radius $r \in [0,1)$ centered at the origin. For all $x \in \partial D_r = \{x \in X : \|x\| = r\}$ and for all $x^* \in J(x)$ we have by (5),

$$\begin{aligned} Re \ \langle g_n(x), x^* \rangle &= \|x\|^2 + \tfrac{t}{n} Re \langle g(x), x^* \rangle - Re \langle y, x^* \rangle \geq r^2 + \tfrac{t}{n} r\alpha(r) - r\|y\| \\ &= r \left(r + \tfrac{t}{n} \alpha(r) - \|y\| \right). \end{aligned} \tag{10}$$

Since $\alpha(1) > 0$, it is easy to see that for sufficiently large n the equation

$$\varphi_n(r) := r + \frac{t}{n} \alpha(r) = 1 \tag{11}$$

has a solution $r_n \in [0, 1)$.

Indeed, $\varphi_n(0) = \frac{t}{n}\,\alpha(0) \le 1$ for $n \ge t|\alpha(0)|$ and $\varphi_n(1) = 1 + \frac{t}{n}\,\omega > 1$. The inequality (5) implies in turn that for such n and r_n, and for all x with $\|x\| = r_n$ and $x^* \in J(x)$, the following inequality holds:

$$Re\,\langle g_n(x), x^*\rangle \ge r_n(1 - \|y\|). \tag{12}$$

Since g_n is bounded on $\bar{D}_{r_n}$ [8], it follows by [3] that the equation

$$g_n(x) = x + \frac{t}{n}\,g(x) - y = 0$$

has a unique solution $x = J_{\frac{t}{n}}(y) := \left(I + \frac{t}{n}\,g\right)^{-1}(y) \in D_{r_n}$ for each $y \in D$. In other words, the resolvent mapping $J_{\frac{t}{n}}$ maps D into D_{r_n}.

It now follows by the Earle-Hamilton fixed point theorem [6] that $J_{\frac{t}{n}}$ has a unique fixed point τ in D. This point is also a null point of g. In addition, repeating the proof of this fixed point theorem we obtain the estimate

$$\rho\left(J_{\frac{t}{n}}(x), J_{\frac{t}{n}}(y)\right) \le \frac{1}{1 + \left(\frac{t}{n}\right)\left(\frac{\alpha(r_n)}{2}\right)}\,\rho(x, y) \tag{13}$$

for each pair of points x and y in D.

Since $\alpha(r)$ is continuous on the interval $[0, 1]$, it follows by (11) that $r_n \to 1$ and $\alpha(r_n) \to \omega$ as $n \to \infty$. Therefore, by using the exponential formula [13, 14]

$$S(t)x = \lim_{n\to\infty} J^n_{\frac{t}{n}}(x)$$

and (13), we get by induction the estimates (7) and (8). Theorem 4 is proved.

Example. Let $D = \Delta$ be the open unit disk in the complex plane $\mathbb{C}$ and let $g \in Hol(\Delta, \mathbb{C})$ be defined by

$$g(z) = a - \bar{a}z^2 + bz\,\frac{1 - cz}{1 + cz}$$

where $a \in \mathbb{C}$, $Re\,b > 0$ and $0 \le c < 1$. If we take

$$\alpha(s) = -|a|(1 - s^2) + (Re\,b)s\,\frac{1 - cs}{1 + cs},$$

then we get

$$Re\,g(z)\bar{z} \ge \alpha(|z|)|z|$$

and $\alpha(1) = Re\,b\,\frac{1-c}{1+c} > 0$. Hence $g(z)$ is a strongly semi-complete vector field on Δ.

Remark. Note that if $g \in Hol(D, X)$ is known to be a semi-complete vector field on D, then condition (6) can be replaced by a slightly more general condition, namely,

$$(6') \qquad \alpha(l) > 0 \quad \text{for some} \quad l \in (0, 1],$$

which will still ensure the validity of assertion (i) of Theorem 4. This implies the following very simple and interesting sufficient condition.

Recall that a bounded linear operator $A : X \to X$ is called strongly accretive if

$$(14) \qquad Re\,\langle Ax, x^* \rangle \geq k\|x\|^2$$

for some $k > 0$ and all $x \in X,\ x^* \in J(x)$.

Corollary 5. *Let $g \in \mathcal{G}(D)$ and suppose that the bounded linear operator $A = g'(0)$ is strongly accretive with constant $k > 0$. If*

$$(15) \qquad k > 4\|g(0)\|,$$

then g is a strongly semi-complete vector field.

Proof. Consider the function $\alpha(s) = -\|g(0)\|(1-s^2) + ks\,\frac{1-s}{1+s}$. Using (15) we see that $\alpha(1) = 0$ and $\alpha'(1) < 0$. Hence there is $l \in (0, 1)$ such that $\alpha(l) > 0$. By [2] we know that

$$Re\,\langle g(x), x^* \rangle \geq \alpha(\|x\|)\|x\|, \quad x \in D.$$

Therefore the result follows from the above Remark.

Note that if $A = g'(0)$ is strongly accretive and $g(0) = 0$, then condition (15) is fulfilled automatically. Hence the origin is an attractive fixed point of the semigroup generated by g. Actually, this fact follows from more general considerations and in this case one can obtain an exponential rate of convergence.

To see this, we shall need the following lemma, the proof of which is omitted because it is similar to part of the proof of Theorem 4.

Lemma 6. *Let D be the open unit ball in a complex Banach space X and let $g \in Hol(D, X)$ satisfy the following condition:*

$$(16) \qquad Re\,\langle g(x), x^* \rangle \geq \alpha(\|x\|)\|x\|$$

for all $x \in D$ and some $x^* \in J(x)$, where α is a real continuous function on $[0, 1]$ such that for all $\mu \in [0, 1]$ and for all sufficiently small $r > 0$ the equation

$$(17) \qquad s + r\alpha(s) = \mu$$

has a unique solution $s(\mu)$ in $[0, 1]$. Then

(i) g is a semi-complete vector-field on D;
(ii) if $\beta(t,s)$ is the solution of the Cauchy problem

$$\left\{\begin{array}{l}\frac{\partial\beta(t,s)}{\partial t}+\alpha(\beta(t,s))=0\\ \beta(0,s)=s\in[0,1)\end{array}\right. \tag{18}$$

and $v(t,x)$ is the solution of (1), then the following estimate holds:

$$\|v(t,x)\|\leq\beta(t,\|x\|),\quad x\in D. \tag{19}$$

Proposition 7. *Let $g\in\mathcal{G}(D)$ be such that $g(0)=0$ and $A=g'(0)$ is strongly accretive with $\operatorname{Re}\langle Ax,x^*\rangle\geq k\|x\|^2$. Suppose that $\{S(t)\}_{t\geq0}$ is the semigroup generated by g. Then the following estimates hold:*

(i)

$$\|S(t)x\|\leq\|x\|e^{-k\frac{1-\|x\|}{1+\|x\|}t},\quad x\in D,\ t\geq0;$$

(ii)

$$\frac{\|S(t)x\|}{(1-\|S(t)x\|)^2}\leq e^{-kt}\,\frac{\|x\|}{(1-\|x\|)^2}.$$

Proof. Both estimates follow directly from Lemma 6 if we set

$$\alpha(s)=ks\,\frac{1-s}{1+s}.$$

In this case

$$\beta(t,s)\leq se^{-k\frac{1-s}{1+s}t},$$

where $\{\beta(t,\cdot)\}$ is the semigroup generated by α.

Note that the estimate (i) is due to Gurganus [7, Proposition 2.5.4] while (ii) was obtained by Poreda [11]. Note also that the condition $g(0)=0$ is essential in their considerations as well as in the above approach.

For the case where X is a Hilbert space we would like to obtain more general estimates when g has an arbitrary null point which is strictly regular.

For a step in this direction we shall need the following general lemma. We omit its simple proof.

Lemma 8. *Let D and Ω be two domains in a complex Banach space X such that $\Omega=f(D)$ for some biholomorphic mapping $f:D\to\Omega$. Then there is a linear invertible operator T from the space $Hol(\Omega,X)$ onto the space $Hol(D,X)$ which takes the set $\mathcal{G}(\Omega)$ onto the set $\mathcal{G}(D)$ (i.e. $\mathcal{G}(D)=T(\mathcal{G}(\Omega))$. In other words, the classes of semi-complete vector fields on Ω and D are linearly isomorphic. Moreover, such an isomorphism $T:\mathcal{G}(\Omega)\to\mathcal{G}(D)$ can be given by the formulae*

$$T(g)(\cdot)=[f'(\cdot)]^{-1}g(f(\cdot)) \tag{20}$$

and

$$T^{-1}(\varphi)(\cdot) = [f'(f^{-1}(\cdot))]\varphi(f^{-1}(\cdot)), \tag{21}$$

where $g \in \mathcal{G}(\Omega)$ *and* $\varphi \in \mathcal{G}(D)$.

Now let $\alpha : [0,1] \to \mathbb{R}^+$ be a continuous function on the interval $[0,1]$ such that for some $\delta > 0$ and for each $r \in [0,\delta)$ the function $s + r\alpha(s)$ is increasing on $[0,1]$. We also assume that α satisfies the following range condition: For each $r \in [0,\delta)$ and for all $p \in [0,1]$, the equation

$$s + r\alpha(s) = p \tag{22}$$

has a (unique) solution $s = s(r,p) \in [0,1]$. This solution is an increasing function of $p \in [0,1]$ for each fixed $r \in [0,\delta)$. Also, for each $t \geq 0$ there exists the limit

$$\beta(t,p) = \lim_{n\to\infty} s^{(n)}\left(\frac{1}{n}t, p\right), \tag{23}$$

where by $s^{(n)}(t,p)$ we denote the n-th fold iterate of $s(r,p)$, i.e., $s^{(0)}(r,p) = p$, $s^{(n)}(r,p) = s^{(n-1)}(r, s(r,p))$, $n = 1, 2, \ldots$ This limit is the solution of the Cauchy problem

$$\begin{cases} \frac{\partial\beta(t,p)}{\partial t} + \alpha(\beta(t,p)) = 0 \\ \beta(0,p) = p. \end{cases} \tag{24}$$

Assume now that X is a Hilbert space H and that g is a holomorphic mapping on $\mathbb{B}$, the open unit ball of H. As we have already mentioned, since g is locally bounded, hence locally Lipschitzian, the Cauchy problem

$$\begin{cases} \frac{\partial u(t,x)}{\partial t} + g(u(t,x)) = 0 \\ u(0,x) = x \end{cases} \tag{25}$$

has a unique local solution $u = u(t,x)$ which is real-analytic in t in some neighborhood of zero, and holomorphic in a neighborhood U_x of x. We would like to compare this solution with the function $\beta(t,p)$ defined by (24) and to find out when $u = u(t,x)$ can be extended to a global solution of (25) defined on $\mathbb{R}^+ \times \mathbb{B}$. To this end, we define, for a given $\tau \in \mathbb{B}$, a function $m_\tau : [0,\delta) \times U_x \to \mathbb{R}^+$ by

$$m_\tau(t,y) = \|M_{-\tau}(u(t,y))\|, \tag{26}$$

where $M_{-\tau}$ is the Möbius transformation which takes τ to zero [6]. It is natural to compare this function with the function $\beta(t, \|M_{-\tau}(y)\|)$. As a matter of fact, we will see below that both local and global relations between these two functions are completely determined by their derivatives at merely one point.

For each $\tau \in \mathbb{B}$, let $N_\tau = \{g \in \mathcal{G}(\mathbb{B}) : g(\tau) = 0\}$.

Theorem 9. *Let $g \in Hol(\mathbb{B}, H)$ and let α, β and m be the functions defined above. Then the following assertions are equivalent:*
(i) for some $\tau \in \mathbb{B}$,

$$m_\tau(t,x) \le \beta(t, \|M_{-\tau}(x)\|)$$

whenever $u(t,x)$ is defined;
(ii) for some $\tau \in \mathbb{B}$,

$$\left.\frac{\partial m_\tau}{\partial t}\right|_{t=0+} \le \left.\frac{\partial \beta}{\partial t}\right|_{t=0+} = -\alpha(\|M_{-\tau}(x)\|);$$

(iii) for some $\tau \in \mathbb{B}$, the mapping g belongs to N_τ and satisfies

$$Re\,\langle g(x), x^*\rangle \ge \frac{\alpha(\|M_{-\tau}(x)\|)\|M_{-\tau}(x)\|}{\sigma(\tau, x)}$$

where $x \in \mathbb{B}$, $\sigma(x,y) = 1 - \|M_{-y}(x)\|^2$ and $z^* = \frac{z}{1-\|z\|^2} - \frac{\tau}{1-\langle \tau, z\rangle}$.

Thus, if one of the above assertions holds, then $u(t,x)$ has a unique extension to all of $\mathbb{R}^+ \times \mathbb{B}$ and the estimate (i) holds globally. The points τ in (i)-(iii) are one and the same.

Proof. Since $m_\tau(0,x) = \beta(0, \|M_{-\tau}(x)\|)$, the implication (i)⇒(ii) is clear. Further, by direct calculations we obtain that

$$\left.\frac{\partial m_\tau}{\partial t}\right|_{t=0+} = -\frac{\sigma(\tau,x)}{\|M_{-\tau}(x)\|}\,Re\,\langle g(x), x^*\rangle\,.$$

Since the inequality in (iii) and [1] imply that $g \in N_\tau$, the implication (ii)⇒(iii) follows because $\left.\frac{\partial \beta(t,\|M_{-\tau}(x)\|}{\partial t}\right|_{t=0+} = -\alpha(\|M_{-\tau}(x)\|)$. Thus it remains to be shown that (iii)⇒(i). To this end, we use Lemma 8. Since $g \in N_\tau$, $u(t,x)$ is well-defined and belongs to $\mathbb{B}$ for all $(t,x) \in \mathbb{R}^+ \times \mathbb{B}$. Hence the operator $T : Hol(\mathbb{B}, H) \to Hol(\mathbb{B}, H)$ defined by

$$T(g)(\cdot) = [(M_\tau)'(\cdot)]^{-1} g(M_\tau(\cdot)) = (M_{-\tau})'(M_\tau(\cdot))g(M_\tau(\cdot)) \tag{27}$$

takes $g \in N_\tau$ to $\varphi \in N_0$. Now we observe that since the explicit expression for the linear operator A $(= A(x)) := (M_{-\tau})'(M_\tau(x))$ is

$$A = \frac{1+\langle x,\tau\rangle}{1-\|\tau\|^2}\left(P_\tau + \sqrt{1-\|\tau\|^2}\,(I-P_\tau) + \langle\cdot,\tau\rangle\, x\right), \tag{28}$$

where P_τ is the orthogonal projection of the Hilbert space H onto the one-dimensional subspace spanned by τ,

we get

$$\langle\varphi(x), x\rangle = \langle Ag(M_\tau(x)), x\rangle = \langle g(M_\tau(x)), A^*x\rangle =$$

$$= \frac{|1+\langle x,\tau\rangle|^2}{1-\|\tau\|^2}\left\langle g(M_\tau(x)),\ M_\tau(x) - \tau\,\frac{1-\|x\|^2}{1+\langle x,\tau\rangle}\right\rangle. \tag{29}$$

Also, if $z = M_\tau(x)$, then

$$1 + \langle x, \tau \rangle = 1 + \langle M_{-\tau}(z), \tau \rangle = \frac{1 - \|\tau\|^2}{1 - \langle z, \tau \rangle}$$

and

$$1 - \|x\|^2 = 1 - \|M_{-\tau}(z)\|^2 = \sigma(\tau, z).$$

Hence by (29) we obtain

$$\begin{aligned}\langle \varphi(x), x \rangle &= \frac{1 - \|\tau\|^2}{|1 - \langle z, \tau \rangle|^2} \left\langle g(z), z - \tau \frac{1 - \|z\|^2}{1 - \langle \tau, z \rangle} \right\rangle = \\ &= \sigma(\tau, z) \left\langle g(z), \frac{z}{1 - \|z\|^2} - \frac{\tau}{1 - \langle \tau, z \rangle} \right\rangle.\end{aligned}$$

Thus (iii) implies that

$$Re \, \langle \varphi(z), z \rangle \geq \alpha(\|z\|)\|z\|, \quad z \in \mathbb{B}. \tag{30}$$

Now, if $v(t, z)$ is the solution of the Cauchy problem

$$\begin{cases} \frac{\partial v}{\partial t} + \varphi(v) = 0 \\ v(0, z) = z, \end{cases}$$

then it follows by Lemma 6 that

$$\|v(t, z)\| \leq \beta(t, \|z\|), \quad z \in \mathbb{B}.$$

But $v(t, M_{-\tau}(x)) = M_{-\tau}(u(t, x))$ and this concludes the proof.

Remark. If $\beta(t, s) \to 0$ as $t \to \infty$ for a fixed $s \in [0, 1)$, then g is seen to be strongly semi-complete and assertion (i) establishes a rate of convergence of the semigroup $S(t) = u(t, \cdot)$ to its stationary point $\tau \in \mathbb{B}$. It would be nice, of course, to find a universal class of functions α such that this convergence will be of exponential type. We intend to study this problem elsewhere.

References

[1] D. Aharonov, M. Elin, S. Reich and D. Shoikhet, Parametric representations of semi-complete vector fields on the unit balls in $\mathbb{C}^n$ and in Hilbert space, *Atti. Accad. Naz. Lincei*, in press.

[2] D. Aharonov, S. Reich and D. Shoikhet, Flow invariance conditions for holomorphic mappings in Banach spaces, *Math. Proc. Royal Irish Academy*, in press.

[3] L. Aizenberg, S. Reich and D. Shoikhet, One-sided estimates for the existence of null points of holomorphic mappings in Banach spaces, *J. Math. Anal. Appl.* **203** (1996), 38-54.

[4] S. Dineen, *The Schwartz Lemma*, Clarendon Press, Oxford, 1989.

[5] T. Franzoni and E. Vesentini, *Holomorphic Maps and Invariant Distances*, North-Holland, Amsterdam, 1980.

[6] K. Goebel and S. Reich, *Uniform Convexity, Hyperbolic Geometry and Nonexpansive Mappings*, Marcel Dekker, New York and Basel, 1984.

[7] K.R. Gurganus, Φ-like holomorphic functions in $\mathbb{C}^n$ and Banach space, *Trans. Amer. Math. Soc.* **205** (1975), 389-406.

[8] L.A. Harris, S. Reich and D. Shoikhet, Dissipative holomorphic functions, Bloch radii, and the Schwarz lemma, *Technion Preprint Series* **1050** (1999), 1-14.

[9] E. Hille and R. S. Phillips, *Functional Analysis and Semigroups*, Amer. Math. Soc., Providence, RI, 1957.

[10] J.M. Isidro and L.L. Stacho, *Holomorphic Automorphism Groups in Banach Spaces: An Elementary Introduction*, North Holland, Amsterdam, 1984.

[11] T. Poreda, On generalized differential equations in Banach space, *Dissert. Math.* **310** (1991), 1-50.

[12] S. Reich and D. Shoikhet, Generation theory for semigroups of holomorphic mappings in Banach spaces, *Abstr. Appl. Anal.* **1** (1996), 1-44.

[13] S. Reich and D. Shoikhet, Semigroups and generators on convex domains with the hyperbolic metric, *Atti. Accad. Naz. Lincei* **8** (1997), 231-250.

[14] S. Reich and D. Shoikhet, Metric domains, holomorphic mappings and nonlinear semigroups, *Abstr. Appl. Anal.* **3** (1998), 203-228.

[15] H. Upmeier, *Jordan Algebras in Analysis, Operator Theory and Quantum Mechanics*, CBMS, Reg. Conf. Ser. in Math., Vol. **67**, Amer. Math. Soc., Providence, RI, 1986.

Mark Elin, Department of Applied Mathematics, Braude College, 21982 Karmiel, ISRAEL

Simeon Reich, Department of Mathematics, Technion-Israel Institute of Technology, 32000 Haifa, ISRAEL
E-mail address: `sreich@techunix.technion.ac.il`

David Shoikhet, Department of Applied Mathematics, Braude College, 21982 Karmiel; Department of Mathematics, Technion-Israel Institute of Technology, 32000 Haifa, ISRAEL

Progress in Nonlinear Differential Equations
and Their Applications, Vol. 42

FRACTIONAL POWERS OF OPERATORS VIA HYPERSINGULAR INTEGRALS

STEFAN SAMKO
UNIVERSIDADE DO ALGARVE

Introduction

The well known Balakrishnan formula represents the fractional power $(-A)^\alpha$ in case of the generator A of a semigroup $T_t, t > 0$, in terms of a (hyper)-singular integral with respect to the variable $t \in R^1_+$, that is,

$$(-A)^\alpha f = \frac{1}{\Gamma(-\alpha)} \int_0^\infty t^{-\alpha-1} (T_t - I) f dt ,$$

where $0 < \alpha < 1, \varphi \in D(A)$, and I is the identity operator. In the case $\alpha > 1$, this formula can be written with the usage of "finite differences" $(T_t - I)^\ell, \ell = 1, 2, 3, ..., \ell > \alpha$:

$$(1) \qquad (-A)^\alpha f = \frac{1}{\varkappa(\alpha, \ell)} \int_0^\infty t^{-\alpha-1} (I - T_t)^\ell f dt , \ell > \alpha,$$

with $\varkappa(\alpha, \ell) = -\Gamma(-\alpha) A_\alpha(\ell)$, where $A_\alpha(\ell) = \sum_{k=0}^\ell (-1)^{k-1} \binom{\ell}{k}$. In particular, the fractional power of the Laplace operator is given by (1) with $T_t = P_t$ where P_t is the Poisson semigroup of operators:

$$P_t f = c_n \int_{R^n} \frac{t f(x-y)}{(|x|^2 + t^2)^{(n+1)/2}} dy , \quad t > 0.$$

On the other hand, positive fractional powers of the Laplace operator can be given also in the form

$$(2) \qquad (-\Delta)^{\frac{\alpha}{2}} f = \frac{1}{d_{n,\ell}(\alpha)} \lim_{\varepsilon \to 0} \int_{|y|>\varepsilon} \frac{(\Delta_y^\ell f)(x)}{|y|^{n+\alpha}} dy ,$$

see [23], p.56, which is also known as the Riesz fractional derivative and denoted as $\mathbb{D}^\alpha f = (-\Delta)^{\frac{\alpha}{2}} f$. Here $d_{n,\ell}(\alpha)$ is the known normalizing constant and the finite difference $(\Delta_y^\ell f)(x)$, generated by the standard shift $\tau_y f = f(x-y)$, may be centered one and then $0 < \alpha < \ell$, see [15], or a non-centered and then $0 < \alpha < 2\left[\frac{\ell}{2}\right]$, where $\left[\frac{\alpha}{2}\right]$ stands for the entire part of $\frac{\alpha}{2}$, see [23], Ch.3, Section 1.

Hypersingular constructions of the type (2) can be used for an effective realization of fractional powers of some differential operators of mathematical physics, such as fractional powers $(I - \Delta)^{\frac{\alpha}{2}}, \Delta$ being the Laplace operator; fractional powers $(-\Delta_x + \frac{\partial}{\partial t})^{\frac{\alpha}{2}}$ of *parabolic (heat) operator* or $(I - \Delta_x + \frac{\partial}{\partial t})^{\frac{\alpha}{2}}$, the Laplace operator being applied in the spatial variable $x = (x_1, \cdots, x_n)$; fractional powers of the *wave operator*, of *Schrödinger operator* and others.

What follows is a brief survey of some of the results on application of hypersingular integrals to the realization of fractional powers of these and other differential operators in partial derivatives. Details of some of the presented results, as well as further applications, may be found in the book [23].

1. The fractional powers $(I-\Delta)^{\frac{\alpha}{2}}$

According to the Balakrishnan formula, the fractional power $(I-\Delta)^{\frac{\alpha}{2}}f$ may be represented as

$$(I-\Delta)^{\frac{\alpha}{2}}f = \frac{1}{\varkappa(\alpha,\ell)}\int_0^\infty t^{-1-\alpha}(I-e^{-t}P_t)^\ell f\,dt\,, \quad \ell>\alpha\,.$$

Keeping applications in mind, we wish to give a construction of this fractional power directly in terms of multidimensional hypersingular integrals.

a) The idea of the construction

We start with the negative power $(I-\Delta)^{-\frac{\alpha}{2}}$, which is the Bessel potential represented by the convolution

$$(I-\Delta)^{-\frac{\alpha}{2}}\varphi = B^\alpha\varphi = \frac{\omega_\alpha(|x|)}{|x|^{n-\alpha}}*\varphi, \qquad \alpha>0, \tag{3}$$

where

$$\omega_\alpha(|x|) = \frac{2^{1-\frac{n+\alpha}{2}}}{\pi^{\frac{n}{2}}\Gamma(\frac{\alpha}{2})}|x|^{\frac{n-\alpha}{2}}K_{\frac{n-\alpha}{2}}(|x|) = \frac{1}{\pi^{\frac{n}{2}}2^\alpha\Gamma(\frac{\alpha}{2})}\int_0^\infty t^{\frac{n-\alpha}{2}-1}e^{-t-\frac{|x|^2}{4t}}dt.$$

Since the operators B^α form a semigroup, it is natural to expect that the inverse operator should be formally given by

$$(B^\alpha)^{-1}f = \frac{\omega_{-\alpha}(|x|)}{|x|^{n+\alpha}}*f\,, \tag{4}$$

under the appropriate interpretation of this convolution.

We note that the function $\omega_{-\alpha}(|x|)$ is differentiable, exponentially decays at infinity and stabilizes at the origin to $\omega_{-\alpha}(0)=\frac{2^{\alpha-1}\Gamma(\frac{n+\alpha}{4})}{\pi^{\frac{n}{2}}\Gamma(-\frac{\alpha}{2})}$. We also observe that formally $\omega_{-\alpha}(|x|)\equiv 0$ for $\alpha=2,4,6,\cdots$, but $\frac{1}{|x|^{n+\alpha}}$, as a distribution, has poles at the same points $\alpha=2,4,6,\cdots$. We describe an effective realization of the convolution (4) with the distribution $\frac{\omega_{-\alpha}(|x|)}{|x|^{n+\alpha}}$.

b) The operator $(I-\Delta)^{\frac{\alpha}{2}}$ as the convolution (4)

Theorem 1. *Let $f(x)\in\mathcal{S}(R^n)$ and $\Re\alpha>0$. Then*

$$f.p.\frac{\omega_{-\alpha}(|x|)}{|x|^{n+\alpha}}*f = \sum_{k=0}^{[\frac{m-1}{2}]}(-1)^k\binom{\alpha/2}{k}\Delta^k f+\int_{R^n}\frac{f(x-y)-T_{m-1}(x,y)}{|y|^{n+\alpha}}\omega_{-\alpha}(|y|)dy\,, \tag{5}$$

where $m>\Re\alpha$ and $T_m(x,y)=\sum_{|j|\le m}\frac{(-y)^j}{j!}(D^jf)(x)$ is the Taylor polynomial.

We refer to [23], p.274, for the proof of Theorem 1,

Basing on (3), we may write

$$(I-\Delta)^{\frac{\alpha}{2}} f = \sum_{k=0}^{[\frac{m-1}{2}]} (-1)^k \binom{\alpha/2}{k} \Delta^k f + \int_{R^n} \frac{f(x-y) - T_{m-1}(x,y)}{|y|^{n+\alpha}} \omega_{-\alpha}(|y|) dy \,, \tag{6}$$

where $m > \alpha$. We remind that $\omega_{-\alpha}(|y|) \equiv 0$ in the case when $\alpha = 2, 4, 6, \cdots$, so that in this case the right-hand side of (6) turns into just an integer power of $I - \Delta$. Now, having obtained the exact expression for the fractional power, we may justify it directly, firstly for nice function again.

c) Direct justification of the formula (6) in case of nice functions.

Let $B^\alpha \varphi = F^{-1}(1+|\xi|^2)^{-\frac{\alpha}{2}} F\varphi$, $\alpha > 0$, be the Bessel potential operator. A justification of the formula (6) may be given in the following form.

Theorem 2. *Let $f(x) \in \mathcal{S}(R^n)$. Then*

$$(I-\Delta)^{\frac{\alpha}{2}} B^\alpha f = B^\alpha (I-\Delta)^{\frac{\alpha}{2}} f = f \,, \quad \alpha > 0 \,.$$

We consider specially the case $0 < \alpha < 2$. Let $0 < \alpha < 1$ first. The formula (6) turns into

$$(I-\Delta)^{\frac{\alpha}{2}} f := f(x) + \int_{R^n} \frac{f(x-y) - f(x)}{|y|^{n+\alpha}} \omega_{-\alpha}(|y|) dy \,. \tag{7}$$

The integral here converges absolutely, for example, for $f \in \mathcal{S}(R^n)$ if $0 < \alpha < 1$. It converges as the limit of truncated integrals over $\{y \in R^n : |y| > \varepsilon\}$ for all $0 < \alpha < 2$. Indeed, since $\omega_{-\alpha}(|y|)$ is even, this follows from the relation:

$$\lim_{\epsilon\to 0} \int_{|y|>\epsilon} \frac{f(x-y) - f(x)}{|y|^{n+\alpha}} \omega_{-\alpha}(|y|) dy$$

$$= -\int_{R^n} \frac{(\Delta_y^2 f)(x)}{|y|^{n+\alpha}} \omega_{-\alpha}(|y|) dy - \frac{1}{2} \int_{R^n} \frac{(\Delta_y^2 f)(x+y)}{|y|^{n+\alpha}} \omega_{-\alpha}(|y|) dy \tag{8}$$

So, we arrive at the following theorem.

Theorem 3. *Let $0 < \alpha < 2$. The fractional power $(I-\Delta)^{\frac{\alpha}{2}} f$, interpreted as (6), or which is the same, as the operator, inverse to the Bessel potential operator B^α within the framework of the space $\mathcal{S}$, has the form,*

$$(I-\Delta)^{\frac{\alpha}{2}} f := f(x) + \int_{R^n} \frac{f(x-y) - f(x)}{|y|^{n+\alpha}} \omega_{-\alpha}(|y|) dy \,, \tag{9}$$

where the integral converges absolutely in the case $0 < \alpha < 1$ and conditionally in the case $1 \le \alpha < 2$.

Corollary. *The corresponding realization of* $(\lambda^2 I - \Delta)^{\frac{\alpha}{2}}$ *with* $\lambda > 0$ *is*

$$(\lambda^2 I - \Delta)^{\frac{\alpha}{2}} f \ := \lambda^\alpha f(x) + \int_{R^n} \frac{f(x-y) - f(x)}{|y|^{n+\alpha}} \omega_{-\alpha}(\lambda |y|) dy \ , \quad 0 < \alpha < 2 \ .$$

In the case $\alpha = 1$, we arrive at the following interpretation of the square root $\sqrt{\lambda I - \Delta}$:

$$(10) \ \sqrt{\lambda^2 I - \Delta} \ f \ = \ \lambda f(x) \ - \frac{2}{(2\pi)^{\frac{n+1}{2}}} \int_{R^n} \frac{f(x-y) - f(x)}{|y|^{\frac{n+1}{2}}} K_{\frac{n+1}{2}}(\lambda |y|) dy \ .$$

This operator is well known under the name of square root Klein-Gordon operator in mathematical aspects of quantum physics, see e.g. [26], [11] and [22].

In particular, in the planar case $n = 2$ we have $K_{\frac{3}{2}}(z) = \sqrt{\frac{\pi}{2}} \frac{z-1}{z^{\frac{3}{2}}} e^{-z}$ and the formula (10) turns into

$$\sqrt{\lambda^2 I - \Delta} \ f \ = \ \lambda f(x) \ - \frac{1}{2\pi \lambda^{3/2}} \int_{R^2} \frac{f(x-y) - f(x)}{|y|^3} (1 - \lambda |y|) e^{-\lambda |y|} dy \ .$$

In the case $1 \leq \Re\alpha < 2$ we may also use the form (6) which is

$$(11) \ (I - \Delta)^{\frac{\alpha}{2}} f, \ = \ f(x) + \int_{R^n} \frac{f(x-y) - f(x) + y \cdot grad \ f(x)}{|y|^{\frac{n+1}{2}}} \omega_{-\alpha}(|y|) dy \ .$$

Here the integral is already absolutely convergent (in case of nice functions).

The following lemma presents another version of the construction (11).

Lemma 4. *Let* $1 < \Re\alpha < 2$. *The construction (11) may be represented in terms of hypersingular integral*

$$(12) \ (I - \Delta)^{\frac{\alpha}{2}} f \ = \ f(x) \ + \int_{R^n} \frac{f(x) - 2f(x-y) + f(x-2y)}{|y|^{\frac{n+1}{2}}} \mu_\alpha(|y|) dy$$

where $\mu_\alpha(r) \ = \ \frac{1}{2} \sum_{k=1}^{\infty} 2^{(1-\alpha)k} \omega_{-\alpha}(2^k r)$.

We note that $\mu_\alpha(r) \leq c r^{\frac{n+\Re\alpha-1}{2}} e^{-2r}$, $r \geq 1$, and $|\mu_\alpha(r) - \mu_\alpha(0)| \leq c r^\beta$, $0 < r \leq 1$, where $\beta \in (0, \Re\alpha - 1)$, the constant c not depending on r.

d) Justification of the formula within the framework of the spaces L_p; the cases $0 < \Re\alpha < 1$ and $1 < \Re\alpha < 2$

For nice functions $f(x)$ the operator $(I - \Delta)^{\frac{\alpha}{2}}$ was obtained in (7) which implies a non-absolutely convergent integral in the case $1 \leq \Re\alpha < 2$, even in the case of "nice" functions $f(x)$. To deal with "not so nice" functions $f(x)$ in the range $B^\alpha(L_p)$, we shall use only absolutely convergent constructions, that is, the construction (7) in the case $0 < \Re\alpha < 1$ and the construction (12) in the case $1 < \Re\alpha < 2$. But on the whole range $B^\alpha(L_p)$ they will not already be absolutely convergent and will be treated as

$$(13) \qquad (I - \Delta)^{\frac{\alpha}{2}} f \ = \ f(x) \ + \ \lim_{\substack{\epsilon \to 0 \\ (L_p)}} T_\epsilon^\alpha f$$

with $T_\epsilon^\alpha f = \int_{|y|>\epsilon} \frac{(\Delta_y^\ell f)(x)}{|y|^{n+\alpha}} \mu_\alpha(|y|)dy$, where $\ell = 2$ and $\mu_\alpha(|y|)$ is a function from (12), if $1 < \Re\alpha < 2$, and $\ell = 1$ and $\mu_\alpha(|y|) = \omega_{-\alpha}(|y|)$ if $0 < \Re\alpha < 1$.

Theorem 5. *Let $f(x) = B^\alpha\varphi, \varphi \in L_p(R^n), 1 < p < \infty, 0 < \Re\alpha < 1$ or $1 < \Re\alpha < 2$. Then*

$$(I - \Delta)^{\frac{\alpha}{2}} f = \varphi \tag{14}$$

with $(I-\Delta)^{\frac{\alpha}{2}} f$ interpreted according to (13).

2. Parabolic (heat) hypersingular integrals

We consider the fractional powers $\left(\frac{\partial}{\partial t} - \Delta_x\right)^{\frac{\alpha}{2}}$ introduced via the corresponding Fourier multipliers. The negative fractional powers are known as parabolic fractional integrals. The positive fractional powers will be realized as hypersingular integrals. We refer to original papers [17] and [18], see also [23], Chapter 9, Section 2.

a) Parabolic fractional potentials

The fractional parabolic potentials $H^\alpha\varphi$ are introduced via Fourier transforms by the relations

$$F(H^\alpha\varphi) = \left(|\xi|^2 - i\tau\right)^{-\frac{\alpha}{2}} \widehat{\varphi}(\xi,\tau), \tag{15}$$

where

$$\widehat{\varphi}(\xi,\tau) = (F\varphi)(\xi,\tau) = \int_{R^{n+1}} e^{i\xi x + i\tau t} dx dt \tag{16}$$

and $(|\xi|^2 - i\tau)^{-\frac{\alpha}{2}} := (|\xi|^4 + \tau^2)^{-\frac{\alpha}{4}} e^{-\frac{i\alpha}{2}\arg(|\xi|^2 - i\tau)}$ under the choice $\arg(|\xi|^2 - i\tau) \in \left(-\frac{\pi}{2}, \frac{\pi}{2}\right)$.

The potentials $H^\alpha\varphi$ which are, in fact, negative fractional powers of parabolic differential operators $\left(\frac{\partial}{\partial t} - \Delta_x\right)^{-\frac{\alpha}{2}}$, were introduced in [13] and [25]. It is known that the convolution operator H^α has the form

$$(H^\alpha\varphi)(x,t) = \frac{1}{\Gamma\left(\frac{\alpha}{2}\right)} \int_{R_+^{n+1}} \tau^{\frac{\alpha}{2}-1} W(y,\tau)\varphi(x-y, t-\tau) dy d\tau\ , \tag{17}$$

where $W(y,\eta) = (4\pi t)^{-\frac{n}{2}} e^{-\frac{|x|^2}{4t}}$ is the Gauss-Weierstrass kernel.

Remark 6. *The fractional parabolic potential $H^\alpha\varphi$ may be interpreted as a result of the one-dimensional fractional integration applied in the time variable t to the Gauss-Weierstrass operator $W_t\varphi$, that is,*

$$(H^\alpha\varphi)(x,t) = \frac{1}{\Gamma\left(\frac{\alpha}{2}\right)} \int_0^\infty \tau^{\frac{\alpha}{2}-1} (W_\tau\varphi)(x, t-\tau) d\tau\ . \tag{18}$$

These formulas evidently generate the corresponding Balakrishnan formula when we pass to positive fractional powers. But we remind that we are now interested in the realization of positive powers in terms of multidimensional hypersingular type integrals.

b) Positive fractional powers

$\left(\frac{\partial}{\partial t} - \Delta_x\right)^{\frac{\alpha}{2}}$, $\alpha > 0$. These powers may be constructed effectively as some hypersingular integral operators inverse to the parabolic potential operators H^α. They will contain non-standard finite differences which take into account different behaviour of potentials with respect to the space variable $x \in R^n$ and the time variable $t \in R^1$.

To arrive at the idea of the construction of these fractional powers, we apply the Fourier transform F_x in x to both parts of the equality $f(x,t) = (H^\alpha\varphi)(x,t)$. We get

$$I_+^{\frac{\alpha}{2}}\left[e^{\tau|x|^2}(F_x\varphi)(x,\tau)\right] = e^{t|x|^2}(F_x f)(x,t) ,$$

where I_+^α is the one-dimensional fractional integration operator applied in the time variable. Inverting the operator $I_+^{\frac{\alpha}{2}}$ according to the well known Marchaud formula for fractional derivatives, we get

$$e^{t|x|^2}(F_x\varphi)(x,t) = \frac{1}{\varkappa\left(\frac{\alpha}{2},\ell\right)}\int_0^\infty \sum_{k=0}^{\ell}(-1)^k\binom{\ell}{k}(F_x f)(x,t-k\tau)e^{(t-k\tau)|x|^2}\frac{d\tau}{\tau^{1+\frac{\alpha}{2}}} .$$

Multiplying this by $e^{-t|x|^2}$ and then applying the inverse Fourier transform in x, we arrive at

$$\varphi(x,t) =$$

$$\frac{1}{\varkappa\left(\frac{\alpha}{2},\ell\right)}\int_0^\infty\left[f(x,t)+\sum_{k=0}^{\ell}(-1)^k\binom{\ell}{k}\int_{R^n} f(x-y,t-k\tau)W(y,k\tau)dy\right]\frac{d\tau}{\tau^{1+\frac{\alpha}{2}}} .$$

Hence, after the change of variables $y \to \sqrt{k\tau}y$ and easy transformations, we arrive at the following

Conclusion 7. *The formal solution of the equation* $(H^\alpha\varphi)(x,t) = f(x,t)$ *is given by the formula*

$$\varphi(x,t) = \frac{1}{(4\pi)^{\frac{n}{2}}\varkappa\left(\frac{\alpha}{2},\ell\right)}\int_{R_+^{n+1}}\frac{(\Delta_{y,\tau}^\ell f)(x,t)}{\tau^{1+\frac{\alpha}{2}}}e^{-\frac{|y|^2}{4}}dyd\tau = \left(\frac{\partial}{\partial t} - \Delta_x\right)^{\frac{\alpha}{2}} f , \tag{19}$$

where

$$(\Delta_{y,\tau}^\ell f)(x,t) = \sum_{k=0}^{\ell}(-1)^k\binom{\ell}{k}f(x-y\sqrt{k\tau},t-k\tau) , \quad \ell > \frac{\alpha}{2} . \tag{20}$$

In particular, in the case $0<\alpha<2$ *we have*

$$\left(\frac{\partial}{\partial t} - \Delta_x\right)^{\frac{\alpha}{2}} f = \frac{1}{(4\pi)^{\frac{n}{2}}\Gamma\left(-\frac{\alpha}{2}\right)}\int_{R_+^{n+1}}\frac{f(x,t)-f(x-y\sqrt{\tau},t-\tau)}{\tau^{1+\frac{\alpha}{2}}}e^{-\frac{|y|^2}{4}}dyd\tau. \tag{21}$$

Remark 8. *The constructions (19) and (22) were obtained formally. The question which arise naturally, is whether the integrals in (19) and (22) converge. For $f \in \mathcal{S}$ it is not hard to show that they converge absolutely if we take $\ell > \alpha$. In the case $\ell > \frac{\alpha}{2}$ they converge non absolutely, as the limits of the corresponding truncated integrals:*

$$H^{-\alpha}f = \lim_{\epsilon \to 0} H_\epsilon^{-\alpha} f \tag{22}$$

where

$$H_\epsilon^{-\alpha}f = \frac{1}{(4\pi)^{\frac{n}{2}} \varkappa\left(\frac{\alpha}{2}, \ell\right)} \int_{R^{n+1}_{\epsilon+}} \frac{(\Delta^\ell_{y,\tau} f)(x,t)}{\tau^{1+\frac{\alpha}{2}}} e^{-\frac{|y|^2}{4}} dy d\tau \tag{23}$$

and $R^{n+1}_{\epsilon+} = \{(y,\tau) : y \in R^n, \tau > \epsilon\}, \epsilon > 0$.

c) Justification of the inversion in case of nice functions

Below Φ is the Lizorkin test function space, invariant with respect to fractional powers of the Laplace operator, see [23], p.39. The following auxiliary statement is a matter of direct verification, from which Theorem 10 follows.

Lemma 9. *Let $f = H^\alpha \varphi$, $\alpha > 0$ where $\varphi \in \Phi$. For the truncated construction $H_\epsilon^{-\alpha} f$ the representation*

$$H_\epsilon^{-\alpha} f = \int_{R^n} W(y,\tau) \mathcal{K}^+_{\ell,\frac{\alpha}{2}}(\tau) \varphi\left(x - \sqrt{\varepsilon} y, t/\varepsilon\tau) dy d\tau\right) \tag{24}$$

is valid, where $\mathcal{K}^+_{\ell,\alpha}(\tau)$ is some integrable kernel of the type of the identity approximation.

Theorem 10. *Let $f = H^\alpha \varphi$, $\alpha > 0$, where $\varphi \in \Phi$. Then*

$$\left(\frac{\partial}{\partial t} - \Delta_x\right)^{\frac{\alpha}{2}} f := \lim_{\epsilon \to 0} H_\epsilon^{-\alpha} f = \varphi . \tag{25}$$

d) The case of functions in L_p

Theorem 11. *Let $0 < \alpha < \frac{n+2}{p}$, $1 < p < \infty$, and $f = H^\alpha \varphi$ with $\varphi \in L_p(R^{n+1})$. Then*

$$\lim_{\substack{\epsilon \to 0 \\ (L_p)}} H_\epsilon^{-\alpha} f = \varphi .$$

e) The spaces of parabolic potentials

Let

$$H^\alpha(L_p) = \left\{ f : f = H^\alpha \varphi, \varphi \in L_p(R^{n+1}) \right\}, \quad 1 \le p < \infty , \quad \alpha > 0.$$

The range $H^\alpha(L_p)$ is understood in the usual sense only in the case $0 < \alpha < \frac{n+2}{p}$. In the case $\alpha \ge \frac{n+2}{p}$ the potential $H^\alpha \varphi$, with $\varphi \in L_p$, is understood as the convolution with the kernel $\frac{\tau_+^{\frac{\alpha}{2}-1}}{\Gamma(\frac{\alpha}{2})}W(y,\tau)$ in the Lizorkin space $\Phi'(R^{n+1})$ of distributions. The space $H^\alpha(L_p)$ is a Banach space with respect to the norm $\|f\|_{H^\alpha(L_p)} = \|\varphi\|_p$.

Theorem 12 below provides a characterization of the range $H^\alpha(L_p)$. Naturally, this range can be described in terms of convergence of the hypersingular integral $H^{-\alpha}f$. But the heat operator is a quasihomogeneous operator with a non-degenerate symbol. Therefore, it is also natural to expect that the range $H^\alpha(L_p)$ may be also characterized in terms of convergence of the anisotropic hypersingular integral. The corresponding anisotropic distance may be obtained in terms of elementary functions in this case:

$$\rho = \rho(y,\eta) = \left(\frac{\sqrt{|y|^4+4|\eta|^2}+|y|^2}{2}\right)^{\frac{n+2}{2(n+1)}} .$$

Let

$$(T^\alpha f)(x,t) = \lim_{\substack{\epsilon\to 0\\ (L_p(R^{n+1}))}} \int_{\rho(y,\eta)>\epsilon} \frac{(\Delta^{2\ell}_{y,\eta}f)(x,t)}{[\rho(y,\eta)]^{n+\frac{\alpha(n+1)}{n+2}}}dyd\eta\ , \quad \ell>\frac{\alpha}{2} .$$

Theorem 12. *Let* $0<\alpha<\frac{n+2}{p}$, $1<p<\infty$, $\frac{1}{q}=\frac{1}{p}-\frac{\alpha}{n+2}$. *Then*

$$H^\alpha(L_p) = \{f\in L_q:\ H^{-\alpha}f\in L_p\} = \{f\in L_q:\ T^\alpha f\in L_p\} . \tag{26}$$

For the proof of this theorem we refer to [17] for the first of the equalities in (26) and [10] for the second one.

3. Fractional powers of the Schrödinger operator

The results shown below represent a slight modification of what was obtained in [20]. The Schrödinger operator $S=\Delta_x+i\frac{\partial}{\partial t}$, $x\in R^n$, $t\in R^1$ has the symbol $\tau-|\xi|^2$, so that its fractional power may be defined via Fourier transforms as

$$S^{\frac{\alpha}{2}}\varphi = F^{-1}\left[(\tau-|\xi|^2)^{\frac{\alpha}{2}}(F\varphi)(\xi,\tau)\right] , \tag{27}$$

at least, formally. We should make sense to the construction (27).

a) Fractional Schrödinger potential operator

Negative fractional powers are realized as integral operators

$$I^\alpha_{Schr}\varphi = S^{-\frac{\alpha}{2}}\varphi = \int_{R^{n+1}_+} s_\alpha(y,\eta)\varphi(x-y,t-\eta)dyd\eta , \tag{28}$$

where $R^{n+1}_+ = \{(y,\eta)\in R^{n+1} : y\in R^n, \eta\in R^1_+\}$ and

$$s_\alpha(y,\eta) = \frac{e^{-\frac{\pi i(n+\alpha)}{4}}}{(4\pi)^{\frac{n}{2}}\Gamma\left(\frac{\alpha}{2}\right)}\eta^{\frac{\alpha-n-2}{2}}e^{i\frac{|y|^2}{4\eta}} , \quad \widehat{s}_\alpha(\xi,\tau) = (\tau-|\xi|^2+i0)^{-\frac{\alpha}{2}} . \tag{29}$$

The integral (28) may be referred to as *the Schrödinger fractional potential.* The integral in (64) exists in the usual sense only when $\Re\alpha > n$. If $\Re\alpha \leq n$, the kernel $s_\alpha(y,\eta)$ has a non integrable singularity in η at $\eta = 0$. We shall treat the Shrödinger fractional potential with $0 < \Re\alpha \leq n$ as analytical continuation in α. But before that we wish to justify the passage of the type (27), even if in the case of nice functions φ.

Let

$$P = \{(\xi,\tau) \in R^{n+1} : \ \xi \in R^n, \tau \in R^1_+, \tau = |\xi|^2\}$$

be the paraboloid surface at which the symbol $\tau - |\xi|^2$ of the Schrödinger operator vanishes, and let Φ_P be the Lizorkin test function space generated by the set P, so that it is invariant with respect to the operator I^α_{Schr}.

For functions in Φ_P the passage to Fourier transforms may be justified. Namely, the following lemma is valid.

Lemma 13. *Let* $\Re\alpha > n, \varphi \in \Phi_P$. *Then*

$$(FI^\alpha_{Schr}\varphi)(\xi,\tau) = \frac{(F\varphi)(\xi,\tau)}{(\tau - |\xi|^2 + i0)^{\frac{\alpha}{2}}} \ .$$

To deal with the operator I^α_{Schr} in the case $0 < \Re\alpha \leq n$, when it does not exist under the direct approach (28), we shall use its analytic continuation which is constructively realized from the half-space $\Re\alpha > n$ into the strip $-2\ell < \Re\alpha \leq n$ with an arbitrary $\ell \in \mathbb{N}$ except for some points in this strip, where it may have poles.

b) Analytical continuation of the Schrödinger fractional potential

Let $\varphi(x,t) \in \mathcal{S}(R^{n+1})$. To organize convergence in (28) for $\Re\alpha \leq n$ by the familiar usage of finite differences, it is convenient to single out the factor $\frac{1}{\eta^{1-\frac{\alpha}{2}}}$ from $s_\alpha(y,\eta)$, and we introduce the following hypersingular construction

$$I^\alpha_{Schr}\varphi = \lim_{\varepsilon\to 0} \frac{e^{-\frac{\alpha\pi i}{4}}}{d_{n,\ell}(-\alpha)} \int_{R^{n+1}_{+,\varepsilon}} \frac{(\widetilde{\Delta}^\ell_{y,\eta}\varphi)(x,t)}{\eta^{1-\frac{\alpha}{2}}} M(y,\tau)dyd\tau \tag{30}$$

where $R^{n+1}_{+,\epsilon} = \{(y,\eta) \in R^{n+1} : y \in R^n, \eta > \epsilon\}, \epsilon > 0$,

$$M(y,\eta) = \frac{e^{-\frac{n\pi i}{4}}}{(4\pi\eta)^{\frac{n}{2}}} e^{i\frac{|y|^2}{4\eta}},$$

and $\widetilde{\Delta}^\ell_{y,\eta}\varphi$ is the Marchaud type generalized difference (see [23], p. 78) with different kind of steps in y and η, which reflects the non-homogeneity of the Schrödinger operator:

$$(\widetilde{\Delta}^\ell_{y,\eta}\varphi)(x,t) = \sum_{j=0}^{\ell} (-1)^j C^{(\ell)}_j \varphi(x - a^{\frac{j}{2}}y, t - a^j\eta) \ , \quad \ell \in \mathbb{N} \ ,$$

where $a > 1$ may be chosen arbitrarily. Here the $C_j^{(\ell)}$ are given by the formulas

$$C_i^{(\ell)} = a^{i(\frac{i+1}{2}-\ell)} \frac{\prod_{k=M_i+1}^{\ell}(a^k-1)}{\prod_{k=1}^{m_i}(a^k-1)}, i = 1, 2, ..., \ell-1, \tag{31}$$

$$C_i^{(\ell)} = a^{i(\frac{i+1}{2}-\ell)} \frac{\prod_{k=M_i+1}^{\ell}(a^k-1)}{\prod_{k=1}^{m_i}(a^k-1)}, i = 1, 2, ..., \ell-1, \tag{32}$$

with $m_i = \min(i, \ell - i)$ and $M_i = \max(i, \ell - i)$ and the normalizing constant in (30) is equal to

$$d_{n,\ell}(-\alpha) = \Gamma\left(\frac{\alpha}{2}\right) \sum_{j=0}^{\ell} (-1)^j C_j^{(\ell)} a^{-\frac{\alpha}{2}j} .$$

Lemma 14. *Let $\varphi(x,t) \in \mathcal{S}(R^{n+1})$. The limit in (30) exists and represents the analytical continuation of the Schrödinger fractional potential $I_{Schr}^{\alpha}\varphi$ from the half-plane $\Re\alpha > n$ to the half-plane $\Re\alpha > -2\ell$, except, probably, for the points*

$$\alpha = -2m + \frac{4k\pi i}{ln\ a}, \quad m = 0, 1, 2, ..., \ell - 1, \quad k \in \mathbb{Z}\backslash\{0\}.$$

c) Positive fractional powers of the Schrödinger operator

Lemma 14 already represents such positive powers since $\Re\alpha$ may be negative in the construction (30). So we define it as

$$\left(\Delta_x + i\frac{\partial}{\partial t}\right)^{\frac{\alpha}{2}} f = J_{Schr}^{-\alpha} f$$

and rewrite this in the convenient way, assuming that $\ell > \frac{\Re\alpha}{2}$:

$$\left(\Delta_x + i\frac{\partial}{\partial t}\right)^{\frac{\alpha}{2}} f = \frac{e^{\frac{\alpha\pi i}{4}}}{d_{n,\ell}(\alpha)} \int_{R_+^{n+1}} \frac{\left(\widetilde{\Delta}_{y,\eta}^{\ell} f\right)(x,t)}{\eta^{1+\frac{\alpha}{2}}} M(y,\eta)dyd\eta , \tag{33}$$

where the finite difference is defined by $a > 1$ such that

$$\alpha \neq 2m + \frac{4k\pi i}{ln\ a}, \quad m = 0, 1, 2, ..., \ell - 1, \quad k \in \mathbb{Z}\backslash\{0\}. \tag{34}$$

In particular,

$$\sqrt{\Delta_x + i\frac{\partial}{\partial t}}\, f = \tag{35}$$

$$\frac{1+i}{2\sqrt{2\pi}(\sqrt{a}-1)} \int_{R_+^{n+1}} \frac{f(x-y, t-\eta) - f(x-\sqrt{a}y, t-a\eta)}{\eta^{\frac{3}{2}}} M(y,\eta)dyd\eta ,$$

where $a > 1$ may be arbitrary.

The following theorem states that the positive fractional power of the Schrödinger operator is the inverse operator to the Schrödinger fractional potential operator, at least on "nice" functions.

Theorem 15. *Let $\varphi \in \Phi_P, \Re\alpha > 0$. Then*

$$\left(\Delta_x + i\frac{\partial}{\partial t}\right)^{\frac{\alpha}{2}} I^{\alpha}_{Schr}\varphi = I^{\alpha}_{Schr}\left(\Delta_x + i\frac{\partial}{\partial t}\right)^{\frac{\alpha}{2}} \varphi = \varphi, \tag{36}$$

where $\left(\Delta_x + i\frac{\partial}{\partial t}\right)^{\frac{\alpha}{2}}$ is the operator (33) constructed under the choice of $a > 1$, satisfying the condition (34) and $I^{\alpha}_{Schr}\varphi$ is defined by (27) in the case $\Re\alpha > n$ and treated as its analytical continuation (30) when $0 < \Re\alpha \leq n$.

4. Fractional powers of some other differential operators

We give some brief indications on investigations of fractional powers of differential operators with homogeneous non-constant coefficients (which thereby are not invariant with respect to the translation operator) or of some non-homogeneous differential operators.

The fractional powers

$$\left[-\sum_{k=1}^{n}\left(x_k\frac{\partial}{\partial x_k}\right)^2\right]^{\frac{\alpha}{2}} \tag{37}$$

were studied in [7] - [10].

The negative powers are realized as the Riesz-type potential operator

$$\mathcal{I}^{\alpha}\varphi = \frac{1}{\gamma_n(\alpha)}\int_{R^n_+} \varphi(x \circ y)\left|\overrightarrow{ln}\ y\right|^{\alpha-n} dy\ ,$$

where $\overrightarrow{ln}\ y = (ln\ y_1, ..., ln\ y_n)$ and $R^n_+ = \{y \in R^n : y_1 > 0, ..., y_n > 0\}$. The positive fractional powers are realized as the corresponding modification of hypersingular integrals, adjusted to the n-tant R^n_+, but with a non-standard truncation of the hypersingular integral when it is interpreted on "not very nice" functions.

The fractional powers of the type

$$\mathbb{I}^{\alpha}_{\pm}\varphi = \left[\pm\sum_{k=0}^{n} x_k^2\frac{\partial^2}{\partial x_k^2}\right]^{-\frac{\alpha}{2}}, \ \Re\alpha > 0, \tag{38}$$

were considered in [1] with the realization $\mathbb{I}^{\alpha}_{\pm}\varphi = \int_{R^n_+} i_{\alpha}(t)\varphi\left(\frac{x}{t}\right) dt$, where

$$i^{\alpha}(t) = \frac{n^{\frac{n-\alpha}{4}}}{2^{n-1}\pi^{\frac{n}{2}}\Gamma\left(\frac{\alpha}{2}\right)}\left|\overrightarrow{ln}\ t\right|^{\frac{\alpha-n}{4}} K_{\frac{n-\alpha}{4}}\left(\frac{\sqrt{n}}{2}\left|\overrightarrow{ln}\ t\right|\right)\sqrt{t_1...t_n}.$$

Positive powers were realized by means of weighted hypersingular integrals.

Another dilatation-invariant version

$$\left(-|x|^2\Delta\right)^{\frac{\alpha}{2}}$$

was considered in [5] with the corresponding realization as potentials in case of negative powers and in the form of non-standard hypersingular integrals in case of positive powers.

The following fractional powers of the operator with constant coefficients, but non-homogeneous one,

$$\left(-P(D,D)+\sum_{k=0}^{n} a_k \frac{\partial}{\partial x_k}\right)^{\frac{\alpha}{2}}, \quad a=(a_1,...,a_n)\in R^n, \tag{39}$$

where $P(x,x)$ is an arbitrary positive quadratic form with real coefficients, were treated in [2] but in case of positive fractional powers, instead of the method of hypersingular integrals, there was used a modification of this method in the form of the limit of approximative inverse operators (AIO).

By the same method in [3] and [4] the complex powers (39) were constructed under a weaker assumption that $P(x,x) - ia\cdot x$ is hypoelliptic; a more general case when $1 \le rang\ P \le n-1$ was also treated there.

The case of complex a_k in (39) has a different nature in comparison with that of real coefficients, because the symbol of the operator $-P(D,D)+a\cdot D$ with complex a degenerates on some ellipsoid if $\Re\alpha = 0$ and on the intersection of the ellipsoid and a hyperplane, if $\Re\alpha = 0$. This more difficult case was treated in [14].

Negative fractional powers $(-\Delta - I)^{-\frac{\alpha}{2}}$ are known as fractional acoustic potentials. They have the form

$$(-\Delta - I)^{-\frac{\alpha}{2}}\varphi = \frac{i\,2^{-\frac{n+\alpha}{2}}}{\pi^{\frac{\alpha}{2}}}\int_{R^n}\frac{H^{(1)}_{\frac{n-\alpha}{2}}(|y|)}{|y|^{\frac{n-\alpha}{2}}}\varphi(x-y)dy\,,$$

where $0 < \Re\alpha < n+1$ and $H^{(1)}_\nu$ is the Hankel function of the first kind. The positive powers can be realized by means of AIO. Fractional powers

$$\left(\frac{\partial^2}{\partial x_1^2} - \frac{\partial^2}{\partial x_2^2} - \dots - \frac{\partial^2}{\partial x_n^2}\right)^{\frac{\alpha}{2}} \tag{40}$$

of the wave operator by the method of AIO were constructed in [19].

The powers of the Klein-Gordon-Fock operator

$$\left(m^2 I + \frac{\partial^2}{\partial x_1^2} - \frac{\partial^2}{\partial x_2^2} - \dots - \frac{\partial^2}{\partial x_n^2}\right)^{\frac{\alpha}{2}}, \quad m>0, \tag{41}$$

were treated in [20] and the powers

$$\left(m^2 I + \frac{\partial}{\partial x_1} + \frac{\partial^2}{\partial x_1^2} - \frac{\partial^2}{\partial x_2^2} - \dots - \frac{\partial^2}{\partial x_n^2}\right)^{\frac{\alpha}{2}}, \quad 0 \le m \le \frac{1}{2} \tag{42}$$

including the case of the telegraph operator ($m = 0$) were dealt with in [16]. The negative fractional power ($\Re\alpha < 0$) in (42) proves to be the following potential type operator

$$(H^{\alpha}_{m}\varphi)(x) = c_{n,m}(\alpha) \int\limits_{K^+_+} \frac{e^{-\frac{y_1}{2}} I_{\frac{\alpha-n}{2}}(\lambda r(y))}{r^{\frac{n-\alpha}{2}}(y)} \varphi(x-y)dy,$$

with $r^2(y) = y_1^2 - y_2^2 - \cdots - y_n^2$ and $\lambda = \frac{1}{2} - 2m^2$, $0 \leq m < \frac{1}{2}$, and the Bessel function replaced by 1 in the case $m = \frac{1}{2}$, $c_{n,m}(\alpha)$ is some constant. The kernel of this potential has locally the same behaviour as the kernel of the Riesz hyperbolic potential $I^{\alpha}_{\square}\varphi$, but exponentially decreases at infinity.

Bibliography

[1] Abramyan, A.V. and Nogin, V.A. (1994) *On fractional powers of some differential operators commuting with rotations* (Russian). Izv. Vysh. Uchebn. Zaved., no 4, 79–83.

[2] Abramyan, A.V. and Nogin, V.A. (1994) Integral transforms, connected with fractional powers of nonhomogeneous differential operators in L_p-spaces. *Integr. Transf. and Special Funct.*, **2**, no 1, 1–14.

[3] Abramyan, A.V. and Nogin, V.A. (1996) *Complex powers of second order hypoelliptic operators with constant coefficients in L_p-spaces* (Russian). Deponierted in VINITI, Moscow, no 382–396, 48pp.

[4] Abramyan, A.V. and Nogin, V.A. (1998) Complex powers of second order hypoelliptic operators with constant coefficients in L_p-spaces. *Differen. Uravnenia*, **33**, no 8, 1134–1135.

[5] Abramyan, A.V. and Nogin, V.A. (1996) Fractional powers of the operators $-|x|^2\Delta$ in L_p-spaces (Russian). *Differen. Uravnenia*, **32**, no 2, 275–276.

[6] Bagby, R. (1971) Lebesgue spaces of parabolic potentials. *Illinois J. Math.*, **5**, 610–634.

[7] Emgusheva, G.P. and Nogin, V.A. (1987) *On convergence of hypersingular integrals with a non-standard truncation* (Russian). Deponierte in VINITI, Moscow, no 3714–3787, 39pp.

[8] Emgusheva, G.P. and Nogin, V.A. (1988) Riesz derivatives with a non-standard truncation and their application to the inversion and characterization of potentials commuting with dilatations (Russian). *Dokl. Akad. Nauk SSSR*, **300**, no 2, 277–280.

[9] Emgusheva, G.P. and Nogin, V.A. (1988) *On anisotropic spaces of differentiable functions* (Russian). Deponierted in VINITI, Moscow, no 1173–1188, 55pp.

[10] Emgusheva, G.P. and Nogin, V.A. (1991) *On convergence in $L_p(R^n)$ of hypersingular integrals with a non-standard truncation* (Russian). Izv. Vysch. Uchebn. Zaved., no 7, 71–74.

[11] Friedrichs, K.O. (1953) *Mathematical aspects of quantum theory of fields.* Interscience Publishers, Inc., New York.

[12] Gopala Rao, V.R. (1977) A characterization of parabolic function spaces. *Amer. J. Math.*, **99**, 985–993.

[13] Jones, B.F. (1968) Lipschitz spaces and heat equation. *J. Math. Mech.*, **18**, 379–410.

[14] Karapetyants, A.N. and Nogin, V.A. (1998) Complex powers of second order elliptic operators with degenerate symbols in the space $L_p(R^n)$ (Russian). *Dokl. Ross. Acad. Nauk*, **358**, no 1, 10–12.

[15] Lizorkin, P.I. (1970) Characterization of the spaces $L_p^r(R^n)$ in terms of difference singular integrals (Russian). *Mat. Sb. (N.S.)*, **81**, no 1, 79–91.

[16] Nogin, V.A. and Chegolin, A.P. (1996) *Inversion of some integral operators with degenerating and oscillating symbols* (Russian). Izv. Vysch. Uchebn. Zaved., no 10, 36–39.

[17] Nogin, V.A. and Rubin, B.S. (1986) Inversion of parabolic potentials with L_p-densities (Russian). *Mat. Zametki*, **39**, no 6, 831–840.

[18] Nogin, V.A. and Rubin, B.S. (1987) The spaces $\mathcal{L}_{p,r}^{\alpha}(R^{n+1})$ of parabolic potentials. *Analysis Mathematica*, **13**, no 4, 321–328.

[19] Nogin, V.A. and Sukhinin, E.V. (1992) *Inversion and characterization of hyperbolic potentials in L_p-spaces* (Russian). Deponierted in VINITI, Moscow, no 2512–2592, 50pp.

[20] Nogin, V.A. and Sukhinin, E.V. (1995) Fractional powers of the Klein-Gordon-Fock operator in L_p-spaces (Russian). *Dokl. Acad. Nauk*, **341**, no 2, 166–168.

[21] Samarov, K.L. (1984) On the pseudodifferential Schrödinger equation (Russian). *Dokl. Akad. Nauk SSSR*, **279**, 83–87. (Transl. in: Sov. Phys. Dokl. (1984) 29(11), 909–910.)

[22] Samko, S.G. (1976) Spaces of Riesz potentials (Russian). *Izv. Akad. Nauk SSSR, ser. Mat.*, **40**, no 5, 1143–1172.

[23] Samko, S.G. (1999) *Hypersingular Integrals and Their Applications*, Gordon and Breach Sci. Publ., Serie "Analytic Methods and Special Functions", vol.3, 370pp.

[24] Samko, S.G., Kilbas, A.A. and Marichev, O.I. (1993) *Fractional Integrals and Derivatives. Theory and Applications.* Gordon & Breach. Sci. Publ., London-New-York, 1012pp.

[25] Sampson, C.H. (1968) *A characterization of parabolic potentials.* Thesis, Rice University.

[26] Weyl, H. (1927) Quantummechanik und Gruppentheorie. *Zeit. für Physik*, **46**, 1–46.

STEFAN SAMKO, UNIVERSIDADE DO ALGARVE, UCEH, CAMPUS DE GAMBELAS, FARO 8000, PORTUGAL

E-mail address: `ssamko@ualg.pt`

Progress in Nonlinear Differential Equations
and Their Applications, Vol. 42

EXISTENCE OF SOLUTIONS FOR CAUCHY PROBLEMS AND PERIODIC PROBLEMS WITH MULTIVALUED PSEUDO MONOTONE OPERATORS

NAOKI SHOIJI
YOKOHAMA NATIONAL UNIVERSITY

Abstract. In this paper, we study the existence of T-periodic solutions for the problem

$$u'(t) + A(t)u(t) = 0, \quad t \in \mathbb{R},$$

where $A(t)$ is a T-periodic, pseudo monotone mapping from a reflexive Banach space into its topological dual.

1. Introduction

Let V be a reflexive Banach space which is densely and continuously imbedded in a Hilbert space, let $T > 0$ and let $\{A(t) : t \in [0, T]\}$ be multivalued operators from V into its topological dual V'. We study the existence of solutions for Cauchy problems and periodic problems to a class of a nonlinear evolution equations of the form

$$u'(t) + A(t)u(t) \ni 0 \qquad \text{for } t \in [0, T].$$

In the case when $A(t)$ is single valued, Hirano [12], Ahmed and Xiang [1], Berkovits and Mustonen [3], and the author [15] studied the problems of this kind. Hirano and Ahmed and Xiang studied the Cauchy problems in the case when $A(t)$ is a monotone operator with nonmonotonic perturbations. Berkovits and Mustonen generalized their results to the case that $A(t)$ is pseudo monotone and they also studied the periodic problems. The author also studied the periodic problems in the case when $A(t)$ is pseudo monotone.

In this paper, we study the Cauchy problems and periodic problems in the case when $A(t)$ is a multivalued pseudo monotone operator. The conditions (A1)–(A4) in our theorems are more general than those in [1, 3, 12, 15] even in the case when $A(t)$ is single valued. To prove our results, we employ the method employed in [10]. We use a topological property of a solution set for a differential inclusion in Euclidean space[7], a topological fixed point theorem [9], Filippov's type implicit function theorem [16] and a minimax theorem [8].

The next section is devoted to some preliminaries and notations. In Section 3, we state our main results and we prove them in Section 4. In the final section, we study some applications.

2. Preliminaries

Throughout this paper, all vector spaces are real, and we denote by $\mathbb{N}$, $\mathbb{R}$ and $\mathbb{R}_+$ the set of positive integers, the set of real numbers and the set of nonnegative real numbers, respectively. Let X and Y be topological spaces. A multivalued mapping $F : X \to 2^Y$ is said to be *upper semicontinuous* if for every $x_0 \in X$ and open set $V \subset Y$ with $Fx_0 \subset V$, there exists an open neighborhood U of x_0 such that $Fx \subset V$ for every $x \in U$. Let E be a measure space. A multivalued mapping $G : E \to 2^Y$ is said to be *measurable* if for every closed set $C \subset Y$, the set $\{t \in E : G(t) \cap C \neq \emptyset\}$ is measurable. A multivalued mapping $H : E \times X \to 2^Y$ is said to be *Carathéodory* if $H(\cdot, x) : E \to 2^Y$ is measurable for every $x \in X$ and $H(t, \cdot) : X \to 2^Y$ is upper semicontinuous for almost every $t \in E$. Let V be a reflexive Banach space. We denote by V' the topological dual of V. The value of $y \in V'$ at $x \in V$ will be denoted by (y, x). Let V be densely and continuously imbedded in a Hilbert space H. Since we identify H with its topological dual, we have $V \subset H \subset V'$. Let p, q and T be positive constants such that $1/p + 1/q = 1$. For every $u \in L^p(0, T; V)$ and $v \in L^q(0, T; V')$, $\langle v, u\rangle$ is defined by $\int_0^T (v(t), u(t))\, dt$. We denote by $W_p^1(0, T; V, H)$ the Banach space

$$W_p^1(0, T; V, H) = \{u \in L^p(0, T; V) : u' \in L^q(0, T; V')\}$$

with the norm $\|u\| + \|u'\|_*$, where u' is the generalized derivative [2, 17] of u and $\|\cdot\|$ and $\|\cdot\|_*$ are the norms of $L^p(0, T; V)$ and $L^q(0, T; V')$, respectively. We know from [17] that $W_p^1(0, T; V, H)$ is a reflexive Banach space and that $W_p^1(0, T; V, H)$ is continuously imbedded in $C(0, T; H)$.

Let V be a reflexive Banach space and let $A : V \to 2^{V'}$. A is said to be *monotone* if $(w - z, x - y) \geq 0$ for every $x, y \in V$, $w \in Ax$ and $z \in Ay$. A is said to be *pseudo monotone* if for every sequence $\{x_n\}$ in V and $\{w_n\}$ in V' such that $\{x_n\}$ converges weakly to $x \in V$, $w_n \in Ax_n$ for every $n \in \mathbb{N}$ and $\overline{\lim}(w_n, x_n - x) \leq 0$, there exists $w_y \in Ax$ such that

$$(w_y, x - y) \leq \varliminf_{n\to\infty} (w_n, x_n - y)$$

for every $y \in V$. It is well known from [4, Proposition 7.4] that if A is monotone, Ax is a nonempty, closed convex subset of V' for every $x \in V$ and A is upper semicontinuous from every line segment in V to the weak topology of V', then A is pseudo monotone.

To prove our theorems in the next section, we use the following; see [7, Theorem and Lemma 5], [9, Example 1 in Section III.2 and Corollary V.3.8], [16, p. 864, Theorem 4.1 and Theorem 7.4] and [8, Theorem 2], respectively:

Proposition 1 (De Blasi and Myjak). Let $T > 0$ and let $g : [0, T] \times \mathbb{R}^n \to 2^{\mathbb{R}^n}$ be a Carathéodory mapping such that for almost every $t \in [0, T]$ and for every $y \in \mathbb{R}^n$, $g(t, y)$ is a nonempty, compact, convex subset of $\mathbb{R}^n$, and there exists $\mu \in L^1(0, T; \mathbb{R}_+)$ such that $\sup_{y\in\mathbb{R}^n} |g(t, y)| \leq \mu(t)$ for almost every $t \in [0, T]$. For every $y \in \mathbb{R}^n$, let

$$S_g(y) = \{x \in C(0,T;\mathbb{R}^n) : \text{absolutely continuous,} \\ x(0) = y,\ x'(t) \in g(t,x(t)) \text{ for almost every } t \in [0,T]\}.$$

Then for every $y \in \mathbb{R}^n$, $S_g(y)$ is the intersection of a decreasing sequence of nonempty, compact, contractible subsets of $C(0,T;\mathbb{R}^n)$.

Proposition 2 (Górniewicz). Let X be a convex subset of a normed linear space and let $H : X \to 2^X$ be an upper semicontinuous mapping such that $\cup_{x \in X} Hx$ is contained in a compact subset of X and for every $x \in X$, Hx is a nonempty, acyclic, compact subset of X, where acyclic is in the sense of Čech homology with rational coefficients. Then H has a fixed point, i.e., there is an element x of X such that $x \in Hx$.

Proposition 3 (Wagner). Let $\mathcal{T}$ be a complete measure space, let V be a complete, separable metric space and let Y be a separable metric space. Let $F : \mathcal{T} \to 2^V$ be a measurable mapping such that $F(t)$ is a nonempty, closed subset of V for almost every $t \in \mathcal{T}$, let $g : \mathcal{T} \times V \to Y$ be a Carathéodory mapping and let $h : \mathcal{T} \to Y$ be a measurable mapping such that there exists $v_t \in F(t)$ with $h(t) = g(t, v_t)$ for almost every $t \in \mathcal{T}$. Then there exists a measurable mapping $f : \mathcal{T} \to V$ such that $f(t) \in F(t)$ and $h(t) = g(t, f(t))$ for almost every $t \in \mathcal{T}$.

Proposition 4 (Fan). Let U be a compact, convex subset of a Hausdorff topological vector space and let V be a convex subset of a vector space. Let $f : U \times V \to \mathbb{R}$ such that $f(\cdot, v)$ is lower semicontinuous and convex for every $v \in V$ and $f(u, \cdot)$ is concave for every $u \in U$. Then

$$\min_{u \in U} \sup_{v \in V} f(u,v) = \sup_{v \in V} \min_{u \in U} f(u,v).$$

3. Main Results

Now we state our main results.

Theorem 1. Let T, p and q be positive constants such that $1/p + 1/q = 1$. Let $(V, \|\cdot\|)$ be a separable, reflexive Banach space which is densely and continuously imbedded in a Hilbert space $(H, |\cdot|)$ and let $\{A(t) : 0 \le t \le T\}$ be a family of multivalued mappings such that

(A1) for almost every $t \in [0,T]$, $A(t) : V \to 2^{V'}$ is pseudo monotone and $A(t)x$ is a nonempty, closed, convex subset of V' for every $x \in V$;

(A2) for every $u \in L^p(0,T;V)$, $t \mapsto A(t)u(t) : [0,T] \to 2^{V'}$ is measurable;

(A3) there exist $C_0, C_2 \in L^1(0,T;\mathbb{R}_+)$ and $C_1 > 0$ such that for almost every $t \in [0,T]$, for every $x \in V$ and for every $y \in A(t)x$,

$$(y, x) + C_0(t)|x|^2 \ge C_1\|x\|^p - C_2(t); \tag{3.1}$$

(A4) there exist an increasing function $C_3 : \mathbb{R}_+ \to \mathbb{R}_+$ and a function $C_4 : [0,T] \times \mathbb{R}_+ \to \mathbb{R}_+$ such that $C_4(\cdot, a) \in L^q(0,T;\mathbb{R}_+)$ for every $a \in \mathbb{R}_+$, $C_4(t,\cdot)$ is increasing for almost every $t \in [0,T]$, and

$$\sup\{\|y\|_* : y \in A(t)x\} \le C_3(|x|)\|x\|^{p-1} + C_4(t, |x|)$$

for almost every $t \in [0,T]$ and for every $x \in V$, where $\|\cdot\|_*$ is the norm of V'.

Then for every $u_0 \in H$, there exists $u \in W_p^1(0,T;V,H)$ such that

$$u(0) = u_0, \quad u'(t) + A(t)u(t) \ni 0 \quad \text{for almost every } t \in [0,T].$$

Theorem 2. Let T, p, q, V, H and $\{A(t) : 0 \le t \le T\}$ be as in Theorem 1. Assume $p > 2$ or $C_0(t) \equiv 0$. Then there exists $u \in W_p^1(0,T;V,H)$ such that

$$u(0) = u(T), \quad u'(t) + A(t)u(t) \ni 0 \quad \text{for almost every } t \in [0,T].$$

Remark 1. In both theorems, if V is compactly imbedded in H or $C_0(t) \equiv 0$, it is sufficient to assume the following (A5) and (A6) instead of (A3) and (A4):

(A5) there exist $C_2 \in L^1(0,T;\mathbb{R}_+)$ and $C_1 > 0$ such that for almost every $t \in [0,T]$ and for every $x \in V$, there exists $y \in A(t)x$ which satisfies (3.1) with $C_0(t) \equiv 0$ or some $C_0 \in L^1(0,T;\mathbb{R}_+)$ in the case when V is compactly imbedded into H;

(A6) there exist an increasing function $C_3 : \mathbb{R}_+ \to \mathbb{R}_+$ and a function $C_4 : [0,T] \times \mathbb{R}_+ \to \mathbb{R}_+$ such that $C_4(\cdot, a) \in L^q(0,T;\mathbb{R}_+)$ for every $a \in \mathbb{R}_+$, $C_4(t,\cdot)$ is increasing for almost every $t \in [0,T]$, and

$$\sup\{\|y\|_* : \ y \in A(t)x \text{ satisfes (3.1) } \} \le C_3(|x|)\|x\|^{p-1} + C_4(t, |x|)$$

for almost every $t \in [0,T]$ and for every $x \in V$.

For details, see Remark 2 in Section 4.

4. Proofs of Theorems

First, we give the proof of Theorem 1. We assume that T, p, q, V, H, $\{A(t)\}$ and u_0 are as in Theorem 1.

Lemma 1. Let $t \in [0,T]$ such that $A(t)$ is pseudo monotone and $A(t)x$ is a nonempty, closed, convex subset of V' for every $x \in V$. Let $x \in V$ and let $\{x_n\}$ be a sequence in V which converges weakly to x. Let $y \in V'$ and let $\{y_n\}$ be a sequence in V' such that it converges weakly to y, $y_n \in A(t)x_n$ for every $n \in \mathbb{N}$ and $\overline{\lim}_n (y_n, x_n - x) \le 0$. Then $y \in A(t)x$. Especially, for almost every

$t \in [0,T]$, $A(t)$ is upper semicontinuous from the strong topology of V to the weak topology of V'.

Proof. Since $A(t)$ is pseudo monotone, there exists $y_x \in A(t)x$ such that $0 = (y_x, x-x) \leq \underline{\lim}_n (y_n, x_n - x)$. So we have $\lim_n (y_n, x_n - x) = 0$. For every $z \in V$, there also exists $y_z \in A(t)x$ such that $(y_z, x-z) \leq \underline{\lim}_n (y_n, x_n - z) = (y, x-z)$. So we have $\sup_{z \in V} \min_{w \in A(t)x} (w - y, x - z) \leq 0$. By Proposition , there exists $w \in A(t)x$ such that $(w - y, x - z) \leq 0$ for all $z \in V$, which implies $w = y$. Hence we obtain $y \in A(t)x$. □

We denote by $\mathcal{V}$ and $\mathcal{V}'$, the spaces $L^p(0,T;V)$ and $L^q(0,T;V')$, respectively and the norms of these spaces are also denoted by $\|\cdot\|$ and $\|\cdot\|_*$, respectively. By $\mathcal{W}$ we mean $W^1_p(0,T;V,H)$. For $u \in \mathcal{V}$ and $t \in [0,T]$, we write $\mathcal{A}u(t)$ instead of $A(t)u(t)$.

To prove the following, we use the method employed in [11, 12, 15].

Lemma 2. Let $\{u_n\}$ be a sequence in $\mathcal{W}$ and let $\{w_n\}$ be a sequence in $\mathcal{V}'$ such that $\{u_n\}$ converges weakly to u in $\mathcal{W}$, $w_n \in \mathcal{A}u_n$ for every $n \in \mathbb{N}$ and $\overline{\lim}_n \langle w_n, u_n - u\rangle \leq 0$. Then for every $v \in \mathcal{V}$, there exists $w_v \in \mathcal{A}u$ such that $\langle w_v, u - v\rangle \leq \underline{\lim}_n \langle w_n, u_n - v\rangle$. Further, if $\{w_n\}$ converges weakly to w in $\mathcal{V}'$, then $w \in \mathcal{A}u$.

Proof. Since $\{u_n\}$ converges weakly to u in $\mathcal{W}$, it also converges weakly to u in $C(0,T;H)$. So $\{u_n\}$ is bounded in $C(0,T;H)$. By (A3) and (A4), there exist $K_1 > 0$, $K_2 > 0$ and $K_3 \in L^1(0,T;\mathbb{R}_+)$ such that

$$(w_n(t), u_n(t) - v(t)) \geq K_1\|u_n(t)\|^p - K_2\|v(t)\|^p - K_3(t) \tag{4.1}$$

for almost every $t \in [0,T]$, for every $n \in \mathbb{N}$ and for every $v \in \mathcal{V}$. We will show that

$$\underline{\lim}_{n\to\infty} (w_n(t), u_n(t) - u(t)) \geq 0 \quad \text{for almost every } t \in [0,T]. \tag{4.2}$$

Suppose not. Then the set

$$\{t \in [0,T] : \underline{\lim}_{n\to\infty} (w_n(t), u_n(t) - u(t)) < 0,$$
$$(w_n(t), u_n(t) - u(t)) \geq K_1\|u_n(t)\|^p - K_2\|u(t)\|^p - K_3(t) \quad \text{for all } n \in \mathbb{N}\}$$

has a positive measure. Let t be an element of the set. Since $\{u_n(t)\}$ is bounded in V from (4.1), $\{u_n(t)\}$ converges weakly to $u(t)$ in V. By (A1), we have $\underline{\lim}_n (w_n(t), u_n(t) - u(t)) = 0$, which contradicts that t is an element of the above set. Hence we have (4.2). By (4.1) and Fatou's lemma, we have

$$0 \leq \int_0^T \underline{\lim}_{n\to\infty} (w_n(t), u_n(t) - u(t))\, dt \leq \underline{\lim}_{n\to\infty} \langle w_n, u_n - u\rangle \leq \overline{\lim}_{n\to\infty} \langle w_n, u_n - u\rangle \leq 0.$$

Hence we obtain $\lim_n \langle w_n, u_n - u\rangle = 0$. Next we will show that there exists a subsequence $\{n_i\}$ of $\{n\}$ such that

$$\lim_{i\to\infty} (w_{n_i}(t), u_{n_i}(t) - u(t)) = 0 \quad \text{for almost every } t \in [0,T]. \tag{4.3}$$

Put $h_n(t) = (w_n(t), u_n(t) - u(t))$ for $t \in [0,T]$. We know that $\underline{\lim}_n h_n(t) \geq 0$ for almost every $t \in [0,T]$ and $\lim_n \int_0^T h_n(t)\,dt = 0$. Set $h_n^-(t) = -\min\{h_n(t), 0\}$ for $t \in [0,T]$. Since $h_n^-(t) \leq K_2\|u(t)\|^2 + K_3(t)$, by Lebesgue's dominated convergence theorem, we get $\lim_n \int_0^T h_n^-(t)\,dt = 0$. So we obtain $\lim_n \int_0^T |h_n(t)|\,dt = 0$. Hence we can choose a subsequence $\{h_{n_i}\}$ of $\{h_n\}$ which satisfies (4.3).

Let $v \in \mathcal{V}$. By the preceding, there exists a subsequence $\{n_j\}$ of $\{n\}$ such that $\lim_j \langle w_{n_j}, u_{n_j} - v\rangle = \underline{\lim}_n \langle w_n, u_n - v\rangle$ and $\lim_j (w_{n_j}(t), u_{n_j}(t) - u(t)) = 0$ for almost every $t \in [0,T]$. Since $\{u_{n_j}(t)\}$ is bounded in V by (4.1), $\{u_{n_j}(t)\}$ converges weakly to $u(t)$ in V for almost every $t \in [0,T]$. We know $\underline{\lim}_j (w_{n_j}(t), u_{n_j}(t) - v(t))$ is measurable and from Lemma 1, there exists $y_t \in A(t)u(t)$ with $\underline{\lim}_j (w_{n_j}(t), u_{n_j}(t) - v(t)) = (y_t, u(t) - v(t))$ for almost every $t \in [0,T]$. By Proposition 3, there exists $w_v \in \mathcal{V}'$ such that $w_v(t) \in A(t)u(t)$ and $\underline{\lim}_j (w_{n_j}(t), u_{n_j}(t) - v(t)) = (w_v(t), u(t) - v(t))$ for almost every $t \in [0,T]$. Hence by (4.1) and Fatou's lemma, we have

$$\langle w_v, u - v\rangle = \int_0^T \underline{\lim}_{j\to\infty} (w_{n_j}(t), u_{n_j}(t) - v(t))\,dt \leq \underline{\lim}_{j\to\infty} \langle w_{n_j}, u_{n_j} - v\rangle$$
$$= \underline{\lim}_{n\to\infty} \langle w_n, u_n - v\rangle.$$

Now, assume that $\{w_n\}$ converges weakly to w in $\mathcal{V}'$. From the inequality above, we have $\lim_n \langle w_n, u_n - u\rangle = 0$. For every $v \in \mathcal{V}$, there exists $w_v \in \mathcal{A}u$ such that $\langle w_v, u - v\rangle \leq \lim_n \langle w_n, u_n - v\rangle = \langle w, u - v\rangle$, which implies $\sup_{v\in\mathcal{V}} \min_{y\in\mathcal{A}u} \langle y - w, u - v\rangle \leq 0$. By Proposition 4, there exists $y \in \mathcal{A}u$ such that $\langle y - w, u - v\rangle \leq 0$ for all $v \in \mathcal{V}$. So we have $y = w$, and hence $w \in \mathcal{A}u$. □

Let $\{e_1, e_2, \cdots\}$ be a subset of V such that the subspace spanned by $\{e_1, e_2, \cdots\}$ is dense in V and

$$(e_i, e_j) = \begin{cases} 1 & \text{if } i = j, \\ 0 & \text{if } i \neq j, \end{cases}$$

i.e., $\{e_i\}$ is a complete orthonormal basis of H with $\{e_i\} \subset V$. For every $n \in \mathbb{N}$, we denote by F_n the subspace of V spanned by $\{e_1, \cdots, e_n\}$.

We set $u_0^n = \sum_{i=1}^n (u_0, e_i)e_i$ for every $n \in \mathbb{N}$ and

$$K = \left(|u_0|^2 + 2\int_0^T C_2(s)\,ds\right) \exp\left(2\int_0^T C_0(s)\,ds\right).$$

Lemma 3. For every $n \in \mathbb{N}$, there exist an absolutely continuous function $u_n : [0,T] \to F_n$ and $w_n \in \mathcal{V}'$ such that $u_n' \in L^q(0,T;F_n)$, $w_n(t) \in A(t)u_n(t)$ for almost every $t \in [0,T]$, $|u_n(t)| \leq \sqrt{K}$ for every $t \in [0,T]$ and

$$u_n(0) = u_0^n, \qquad (u_n'(t) + w_n(t), v) = 0 \tag{4.4}$$

for almost every $t \in [0,T]$ and for every $v \in F_n$.

Proof. Let $n \in \mathbb{N}$. Let $M > 0$ with $M|\cdot| \le \|\cdot\|$. Since F_n is finite dimensional, there exists $L_n > 0$ such that $\|v\| \le L_n|v|$ for all $v \in F_n$. Define $x_0 \in \mathbb{R}^n$ and $f : [0,T] \times \mathbb{R}^n \to 2^{\mathbb{R}^n}$ by

$$x_0 = \begin{pmatrix} (u_0^n, e_1) \\ \vdots \\ (u_0^n, e_n) \end{pmatrix},$$

$$f(t,x) = -\begin{pmatrix} \left(A(t)\left(\sum_{i=1}^n x_i e_i\right), e_1\right) \\ \vdots \\ \left(A(t)\left(\sum_{i=1}^n x_i e_i\right), e_n\right) \end{pmatrix}, \quad (t,x) \in [0,T] \times \mathbb{R}^n.$$

By (A2) and Lemma 1, f is a Carathéodory mapping. By (A4), we have

$$\begin{aligned} |f(t,x)| &\le n \sum_{j=1}^n \left\| A(t)\left(\sum_{i=1}^n x_i e_i\right)\right\|_* \|e_j\| \\ &\le n^2 L_n \left(C_3\left(\left|\sum_{i=1}^n x_i e_i\right|\right) \left\|\sum_{i=1}^n x_i e_i\right\|^{p-1} + C_4\left(t, \left|\sum_{i=1}^n x_i e_i\right|\right)\right) \qquad (4.5) \\ &\le n^2 L_n \left(C_3(|x|) L_n^{p-1} |x|^{p-1} + C_4(t,|x|)\right) \end{aligned}$$

for almost every $t \in [0,T]$ and for every $x \in \mathbb{R}^n$. By (A3), we have

$$\begin{aligned} (x, f(t,x)) &= -\left(A(t)\left(\sum_{i=1}^n x_i e_i\right), \sum_{j=1}^n x_j e_j\right) \\ &\le -C_1 \left\|\sum_{i=1}^n x_i e_i\right\|^p + C_0(t)\left|\sum_{i=1}^n x_i e_i\right|^2 + C_2(t) \qquad (4.6) \\ &\le -M^p C_1 |x|^p + C_0(t)|x|^2 + C_2(t) \end{aligned}$$

for almost every $t \in [0,T]$ and for every $x \in \mathbb{R}^n$. We show an a priori estimate for an absolutely continuous function $x : [0,T] \to \mathbb{R}^n$ which satisfies $x(0) = x_0$ and $x'(t) \in f(t,x(t))$ for $t \in [0,T]$. Since from (4.6), $(|x(t)|^2)'/2 \le C_0(t)|x(t)|^2 + C_2(t)$ for almost every $t \in [0,T]$, we have $|x(t)|^2 \le |x_0|^2 + 2\int_0^T C_2(s)\,ds + 2\int_0^t C_0(s)|x(s)|^2\,ds$ for every $t \in [0,T]$. So by Gronwall's inequality, we get

$$|x(t)|^2 \le \left(|x_0|^2 + 2\int_0^T C_2(s)\,ds\right)\exp\left(2\int_0^T C_0(s)\,ds\right) \quad \text{for every } t \in [0,T].$$

By the standard fixed point argument, there exists an absolutely continuous function $x : [0,T] \to \mathbb{R}^n$ such that $x'(t) \in f(t,x(t))$ for almost every $t \in [0,T]$

and $x(0) = x_0$. Let $u_n : [0,T] \to F_n$ be the absolutely continuous function defined by $u_n(t) = \sum_{i=1}^n x_i(t)e_i$ for every $t \in [0,T]$. It is easy to see that $u_n(0) = u_0^n$, $|u_n(t)| \le \sqrt{K}$ for every $t \in [0,T]$ and $u_n' \in L^q(0,T;F_n)$. Since the mapping

$$t \mapsto \begin{pmatrix} (u_n'(t), e_1) \\ \vdots \\ (u_n'(t), e_n) \end{pmatrix} : [0,T] \to \mathbb{R}^n$$

is measurable, by Proposition , there exists a measurable mapping $w_n \in \mathcal{V}'$ such that $w_n(t) \in A(t)u_n(t)$ and $(u_n'(t) + w_n(t), e_j) = 0$ for almost every $t \in [0,T]$ and for every $j = 1, \ldots, n$. □

Lemma 4. $\{u_n\}$ is bounded in $\mathcal{W}$.

Proof. Let $n \in \mathbb{N}$. From (4.4) and (A3), we have

$$\int_0^T \|u_n(t)\|^p \, dt \le \frac{1}{C_1}\left(K \int_0^T C_0(t)\, dt + \int_0^T C_2(t)\, dt + \frac{|u_0|^2}{2}\right).$$

Since $u_n' \in L^q(0,T;F_n)$, there exists $v_n \in L^p(0,T;F_n)$ such that $\langle u_n', v_n\rangle = \|v_n\|^2 = \|u_n'\|_*^2$. So, by (4.4) and (A4), we get

$$\|u_n'\|_* \le C_3(\sqrt{K})\|u\|^{\frac{p}{q}} + \left(\int_0^T |C_4(t, \sqrt{K})|\, dt\right)^{\frac{1}{q}}.$$

Hence $\{u_n\}$ is bounded in $\mathcal{W}$. □

Since $\{u_n\}$ and $\{w_n\}$ are bounded in $\mathcal{W}$ and $\mathcal{V}'$, respectively, we may assume that $\{u_n\}$ converges weakly to u in $\mathcal{W}$ and $\{w_n\}$ converges weakly to w in $\mathcal{V}'$.

Lemma 5. $u' + w = 0$ and $u(0) = u_0$.

Proof. First, we will show $u' + w = 0$. Let $\varphi \in C_0^\infty(0,T)$ and let $n \in \mathbb{N}$. By (4.4), we have for every $m \ge n$,

$$\begin{aligned} 0 &= (u_m(T), \varphi(T)e_n) - (u_m(0), \varphi(0)e_n) \\ &= \int_0^T \left((u_m'(t), \varphi(t)e_n) + (\varphi'(t)e_n, u_m(t))\right) dt \\ &= \int_0^T \left((-w_m(t), \varphi(t)e_n) + (\varphi'(t)e_n, u_m(t))\right) dt \\ &= \langle -w_m, \varphi e_n\rangle + \langle \varphi' e_n, u_m\rangle. \end{aligned}$$

So we get $0 = \langle -\varphi w + \varphi' u, e_n\rangle$ for all $n \in \mathbb{N}$ and $\varphi \in C_0^\infty(0,T)$. Hence we obtain $u' + w = 0$. Since $\{u_n\}$ converges weakly to u in $\mathcal{W}$ and $\mathcal{W}$ is continuously imbedded in $C(0,T;H)$, $\{u_n(0)\}$ converges weakly to $u(0)$ in H. So we obtain $u(0) = u_0$. □

Proof of Theorem 1. By Lemmas 3 and 5, we have $\langle w_n, u_n\rangle = 1/2|u_0^n|^2 - 1/2|u_n(T)|^2$ and $\langle w, u\rangle = 1/2|u_0|^2 - 1/2|u(T)|^2$. So we get

$$\overline{\lim_{i\to\infty}}\,\langle w_n, u_n - u\rangle = \frac{1}{2}\Big(\lim_{n\to\infty}|u_0^n|^2 - |u_0|^2 + |u(T)|^2 - \underline{\lim}_{i\to\infty}|u_n(T)|^2\Big) \le 0.$$

By Lemma 2, we have $w \in \mathcal{A}u$. Hence we obtain $u \in \mathcal{W}$ and $w \in \mathcal{A}u$ such that $u(0) = u_0$ and $u' + w = 0$. □

Next, we give the proof of Theorem 2. Till the end of this section, we assume $p > 2$ or $C_0(t) \equiv 0$. We fix $r > 0$ which satisfies

$$-C_1 M^p r^p T + \int_0^T C_2(s)\,ds + \big((r^2+a)b + a\big)b\int_0^T C_0(s)\,ds < 0,$$

where $a = 2\int_0^T C_2(s)\,ds$ and $b = \exp\big(2\int_0^T C_0(s)\,ds\big)$. We set $\rho = \sqrt{(r^2+a)b}$ and $R = \sqrt{(\rho^2+a)b}$.

Lemma 6. For every $n \in \mathbb{N}$, there exist an absolutely continuous function $u_n : [0,T] \to F_n$ and $w_n \in \mathcal{V}'$ such that $u_n' \in L^q(0,T;F)$, $w_n(t) \in A(t)u_n(t)$ for almost every $t \in [0,T]$, $|u_n(t)| \le R$ for every $t \in [0,T]$ and

$$u_n(0) = u_n(T), \quad (u_n'(t) + w_n(t), v) = 0$$

for almost every $t \in [0,T]$ and for every $v \in F_n$.

Proof. Let $n \in \mathbb{N}$. Let M, L_n and f be as in the proof of Lemma 3. Let $x : [0,T] \to \mathbb{R}^n$ be an absolutely continuous function which satisfies $x'(t) \in f(t,x(t))$ for almost every $t \in [0,T]$ and $|x(0)| \le \rho$. We will show $|x(T)| \le \rho$. By (4.6), it is easy to see that if there exists $t_0 \in [0,T]$ with $|x(t_0)| \le r$ then $|x(T)| \le \rho$. So we may assume that $|x(t)| > r$ for every $t \in [0,T]$. Since $|x(t)| \le R$ for every $t \in [0,T]$ from (4.6), we get

$$\frac{1}{2}|x(T)|^2 - \frac{1}{2}|x(0)|^2 \le -C_1 M^p r^p T + \int_0^T C_2(s)\,ds + R^2\int_0^T C_0(s)\,ds < 0$$

from (4.6). So we obtain $|x(T)| \le \rho$.

Let $g : [0,T] \times \mathbb{R}^n \to 2^{\mathbb{R}^n}$ be a Carathéodory mapping defined by

$$g(t,y) = \begin{cases} f(t,y) & \text{if } |y| \le R, \\ f(t, Ry/|y|) & \text{if } |y| > R \end{cases} \qquad \text{for } (t,y) \in [0,T] \times \mathbb{R}^n.$$

We remark that $S_f(y) = S_g(y)$ for $y \in \mathbb{R}^n$ with $|y| \le \rho$, where S_f and S_g are as in Proposition 1. From Proposition 1 and the continuity property of Čech homology, $S_g(y)$ is acyclic in the sense of Čech homology with rational coefficients for every $y \in \mathbb{R}^n$. We set $X = \{x \in C(0,T;\mathbb{R}^n) : |x(T)| \le \rho\}$. For every $x \in X$, put $Hx = S_f(x(T))$. Then $H : X \to 2^X$ is an upper semicontinuous mapping such that $\cup_{x\in X} H(x)$ is contained in a compact subset of X and Hx is acyclic and compact for every $x \in X$. Hence, by Proposition 2, there exists a fixed point of H, i.e., there exists an absolutely continuous function $x : [0,T] \to \mathbb{R}^n$ such that $x'(t) \in f(t,x(t))$ for almost every $t \in [0,T]$,

$x(0) = x(T)$ and $|x(t)| \le R$ for every $t \in [0,T]$. Let $u_n : [0,T] \to F_n$ be the absolutely continuous function defined by $u_n(t) = \sum_{i=1}^n x_i(t)e_i$ for every $t \in [0,T]$. By the similar argument as in the proof of Lemma 3, we finish the proof. □

Lemma 7. $\{u_n\}$ is bounded in $\mathcal{W}$.

Proof. Let $n \in \mathbb{N}$. From Lemma 6 and (A3), we have

$$\begin{aligned} 0 &= \int_0^T (w_n(t), u_n(t))\,dt + \frac{1}{2}|u_n(T)|^2 - \frac{1}{2}|u_n(0)|^2 \\ &\ge C_1 \int_0^T \|u_n(t)\|^p\,dt - R^2 \int_0^T C_0(t)\,dt - \int_0^T C_2(t)\,dt. \end{aligned}$$

Hence by the same argument as in the proof of Lemma 4, $\{u_n\}$ is bounded in $\mathcal{W}$. □

Since $\{u_n\}$ and $\{w_n\}$ are bounded in $\mathcal{W}$ and $\mathcal{V}'$, respectively, we may assume that $\{u_n\}$ converges weakly to u in $\mathcal{W}$ and $\{w_n\}$ converges weakly to w in $\mathcal{V}'$.

By the same lines as those in the proof of Lemma 5, we obtain the following:

Lemma 8. $u' + w = 0$ and $u(0) = u(T)$.

Proof of Theorem 2. By Lemmas 6 and 8, we have $\langle w_n, u_n\rangle = 0$ for every $n \in \mathbb{N}$ and $\langle w, u\rangle = 0$. So we get $\lim_n \langle w_n, u_n - u\rangle = 0$. By Lemma 2, we have $w \in \mathcal{A}u$. Hence we obtain $u \in \mathcal{W}$ and $w \in \mathcal{A}u$ such that $u(0) = u(T)$ and $u' + w = 0$. □

Remark 2. We give the proof of Remark . We set

$$\tilde{A}(t)x = \{y \in A(t)x : \ y \text{ satisfies } (3.1)\} \qquad \text{for } (t,x) \in [0,T] \times V.$$

Since it is easy to see that $\{\tilde{A}(t)\}$ satisfies (A2), (A3) and (A4), it is sufficient to show the following:

Lemma 9. $\tilde{A}(t)$ is pseudo monotone for almost every $t \in [0,T]$.

Proof. Let $t \in [0,T]$ such that $A(t)$ is pseudo monotone and $A(t)x$ is a nonempty, closed, convex subset of V' for every $x \in V$. Let $x \in V$ and let $\{x_n\}$ be a sequence in V which converges weakly to x in V. Let $\{y_n\}$ be a sequence in V' such that $y_n \in \tilde{A}(t)x_n$ for every $n \in \mathbb{N}$ and $\overline{\lim}_n (y_n, x_n - x) \le 0$. Fix $z \in V$. Then there exists a subsequence $\{n_i\}$ of $\{n\}$ such that $\{y_{n_i}\}$ converges weakly to y in V' and $\underline{\lim}_n (y_n, x_n - z) = \lim_i (y_{n_i}, x_{n_i} - z)$. Since $(y, x - z) = \underline{\lim}_n (y_n, x_n - z)$, it is sufficient to show that $y \in \tilde{A}(t)x$. From Lemma 1, we have $y \in A(t)x$. Since $\{y_{n_i}\}$ converges weakly to y, $y_{n_i} \in \tilde{A}(t)x_{n_i}$ and V is compactly imbedded in H or $C_0(t) \equiv 0$, we obtain $y \in \tilde{A}(t)x$ from (3.1). This completes the proof. □

5. Application

Throughout this section, $T > 0$, $p \geq 2$ with $1/p + 1/q = 1$, Ω is an bounded domain in $\mathbb{R}^N$ with smooth boundary, $Q = (0,T) \times \Omega$, and m is a positive natural number. For a real valued function u on Ω, we mean $\eta(u)$ and $\zeta(u)$ as follows:

$$\eta(u) = \{D^\alpha u : |\alpha| \leq m-1\} \in \mathbb{R}^{N_1}, \qquad \zeta(u) = \{D^\alpha u : |\alpha| = m\} \in \mathbb{R}^{N_2}.$$

Let $\{A_\alpha : |\alpha| = m\}$ be functions from $Q \times \mathbb{R}^{N_1} \times \mathbb{R}^{N_2}$ into $\mathbb{R}$, let $\{A_\alpha : |\alpha| \leq m-1\}$ be multivalued functions $Q \times \mathbb{R}^{N_1} \times \mathbb{R}^{N_2}$ into $2^{\mathbb{R}}$, and let h be a function from Q into $\mathbb{R}$.

We consider the following nonlinear differential inclusion

$$(5.1)\qquad \begin{cases} \dfrac{\partial u}{\partial t}(t,x) + \displaystyle\sum_{|\alpha|=m} (-1)^{|\alpha|} D^\alpha A_\alpha(t,x,\eta(u),\zeta(u)) \\ \qquad\qquad + \displaystyle\sum_{|\alpha|\leq m-1} (-1)^{|\alpha|} D^\alpha w_\alpha(t,x) = h(t,x) \quad \text{on } Q, \\ w_\alpha(t,x) \in A_\alpha(t,x,\eta(u),\zeta(u)) \quad \text{on } Q \text{ for } |\alpha| \leq m-1 \end{cases}$$

with Dirichlet boundary condition

$$(5.2)\qquad D^\alpha u = 0 \qquad \text{on } [0,T] \times \partial\Omega \text{ for } |\alpha| \leq m-1.$$

Theorem 3. Assume the following:

(i) for every α with $|\alpha| = m$, $A_\alpha(\cdot,\cdot,\eta,\zeta)$ is measurable for every $(\eta,\zeta) \in \mathbb{R}^{N_1} \times \mathbb{R}^{N_2}$ and $A_\alpha(t,x,\cdot,\cdot)$ is continuous for almost every $(t,x) \in Q$;

(ii) for every α with $|\alpha| \leq m-1$, $A_\alpha(t,x,\eta,\zeta)$ is a nonempty, closed, convex subset of $\mathbb{R}$ for every $(t,x,\eta,\zeta) \in Q \times \mathbb{R}^{N_1} \times \mathbb{R}^{N_2}$, $A_\alpha(\cdot,\cdot,\eta,\zeta)$ is measurable for every $(\eta,\zeta) \in \mathbb{R}^{N_1} \times \mathbb{R}^{N_2}$ and $A_\alpha(t,x,\cdot,\cdot)$ is upper semicontinuous for almost every $(t,x) \in Q$;

(iii) there exist $c_3 > 0$ and $c_4 \in L^q(Q;\mathbb{R}_+)$ such that

$$\sup |A_\alpha(t,x,\eta,\zeta)| \leq c_3\big(|\zeta|^{p-1} + |\eta|^{p-1}\big) + c_4(t,x)$$

for every $|\alpha| \leq m$ and $(t,x,\eta,\zeta) \in Q \times \mathbb{R}^{N_1} \times \mathbb{R}^{N_2}$;

(iv) for every $(t,x,\eta) \in Q \times \mathbb{R}^{N_1}$ and $\zeta,\zeta' \in \mathbb{R}^{N_2}$ with $\zeta \neq \zeta'$,

$$\sum_{|\alpha|=m} \big(A_\alpha(t,x,\eta,\zeta) - A_\alpha(t,x,\eta,\zeta')\big)(\zeta_\alpha - \zeta'_\alpha) > 0;$$

(v) there exist $c_1 > 0$, $c_0 \in L^1(0,T;\mathbb{R}_+)$ and $c_2 \in L^1(Q;R_+)$ such that

$$\sum_{|\alpha|\le m} A_\alpha(t,x,\eta,\zeta)\zeta_\alpha + c_0(t)|\eta_0|^2 \ge c_1|\zeta|^p - c_2(t,x)$$

for every $(t,x,\eta,\zeta) \in Q \times \mathbb{R}^{N_1} \times \mathbb{R}^{N_2}$, where $\eta_0 = \{\eta_\beta : |\beta| = 0\}$;

Then for every $h \in L^q(0,T;W^{-1,q}(\Omega))$ and $u_0 \in L^2(\Omega)$, there exists $u \in W_p^1(0,T;W_0^{1,p}(\Omega),L^2(\Omega))$ which satisfies (5.1) and (5.2) with $u(0) = u_0$, and if $p > 2$ or $c_0 \equiv 0$, then for every $h \in L^q(0,T;W^{-1,q}(\Omega))$, there exists $u \in W_p^1(0,T;W_0^{1,p}(\Omega),L^2(\Omega))$ which satisfies (5.1) and (5.2) with $u(0) = u(T)$.

Proof. For every $t \in [0,T]$, we define an operator $A(t) : W_0^{m,p}(\Omega) \to 2^{W^{-m,q}(\Omega)}$ by $w \in A(t)u$ if for almost every $x \in \Omega$ and $|\alpha| \le m$, $w_\alpha(x) \in A_\alpha(t,x,\eta(u),\zeta(u))$ and for every $v \in W_0^{m,p}(\Omega)$, $(w,v) = \int_\Omega \sum_{|\alpha|\le m} w_\alpha(x)D^\alpha v\,dx$. First, we will show that $A(t)$ is pseudo monotone for almost every $t \in [0,T]$. We use the method employed in [5, 14]. Fix $t \in [0,T]$. Let $\{(u^n,w^n)\}$ be a sequence in $W_0^{m,p}(\Omega) \times W^{-m,q}(\Omega)$ such that $\{u^n\}$ converges weakly to u, $w^n \in Au^n$ and $\overline{\lim}_n(w^n,u^n-u) \le 0$. Fix $v \in W_0^{m,p}(\Omega)$. Taking a subsequence, if necessary, we may assume that $\underline{\lim}_n(w^n,u^n-v) = \lim_n(w^n,u^n-v)$, $\{D^\alpha u^n\}$ converges strongly in $L^p(\Omega)$ and almost everywhere to $D^\alpha u$ for $|\alpha| \le m-1$ and $\{w_\alpha^n\}$ converges weakly to w_α in $L^q(\Omega)$ for $|\alpha| \le m$. We show $w_\alpha(x) \in A_\alpha(t,x,\eta(u),\zeta(u))$ for almost every $x \in \Omega$ and for every $|\alpha| \le m-1$. Fix α with $|\alpha| \le m-1$. We may assume that there exists $\omega_\alpha^n \in \mathrm{co}\{w_\alpha^n, w_\alpha^{n+1},\cdots\}$ and $\omega_\alpha^n(x)$ converges to $w_\alpha(x)$ almost everywhere. Fix $x \in \Omega$. Let (β,γ) be an open interval with $A_\alpha(t,x,\eta(u),\zeta(u)) \subset (\beta,\gamma)$. From the upper semicontinuity of $A_\alpha(t,x,\cdot,\cdot)$, we have $w_\alpha(x) \in (\beta,\gamma)$. Since (β,γ) is an arbitrary open interval which contains $A_\alpha(t,x,\eta(u),\zeta(u))$ and $A_\alpha(t,x,\eta(u),\zeta(u))$ is closed and convex, we have $w_\alpha(x) \in A_\alpha(t,x,\eta(u),\zeta(u))$ almost everywhere. Set

$$\begin{aligned} p_n(x) &= \textstyle\sum_{|\alpha|=m} A_\alpha(t,x,\eta(u^n),\zeta(u^n))(D^\alpha u^n - D^\alpha u) \\ &\qquad + \textstyle\sum_{|\alpha|\le m-1} w_\alpha^n(x)(D^\alpha u^n - D^\alpha u); \\ q_n(x) &= \textstyle\sum_{|\alpha|=m}(A_\alpha(t,x,\eta(u^n),\zeta(u^n)) - A_\alpha(t,x,\eta(u^n),\zeta(u))) \\ &\qquad \cdot(D^\alpha u^n - D^\alpha u); \\ r_n(x) &= \textstyle\sum_{|\alpha|=m} A_\alpha(t,x,\eta(u^n),\zeta(u))(D^\alpha u^n - D^\alpha u); \\ s_n(x) &= \textstyle\sum_{|\alpha|\le m-1} w_\alpha^n(x)(D^\alpha u^n - D^\alpha u). \end{aligned}$$

It is easy to see $q_n = p_n - r_n - s_n$ and $\overline{\lim}_n \int q_n \le 0$. From $q_n \ge 0$, $q_n(x) \to 0$ almost everywhere. By [13, Lemma 6], $D^\alpha u^n(x) \to D^\alpha u(x)$ almost everywhere for $|\alpha| = m$, and hence $p_n(x) \to 0$ almost everywhere. Since

$$\begin{aligned} p_n(x) \ \ge\ & c_1|\zeta(u^n(x))|^p - c_2(t,x) - c_0(t)|u^n(x)|^2 \\ & - \sum_{|\alpha|\le m} \Big(c_3\big(|\eta(u^n(x))|^{p-1} + |\zeta(u^n(x))|^{p-1}\big) + c_4(t,x)\Big)|D^\alpha u(x)| \end{aligned}$$

and $\{D^\alpha u^n\}$ converges strongly to $D^\alpha u$ in $L^p(\Omega)$ for every α with $|\alpha| \leq m-1$, $\{p_n^-\}$ is uniformly integrable. By Vitali's convergence theorem, we have $\lim_n \int p_n^- = 0$. So we obtain $\lim_n \int p_n^+ = 0$ and

$$\lim_{n\to\infty}(w^n, u^n - v) = \lim_{n\to\infty}\left(\int_\Omega p_n(x)\,dx + (w^n, u - v)\right) = (w, u - v).$$

Hence $A(t)$ is pseudo monotone, which implies (A1). By Poincare's inequality, we have (A3). Hence by our theorems, we obtain the conclusion. □

As a direct consequence, we have the following, which improves [6, Theorem 4.1] and [15, Theorem 2].

Corollary. Let $b_i, a : \mathbb{R} \to \mathbb{R}$ be bounded and continuous functions and let $g : Q \times \mathbb{R} \times \mathbb{R}^n \to 2^{\mathbb{R}}$ be a Carathéodory mapping such that there exist $c_3 > 0$, $c_4 \in L^q(Q; \mathbb{R}_+)$ and $c_0 \in L^1(0, T; \mathbb{R}_+)$ which satisfy $\sup|g(t, x, \eta, \zeta)| \leq c_3(|\eta|^{p-1} + |\zeta|^{p-1}) + c_4(t, x) + c_0(t)|\eta|$ for every $(t, x, \eta, \zeta) \in Q \times \mathbb{R} \times \mathbb{R}^n$. If $p > 2$, then the problem

$$\frac{\partial u}{\partial t} - \sum_{i=1}^n \frac{\partial}{\partial x_i}\left(\left|\frac{\partial u}{\partial x_i}\right|^{p-2}\frac{\partial u}{\partial x_i}\right) + \sum_{i=1}^n b_i(x)\frac{\partial u}{\partial x_i}$$

$$+\ a(x)|u(x)|^{p-2}u(x) + g(t, x, u(x), \nabla u(x)) \ni 0$$

with Dirichlet boundary condition has a T-periodic, weak solution $u \in W_p^1(0, T; W_0^{1,p}(\Omega), L^2(\Omega))$.

References

[1] N. U. Ahmed and X. Xiang, *Existence of solutions for a class of nonlinear evolution equations with nonmonotone perturbations,* Nonlinear Anal. **13** (1994), 81-89.

[2] V. Barbu, *Nonlinear semigroups and differential equations in Banach spaces,* Noordhoff, Leyden, 1976.

[3] J. Berkovits and V. Mustonen, *Monotonicity methods for nonlinear evolution equations,* Nonlinear Anal. **27** (1996), 1397-1405.

[4] F. E. Browder, *Nonlinear operators and nonlinear equations of evolution in Banach spaces,* Proc. Symp. Pure Math. **18**, part 2, 1976.

[5] F. E. Browder, *Pseudo-monotone operators and nonlinear elliptic boundary value problems on unbounded domains,* Proc. Nat. Acad. Sci. USA **74** (1977), 2659-2661.

[6] J. P. Caşcaval and I. I. Vrabie, *Existence of periodic solutions for a class of nonlinear evolution equations,* Rev. Mate. Univ. Complutense Madrid **7** (1994), 325-338.

[7] F. S. De Blasi and J. Myjak, *On the solutions sets for differential inclusions,* Bull. Pol. Acad. Sci. Math. **33** (1985), 17-23.

[8] K. Fan, *Minimax theorems,* Proc. Nat. Acad. Sci. USA **39** (1953), 42-47.

[9] L. Górniewicz, *Homological methods in fixed point theory of multi-valued maps,* Dissertationes Math. **129** (1976), 1-66.

[10] L. Górniewicz and S. Plaskacz, *Periodic solutions of differential inclusions in R^n,* Boll. U. M. I. (7) **7-A** (1993), 409-420.

[11] N. Hirano, *Nonlinear Volterra equations with positive kernels,* in Nonlinear and Convex Analysis (Santa Barbara, Calif., 1985), Lecture Notes in Pure and Appl. Math. **107**, 83-98, 1987.

[12] N. Hirano, *Nonlinear evolution equations with nonmonotonic perturbations,* Nonlinear Anal. **13** (1989), 599-609.

[13] R. Landes, *On Galerkin's method in the existence theory of quasilinear elliptic equations,* J. Funct. Anal. **39** (1980), 123-148.

[14] R. Landes and V. Mustonen, *On pseudo-monotone operators and nonlinear noncoercive variational problems on unbounded domains,* Math. Ann. **248** (1980), 241-246.

[15] N. Shioji, *Existence of periodic solutions for nonlinear evolution equations with nonmonotonic perturbations,* Proc. Amer. Math. Soc. **125** (1997), 2921-2929.

[16] D. H. Wagner, *Survey of measurable selection theorems,* SIAM J. Control Optimization **15** (1977), 859-903.

[17] E. Zeidler, *Nonlinear functional analysis and its applications II/A,B, Linear monotone operators,* Springer-Verlag, New York, 1990.

Naoki Shioji, Department of Mathematics, Faculty of Engineering, Yokohama National University, Tokiwadai, Hodogaya-ku, Yokohama 240-8501, JAPAN

E-mail address: shioji@math.sci.ynu.ac.jp

Progress in Nonlinear Differential Equations
and Their Applications, Vol. 42

NONHOMOGENEOUS DAMPED STRING: RIESZ BASIS PROPERTY OF ROOT VECTORS VIA TRANSFORMATION OPERATORS METHOD

MARIANNA A. SHUBOV
TEXAS TECH UNIVERSITY

Abstract. We extend the classical concept of transformation operators to the 1-dim wave equation with spacially nonhomogeneous coefficients containing the first order damping term. The equation governs the vibrations of a damped string. Transformation operators were introduced in the fifties by I.M. Gelfand, B.M. Levitan and V.A. Marchenko in connection with the inverse scattering problem for the 1-dim Schrödinger equation. In the classical case, the transformation operator maps the exponential function (stationary wave function of a free particle) into the so-called Jost solution of the perturbed Schrödinger equation. In our case, it is natural to introduce two transformation operators (*outgoing* and *incoming* transformation operators). The terminology is motivated by an analog with the Lax-Phillips scattering theory. Our main results are the following. First, we prove the existence of transformation operators (which have the forms of the identity operator plus certain Volterra integral operators) and estimates for their kernels. Secondly, using the transformation operators, we prove that the dynamics generator of a finite string with damping both in the equation and in the boundary conditions is a Riesz spectral operator. The latter result provides a class of nontrivial examples of nonselfadjoint operators which admit an analog of the spectral decomposition. The result also has a significant applications in the control theory of distributed parameter systems.

I.

The goal of the present paper is to formulate the results obtained in our works [1,2]. We extend the classical concept of the transformation operator to the equation of a nonhomogeneous damped string

$$w_{tt} + 2\mathbf{d}(x)w_t + Lw = 0, \tag{1}$$

where L is the Sturm-Liouville operator defined on a smooth function φ by the formula

$$(L\varphi)(x) = -\frac{1}{\rho(x)}\frac{d}{dx}\left(p(x)\frac{d\varphi(x)}{dx}\right) + q(x)\varphi(x). \tag{2}$$

We assume that $w(x,t)$ is defined for $\infty > a \geq x \geq 0, t \geq 0$. The coefficients $\mathbf{d}(x)$ (the viscous damping), $\rho(x)$ (the density of the string), $p(x)$ (the modulus of elasticity) and $q(x)$ (the rigidity of an external harmonic force) are subject to the following conditions:

$$\rho, p \in W_1^2(0,a); \mathbf{d} \in W_1^1(0,a); q \in L^\infty(0,a); \rho(x), p(x) \geq c_0 > 0 \quad \text{for } x \in [0,a]. \tag{3}$$

We show that there are two transformation operators naturally associated to Eq.(1). In the case when $\mathbf{d}(x) = 0$, these operators coincide. If in addition $\rho(x) = p(x) = 1$, then they coincide with the classical transformation operator [3]. We impose the following boundary conditions:

$$(w_x + kw_t)(0,t) = (w_x + hw_t)(a,t) = 0; \quad h, k \in \mathbb{C} \cup \{\infty\}. \tag{4}$$

Depending on the values of the parameters h and k, conditions (4) describe different physical phenomena. Namely, in the case $h = k = \infty$,we have the homogeneous Dirichlet boundary conditions (the string is clamped at both ends). If $h = k = 0$,we have the Neumann boundary conditions (the string has free ends). If $k = \infty$ and $h = 1$, we have the Sommerfeld boundary condition at the right end. This condition is used to describe the resonance phenomena in the scattering of acoustical waves on the semi-infinite string. The boundary conditions with $h = -k = 1$ are used in the description of the resonance phenomena in the scattering of acoustical waves on the entire axis by an obstacle concentrated on the interval $[0, a]$.

Before we formulate our main results, we breifly describe possible applications. 1) One of the most important applications is, certainly, the inverse scattering problem for Eq.(1). However, in this case, the derivation of an analog of the Gelfand-Levitan-Marchenko equation is a nontrivial open problem. 2) Another application of the transformation operators is given in our paper [2], where we apply them to the spectral analysis of a class of nonselfadjoint operators generated by Eq.(1) with linear dissipative boundary conditions. The main result of [2] (stated below in Theorems 4-7) is the proof that these operators belong to the class of spectral operators in the sense of Dunford [4]. Significance of the results from [2] consists of the following. It is well known that at the present moment, there is no general spectral theory of nonselfadjoint operators in a Hilbert space. Such a theory has been developed only for some special classes of nonselfadjoint operators, and one of these classes is the class of spectral operators [4]. While the general spectral theory of such operators has been developed a long time ago, there is still a problem of finding specific examples of the operators of this class. In particular, it turned out that for many natural nonselfadjoint differential operators, it is very difficult to verify the spectral property. The results of [2] provide a class of nontrivial examples of spectral operators. 3) Another important aspect of the spectral results obtained in [2] is their applications to control theory. Namely, the spectral information allows us to solve the controllability problem for the damped string using the *spectral decomposition method.* This method in control theory was suggested by D. Russell [5] in the late sixties for the equation of the undamped string. His results were extended to many systems governed by undamped hyperbolic equations. However, they have never been extended to the damped string because of the absence of the necessary spectral information.

We expect that the method developed in this work can be extended to new classes of damped hyperbolic equations including various beam models and systems of N coupled strings. These results, in turn, could be applied to the solution of a wide class of control problems important in structural engineering.

Now we introduce the nonselfadjoint operators which are our main objects of interest.

Eq.(1) can be represented as the first order evolution equation for a 2-component function

$$\mathcal{W}_t = i\mathcal{L}\mathcal{W}, \quad \mathcal{W}(x,t) = (w(x,t),\ w_t(x,t))^T, \tag{5}$$

where the superscript 'T' means the transposition and the matrix differential expression $\mathcal{L}$ is given by

$$\mathcal{L} = -i\begin{pmatrix} 0 & 1 \\ -L & -2\mathbf{d}(x) \end{pmatrix}. \tag{6}$$

Eq.(5) with boundary conditions (4) defines a strongly continuous semigroup of transformations in a Hilbert space of initial data. This Hilbert space (the energy space) $\mathcal{H}$ is a closure of smooth 2-component functions $\mathcal{W}(x) = (w(x,t), w_t(x,t))^T$ in the standard energy norm

$$\|\mathcal{W}\|^2_{\mathcal{H}} = \frac{1}{2}\int_0^a \left[p(x)|w_0'|^2 + q(x)\rho(x)|w_0|^2 + \rho(x)|w_1|^2\right] dx. \tag{7}$$

The generator $\mathcal{L}_{hk}$ of the aforementioned semigroup is defined by (6) ($\mathcal{L}_{hk} = \mathcal{L}$) on the domain:

$$\mathcal{D}(\mathcal{L}_{hk}) = \{\mathcal{W}\in\mathcal{H} : \mathcal{L}\mathcal{W}\in\mathcal{H};\ (w_0' + kw_1)(0) = 0,\ (w_0' + hw_1)(a) = 0\}. \tag{8}$$

The following facts are valid: $\mathcal{L}_{hk}$ is a *nonselfadjoint, simple, maximal (dissipative* for *Re* $h \geq 0$ and *Re* $k \leq 0$) operator in $\mathcal{H}$. The resolvent of each operator is compact and, therefore, the spectrum is discrete.

The main result of [2] is the fact that operators $\mathcal{L}_{hk}$ are spectral (Riesz spectral).

II.

To introduce the transformation operators, we have to consider two ordinary differential equations containing the spectral parameter $\lambda \in \mathbb{C}$. These equations are naturally associated with the original wave equation (1). Namely, let us substitute into (1) the function $w(x,t)$ in one of the following two forms: $w(x,t) = e^{\pm i\lambda t}w(x)$, where for each sign of the exponent, $w(x)$ is a new function of x only. As a result, we obtain two equations for the functions $w(x)$:

$$(p(x)w')' + \lambda^2\rho(x)w - 2i\lambda\mathbf{d}(x)\rho(x)w - q(x)\rho(x)w = 0, \tag{9}$$

$$(p(x)w')' + \lambda^2\rho(x)w + 2i\lambda\mathbf{d}(x)\rho(x)w - q(x)\rho(x)w = 0. \tag{10}$$

Note, the only difference between Eqs.(9) and (10) is the sign of the term linear with respect to λ. First,we formulate auxiliary results about special solutions of Eqs.(9) and (10). To this end, we need new notations. Let $E_d^{\pm}$ and $E_{-d}^{\pm}$ be the following functions:

(11) $\overset{\circ}{E}{}_d^{\pm}(\lambda,x) = (\rho(x)p(x))^{-\frac{1}{4}} e^{\pm i\nu(\lambda,x)};\ \overset{\circ}{E}{}_{-d}^{\pm}(\lambda,x) = (\rho(x)p(x))^{-\frac{1}{4}} e^{\pm i\nu^*(\lambda,x)},$

where ν and ν^* are given by the formulas

$$\nu(\lambda,x) = \lambda \int_0^x \sqrt{\tau(t)}dt - i \int_0^x \mathbf{d}(t)\sqrt{\tau(t)}dt, \tag{12}$$

$$\nu^*(\lambda,x) = \overline{\nu(\overline{\lambda},x)}, \quad \tau(x) = \rho(x)/p(x).$$

Theorem 1. *Eq. (9) has two linearly independent solutions* $E_d^{\pm}(\lambda,x)$ *with the following approximations:*

$$\left|E_d^{\pm}(\lambda,x) - \overset{\circ}{E}{}_d^{\pm}(\lambda,x)\right| \le C\,|\lambda|^{-1}\left|\overset{\circ}{E}{}_d^{\pm}(\lambda,x)\right|. \tag{13}$$

Eq.(10) has two linearly independent solutions $E_{-d}^{\pm}(\lambda,x)$ *with the following approximations:*

$$\left|E_{-d}^{\pm}(\lambda,x) - \overset{\circ}{E}{}_{-d}^{\pm}(\lambda,x)\right| \le C|\lambda|^{-1}\left|\overset{\circ}{E}{}_{-d}^{\pm}(\lambda,x)\right|. \tag{14}$$

For each x*, these solutions are holomorphic functions of* λ *for* $Im\lambda > 0$ *and are continuous functions of* λ *for* $Im\lambda \ge 0$.

Let us extend the coefficients of Eq.(9) to the interval $(0, a+\mathcal{M})$,where $\mathcal{M} = \int_0^a \sqrt{\tau(t)}dt$ in the following fashion:

$$\rho(x) = p(x) = 1 \quad \mathbf{d}(x) = q(x) = 0 \quad \text{for } x \in (a, a+\mathcal{M}), \tag{15}$$

and consider the problem in the weighted space $L^2_{\rho/p}(0,a+\mathcal{M})$.

Now we are in a position to define the transformation operators.

Definition 1. *Linear operators* $\mathbb{K}$ *and* $\hat{\mathbb{K}}$ *in the space* $H = L^2_{\rho/p}(0,a+\mathcal{M})$ *which transform the functions* $\overset{\circ}{E}{}_d^{\pm}(\lambda,x)$ *into the solutions* $E_d^{\pm}(\lambda,x)$ *of Eq.(9) are called* outgoing *and* incoming *transformation operators respectively.*

To formulate our main statement on the existence and properties of the transformation operators, we need the following functions:

$$\hat{\varphi} = q + (p')^2p^{-1} - p^{3/2}(p^{-1/2})'' - 0.8\tau'\tau^{-1} + 0.5\tau''\tau^{-2};$$

$$\varphi = \hat{\varphi} - \mathbf{d}^2 - \mathbf{d}';$$

$$\psi = \int_0^x \mathbf{d}(t)\sqrt{\tau(t)}dt; \quad g = \varphi\exp(2\psi), \quad \hat{g} = \hat{\varphi} - \mathbf{d}^2 + \mathbf{d}'. \tag{16}$$

We also introduce the functions $\sigma, \sigma_1, \hat{\sigma}$ and $\hat{\sigma}_1$ by formulas

$$\sigma(x) = \int_x^a |g(t)|\sqrt{\tau(t)}dt, \quad \sigma_1(x) = \int_x^a \sigma(t)\sqrt{\tau(t)}dt. \tag{17}$$

To obtain formulas for $\hat{\sigma}$ and σ_1, it sufficies to replace g and σ in (17) with $\hat{g}$ and $\hat{\sigma}$ respectively.

Theorem 2. *1) There exist bounded and boundedly invertible operators $\mathbb{K}$ and $\hat{\mathbb{K}}$ in the space H such that the following relations are valid:*

$$E_d^+(\lambda, x) = (\mathbb{K}\, \mathring{E}_d^+)(\lambda, x), \quad E_d^-(\lambda, x) = (\hat{\mathbb{K}}\, \mathring{E}_d^-)(\lambda, x). \tag{18}$$

$\mathbb{K}$ and $\hat{\mathbb{K}}$ are precisely the outgoing and incoming transformation operators, respectively. Similar relations are valid for the solutions of the adjoint equation

$$E_{-d}^+(\lambda, x) = (\hat{\mathbb{K}}\mathring{E}_{-d}^+)(\lambda, x), \quad E_{-d}^-(\lambda, x) = (\mathbb{K}\, \mathring{E}_{-d}^-)(\lambda, x). \tag{19}$$

2) The operators $\mathbb{K}$ and $\hat{\mathbb{K}}$ can be represented in the forms:

$$\mathbb{K} = I + \mathbf{K}, \quad \hat{\mathbb{K}} = I + \hat{\mathbf{K}}, \tag{20}$$

where $\mathbf{K}$ and $\hat{\mathbf{K}}$ are compact. Both $\mathbf{K}$ and $\hat{\mathbf{K}}$ are Volterra integral operators

$$\mathbf{K}\varphi = \int_x^\infty \mathfrak{K}(x,t)\varphi(t)dt, \quad \hat{\mathbf{K}}\varphi = \int_x^\infty \hat{\mathfrak{K}}(x,t)\varphi(t)dt. \tag{21}$$

The kernels $\mathfrak{K}(x,t)$ and $\hat{\mathfrak{K}}(x,t)$ are differentiable functions with the following properties:

$$a) \quad \mathfrak{K}(x,t) = \hat{\mathfrak{K}}(x,t) = 0 \quad \text{for } x \geq t;$$

$$b) \quad |\mathfrak{K}(x,t)| \leq C_0 \ \sigma\left(\frac{x+t}{2}\right) \exp\left\{\sigma_1(x) - \sigma_1\left(\frac{x+t}{2}\right)\right\}, \tag{22}$$

with C_0 being an absolute constant. Similar estimates are valid for $\hat{\mathfrak{K}}(x,t)$ if we replace σ and σ_1 with $\hat{\sigma}$ and $\hat{\sigma}_1$.

$$c) \quad \mathfrak{K}(x,x) = \frac{1}{2}\int_x^\infty \varphi(t)\sqrt{\tau(t)}dt, \quad \hat{\mathfrak{K}}(x,x) = \frac{1}{2}\int_x^\infty \hat{g}(t)\sqrt{\tau(t)}dt. \tag{23}$$

III.

Now we turn to the statement of our spectral results from [2]. To obtain the Riesz basis property of the root vectors of the operator $\mathfrak{L}_{hk}$, we need a detailed information on the spectral asymptotics, which is given in the next statement. We assume that the following conditions on h and k are satisfied:

$$\text{if } \mathcal{I}m\ h = 0, \ \text{ then } \sqrt{\frac{p(a)}{\rho(a)}} \neq |h|; \quad \text{if } \mathcal{I}m\ k = 0, \ \text{ then } \sqrt{\frac{\rho(0)}{p(0)}} \neq |k|. \tag{24}$$

Theorem 3. *a) The operator* $\mathfrak{L}_{hk}$ *has a countable set of eigenvalues* $\{\lambda_n^{hk}\}_{n\in\mathbb{Z}}$. *If both parameters h and k satisfy conditions (24), then the set of eigenvalues is located in a strip parallel to the real axis and has only two points of accumulation:* $+\infty$ *and* $-\infty$ *in the sense that* $Re\lambda_n^{hk} \to \pm\infty$ *as* $n \to \pm\infty$ *and* $Im\,\lambda_n^{hk} \to const$ *as* $n \to \pm\infty$ *(see (25) below).*

b) For each pair (h, k), satisfying conditions (24), the following asymptotic formula is valid:

$$\lambda_n^{hk} = \mathcal{M}^{-1}[(n + 0.5\ sgn\ n)\pi + i\mathcal{N}] + \mathcal{D}_{hk} + O(|n|^{-\alpha}), \quad |n| \to \infty, \tag{25}$$

where $\alpha = 1$ *if* $p, \rho \in C^2[0, a]$ *and, in general,* $0 < \alpha < 1$ *if only conditions (3) are satisfied;* $\mathcal{N} = \int_0^a \mathbf{d}(t)\sqrt{\tau(t)}dt$, *and* $\mathcal{D}_{hk}$ *is defined by the formula*

$$\mathcal{D}_{hk} = \mathcal{M}^{-1}\left[(n + \frac{1}{2}\ sgn\ n)\pi + i\left(\mathcal{N} + \frac{1}{2}\ \ell n\ \frac{\sqrt{p(a)/\rho(a)} + h}{\sqrt{p(a)/\rho(a)} - h}\right.\right.$$

$$\left.\left. + \frac{1}{2}\ \ell n\ \frac{\sqrt{\rho(0)/p(0)} - k}{\sqrt{\rho(0)/p(0)} + k}\right)\right]. \tag{26}$$

c) All eigenvalues of $\mathfrak{L}_{hk}$ *have geometric multiplicities 1; i.e., for each* λ_n^{hk} *there exists only one normalized eigenvector* Ψ_n. *However, a finite number of eigenvalues* $\{\lambda_n^{hk},\ n \in R \in \mathbb{Z}\}$ *may have finite algebraic multiplicities* $\{m_n\}$; *i.e, for such* λ_n^{hk} *there exists a finite chain of associate vectors* $\{\Psi_n^j\}_{j=1}^{m_n-1}$:

$$(\mathfrak{L}_{hk} - \lambda_n^{hk} I)\Psi_n^j = \Psi_n^{j-1}, \quad \Psi_n^0 = \Psi_n, \quad \Psi_n^{-1} = 0. \tag{27}$$

Let $\mathcal{J}_d(\lambda, x)$ be the solution of Eq.(9) satisfying the boundary condition at $x = a$. Let

$$\delta_\pm = \sqrt{p(a)/\rho(a)} \pm h, \quad \sqrt{p(a)/\rho(a)} \neq |h| \quad \text{for real } h. \tag{28}$$

For every $h \in \mathbb{C} \cup \{\infty\}$, there exists a pair of coefficients $A_h(\lambda)$ and $B_h(\lambda)$ such that

$$\mathcal{J}_d(\lambda, x) = A_h(\lambda)E_d^+(\lambda, x) + B_h(\lambda)E_d^-(\lambda, x). \tag{29}$$

The coefficients $A_h(\lambda)$ and $B_h(\lambda)$ are holomorphic for $Im\lambda > 0$, are continuous for $\mathrm{Im}\lambda \geq 0$, and have the following asymptotic approximations for $Im\,h \geq 0$:

$$A_h(\lambda) = \delta_-(1 + O(|\lambda|^{-\alpha})) - \delta_+ e^{2i\lambda\mathcal{M}+2\mathcal{N}}(1 + O((|\lambda|^{-\alpha})),$$

$$B_h(\lambda) = \delta_+ e^{i\lambda\mathcal{M}+\mathcal{N}}(1 + O(|\lambda|^{-\alpha}), \quad 0 < \alpha \leq 1, \tag{30}$$

Note that if $\lambda = \lambda_{hk}$, then $\mathcal{J}_d(\lambda, x)$ becomes an eigenfunction of the boundary-value problem given by Eq.(9) and both boundary conditions (4). As follows from (29), it is important to know the geometric properties of the components $E_d^\pm(\lambda, x)$. The next result is an immediate corollary of the properties of the transformation operators.

Theorem 4. *a) Each set of exponential-like functions* $\{\overset{\circ}{E}{}^{\pm}_d(\lambda_n^{hk},x)\}_{n\in\mathbb{Z}}$ *(see (11)) forms a Riesz basis in its closed linear span in* $L^2_{\rho/p}(0,\mathcal{A})$*, where* $\mathcal{A}=a+\mathcal{M}$.

b) If the set of exponentials $\{\exp(i\lambda_n^{hk}x)\}_{n\in\mathbb{Z}}$ *is complete in* $L^2(0,2\mathcal{M})$*, then the set* $\{\overset{\circ}{E}{}^{+}_d(\lambda_n^{hk},x)\}_{n\in\mathbb{Z}}$ *is complete in* $L^2_{\rho/p}(0,\mathcal{A})$*, and, therefore, it is a Riesz basis in* $L^2_{\rho/p}(0,\mathcal{A})$.

c) Each set of solutions $\{E_d^{\pm}(\lambda_n^{hk},x)\}_{n\in\mathbb{Z}}$ *of Eq. (9) forms a Riesz basis in its closed linear span* $\mathcal{E}^+$ *(or* $\mathcal{E}^-$*) in the space* $L^2_{\rho/p}(0,\mathcal{A})$*. If the set of nonharmonic exponentials* $\{e^{i\lambda_n^{hk}x}\}_{n\in\mathbb{Z}}$ *is complete in* $L^2(0,2\mathcal{M})$*, (e.g., if the spectrum of the operator* $\mathcal{L}_{hk}$ *is simple), then* $\mathcal{E}^+$ *(or* $\mathcal{E}^-$*) coincides with* $L^2_{\rho/p}(0,\mathcal{A})$.

Theorem 4 is, in fact, a particular case of the following more general result.

Theorem 5. *If the set of complex points* $\{\kappa_n\}_{n\in\mathbb{Z}}$ *is such that the system of exponentials* $\{e^{i\kappa_n x}\}_{n\in\mathbb{Z}}$ *forms a Riesz basis in* $L^2(0,2\mathcal{M})$*, then the set of solutions* $\{E_d^{+}(\kappa_n,x)\}_{n\in\mathbb{Z}}$ *(or* $\{E_d^{-}(\kappa_n,x)\}_{n\in\mathbb{Z}}$*) of Eq. (9) forms a Riesz basis in* $L^2_{\rho/p}(0,\mathcal{A})$.

Theorem 5 can be understood as the stability result for a special set of solutions of Eq.(9). Namely, with a given sequence of complex points $\{\kappa_n\}_{n\in\mathbb{Z}}$, the set of the corresponding solutions $\{E_d^{+}(\kappa_n,x)\}_{n\in\mathbb{Z}}$ forms a Riesz basis in $L^2_{\rho/p}(0,\mathcal{A})$ if and only if the set of nonharmonic exponentials $\{e^{i\kappa_n x}\}_{n\in\mathbb{Z}}$ forms a Riesz basis in $L^2(0,2\mathcal{M})$. Since there exists an extensive literature on the geometry of nonharmonic exponentials (see [6,7]), we obtain a lot of information on the geometry of the sets of solutions $\{E_d^{\pm}(\kappa_n,x)\}_{n\in\mathbb{Z}}$. Moreover, if for a particular set $\{\kappa_n\}_{n\in\mathbb{Z}}$, the set of exponentials $\{e^{i\kappa_n x}\}_{n\in\mathbb{Z}}$ is minimal (linearly independent in the case of infinitely many vectors), but not uniformly minimal, let alone the Riesz basis property, then the same is valid for each of the sets $\{E_d^{\pm}(\kappa_n,x)\}_{n\in\mathbb{Z}}$ in $L^2_{\rho/p}(0,\mathcal{A})$. If the set $\{e^{i\kappa_n x}\}_{n\in\mathbb{Z}}$ is only complete in $L^2(0,2\mathcal{M})$, then the same fact holds for $\{E_d^{\pm}(\kappa_n,x)\}_{n\in\mathbb{Z}}$. Obviously, all of the above takes place for the sets $\{E_{-d}^{\pm}(\kappa_n,x)\}_{n\in\mathbb{Z}}$ in the space $L^2_{\rho/p}(0,\mathcal{A})$.

Next we describe the Riesz basis which is biorthogonal to the Riesz basis $\{E_d^{\pm}(\lambda_n^{hk},x)\}_{n\in\mathbb{Z}}$.

Theorem 6. *Let* $\{\mu_n\}_{n\in\mathbb{Z}}$ *be any sequence of complex numbers for which* $\{\exp(i\mu_n x)\}_{n\in\mathbb{Z}}$ *is a Riesz basis in* $L^2(0,2\mathcal{M})$*. Hence, the set* $\{\overset{\circ}{E}{}^{+}_d(\mu_n,x)\}_{n\in\mathbb{Z}}$ *is a Riesz basis in* $L^2_{\rho/p}(0,\mathcal{A})$*. Let* $\{W_n^0(x)\}_{n\in\mathbb{Z}}$ *be a Riesz basis biorthogonal to the basis* $\{\overset{\circ}{E}{}^{+}_d(\mu_n,x)\}_{n\in\mathbb{Z}}$ *in* $L^2_{\rho/p}(0,\mathcal{A})$*. Then the following formula is valid for the system* $\{W_n(x)\}_{n\in\mathbb{Z}}$ *which is biorthogonal to the set* $\{E_d^{+}(\mu_n,x)\}_{n\in\mathbb{Z}}$*:*

$$W_n(x)=W_n^0(x)+\int_0^x \mathcal{L}(t,x)W_n^0(t)\,dt, \tag{31}$$

where $\mathcal{L}(t,x)$ is the kernel of the integral operator $\mathbb{L}$ satisfying the equation

$$(I+\mathbb{K})^{-1} = I + \mathbb{L}. \tag{32}$$

In (32) $\mathbb{K}$ is the Volterra integral operator introduced in (18). It is clear that Eq. (31) can be written in the form: $W_n = (I+\mathbb{L}^*)W_n^0$.

In the last statement, we give our main results on the Riesz basis property of the root vectors of the operator $\mathcal{L}_{hk}$.

Theorem 7. *1) Let $\{\lambda_n^{hk}\}_{n\in\mathbb{Z}}$ and $\{\varphi_n^{hk}\}_{n\in\mathbb{Z}}$ be the spectrum and the set of eigenvectors of the problem defined by Eq.(9) and the boundary conditions $(w'+i\lambda kw)(\lambda,0) = (w'+i\lambda hw)(\lambda,a) = 0$. Then for any $h,k \in \mathbb{C}\cup\{\infty\}$, there exist bounded linear operators S_{hk} and T_{hk} in the space $L^2_{\rho/p}(0,\mathcal{A})$ which act in the following ways:*

$$\varphi_n^{hk}(x) = (S_{hk}\overset{\circ}{E}{}_d^{+})(\lambda_n^{hk},x), \quad (i\lambda_n^{hk})^{-1}(\varphi_n^{kk}(x))' = (T_{hk}\overset{\circ}{E}{}_d^{+})(\lambda_n^{hk},x). \tag{33}$$

2) Let $\{\hat{\varphi}_n^{hk}\}_{n\in\mathbb{Z}}$ be the set of eigenvectors of the adjoint problem defined by Eq.(10) and the boundary conditions $(w'-i\lambda\overline{k}w)(\lambda,0) = (w'-i\lambda\overline{h}w)(\lambda,a) = 0$. Then there exist bounded linear operators $\hat{S}_{hk}$ and $\hat{T}_{hk}$ in the space $L^2_{\rho/p}(0,\mathcal{A})$ which act in the following ways:

$$\hat{\varphi}_n^{hk}(x) = (\hat{S}_{hk}\overset{\circ}{E}{}_d^{+})(\lambda_n^{hk},x), \quad (i\overline{\lambda}_n^{hk})^{-1}(\hat{\varphi}_n^{hk}(x))' = (\hat{T}_{hk}\overset{\circ}{E}{}_d^{+})(\lambda_n^{hk},x). \tag{34}$$

3) Let $\{\Phi_n^{hk}\}_{n\in\mathbb{Z}}$ and $\{\hat{\Phi}_n^{hk}\}_{n\in\mathbb{Z}}$ be the sets of eigenvectors of the operators $\mathfrak{L}_{hk}$ and $\mathfrak{L}_{hk}^*$ respectively. The following formulas are valid:

$$\Phi_n^{hk}(x) = \begin{pmatrix} \frac{1}{i\lambda_n^{hk}} & \varphi_n^{hk}(x) \\ & \varphi_n^{hk}(x) \end{pmatrix}, \quad \hat{\Phi}_n^{hk}(x) = \begin{pmatrix} \frac{1}{i\overline{\lambda}_n^{hk}} & \hat{\varphi}_n^{hk}(x) \\ & \hat{\varphi}_n^{hk}(x) \end{pmatrix}, \tag{35}$$

$$x \in (0,a), \quad n \in \mathbb{Z}.$$

The systems in (35) are biorthogonal in $\mathcal{H}$.

4) Both the systems $\{\Phi_n^{hk}\}_{n\in\mathbb{Z}}$ and $\{\hat{\Phi}_n^{hk}\}_{n\in\mathbb{Z}}$ are biorthogonal Riesz bases in $\mathcal{H}$.

Acknowledgments. Partial support by the National Science Foundation Grant DMS-9706882 and Advanced Research Program -97 of Texas Grant #0036-44-045 is highly appreciated.

References

[1] Marianna A. Shubov, Transformation operators for class of damped hyperbolic equations, Preprint TTU, 1999.

[2] Marianna A. Shubov, The Riesz basis property of the system of root vectors for the equation of nonhomogenous damped string. Transformation operators method, to appear in *Methods and Applications of Analysis.*

[3] Z.S. Agranovich, V.A. Marchenko, *The Inverse Problem of Scattering Theory*, Gordon and Breach, New York, 1967.

[4] N. Dunford, J.T. Schwartz, *Linear Operators, Part III: Spectral Operators.* New York – London, Toronto, (1971).

[5] D.L. Russell, Nonharmonic Fourier series in the control theory of distributed parameter systems, *J. Math. Anal. Appl.*, **18**, (1967), p. 542-559.

[6] N.K. Nikol'skii, *Treatise on the Shift Operator*, Springer-Verlag, Berlin, (1986).

[7] S.V. Hrščev, N.K. Nikolskii, B.S. Pavlov, Unconditional bases of exponentials and reproducing kernels, *Lecture Notes in Math.* **864** (1981), 214-335.

Marianna A. Shubov, Department of Mathematics and Statistics, Texas Tech University, Lubbock, TX 79409-1042, U.S.A.
E-mail address: mshubov@math.ttu.edu

Progress in Nonlinear Differential Equations
and Their Applications, Vol. 42

M-ACCRETIVE QUANTIZATION OF THE VORTICITY EQUATION

V. BARBU
UNIVERSITY OF IASI

S.S. SRITHARAN
SPACE AND NAVAL WAREFARE SYSTEMS CENTER

Abstract. We will introduce a new quantization method for the vorticity equation which converts the sum of the linear and nonlinear operators in this equation to be α−m-accretive and thus allows us to use nonlinear semigroup theory to establish solvability. This method also sets the stage for the study of vorticity dynamics with state constraints and control forcing, which can be formulated as a control problem for parabolic variational inequalities. Applications of this mathematical work include atmospheric weather prediction, flow and combustion control and control of information flow in massive command-control networks.

1. Introduction

Let us consider the three dimensional vorticity equation [1] (for the curl of the velocity field) $\mathbf{w} = \nabla \times \mathbf{u}$,

$$\partial_t \mathbf{w} - \nu \Delta \mathbf{w} + \mathbf{u} \cdot \nabla \mathbf{w} - \mathbf{w} \cdot \nabla \mathbf{u} = 0, \quad \text{in } \Re^3 \times \Re_+, \tag{1.1}$$

$$\nabla \cdot \mathbf{w} = 0, \text{ in } \Re^3 \times \Re_+, \tag{1.2}$$

$$\mathbf{w}(\mathbf{x}, 0) = \mathbf{w}_0(\mathbf{x}), \ \mathbf{x} \in \Re^3. \tag{1.3}$$

Here $\nu > 0$ is the coefficient of kinematic viscosity. Velocity is related to vorticity by the integral equation,

$$\mathbf{u}(\mathbf{x}, t) = [\mathcal{K} * \mathbf{w}](\mathbf{x}, t) := -\frac{1}{4\pi} \int_{\Re^3} \frac{\mathbf{x} - \mathbf{y}}{|\mathbf{x} - \mathbf{y}|^3} \times \mathbf{w}(\mathbf{y}, t) d\mathbf{y}. \tag{1.4}$$

Let us define spaces,

$$\mathbf{H} := \left\{ \mathbf{u} \in (L^2(\Re^3))^3; \nabla \cdot \mathbf{u} = 0 \right\},$$

$$\mathbf{V} := \left\{ \mathbf{u} \in (H^1(\Re^3))^3; \nabla \cdot \mathbf{u} = 0 \right\},$$

$$D(\mathbf{A}) := (H^2(\Re^3))^3 \cap \mathbf{H}.$$

Thus we have an abstract semilinear equation in $\mathbf{H}$,

$$\frac{d\mathbf{w}}{dt} + \nu \mathbf{A} \mathbf{w} + \mathbf{\Lambda}(\mathbf{w}) = 0, \tag{1.5}$$

where $\mathbf{\Lambda}(\mathbf{w}) := \widetilde{\mathbf{B}}(\mathcal{K} * \mathbf{w}, \mathbf{w})$ with,

$$\widetilde{\mathbf{B}}(\mathbf{u}, \mathbf{v}) = \mathbf{B}(\mathbf{u}, \mathbf{v}) - \mathbf{B}(\mathbf{v}, \mathbf{u}). \tag{1.6}$$

The bilinear operator $\mathbf{B}$ is defined through,

$$< \mathbf{B}(\mathbf{u}, \mathbf{v}), \mathbf{z} >= b(\mathbf{u}, \mathbf{v}, \mathbf{z}) := \int_{\Re^3} (\mathbf{u}(\mathbf{x}) \cdot \nabla \mathbf{v}(\mathbf{x})) \cdot \mathbf{z}(\mathbf{x}) d\mathbf{x} \tag{1.7}$$

and the Riesz representation theorem.

We will introduce a quantization $\mathbf{\Lambda}_N$ for the nonlinearity $\mathbf{\Lambda}$ by modifying it outside an H^1-ball of radius N so that the map,

$$\mathbf{w} \to \alpha \mathbf{w} + \nu \mathbf{A} \mathbf{w} + \mathbf{\Lambda}_N(\mathbf{w}) \tag{1.8}$$

would be m-accretive (maximal monotone) in $\mathbf{H}$. This allows us to establish the solvability of the quantized equation,

$$\frac{d\mathbf{w}}{dt} + \nu \mathbf{A} \mathbf{w} + \mathbf{\Lambda}_N(\mathbf{w}) = 0 \tag{1.9}$$

using Komura type nonlinear semigroup generation theorem. We then show that the solution trajectories of this equation actually lies (upto a maximal time) inside an H^1-ball of radius independent of N and hence by choosing N suitably large we automatically deduce the solvability of the original equation 1.5. This procedure also allows us to treat the "inclusion" problem of the type,

$$\frac{d\mathbf{w}}{dt} + \nu \mathbf{A} \mathbf{w} + \mathbf{\Lambda}(\mathbf{w}) + \sum_{i=1}^{m} \mathbf{N}_{\mathbf{K}_i}(\mathbf{w}) \ni 0 \tag{1.10}$$

where $\mathbf{N}_{\mathbf{K}_i}(\mathbf{w})$ are normal cones (representing feedback controls $\mathbf{U}_i = -\mathbf{N}_{\mathbf{K}_i}(\mathbf{w})$ to ensure that $\mathbf{w}(t) \in \mathbf{K}_i$, $\forall t \geq 0$) to sets $\mathbf{K}_i$. Noting the fact that when $\mathbf{K}_i$ are closed convex set, $\mathbf{N}_{\mathbf{K}_i}(\mathbf{w})$ would be maximal monotone and thus under suitable conditions, perturbation theorems on maximal monotone nonlinear operators will ensure that the operator

$$\mathbf{w} \to \nu \mathbf{A} \mathbf{w} + \mathbf{\Lambda}_N(\mathbf{w}) + \sum_{i=1}^{m} \mathbf{N}_{\mathbf{K}_i}(\mathbf{w}) \tag{1.11}$$

would be maximal monotone. Such analysis for the Navier-Stokes equation is given in [4]. For a general exposition on control of viscous flow and recent advances see [6].

2. M-Accretive Quantization

Let us now consider a quantization of the nonlinearity $\mathbf{\Lambda}$ in the following form,

$$\mathbf{\Lambda}_N(\mathbf{w}) := \begin{cases} \mathbf{\Lambda}(\mathbf{w}) & \text{if } \ ||\mathbf{w}||_1 \leq N \\ \frac{N^2}{||\mathbf{w}||_1^2} \mathbf{\Lambda}(\mathbf{w}) & \text{if } \ ||\mathbf{w}||_1 > N \end{cases} \tag{2.1}$$

where $\|\cdot\|_1$ denotes the H^1-norm (**V**-norm).

Consider the operator $\mathbf{\Gamma}_N : D(\mathbf{\Gamma}_N) \to H$ defined by,

$$\mathbf{\Gamma}_N = \nu\mathbf{A} + \mathbf{\Lambda}_N,\ D(\mathbf{\Gamma}_N) = D(\mathbf{A}). \tag{2.2}$$

We will show first that for $\mathbf{\Gamma}_N(\mathbf{w})$ is well defined in **H** for all $\mathbf{w} \in D(\mathbf{A})$. To see this first note that,

$$|\mathbf{\Gamma}_N(\mathbf{w})| \le \nu|\mathbf{A}\mathbf{w}| + |\mathbf{\Lambda}_N(\mathbf{w})| \tag{2.3}$$

$$\le \nu|\mathbf{A}\mathbf{w}| + |\mathbf{B}(\mathcal{K} * \mathbf{w}, \mathbf{w})| + |\mathbf{B}(\mathbf{w},\mathcal{K} * \mathbf{w})|. \tag{2.4}$$

Here $|\cdot|$ denotes the **H**-norm. Let us estimate the second and third terms on the right. We consider,

$$|\,(\mathbf{B}(\mathcal{K} * \mathbf{w}, \mathbf{w}), \mathbf{z})\,| = |b(\mathcal{K} * \mathbf{w}, \mathbf{w}, \mathbf{z})| \tag{2.5}$$

$$\le |\mathcal{K} * \mathbf{w}|_6 |\nabla\mathbf{w}|_3 |\mathbf{z}| \le C|\mathbf{w}||\nabla\mathbf{w}|_3|\mathbf{z}|$$

using the estimate for the Riesz potentials [7], Theorem 1, page 119):

$$|\int_{R^3} \frac{f(y)}{|x-y|^2} dy|_q \le C|f|_p,\quad \frac{1}{q} = \frac{1}{p} - \frac{1}{3}. \tag{2.6}$$

Thus,

$$|\mathbf{B}(\mathcal{K} * \mathbf{w}, \mathbf{w})| \le C|\mathbf{w}||\nabla\mathbf{w}|_3 \le C|\mathbf{w}||\nabla\mathbf{w}|^{1/2}|\mathbf{A}\mathbf{w}|^{1/2}. \tag{2.7}$$

Similarly,

$$|\,(\mathbf{B}(\mathbf{w},\mathcal{K} * \mathbf{w}), \mathbf{z})\,| = |b(\mathbf{w}, \mathcal{K} * \mathbf{w}, \mathbf{z})| \le |\mathbf{w}|_3 |\nabla(\mathcal{K}*\mathbf{w})|_6 |\mathbf{z}|$$

$$\le |\mathbf{w}|_3 |\mathcal{K}*(\nabla\mathbf{w})|_6 |\mathbf{z}| \le C|\mathbf{w}|_3 |\nabla\mathbf{w}||\mathbf{z}|,$$

using 2.6. Thus,

$$|\mathbf{B}(\mathbf{w},\mathcal{K} * \mathbf{w})| \le C|\mathbf{w}|_3|\nabla\mathbf{w}| \le C|\mathbf{w}|^{1/2}|\nabla\mathbf{w}|^{3/2}. \tag{2.8}$$

Combining (2.4),(2.7) and (2.8) we get,

$$|\mathbf{\Gamma}_N(\mathbf{w})| \le \nu|\mathbf{A}\mathbf{w}| + C|\mathbf{w}|^{1/2}|\nabla\mathbf{w}|^{3/2} + C|\mathbf{w}||\nabla\mathbf{w}|^{1/2}|\mathbf{A}\mathbf{w}|^{1/2}. \tag{2.9}$$

Hence for $\mathbf{w} \in D(\mathbf{A})$, $\mathbf{\Gamma}_N(\mathbf{w})$ is well defined in **H**. We will now prove the α_N-accretivity of $\mathbf{\Gamma}_N$.

Theorem 1. *There exists* $\alpha_N > 0$ *such that* $\mathbf{\Gamma}_N + \alpha_N\mathbf{I}$ *is m-accretive (maximal monotone) in* $\mathbf{H} \times \mathbf{H}$.

Proof. Let us first establish monotonicity and coercivity. We will show in particular that, $\forall \mathbf{w}_1, \mathbf{w}_2 \in D(\mathbf{\Gamma}_N)$,

$$((\mathbf{\Gamma}_N + \lambda)\,\mathbf{w}_1 - (\mathbf{\Gamma}_N + \lambda)\,\mathbf{w}_2, \mathbf{w}_1 - \mathbf{w}_2) \ge \frac{\nu}{2}||\mathbf{w}_1 - \mathbf{w}_2||_1, \tag{2.10}$$

for $\lambda \ge C_N$.

Note that,

$$((\mathbf{\Gamma}_N+\lambda)\,\mathbf{w}_1-(\mathbf{\Gamma}_N+\lambda)\,\mathbf{w}_2,\mathbf{w}_1-\mathbf{w}_2)=\nu|\nabla(\mathbf{w}_1-\mathbf{w}_2)|^2$$

$$+\lambda|\mathbf{w}_1-\mathbf{w}_2|^2+(\mathbf{\Lambda}_N(\mathbf{w}_1)-\mathbf{\Lambda}_N(\mathbf{w}_2),\mathbf{w}_1-\mathbf{w}_2). \tag{2.11}$$

We will now show that,

$$|(\mathbf{\Lambda}_N(\mathbf{w}_1)-\mathbf{\Lambda}_N(\mathbf{w}_2),\mathbf{w}_1-\mathbf{w}_2)|\le\frac{\nu}{2}|\nabla(\mathbf{w}_1-\mathbf{w}_2)|^2+C_N|\mathbf{w}_1-\mathbf{w}_2|^2, \tag{2.12}$$

which will establish (2.10). Let us analyze the different cases for the left hand side of (2.12).

Let us consider the case of $||\mathbf{w}_1||_1, ||\mathbf{w}_2||_1 \le N$. Then,

$$(\mathbf{\Lambda}_N(\mathbf{w}_1)-\mathbf{\Lambda}_N(\mathbf{w}_2),\mathbf{w}_1-\mathbf{w}_2)$$

$$=(\mathbf{B}(\mathbf{u}_1,\mathbf{w}_1)-\mathbf{B}(\mathbf{u}_2,\mathbf{w}_2),\mathbf{w}_1-\mathbf{w}_2)-(\mathbf{B}(\mathbf{w}_1,\mathbf{u}_1)-\mathbf{B}(\mathbf{w}_2,\mathbf{u}_2),\mathbf{w}_1-\mathbf{w}_2)$$

Thus,

$$|(\mathbf{\Lambda}_N(\mathbf{w}_1)-\mathbf{\Lambda}_N(\mathbf{w}_2),\mathbf{w}_1-\mathbf{w}_2)|\le|b(\mathbf{u}_1-\mathbf{u}_2,\mathbf{w}_1,\mathbf{w}_1-\mathbf{w}_2)|$$

$$+|b(\mathbf{w}_1-\mathbf{w}_2,\mathbf{u}_1,\mathbf{w}_1-\mathbf{w}_2)|+|b(\mathbf{w}_2,\mathbf{u}_1-\mathbf{u}_2,\mathbf{w}_1-\mathbf{w}_2)|$$

$$\le|\mathbf{u}_1-\mathbf{u}_2|_6|\nabla\mathbf{w}_1||\mathbf{w}_1-\mathbf{w}_2|_3+|\mathbf{w}_1-\mathbf{w}_2|_3|\nabla\mathbf{u}_1|_6|\mathbf{w}_1-\mathbf{w}_2|$$

$$+|\mathbf{w}_2|_3|\nabla(\mathbf{u}_1-\mathbf{u}_2)|_6|\mathbf{w}_1-\mathbf{w}_2|$$

$$\le C|\mathbf{w}_1-\mathbf{w}_2|_3|\mathbf{w}_1-\mathbf{w}_2|+C|\nabla(\mathbf{w}_1-\mathbf{w}_2)||\mathbf{w}_1-\mathbf{w}_2|$$

Thus,

$$|(\mathbf{\Lambda}_N(\mathbf{w}_1)-\mathbf{\Lambda}_N(\mathbf{w}_2),\mathbf{w}_1-\mathbf{w}_2)|\le\epsilon|\nabla(\mathbf{w}_1-\mathbf{w}_2)|^2+C(\epsilon)|\mathbf{w}_1-\mathbf{w}_2|^2. \tag{2.13}$$

We now consider the case of, $||\mathbf{w}_1||_1 > N, ||\mathbf{w}_2||_1 > N$ with $||\mathbf{w}_1||_1 \ge ||\mathbf{w}_2||_1$ (in fact the case of $||\mathbf{w}_1||_1 > N, ||\mathbf{w}_2||_1 > N$ with $||\mathbf{w}_1||_1 \le ||\mathbf{w}_2||_1$ would be similar). We get,

$$|(\mathbf{\Lambda}_N(\mathbf{w}_1)-\mathbf{\Lambda}_N(\mathbf{w}_2),\mathbf{w}_1-\mathbf{w}_2)|=|\frac{N^2}{||\mathbf{w}_1||_1^2}(\mathbf{B}(\mathbf{u}_1,\mathbf{w}_1)-\mathbf{B}(\mathbf{w}_1,\mathbf{u}_1),\mathbf{w}_1-\mathbf{w}_2)$$

$$-\frac{N^2}{||\mathbf{w}_2||_1^2}(\mathbf{B}(\mathbf{u}_2,\mathbf{w}_2)-\mathbf{B}(\mathbf{w}_2,\mathbf{u}_2),\mathbf{w}_1-\mathbf{w}_2)\,|$$

$$\le\frac{N^2}{||\mathbf{w}_1||_1^2}(|b(\mathbf{u}_1-\mathbf{u}_2,\mathbf{w}_1,\mathbf{w}_1-\mathbf{w}_2)|+|b(\mathbf{w}_1-\mathbf{w}_2,\mathbf{u}_1,\mathbf{w}_1-\mathbf{w}_2)|$$

$$+|b(\mathbf{w}_2,\mathbf{u}_1-\mathbf{u}_2,\mathbf{w}_1-\mathbf{w}_2)|)$$

$$+\left(\frac{N^2}{||\mathbf{w}_1||_1^2}-\frac{N^2}{||\mathbf{w}_2||_1^2}\right)(|b(\mathbf{w}_2,\mathbf{u}_2,\mathbf{w}_1-\mathbf{w}_2)|+|b(\mathbf{u}_2,\mathbf{w}_2,\mathbf{w}_1-\mathbf{w}_2)|)$$

$$\leq C\frac{N^2}{||\mathbf{w}_1||_1^2}(|\mathbf{u}_1-\mathbf{u}_2|_6|\nabla\mathbf{w}_1||\mathbf{w}_1-\mathbf{w}_2|_3+|\mathbf{w}_1-\mathbf{w}_2|_3|\nabla\mathbf{u}_1||\mathbf{w}_1-\mathbf{w}_2|$$

$$+|\mathbf{w}_2|_3|\nabla(\mathbf{u}_1-\mathbf{u}_2)|_6|\mathbf{w}_1-\mathbf{w}_2|)$$

$$+\frac{N^4}{||\mathbf{w}_1||_1^2||\mathbf{w}_2||_1^2}||\mathbf{w}_1-\mathbf{w}_2||_1\left(||\mathbf{w}_1||_1+||\mathbf{w}_2||_1\right)(|\mathbf{w}_2|_3|\nabla\mathbf{u}_2|_6|\mathbf{w}_1-\mathbf{w}_2|$$

$$+|\mathbf{u}_2|_6|\nabla\mathbf{w}_2||\mathbf{w}_1-\mathbf{w}_2|_3)$$

$$\leq C\left(|\mathbf{w}_1-\mathbf{w}_2|_3|\mathbf{w}_1-\mathbf{w}_2|+|\nabla(\mathbf{w}_1-\mathbf{w}_2)||\mathbf{w}_1-\mathbf{w}_2|\right)$$

$$+C||\mathbf{w}_1-\mathbf{w}_2||_1(|\mathbf{w}_1-\mathbf{w}_2|+|\mathbf{w}_1-\mathbf{w}_2|_3)$$

$$\leq \epsilon|\nabla(\mathbf{w}_1-\mathbf{w}_2)|^2+C(\epsilon)|\mathbf{w}_1-\mathbf{w}_2|^2. \tag{2.14}$$

We now consider the case of, $||\mathbf{w}_1||_1>N, ||\mathbf{w}_2||_1\leq N$ (the case of $||\mathbf{w}_1||_1\leq N, ||\mathbf{w}_2||_1>N$ would be similar). We get,

$$|(\mathbf{\Lambda}_N(\mathbf{w}_1)-\mathbf{\Lambda}_N(\mathbf{w}_2),\mathbf{w}_1-\mathbf{w}_2)|=|\frac{N^2}{||\mathbf{w}_1||_1^2}(\mathbf{B}(\mathbf{u}_1,\mathbf{w}_1)-\mathbf{B}(\mathbf{w}_1,\mathbf{u}_1),\mathbf{w}_1-\mathbf{w}_2)$$

$$-(\mathbf{B}(\mathbf{u}_2,\mathbf{w}_2)-\mathbf{B}(\mathbf{w}_2,\mathbf{u}_2),\mathbf{w}_1-\mathbf{w}_2)|$$

$$\leq\frac{N^2}{||\mathbf{w}_1||_1^2}(|b(\mathbf{u}_1-\mathbf{u}_2,\mathbf{w}_1,\mathbf{w}_1-\mathbf{w}_2)|+|b(\mathbf{w}_1-\mathbf{w}_2,\mathbf{u}_1,\mathbf{w}_1-\mathbf{w}_2)|$$

$$+|b(\mathbf{w}_2,\mathbf{u}_1-\mathbf{u}_2,\mathbf{w}_1-\mathbf{w}_2)|)$$

$$+\left(\frac{N^2}{||\mathbf{w}_1||_1^2}-1\right)(|b(\mathbf{w}_2,\mathbf{u}_2,\mathbf{w}_1-\mathbf{w}_2)|+|b(\mathbf{u}_2,\mathbf{w}_2,\mathbf{w}_1-\mathbf{w}_2)|).$$

Noting that $||\mathbf{w}_1||_1>N$ and $||\mathbf{w}_2||_1\leq N$ also implies, $||\mathbf{w}_1||_1^2-N^2\leq||\mathbf{w}_1||_1^2-||\mathbf{w}_2||_1^2$, the rest of the calculation is similar to the case leading to (2.14) and thus we establish (2.12).

Let us now proceed with the proof of m-accretivity. Let us consider the operator,

$$\mathbf{F}_N(\mathbf{w})=\nu\mathbf{A}\mathbf{w}+\mathbf{\Lambda}_N(\mathbf{w})+\alpha_N\mathbf{w},\quad\forall\mathbf{w}\in D(\mathbf{F}_N) \tag{2.15}$$

where,

$$D(\mathbf{F}_N):=\{\mathbf{w}\in\mathbf{V};\,\nu\mathbf{A}\mathbf{w}+\mathbf{\Lambda}_N(\mathbf{w})\in\mathbf{H}\}.$$

Since $\mathbf{F}_N$ is monotone, coercive and continuos from $\mathbf{V}$ to $\mathbf{V}^*$ it is maximal monotone in $\mathbf{H}$ (for example see [5], Chapter II, Example 2.3.7) with domain $D(\mathbf{F}_N)\supseteq D(\mathbf{A})$. We shall show that $\mathbf{F}_N=\mathbf{\Gamma}_N+\alpha_N\mathbf{I}$ is for α_N sufficiently large, m-accretive with domain $D(\mathbf{F}_N)=D(\mathbf{A})$. Note first that by (2.7) and (2.8),

$$|\mathbf{\Lambda}_N(\mathbf{w})|\leq C_N(1+|\mathbf{A}\mathbf{w}|^{1/2})$$

and hence from (2.15),

$$|\mathbf{A}\mathbf{w}|\leq C_N(|\mathbf{F}_N(\mathbf{w})|+|\mathbf{w}|),\ \forall\mathbf{w}\in D(\mathbf{F}_N). \tag{2.16}$$

Next we consider the following two operators.

$$\mathbf{F}_N^1 := \nu(1-\epsilon)A,\ D(\mathbf{F}_N^1) = D(\mathbf{A}) \tag{2.17}$$

and

$$\mathbf{F}_N^2 = \epsilon\nu\mathbf{A} + \mathbf{\Lambda}_N + \alpha_N\mathbf{I},\quad D(\mathbf{F}_N^2) = \{\mathbf{w}\in\mathbf{V}; \epsilon\nu\mathbf{A}\mathbf{w} + \mathbf{\Lambda}_N(\mathbf{w})\in\mathbf{H}\}, \tag{2.18}$$

with α_N is large enough so that $\mathbf{F}_N^2$ is maximal monotone in $\mathbf{H}\times\mathbf{H}$. We have $(\forall\delta > 0)$,

$$\begin{aligned}|\mathbf{F}_N^2(\mathbf{w})| &\le \epsilon\nu|\mathbf{A}\mathbf{w}| + |\mathbf{\Lambda}_N(\mathbf{w})| + \alpha_N|\mathbf{w}|\\ &\le \epsilon\nu|\mathbf{A}\mathbf{w}| + C_N(1+|\mathbf{A}\mathbf{w}|^{1/2}) + \alpha_N|\mathbf{w}| \le \epsilon(1+\delta)|\mathbf{A}\mathbf{w}| + \widetilde{C}_N + \alpha_N|\mathbf{w}|\\ &\le \frac{\epsilon(1+\delta)}{1-\epsilon}|\mathbf{F}_N^1(\mathbf{w})| + \widetilde{C}_N + \alpha_N|\mathbf{w}|,\ \forall\mathbf{w}\in D(\mathbf{F}_N^1) = D(\mathbf{A}).\end{aligned}$$

Thus, taking ϵ such that $\frac{\epsilon(1+\delta)}{1-\epsilon} < 1$, we can use the perturbation theorem on nonlinear accretive operators (Theorem 3.5, Chapter-II of [2]) to conclude that, $\mathbf{F}_N^1 + \mathbf{F}_N^2$ with domain $D(\mathbf{A})$ is m-accretive in $\mathbf{H}\times\mathbf{H}$. Since $\mathbf{F}_N^1 + \mathbf{F}_N^2 = \mathbf{F}_N = \mathbf{\Gamma}_N + \alpha_N\mathbf{I}$, we have proved that $\mathbf{\Gamma}_N + \alpha_N\mathbf{I}$ is m-accretive in $\mathbf{H}\times\mathbf{H}$. □

We will now state the main solvability theorem.

Theorem 2. *Let* $\mathbf{w}_0 \in D(\mathbf{A})$. *Then* $\exists\ T^*(|\mathbf{w}_0|)$ *such that there is a unique solution* $\mathbf{w} \in W^{1,\infty}(0,T^*;\mathbf{H})\cap L^\infty(0,T^*;D(\mathbf{A}))\cap C([0,T^*);\mathbf{V})$ *and*

$$\frac{d^+\mathbf{w}}{dt} + \nu\mathbf{A}\mathbf{w} + \mathbf{\Lambda}(\mathbf{w}) = 0,\ \forall t\in[0,T^*) \tag{2.19}$$

and $\mathbf{w}(0) = \mathbf{w}_0$.

Proof. We have shown earlier that the map

$$\mathbf{w} \to \nu\mathbf{A}\mathbf{w} + \mathbf{\Lambda}_N(\mathbf{w}) + \alpha_N\mathbf{w}$$

is m-accretive in $\mathbf{H}\times\mathbf{H}$ for sufficiently large α_N. Thus by Theorem 1.5, Chapter 4 (section 1) of [3] there is a unique strong solution $\mathbf{w}_N \in W^{1,\infty}(0,T;\mathbf{H})$ (for $T \le +\infty$) to the quantized equation (1.9). Noting that,

$$|\mathbf{A}\mathbf{w}_N| \le \frac{1}{\nu}|\mathbf{\Lambda}_N(\mathbf{w}_N)| + \frac{1}{\nu}|\frac{d\mathbf{w}_N}{dt}|.$$

Hence using (2.7), (2.8) and using the fact that $\mathbf{w}_N \in W^{1,\infty}(0,T;\mathbf{H})$ we deduce that $\mathbf{A}\mathbf{w}_N \in L^\infty(0,T;\mathbf{H})$. This result and the fact that $\frac{d\mathbf{w}_N}{dt} \in L^\infty(0,T;\mathbf{H})$ imply that $\mathbf{w}_N \in C([0,T];\mathbf{V})$.

Moreover, applying Theorem 1.6, Chapter 4 (section 1) of [3] we also deduce that,

$$\frac{d^+\mathbf{w}_N}{dt} + \nu\mathbf{A}\mathbf{w}_N + \mathbf{\Lambda}_N(\mathbf{w}_N) = 0,\ \forall t\in[0,T). \tag{2.20}$$

Let us now take a-priori estimates for the above quantized problem independent of N. Taking inner product by $\mathbf{w}$ with the quantized equation (1.9) we get,

$$\frac{1}{2}\frac{d|\mathbf{w}_N|^2}{dt} + \nu|\nabla \mathbf{w}_N|^2 \leq |b(\mathbf{w}_N, \mathbf{u}_N, \mathbf{w}_N)|$$

$$\leq |\mathbf{w}_N|_3 |\nabla \mathbf{u}_N|_6 |\mathbf{w}_N| \leq C|\mathbf{w}_N|^{3/2} |\nabla \mathbf{w}_N|^{3/2}$$

$$\leq \frac{\nu}{2}|\nabla \mathbf{w}_N|^2 + \frac{1}{2}C_\nu |\mathbf{w}_N|^6.$$

Hence we get,

$$T^* = \frac{1}{C_\nu |\mathbf{w}_0|^4},$$

and $\mathbf{w}_N \in L^\infty(0, T^*; \mathbf{H}) \cap L^2(0, T^*; \mathbf{V})$ uniformly in N. We now take inner product by $\mathbf{A}\mathbf{w}_N$ with (1.9) we get,

$$\frac{1}{2}\frac{d|\nabla \mathbf{w}_N|^2}{dt} + \nu|\mathbf{A}\mathbf{w}_N|^2 \leq |b(\mathbf{w}_N, \mathbf{u}_N, \mathbf{A}\mathbf{w}_N)| + |b(\mathbf{u}_N, \mathbf{w}_N, \mathbf{A}\mathbf{w}_N)|$$

$$\leq |\mathbf{w}_N|_3 |\nabla \mathbf{u}_N|_6 |\mathbf{A}\mathbf{w}_N| + |\nabla \mathbf{w}_N|_3 |\mathbf{u}_N|_6 |\mathbf{A}\mathbf{w}_N|$$

$$\leq C|\mathbf{w}_N|^{1/2} |\nabla \mathbf{w}_N|^{3/2} |\mathbf{A}\mathbf{w}_N| + C|\nabla \mathbf{w}_N|_3 |\mathbf{A}\mathbf{w}_N|$$

$$\leq \frac{\nu}{2}|\mathbf{A}\mathbf{w}_N|^2 + C_\nu(|\mathbf{w}_N||\nabla \mathbf{w}_N| + 1)|\nabla \mathbf{w}_N|^2.$$

Hence noting that the product $|\mathbf{w}_N||\nabla \mathbf{w}_N| \in L^2(0, T^*)$ (uniformly in N), we deduce uniform in N estimates $\mathbf{w}_N \in L^\infty(0, T^*; \mathbf{V}) \cap L^2(0, T^*; D(\mathbf{A}))$. Hence if we take the quantization radius N to be larger than the bound obtained above then the solution $\mathbf{w}_N$ of the quantized equation (1.9) will coincide with the solution $\mathbf{w}$ of the original equation (1.5) in the interval $[0, T^*)$. □

Dedication. Dedicated to the memory of Ralph S. Phillips.

Acknowledgement. Research supported by the office of Naval Research through the probability and statistics program and the Navy laboratory independent research program.

References

[1] G. K. Batchelor, *Fluid Dynamics*, Cambridge University Press, New York, 1988.

[2] V. Barbu, *Nonlinear Semigroups and Differential Equations in Banach Spaces*, Noordhoff, Leyden, 1976.

[3] V. Barbu, *Analysis and Control of Nonlinear Infinite Dimensional Systems*, Academic Press, Boston 1993.

[4] V. Barbu and S. S. Sritharan, Flow Invariance Preserving Feedback Controllers for the Navier-Stokes equation, Submitted for publication, 1999.

[5] H. Brezis, *Operateurs Maximaux et Semi-groupes de Contractions das les Espaces de Hilbert*, North-Holland Publishing Co., New York, 1973.

[6] S. S. Sritharan, Ed. *Optimal Control of Viscous Flow*, SIAM, Philadelphia, 1998.

[7] E. M. Stein, *Singular Integrals and differentiability properties of functions*, Princeton University Press, 1970.

S.S. Sritharan, Surveillance Radar Technology (Code D73H), Joint & National Systems Division, Space & Naval Warfare Systems Center (SPAWAR SSD), San Diego, CA 92152-5001, U.S.A.

E-mail address: srith@spawar.navy.mil

V. Barbu, Department of Mathematics, University of Iasi, Iasi, ROMANIA

E-mail address: barbu@acad.ro

Progress in Nonlinear Differential Equations
and Their Applications, Vol. 42

FELLER SEMIGROUPS GENERATED BY DEGENERATE ELLIPTIC OPERATORS II

KAZUAKI TAIRA
UNIVERSITY OF TSUKUBA

1. Introduction and Results

This paper is devoted to the functional analytic approach to the problem of construction of Feller semigroups with Ventcel' (Wentzell) boundary conditions, generalizing the previous work Taira [11] and Taira-Favini-Romanelli [12].

Let D be a bounded domain of Euclidean space $\mathbf{R}^N$, $N \geq 2$, with smooth boundary ∂D. In this paper we consider a second-order, *degenerate* elliptic differential operator A with real coefficients of the form

$$Au(x) = \sum_{i,j=1}^{N} a^{ij}(x) \frac{\partial^2 u}{\partial x_i \partial x_j}(x) + \sum_{i=1}^{N} b^i(x) \frac{\partial u}{\partial x_i}(x) + c(x)u(x).$$

Here:

(1) $a^{ij} \in C^\infty(\mathbf{R}^N)$, $a^{ij}(x) = a^{ji}(x)$ and

$$\sum_{i,j=1}^{N} a^{ij}(x)\xi_i\xi_j \geq 0, \quad x \in \mathbf{R}^N, \ \xi \in \mathbf{R}^N.$$

(2) $b^i \in C^\infty(\mathbf{R}^N)$.

(3) $c \in C^\infty(\mathbf{R}^N)$ and $c(x) \leq 0$ in D.

The differential operator A is called a diffusion operator which describes analytically a strong Markov process with continuous paths (diffusion process) in the interior D.

In order to state our fundamental hypothesis for A, we introduce a function $b(x')$ on the boundary ∂D by the formula (see [5])

$$b(x') = \sum_{i=1}^{N} \left(b^i(x') - \sum_{j=1}^{N} \frac{\partial a^{ij}}{\partial x_j}(x') \right) n_i, \quad x' \in \partial D,$$

where $\mathbf{n} = (n_1, n_2, \ldots, n_N)$ is the unit interior normal to the boundary ∂D at x'. The function $b(x')$ is called the *Fichera function* for the operator A. We divide the boundary ∂D into the following four disjoint subsets (cf. [5], [7], [8]):

$$\Sigma_3 = \left\{ x' \in \partial D : \sum_{i,j=1}^{N} a^{ij}(x') n_i n_j > 0 \right\}.$$

$$\Sigma_2 = \left\{ x' \in \partial D : \sum_{i,j=1}^{N} a^{ij}(x') n_i n_j = 0, \ b(x') < 0 \right\}.$$

$$\Sigma_1 = \left\{ x' \in \partial D : \sum_{i,j=1}^{N} a^{ij}(x') n_i n_j = 0, \ b(x') > 0 \right\}.$$

$$\Sigma_0 = \left\{ x' \in \partial D : \sum_{i,j=1}^{N} a^{ij}(x') n_i n_j = 0, \ b(x') = 0 \right\}.$$

Then our fundamental hypothesis for the operator A is stated as follows:

(H) $\partial D = \Sigma_0 \cup \Sigma_1 \cup \Sigma_2$ and each set Σ_i, $i = 0, 1, 2$, consists of a *finite* number of connected hypersurfaces.

In other words, we consider the totally characteristic case: $\Sigma_3 = \emptyset$. On the other hand, Cattiaux [2] studied the non-characteristic case ($\partial D = \Sigma_3$) from the point of view of probability theory.

It is worth pointing out here (see [7], [8]) that one may impose a boundary condition only on the set Σ_2, since a Markovian particle reaches the boundary ∂D by means of the drift vector field $X_0(x') = \sum_i (b^i(x') - \sum_j \frac{\partial a^{ij}}{\partial x_j}(x')) \frac{\partial}{\partial x_i}$ and $b(x') = \langle X_0(x'), \mathbf{n} \rangle$. Moreover, in the one-dimensional case ($N = 1$) the four sets Σ_3, Σ_2, Σ_1 and Σ_0 are supposed to correspond to a regular boundary, an exit boundary, an entrance boundary and a natural boundary, respectively (see [4], [6]).

Now let $C(\overline{D})$ be the Banach space of real-valued, continuous functions on the closure $\overline{D} = D \cup \partial D$ equipped with the maximum norm $\|\cdot\|_\infty$. A strongly continuous semigroup $\{T_t\}_{t \geq 0}$ on the space $C(\overline{D})$ is called a *Feller semigroup* if it is non-negative and contractive on $C(\overline{D})$:

$$f \in C(\overline{D}), \ 0 \leq f(x) \leq 1 \quad \text{on } \overline{D} \implies 0 \leq T_t f(x) \leq 1 \quad \text{on } \overline{D}.$$

It is known (see [3], [9]) that if $\{T_t\}_{t \geq 0}$ is a Feller semigroup on $C(\overline{D})$, then there exists a unique Markov transition function $p_t(x, \cdot)$ on $\overline{D}$ such that

$$T_t f(x) = \int_{\overline{D}} p_t(x, dy) f(y), \quad f \in C(\overline{D}).$$

Furthermore it can be shown that the function $p_t(x, \cdot)$ is the transition function of some strong *Markov process*; hence the value $p_t(x, E)$ expresses the transition probability that a Markovian particle starting at position x will be found in the set E at time t.

Let L be a second-order boundary condition such that, in local coordinates

$(x_1, x_2, \ldots, x_{N-1})$ of Σ_2,

$$\begin{aligned} Lu(x') &= \left(\sum_{i,j=1}^{N-1} \alpha^{ij}(x') \frac{\partial^2 u}{\partial x_i \partial x_j}(x') + \sum_{i=1}^{N-1} \beta^i(x') \frac{\partial u}{\partial x_i}(x') \right) \\ &\quad + \gamma(x')u(x') + \mu(x') \frac{\partial u}{\partial \mathbf{n}}(x') - \delta(x') Au(x') \\ &\quad + \int_{\Sigma_2} r(x', y') \left[u(y') - \tau(x', y') \left(u(x') + \sum_{j=1}^{N-1} (y_j - x_j) \frac{\partial u}{\partial x_j}(x') \right) \right] dy'. \end{aligned}$$

Here:

(1) The α^{ij} are the components of a smooth symmetric contravariant tensor of type $\binom{2}{0}$ on Σ_2 satisfying the condition

$$\sum_{i,j=1}^{N-1} \alpha^{ij}(x') \xi_i \xi_j \geq 0, \quad x' \in \Sigma_2, \; \xi' = \sum_{j=1}^{N-1} \xi_j dx_j \in T^*_{x'}(\Sigma_2),$$

where $T^*_{x'}(\Sigma_2)$ is the cotangent space of Σ_2 at x'.

(2) $\gamma \in C^\infty(\Sigma_2)$.

(3) $\mu \in C^\infty(\Sigma_2)$ and $\mu(x') \geq 0$ on Σ_2.

(4) $\delta \in C^\infty(\Sigma_2)$ and $\delta(x') \geq 0$ on Σ_2.

(5) $\mathbf{n} = (n_1, n_2, \ldots, n_N)$ is the unit interior normal to the boundary Σ_2.

(6) The integral kernel $r(x', y')$ is the distribution kernel of a pseudo-differential operator $R \in L^{2-\kappa}_{1,0}(\Sigma_2)$ with $\kappa > 0$, and $r(x', y') \geq 0$ off the diagonal $\Delta_{\Sigma_2} = \{(x', x') : x' \in \Sigma_2\}$. The density dy' is a strictly positive density on Σ_2.

(7) The function $\tau(x', y')$ is a smooth function on $\Sigma_2 \times \Sigma_2$ with compact support in a neighborhood of the diagonal Δ_{Σ_2} such that $0 \leq \tau(x', y') \leq 1$ on $\Sigma_2 \times \Sigma_2$ and, at each point x' of Σ_2, $\tau(x', y') = 1$ for y' in a neighborhood of x'.

(8) The functions $\gamma(x')$, $\tau(x', y')$ and the integral kernel $r(x', y')$ satisfy the condition

$$\gamma(x') + \int_{\Sigma_2} r(x', y') \left[1 - \tau(x', y')\right] dy' \leq 0 \quad \text{on } \Sigma_2. \tag{1.1}$$

The boundary condition L is called a second-order *Ventcel' boundary condition* (see [13]). The five terms of L are supposed to correspond to the diffusion along Σ_2, the absorption phenomenon, the reflection phenomenon, the viscosity phenomenon and the jump phenomenon on Σ_2, respectively.

This paper is devoted to the functional analytic approach to the problem of construction of Feller semigroups with Ventcel' boundary conditions. More precisely we consider the following problem:

Problem. *Given analytic data (A, L), can we construct a Feller semigroup $\{T_t\}_{t \geq 0}$ whose infinitesimal generator $\mathfrak{A}$ is characterized by (A, L) ?*

We say that the boundary condition L is *transversal* on the set Σ_2 if it satisfies the condition

$$\mu(x') + \delta(x') > 0 \quad \text{on } \Sigma_2. \tag{1.2}$$

Intuitively, the transversality condition implies that either reflection or viscosity phenomenon occurs at each point of Σ_2.

The next theorem asserts that there exists a Feller semigroup on $\overline{D}$ corresponding to such a diffusion phenomenon that one of the absorption phenomenon, the reflection phenomenon, the viscosity phenomenon, the diffusion along Σ_2 and the jump phenomenon on Σ_2 occurs at each point of Σ_2:

Main Theorem. *We define a linear operator $\mathfrak{A}$ from the space $C(\overline{D})$ into itself as follows:*

(a) The domain of definition $D(\mathfrak{A})$ is the set

$$D(\mathfrak{A}) = \left\{u \in C(\overline{D}) : Au \in C(\overline{D}),\ Lu = 0 \text{ on } \Sigma_2\right\}. \tag{1.3}$$

(b) $\mathfrak{A}u = Au$, $u \in D(\mathfrak{A})$.

Here Au and Lu are taken in the sense of distributions.

Assume that the boundary condition L is transversal *on the boundary Σ_2. Then the operator $\mathfrak{A}$ generates a Feller semigroup $\{T_t\}_{t\geq 0}$ on $\overline{D}$.*

We remark that Main Theorem is an improvement on the previous work [11] and [12] which only treated the case where $r(x', y') \equiv 0$ on $\Sigma_2 \times \Sigma_2$.

A major part of this work was done at the University of Bologna and the University of Bari in June, 1998 under the auspices of G.N.A.F.A. of C.N.R. while I was on leave from the University of Tsukuba. The research of the author is partially supported by Grant-in-Aid for General Scientific Research (No. 10440050), Ministry of Education, Science and Culture of Japan.

2. Preliminaries

In this section we formulate generation theorems for Feller semigroups which form a functional analytic background for the proof of Main Theorem, and further we state an existence and uniqueness theorem for a class of pseudo-differential operators which enters naturally in the construction of Feller semigroups in Section 3.

2.1. Generation theorems for Feller semigroups

Let K be a compact metric space and let $C(K)$ be the Banach space of real-valued, continuous functions u on K with the maximum norm $\|u\|_\infty = \max_{x\in K} |u(x)|$. A strongly continuous semigroup $\{T_t\}_{t\geq 0}$ on $C(K)$ is called a *Feller semigroup* on K if it is non-negative and contractive on $C(K)$.

The next theorem is a version of the Hille-Yosida theorem adapted to the present context (see [9, Theorem 9.3.1 and Corollary 9.3.2]):

Theorem 2.1. *(i) Let $\{T_t\}_{t\geq 0}$ be a Feller semigroup on K and A its infinitesimal generator. Then we have the following:*

(a) The domain $D(A)$ is dense in the space $C(K)$.

(b) For each $\alpha > 0$, the equation $(\alpha I - A)u = f$ has a unique solution $u \in D(A)$ for any $f \in C(K)$. Hence, for each $\alpha > 0$, the Green operator $(\alpha I - A)^{-1} : C(K) \to C(K)$ can be defined by the formula $u = (\alpha I - A)^{-1} f$, $f \in C(K)$.

(c) For each $\alpha > 0$, the operator $(\alpha I - A)^{-1}$ is non-negative on the space $C(K)$.

(d) For each $\alpha > 0$, the operator $(\alpha I - A)^{-1}$ is bounded on the space $C(K)$ with norm $\|(\alpha I - A)^{-1}\| \leq 1/\alpha$.

(ii) Conversely, if A is a linear operator from $C(K)$ into itself satisfying condition (a) and if there is a constant $\alpha_0 \geq 0$ such that, for all $\alpha > \alpha_0$, conditions (b) through (d) are satisfied, then A is the infinitesimal generator of some Feller semigroup $\{T_t\}_{t\geq 0}$ on K.

Corollary 2.2. *Let A be the infinitesimal generator of a Feller semigroup on K. Assume that the constant function 1 belongs to the domain $D(A)$ and that, for some constant c, we have $A1 \leq -c$ on K. Then the operator $A' = A + cI$ is the infinitesimal generator of some Feller semigroup on K.*

Finally we give useful criteria in terms of the *maximum principle* (see [9, Theorem 9.3.3 and Corollary 9.3.4]):

Theorem 2.3. *Let K be a* compact *metric space. Then we have the following assertions:*

(i) Let B be a linear operator from the space $C(K)$ into itself, and assume that:

(α) The domain $D(B)$ is dense in the space $C(K)$.

(β) There exists an open and dense subset K_0 of K such that if $u \in D(B)$ takes a positive *maximum at a point x_0 of K_0, then we have $Bu(x_0) \leq 0$.*

Then the operator B is closable in the space $C(K)$.

(ii) Let B be as in part (i), and further assume that:

(β') If $u \in D(B)$ takes a positive *maximum at a point x' of K, then we have $Bu(x') \leq 0$.*

(γ) For some $\alpha_0 \geq 0$, the range $R(\alpha_0 I - B)$ of $\alpha_0 I - B$ is dense in the space $C(K)$.

Then the minimal closed extension $\overline{B}$ of B is the infinitesimal generator of some Feller semigroup on K.

Corollary 2.4. *Let A be the infinitesimal generator of a Feller semigroup on a compact metric space K and C a bounded linear operator on the space $C(K)$ into itself. Assume that either C or $A' = A + C$ satisfies condition (β'). Then the operator A' is the infinitesimal generator of some Feller semigroup on K.*

2.2. Unique Solvability Theorem for Pseudo-Differential Operators

The next result, essentially due to Cancelier [1], which will play an important role in the construction of Feller semigroups in Section 3 (cf. [11, Theorem 2.1]):

Theorem 2.5. *Let M be an n-dimensional, compact smooth manifold without boundary, and let T be a classical pseudo-differential operator of second order on M such that $T = P + S$. Here:*

(a) The operator P is a second-order, degenerate elliptic differential operator *on M with non-positive principal symbol.*

(b) The operator S is a classical pseudo-differential operator of order $2-\kappa$, $\kappa > 0$, on M and its distribution kernel $s(x,y)$ is non-negative *off the diagonal $\Delta_M = \{(x,x) : x \in M\}$ in $M \times M$.*

(c) $T1 = P1 + S1 \leq 0$ on M.

Then, for each integer $k \geq 1$, there exists a constant $\lambda = \lambda(k) > 0$ such that, for any $f \in C^{k+\theta}(M)$, one can find a function $\varphi \in C^{k+\theta}(M)$ satisfying the equation

$$(T - \lambda I)\varphi = f \quad \text{on } M,$$

and

$$\|\varphi\|_{C^{k+\theta}(M)} \leq C(\lambda)\|f\|_{C^{k+\theta}(M)}.$$

Here $C(\lambda) > 0$ is a constant independent of f.

3. Proof of Main Theorem

In this section we prove Main Theorem. First we reduce the problem of construction of Feller semigroups to the problem of *unique solvability* for the boundary value problem

$$\begin{cases} (\alpha - A)u = f & \text{in } D, \\ (\lambda - L)u = \varphi & \text{on } \Sigma_2, \end{cases}$$

and then prove existence theorems for Feller semigroups.

3.1. General Existence Theorem for Feller Semigroups

The purpose of this subsection is to give a general existence theorem for Feller semigroups (Theorem 3.10) in terms of boundary value problems.

First we state a fundamental existence and uniqueness theorem for the Dirichlet problem in the framework of *Hölder spaces* (see [7, Theorem 1.8.2], [10, Theorem 2]):

Theorem 3.1. *Assume that hypothesis (H) is satisfied. Then, for each integer $m \geq 2$ one can find a constant $\alpha = \alpha(m) > 0$ such that, for any $f \in C^{2m+2+2\theta}(\overline{D})$ and any $\varphi \in C^{2m+4+2\theta}(\Sigma_2)$, there exists a unique solution $u \in C^{m+\theta}(\overline{D})$ of the Dirichlet problem*

$$(D) \qquad \begin{cases} (\alpha - A)u = f & \text{in } D, \\ u = \varphi & \text{on } \Sigma_2. \end{cases}$$

Now, if we take $m = 2$ in Theorem 3.1 and $\alpha \geq \alpha(2)$, then we can introduce the Green operator

$$G^0_\alpha : C^{6+2\theta}(\overline{D}) \longrightarrow C^{2+\theta}(\overline{D})$$

and the harmonic operator

$$H_\alpha : C^{8+2\theta}(\Sigma_2 \cup \Sigma_3) \longrightarrow C^{2+\theta}(\overline{D})$$

in the following way:

(a) For any $f \in C^{6+2\theta}(\overline{D})$, the function $G^0_\alpha f \in C^{2+\theta}(\overline{D})$ is the unique solution of the problem

$$(3.1) \qquad \begin{cases} (\alpha - A)G^0_\alpha f = f & \text{in } D, \\ G^0_\alpha f = 0 & \text{on } \Sigma_2. \end{cases}$$

(b) For any $\varphi \in C^{8+2\theta}(\Sigma_2)$, the function $H_\alpha \varphi \in C^{2+\theta}(\overline{D})$ is the unique solution of the problem

$$(3.2) \qquad \begin{cases} (\alpha - A)H_\alpha \varphi = 0 & \text{in } D, \\ H_\alpha \varphi = \varphi & \text{on } \Sigma_2. \end{cases}$$

Then we have the following:

Theorem 3.2. *(i) (a) The operator G^0_α can be uniquely extended to a non-negative, bounded linear operator on the space $C(\overline{D})$ into itself, denoted again G^0_α, with norm $\|G^0_\alpha\| = \|G^0_\alpha 1\|_\infty \leq 1/\alpha$.*

(b) For any $f \in C(\overline{D})$, we have $G^0_\alpha f = 0$ on Σ_2.

(c) For all α, β, the resolvent equation *holds:*

$$(3.3) \qquad G^0_\alpha f - G^0_\beta f + (\alpha - \beta)G^0_\alpha G^0_\beta f = 0, \quad f \in C(\overline{D}).$$

(d) For any $f \in C(\overline{D})$, we have $\lim_{\alpha \to +\infty} \alpha G^0_\alpha f = f$ in D. Furthermore, if $f = 0$ on Σ_2, then it follows that $\lim_{\alpha \to +\infty} \alpha G^0_\alpha f = f$ on $\overline{D}$.

(ii) (a′) The operator H_α can be uniquely extended to a non-negative, bounded linear operator on the space $C(\Sigma_2)$ into $C(\overline{D})$, denoted again H_α, with norm $\|H_\alpha\| \leq 1$.

(b′) For any $\varphi \in C(\Sigma_2)$, we have $H_\alpha \varphi = \varphi$ on Σ_2.

(c′) For all α, β, we have

$$(3.4) \qquad H_\alpha \varphi - H_\beta \varphi + (\alpha - \beta)G^0_\alpha H_\beta \varphi = 0, \quad \varphi \in C(\Sigma_2).$$

(d′) For any $\varphi \in C(\Sigma_2)$, we have $\lim_{\alpha \to +\infty} H_\alpha \varphi = 0$ in D.

Now we consider the following boundary value problem (∗) in the framework of the spaces of *continuous functions*.

$$(*) \qquad \begin{cases} (\alpha - A)u = f & \text{in } D, \\ Lu = 0 & \text{on } \Sigma_2. \end{cases}$$

To do so, we introduce three operators associated with problem (∗).

(I) First we introduce a linear operator $\mathcal{A} : C(\overline{D}) \to C(\overline{D})$ as follows.

(a) The domain $D(\mathcal{A})$ is the space $C^{2+\theta}(\overline{D})$.

(b) $\mathcal{A}u = Au$, $u \in D(\mathcal{A})$.

Then we have the following:

Lemma 3.3. *The operator $\mathcal{A}$ has its minimal closed extension $\overline{\mathcal{A}}$ in the space $C(\overline{D})$.*

The extended operators $G_\alpha^0 : C(\overline{D}) \to C(\overline{D})$ and $H_\alpha : C(\Sigma_2) \to C(\overline{D})$ still satisfy formulas (3.1) and (3.2) respectively in the following sense:

Lemma 3.4. *(i) For any $f \in C(\overline{D})$, we have*

$$\begin{cases} G_\alpha^0 f \in D(\overline{\mathcal{A}}), \\ (\alpha I - \overline{\mathcal{A}})G_\alpha^0 f = f \quad \textit{in } D. \end{cases}$$

(ii) For any $\varphi \in C(\Sigma_2)$, we have

$$\begin{cases} H_\alpha \varphi \in D(\overline{\mathcal{A}}), \\ (\alpha I - \overline{\mathcal{A}})H_\alpha \varphi = 0 \quad \textit{in } D. \end{cases}$$

Here $D(\overline{\mathcal{A}})$ is the domain of the closed extension $\overline{\mathcal{A}}$.

Corollary 3.5. *Every u in $D(\overline{\mathcal{A}})$ can be written in the form*

$$u = G_\alpha^0\left((\alpha I - \overline{\mathcal{A}})u\right) + H_\alpha(u|_{\Sigma_2}).$$

(II) Secondly we introduce a linear operator $LG_\alpha^0 : C(\overline{D}) \to C(\Sigma_2)$ as follows.

(a) The domain $D\left(LG_\alpha^0\right)$ is the space $C^{6+2\theta}(\overline{D})$.

(b) $LG_\alpha^0 f = L\left(G_\alpha^0 f\right)$, $f \in D\left(LG_\alpha^0\right)$.

Then we have the following:

Lemma 3.6. *The operator LG_α^0 can be uniquely extended to a non-negative, bounded linear operator $\overline{LG_\alpha^0} : C(\overline{D}) \to C(\Sigma_2)$.*

The next lemma states a fundamental relationship between two operators $\overline{LG_\alpha^0}$ and $\overline{LG_\beta^0}$:

Lemma 3.7. *For any $f \in C(\overline{D})$, we have*

$$\overline{LG_\alpha^0} f - \overline{LG_\beta^0} f + (\alpha - \beta)\overline{LG_\alpha^0}\, G_\beta^0 f = 0. \tag{3.5}$$

(III) Finally we introduce a linear operator $LH_\alpha : C(\Sigma_2) \to C(\Sigma_2)$ as follows.

(a) The domain $D(LH_\alpha)$ is the space $C^{8+2\theta}(\Sigma_2)$.

(b) $LH_\alpha \psi = L(H_\alpha \psi)$, $\psi \in D(LH_\alpha)$.

Then we have the following:

Lemma 3.8. *The operator LH_α has its minimal closed extension $\overline{LH_\alpha}$ in the space $C(\Sigma_2)$.*

The next lemma states a fundamental relationship between two operators $\overline{LH_\alpha}$ and $\overline{LH_\beta}$:

Lemma 3.9. *The domains $D\left(\overline{LH_\alpha}\right)$ do not depend on α; so we denote by $\mathcal{D}$ the common domain. Then we have*

$$\overline{LH_\alpha}\psi - \overline{LH_\beta}\psi + (\alpha - \beta)\overline{LG_\alpha^0}\, H_\beta \psi = 0, \quad \psi \in \mathcal{D}.$$

Now we can give a general existence theorem for Feller semigroups on Σ_2 in terms of boundary value problem $(*)$. The next theorem tells us that the operator $\overline{LH_\alpha}$ is the infinitesimal generator of some Feller semigroup on Σ_2 if and only if problem $(*)$ is solvable for sufficiently *many* functions φ in the space $C(\Sigma_2)$:

Theorem 3.10. *(i) If the operator $\overline{LH_\alpha}$ is the infinitesimal generator of a Feller semigroup on Σ_2, then, for each constant $\lambda > 0$ the boundary value problem*

$$(*') \qquad \begin{cases} (\alpha - A)u = 0 & \text{in } D, \\ (\lambda - L)u = \varphi & \text{on } \Sigma_2 \end{cases}$$

has a solution $u \in C^{2+\theta}(\overline{D})$ for any φ in some dense *subset of $C(\Sigma_2)$.*

(ii) Conversely, if, for some constant $\lambda \geq 0$, problem $(*')$ has a solution $u \in C^{2+\theta}(\overline{D})$ for any φ in some dense subset of $C(\Sigma_2)$, then the operator $\overline{LH_\alpha}$ is the infinitesimal generator of some Feller semigroup on Σ_2.

3.2. Proof of Main Theorem

The next theorem proves Main Theorem:

Theorem 3.11. *We define a linear operator $\mathfrak{A} : C(\overline{D}) \to C(\overline{D})$ as follows (see formula (1.3)).*

(a) The domain $D(\mathfrak{A})$ is the set

$$D(\mathfrak{A}) = \left\{ u \in D(\overline{A}) : u|_{\Sigma_2} \in \mathcal{D},\ Lu = 0 \text{ on } \Sigma_2 \right\}, \tag{3.6}$$

where $\mathcal{D}$ is the common domain of the operators $\overline{LH_\alpha}$.

(b) $\mathfrak{A}u = \overline{A}u$, $u \in D(\mathfrak{A})$.

If the boundary condition L is transversal *on the set Σ_2, then the operator $\mathfrak{A}$ is the infinitesimal generator of some Feller semigroup on $\overline{D}$, and the Green operator $G_\alpha = (\alpha I - \mathfrak{A})^{-1}$ is given by the formula*

$$G_\alpha f = G_\alpha^0 f - H_\alpha \left(\overline{LH_\alpha}^{-1} \left(\overline{LG_\alpha^0 f} \right) \right), \quad f \in C(\overline{D}). \tag{3.7}$$

Proof. We apply part (ii) of Theorem 2.1 to the operator $\mathfrak{A}$ defined by formula (3.6). The proof is divided into several steps.

(1) If we let

$$\begin{aligned} L_0 u(x') &= \sum_{i,j=1}^{N-1} \alpha^{ij}(x') \frac{\partial^2 u}{\partial x_i \partial x_j}(x') + \sum_{i=1}^{N-1} \beta^i(x') \frac{\partial u}{\partial x_i}(x') \\ &\quad + \gamma(x') u(x') + \mu(x') \frac{\partial u}{\partial \mathbf{n}}(x') \\ &\quad + \int_{\Sigma_2} r(x', y') \left[u(y') - \tau(x', y') \left(u(x') + \sum_{j=1}^{N-1} (y_j - x_j) \frac{\partial u}{\partial x_j}(x') \right) \right] dy', \end{aligned}$$

then it follows that

$$\begin{aligned} L_0 H_\alpha \varphi(x') &= \sum_{i,j=1}^{N-1} \alpha^{ij}(x') \frac{\partial^2 \varphi}{\partial x_i \partial x_j}(x') + \sum_{i=1}^{N-1} \beta^i(x') \frac{\partial \varphi}{\partial x_i}(x') \\ &\quad + \gamma(x') \varphi(x') + \mu(x') \frac{\partial}{\partial \mathbf{n}}(H_\alpha \varphi)(x') \\ &\quad + \int_{\Sigma_2} r(x', y') \left[\varphi(y') - \tau(x', y') \left(\varphi(x') + \sum_{j=1}^{N-1} (y_j - x_j) \frac{\partial \varphi}{\partial x_j}(x') \right) \right] dy'. \end{aligned}$$

However we have the following results:

(a) The operator $\dfrac{\partial}{\partial \mathbf{n}}(H_\alpha \varphi)$ is a second-order, degenerate elliptic *differential operator* on Σ_2.

Indeed, if x'_0 is a point of the boundary Σ_2, then we choose a local coordinate system $(y_1, y_2, \dots, y_N)$ in a neighborhood of x'_0 such that

$$x'_0 = 0, \quad D = \{y_N > 0\}, \quad \Sigma_2 = \{y_N = 0\},$$

and assume that, in terms of this coordinate system, the operator A is written in the form

$$A = a^{NN}(y) \frac{\partial^2}{\partial y_N^2} + b^N(y) \frac{\partial}{\partial y_N} + \sum_{i,j=1}^{N-1} a^{ij}(y) \frac{\partial^2}{\partial y_i \partial y_j} + \sum_{i=1}^{N-1} b^i(y) \frac{\partial}{\partial y_i} + c(y).$$

We remark that:

1. $a^{NN}(0) = 0$ and $b^N(0) = b(0) < 0$, since $x'_0 = 0 \in \Sigma_2$.
2. $\sum_{i,j} a^{ij}(y)\xi_i\xi_j \geq 0$, $y \in \mathbf{R}^N$, $\xi \in \mathbf{R}^N$.

Since we have

$$\begin{cases} (A-\alpha)H_\alpha\psi = 0 & \text{in } D, \\ H_\alpha\psi = \psi & \text{on } \Sigma_2, \end{cases}$$

and since the function $H_\alpha\psi$ takes its positive maximum $\psi(0)$ at $x'_0 = 0$, it follows that

$$\begin{aligned} 0 &= a^{NN}(0)\frac{\partial^2}{\partial y_N^2}(H_\alpha\psi)(0) + b^N(0)\frac{\partial}{\partial y_N}(H_\alpha\psi)(0) \\ &\quad + \sum_{i,j=1}^{N-1} a^{ij}(0)\frac{\partial^2\psi}{\partial y_i \partial y_j}(0) + \sum_{i=1}^{N-1} b^i(0)\frac{\partial\psi}{\partial y_i}(0) + (c(0)-\alpha)\,\psi(0) \\ &= b^N(0)\frac{\partial}{\partial y_N}(H_\alpha\psi)(0) + \sum_{i,j=1}^{N-1} a^{ij}(0)\frac{\partial^2\psi}{\partial y_i \partial y_j}(0) + (c(0)-\alpha)\,\psi(0). \end{aligned}$$

This proves that

(3.8)

$$\frac{\partial}{\partial \mathbf{n}}(H_\alpha\psi)(x'_0) = -\frac{1}{b(x'_0)}\left(\sum_{i,j=1}^{N-1} a^{ij}(x'_0)\frac{\partial^2\psi}{\partial y_i\partial y_j}(x'_0) + (c(x'_0)-\alpha)\,\psi(x'_0)\right).$$

(b) The operator $\int_{\Sigma_2} r(x',y')[\varphi(y') - \tau(x',y')(\varphi(x') + \sum(y_j - x_j)\frac{\partial\varphi}{\partial x_j}(x'))]dy'$ is a classical, pseudo-differential operator of order $2-\kappa$ on Σ_2.

(c) Moreover we have by formula (3.8)

$$\frac{\partial}{\partial \mathbf{n}}(H_\alpha 1)(x') = \frac{c(x')-\alpha}{-b(x')} < 0 \quad \text{on } \Sigma_2, \tag{3.9}$$

so that by condition (1.1)

$$\begin{aligned} L_0 H_\alpha 1(x') &= \gamma(x') + \mu(x')\frac{\partial}{\partial \mathbf{n}}(H_\alpha 1)(x') \\ &\quad + \int_{\Sigma_2} r(x',y')\,[1-\tau(x',y')]\,dy' \\ &\leq 0 \quad \text{on } \Sigma_2. \end{aligned}$$

Thus, applying Theorem 2.5 to the operator L_0H_α we find that if $\lambda > 0$ is sufficiently large, then the range $R(L_0H_\alpha - \lambda I)$ contains the space $C^{2+\theta}(\Sigma_2)$. This implies that the range $R(L_0H_\alpha - \lambda I)$ is *dense* in $C(\Sigma_2)$. Therefore, applying part (ii) of Theorem 3.10 to the operator L_0 we obtain that the operator $\overline{L_0H_\alpha}$ is the infinitesimal generator of some Feller semigroup on Σ_2, for any $\alpha \geq \alpha(2)$.

(2) Next, applying Corollary 2.4 with $A := \overline{L_0 H_\alpha}$, $C := -\alpha\delta I$ we obtain that the operator $\overline{LH_\alpha} = \overline{L_0 H_\alpha} - \alpha\delta I$ is the infinitesimal generator of a Feller semigroup on Σ_2, for any $\alpha \geq \alpha(2)$.

(3) Now we prove that the equation $\overline{LH_\alpha}\,\psi = \varphi$ has a unique solution ψ in $D(\overline{LH_\alpha})$ for any $\varphi \in C(\Sigma_2)$; hence the inverse $\overline{LH_\alpha}^{-1}$ can be defined on the whole space $C(\Sigma_2)$. Furthermore the operator $-\overline{LH_\alpha}^{-1}$ is non-negative and bounded on $C(\Sigma_2)$.

Indeed, we have, by formula (3.9) and conditions (1.1) and (1.2),

$$
\begin{aligned}
LH_\alpha 1(x') &= \gamma(x') + \mu(x')\frac{\partial}{\partial \mathbf{n}}(H_\alpha 1)(x') - \alpha\delta(x') \\
&\quad + \int_{\Sigma_2} r(x',y')\,[1-\tau(x',y')]\,dy' \\
&< 0 \quad \text{on } \Sigma_2,
\end{aligned}
\tag{3.10}
$$

and so

$$\ell_\alpha = -\sup_{x' \in \Sigma_2} LH_\alpha 1(x') > 0.$$

Moreover, using Corollary 2.2 with $K := \partial D$, $A := \overline{LH_\alpha}$ and $c := \ell_\alpha$ we obtain that the operator $\overline{LH_\alpha} + \ell_\alpha I$ is the infinitesimal generator of some Feller semigroup on Σ_2. Therefore, since $\ell_\alpha > 0$, it follows from an application of part (i) of Theorem 2.1 with $A := \overline{LH_\alpha} + \ell_\alpha I$ that the equation $-\overline{LH_\alpha}\,\psi = \left(\ell_\alpha I - (\overline{LH_\alpha} + \ell_\alpha I)\right)\psi = \varphi$ has a unique solution $\psi \in D(\overline{LH_\alpha})$ for any $\varphi \in C(\Sigma_2)$, and further that the operator $-\overline{LH_\alpha}^{-1} = \left(\ell_\alpha I - (\overline{LH_\alpha} + \ell_\alpha I)\right)^{-1}$ is non-negative and bounded on the space $C(\Sigma_2)$ with norm

$$\left\| -\overline{LH_\alpha}^{-1} \right\| = \left\| \left(\ell_\alpha I - (\overline{LH_\alpha} + \ell_\alpha I)\right)^{-1} \right\| \leq \frac{1}{\ell_\alpha}.$$

(4) By step (3), we can define the right-hand side of formula (3.7) for all $\alpha \geq \alpha(2)$. We prove that

$$G_\alpha = (\alpha I - \mathfrak{A})^{-1}. \tag{3.11}$$

In view of Lemmas 3.6 and 3.9, it follows that we have, for all $f \in C(\overline{D})$,

$$
\begin{cases}
G_\alpha f = G_\alpha^0 f - H_\alpha\left(\overline{LH_\alpha}^{-1}\left(\overline{LG_\alpha^0} f\right)\right) \in D(\overline{A}), \\
G_\alpha f|_{\Sigma_2} = -\overline{LH_\alpha}^{-1}\left(\overline{LG_\alpha^0} f\right) \in D\left(\overline{LH_\alpha}\right) = \mathcal{D}, \\
LG_\alpha f = \overline{LG_\alpha^0} f - \overline{LH_\alpha}\left(\overline{LH_\alpha}^{-1}\left(\overline{LG_\alpha^0} f\right)\right) = 0,
\end{cases}
$$

and that

$$(\alpha I - \overline{A}) G_\alpha f = f.$$

This proves that

$$G_\alpha f \in D(\mathfrak{A}) \text{ and } (\alpha I - \mathfrak{A}) G_\alpha f = f,$$

that is,

$$(\alpha I - \mathfrak{A})G_\alpha = I \quad \text{on } C(\overline{D}).$$

Therefore, to prove formula (3.11) it suffices to show the injectivity of the operator $\alpha I - \mathfrak{A}$.

Assume that

$$u \in D(\mathfrak{A}) \text{ and } (\alpha I - \mathfrak{A})u = 0.$$

Then, by Corollary 3.5 it follows that the function u can be written as

$$u = H_\alpha\left(u|_{\Sigma_2}\right), \quad u|_{\Sigma_2} \in \mathcal{D} = D\left(\overline{LH_\alpha}\right).$$

Thus we have

$$\overline{LH_\alpha}\left(u|_{\Sigma_2}\right) = Lu = 0 \quad \text{on } \Sigma_2.$$

By step (3), this implies that $u|_{\Sigma_2} = 0$, so that

$$u = H_\alpha\left(u|_{\Sigma_2}\right) = 0 \quad \text{in } D.$$

(5) The non-negativity of G_α follows immediately from formula (3.7), since the operators G_α^0, H_α, $-\overline{LH_\alpha}^{-1}$ and $\overline{LG_\alpha^0}$ are all non-negative.

(6) We prove that the operator G_α is bounded on the space $C(\overline{D})$ with norm

$$\|G_\alpha\| \leq \frac{1}{\alpha}, \quad \alpha \geq \alpha(2).$$

To do so, it suffices to show that

$$G_\alpha 1 \leq \frac{1}{\alpha} \quad \text{on } \overline{D}.$$

since G_α is non-negative on $C(\overline{D})$.

First it follows from the uniqueness property of solutions of problem (D) that

$$\alpha G_\alpha^0 1 + H_\alpha 1 = 1 + G_\alpha^0 c \quad \text{on } \overline{D}. \tag{3.12}$$

Indeed, the both sides have the same boundary value 1 and satisfy the same equation: $(\alpha - A)u = \alpha$ in D.

Applying the operator L to the both hand sides of equality (3.12), we obtain that

$$\begin{aligned} -LH_\alpha 1(x') &= -L1(x') - LG_\alpha^0 c(x') + \alpha LG_\alpha^0 1(x') \\ &= -\gamma(x') - \int_{\Sigma_2} r(x', y')\left[1 - \tau(x', y')\right] dy \\ &\quad - \mu(x')\frac{\partial}{\partial \mathbf{n}}(G_\alpha^0 c)(x') + \alpha LG_\alpha^0 1(x') \\ &\geq \alpha LG_\alpha^0 1(x') \quad \text{on } \Sigma_2, \end{aligned}$$

since $G_\alpha^0 c = 0$ on Σ_2 and $G_\alpha^0 c \leq 0$ on $\overline{D}$. Hence we have, by the non-negativity of $-\overline{LH_\alpha}^{-1}$,

$$(3.13) \qquad -\overline{LH_\alpha}^{-1}\left(LG^0_\alpha 1\right) \leq \frac{1}{\alpha} \quad \text{on } \Sigma_2.$$

Using formula (3.7) with $f := 1$, inequality (3.13) and equality (3.12), we obtain that

$$\begin{aligned} G_\alpha 1 &= G^0_\alpha 1 + H_\alpha(-\overline{LH_\alpha}^{-1}(LG^0_\alpha 1)) \\ &\leq G^0_\alpha 1 + \frac{1}{\alpha} H_\alpha 1 = \frac{1}{\alpha} + \frac{1}{\alpha} G^0_\alpha c \leq \frac{1}{\alpha} \text{ on } \overline{D}, \end{aligned}$$

since the operators H_α and G^0_α are non-negative and since $c(x) \leq 0$ in D.

(7) Finally we prove that:

(3.14) The domain $D(\mathfrak{A})$ is *dense* in the space $C(\overline{D})$.

(7-1) Before the proof, we need some lemmas on the behavior of the operators G^0_α, H_α and $-\overline{LH_\alpha}^{-1}$ as $\alpha \to +\infty$:

Lemma 3.12. *For all $f \in C(\overline{D})$, we have*

$$(3.15) \qquad \lim_{\alpha \to +\infty} \left[\alpha G^0_\alpha f + H_\alpha\left(f|_{\Sigma_2}\right)\right] = f \quad \textit{in } C(\overline{D}).$$

Proof. Choose a constant $\beta > 0$ and let

$$g = f - H_\beta(f|_{\Sigma_2}).$$

Then, using formula (3.4) with $\varphi := f|_{\Sigma_2}$ we obtain that

$$(3.16) \qquad \alpha G^0_\alpha g - g = \left[\alpha G^0_\alpha f + H_\alpha(f|_{\Sigma_2}) - f\right] - \beta G^0_\alpha H_\beta(f|_{\Sigma_2}).$$

However we have, by Theorem 3.2,

$$\lim_{\alpha \to +\infty} G^0_\alpha H_\beta(f|_{\Sigma_2}) = 0 \quad \text{in } C(\overline{D}),$$

and also

$$\lim_{\alpha \to +\infty} \alpha G^0_\alpha g = g \quad \text{in } C(\overline{D}),$$

since $g = 0$ on Σ_2. Therefore formula (3.15) follows by letting $\alpha \to +\infty$ in formula (3.16). □

Lemma 3.13. *If the boundary condition L is* transversal *on the set Σ_2, then we have*

$$\lim_{\alpha \to +\infty} \| -\overline{LH_\alpha}^{-1} \| = 0.$$

Proof. By formulas (3.9) and (3.10), it follows that

$$LH_\alpha 1(x') \leq -\alpha\left(\frac{\mu(x')}{-b(x')} + \delta(x')\right).$$

Hence we have, by transversality condition (1.2),

$$\left\| -\overline{LH_\alpha}^{-1} \right\| = \left\| -\overline{LH_\alpha}^{-1} 1 \right\|_\infty \leq \left\| \frac{1}{LH_\alpha 1} \right\|_\infty \longrightarrow 0 \quad \text{as } \alpha \to +\infty. \quad \square$$

(7-2) *Proof of assertion (3.14)*

In view of formula (3.11), it suffices to prove that

$$\lim_{\alpha \to +\infty} \|\alpha G_\alpha f - f\|_\infty = 0, \quad f \in C^{2+\theta}(\overline{D}), \tag{3.17}$$

since the space $C^{2+\theta}(\overline{D})$ is dense in $C(\overline{D})$.

First we remark that

$$\begin{aligned} \|\alpha G_\alpha f - f\|_\infty &= \|\alpha G_\alpha^0 f - \alpha H_\alpha(\overline{LH_\alpha}^{-1}(LG_\alpha^0 f)) - f\|_\infty \\ &\leq \|\alpha G_\alpha^0 f + H_\alpha(f|_{\Sigma_2}) - f\|_\infty \\ &\quad + \|\alpha H_\alpha(\overline{LH_\alpha}^{-1}(LG_\alpha^0 f)) + H_\alpha(f|_{\Sigma_2})\|_\infty \\ &\leq \|\alpha G_\alpha^0 f + H_\alpha(f|_{\Sigma_2}) - f\|_\infty + \|\alpha \overline{LH_\alpha}^{-1}(LG_\alpha^0 f) + f|_{\Sigma_2}\|_\infty. \end{aligned}$$

Thus, in view of formula (3.15) it suffices to show that

$$\lim_{\alpha \to +\infty} \left[\alpha \overline{LH_\alpha}^{-1} \left(LG_\alpha^0 f \right) + f|_{\Sigma_2} \right] = 0 \quad \text{in } C(\Sigma_2). \tag{3.18}$$

Take a constant β such that $\beta < \alpha$, and write

$$f = G_\beta^0 g + H_\beta \varphi,$$

where

$$g = (\beta - A) f \in C^\theta(\overline{D}), \quad \varphi = f|_{\Sigma_2} \in C^{2+\theta}(\Sigma_2).$$

Then, using equations (3.3) (with $f := g$) and (3.4) we obtain that

$$G_\alpha^0 f = G_\alpha^0 G_\beta^0 g + G_\alpha^0 H_\beta \varphi = \frac{1}{\alpha - \beta} \left(G_\beta^0 g - G_\alpha^0 g + H_\beta \varphi - H_\alpha \varphi \right).$$

Hence we have

$$\begin{aligned} &\left\| \alpha \overline{LH_\alpha}^{-1} \left(LG_\alpha^0 f \right) + f|_{\Sigma_2} \right\|_\infty \\ &= \left\| \frac{\alpha}{\alpha - \beta} \left(-\overline{LH_\alpha}^{-1} \right) \left(LG_\beta^0 g - LG_\alpha^0 g + LH_\beta \varphi \right) + \frac{\alpha}{\alpha - \beta} \varphi - \varphi \right\|_\infty \\ &\leq \frac{\alpha}{\alpha - \beta} \left\| -\overline{LH_\alpha}^{-1} \right\| \cdot \left\| LG_\beta^0 g + LH_\beta \varphi \right\|_\infty \\ &\quad + \frac{\alpha}{\alpha - \beta} \left\| -\overline{LH_\alpha}^{-1} \right\| \cdot \left\| LG_\alpha^0 \right\| \cdot \|g\|_\infty + \frac{\beta}{\alpha - \beta} \|\varphi\|_\infty. \end{aligned}$$

It follows from an application of Lemma 3.13 that the first term on the last inequality converges to 0 as $\alpha \to +\infty$. For the second term, using formula (3.5) with $f := 1$ and the non-negativity of G_β^0 and LG_α^0 we find that

$$\|LG_\alpha^0\| = \|LG_\alpha^0 1\|_\infty = \|LG_\beta^0 1 - (\alpha - \beta) LG_\alpha^0 G_\beta^0 1\|_\infty \leq \|LG_\beta^0 1\|_\infty.$$

Hence the second term also converges to 0 as $\alpha \to +\infty$. It is clear that the third term converges to 0 as $\alpha \to +\infty$. This completes the proof of assertion (3.18) and hence that of assertion (3.17).

The proof of Theorem 3.11 and hence that of Main Theorem is now complete. □

References

[1] C. Cancelier. *Problèmes aux limites pseudo-différentiels donnant lieu au principe du maximum*, Comm. Partial Differential Equations, **11** (1986), 1677–1726.

[2] P. Cattiaux. *Stochastic calculus and degenerate boundary value problems*, Ann. Inst. Fourier (Grenoble), **42** (1992), 541–624.

[3] E.B. Dynkin. *Markov processes I, II.* Springer-Verlag, Berlin Göttingen Heidelberg, 1965.

[4] W. Feller. *The parabolic differential equation and the associated semigroups of transformations*, Ann. of Math. **55** (1952), 468–519.

[5] G. Fichera. *Sulla equazioni differenziali lineari ellittico-paraboliche del secondo ordine*, Atti Accad. Naz. Lincei Mem. **5** (1956), 1–30.

[6] P. Mandl. *Analytical treatment of one-dimensional Markov processes.* Springer-Verlag, Berlin Heidelberg New York, 1968.

[7] O.A. Oleĭnik, E.V. Radkevič, *Second order equations with nonnegative characteristic form* (in Russian). Itogi Nauki, Moscow, 1971. English translation, Amer. Math. Soc., Providence, Rhode Island and Plenum Press, New York, 1973.

[8] D.W. Stroock, S.R.S. Varadhan. *On degenerate elliptic-parabolic operators of second order and their associated diffusions*, Comm. Pure Appl. Math. **25** (1972), 651–713.

[9] K. Taira. *Diffusion processes and partial differential equations.* Academic Press, San Diego New York London Tokyo, 1988.

[10] K. Taira. *On the existence of Feller semigroups with Dirichlet condition*, Tsukuba J. Math. **17** (1993), 377–427.

[11] K. Taira. *On the existence of Feller semigroups with boundary conditions II*, J. Functional Analysis **129** (1995), 108–131.

[12] K. Taira, A. Favini and S. Romanelli. *Feller semigroups generated by degenerate elliptic operators*, Semigroup Forum (1999), to appear.

[13] A.D. Wentzell (Ventcel'). *On boundary conditions for multidimensional diffusion processes* (in Russian), Teoriya Veroyat. i ee Primen. **4** (1959), 172–185. English translation, Theory Prob. and its Appl. **4** (1959), 164–177.

KAZUAKI TAIRA, INSTITUTE OF MATHEMATICS, UNIVERSITY OF TSUKUBA, TSUKUBA 305-8571, JAPAN

E-mail address: `taira@math.tsukuba.ac.jp`

Progress in Nonlinear Differential Equations
and Their Applications, Vol. 42

INCOMPRESSIBLE FLUID FLOWS ON ROUGH DOMAINS

MICHAEL E. TAYLOR
UNIVERSITY OF NORTH CAROLINA

Introduction

There is a substantial literature on the existence of solutions to the Euler and Navier-Stokes equations for incompressible flows in bounded domains. Most papers concentrate on the case of domains with fairly smooth boundary, typically at least $C^{1,1}$ regular. However, it is natural to consider fluid flows in rougher domains. Our discussion here has two parts. In Part I we consider the Navier-Stokes equations on Lipschitz domains in $\mathbb{R}^2$ and $\mathbb{R}^3$. In Part II we consider the Euler equations in a more restricted class of nonsmooth domains in the plane.

P. Deuring and W. von Wahl [7] have established the local existence of strong solutions to the Navier-Stokes equations on a bounded Lipschitz domain (with connected boundary) $\Omega \subset \mathbb{R}^3$. They made use of estimates on the Stokes operator A established in [9] and in [23], as well as a further analysis of A, showing that

$$\mathcal{D}(A^{3/4+\varepsilon}) \subset H^{3/2-\delta}(\Omega), \quad \forall\, \varepsilon, \delta > 0. \tag{0.1}$$

They then obtained local solutions to the Navier-Stokes equations with initial data in $\mathcal{D}(A^{1/4+\varepsilon})$, making use also of techniques developed in [11] to treat smooth domains.

R. Brown and Z. Shen [5] have obtained some refined estimates on the Stokes operator. They show that, if $\Omega \subset \mathbb{R}^3$ is a Lipschitz domain (with connected boundary), then

$$\mathcal{D}(A) \subset H^{1,p}(\Omega), \quad \text{for some } \ p = p(\Omega) > 3. \tag{0.2}$$

They also show that

$$\mathcal{D}(A) \subset H^{3/2}(\Omega). \tag{0.3}$$

The spaces in (0.1)–(0.3) are Sobolev spaces. The analysis in [5] makes use of the results of [9] and [24], and also further estimates of "area integral" type.

Here we show that the estimate (0.2) of [5] leads to a very short proof of local solvability of the Navier-Stokes equations on such Lipschitz domains in $\mathbb{R}^3$. We obtain local solutions with initial data in $\mathcal{D}(A^{s/2})$, for s in some interval $(s_0, 2)$. While this argument requires more regularity of the initial data than [7], we think its simplicity makes it worth pointing out.

In §1 we present a general set-up for obtaining a local solution to the Navier-Stokes equations, converting the system to an integral equation in a standard fashion and casting the problem as a fixed-point problem. This method is familiar; the presentation here is similar to that used in several contexts in Chapters 15 and 17 of [25]. The approach works provided a few conditions on $\mathcal{D}(A)$ can be verified. Such conditions are basically obvious if $\Omega = \mathbb{R}^3$. If Ω is a bounded domain in $\mathbb{R}^3$ with smooth boundary, the conditions follow from the well known inclusion $\mathcal{D}(A) \subset H^2(\Omega)$. In §2, we show that the results of [9] imply that the set-up of §1 works if Ω is a bounded planar Lipschitz domain (with connected boundary). While the argument here is superseded by the even shorter argument of §3, we include it to show that the two-dimensional version of this result is a simple consequence of the result of [9] alone, not requiring further hard analysis. Then, in §3, we show that the results of [5] mentioned above imply that the set-up of §1 works if Ω is a bounded Lipschitz domain (with connected boundary) in $\mathbb{R}^3$. In §4 we mention further estimates and some conjectures about the behavior of the Stokes operator that could potentially be valuable in the study of the Navier-Stokes equations.

We make a comment on the hypothesis that $\partial\Omega$ be connected. This is an artifact of the way in which the method of layer potentials is used in [9], as in many other papers on PDE on Lipschitz domains. Work done in [19]–[22] and in [18] on various PDE with variable coefficients in Lipschitz domains gives a way to eliminate this unnecessary hypothesis. While we have not specifically applied this to the Stokes operator yet, we believe this approach will also work in this case.

In Part II we present some results on existence of solutions to the Euler equations on nonsmooth planar domains, in particular convex planar domains, though the analysis also works for planar domains satisfying a uniform exterior disk condition. This is a context in which initial data with bounded vorticity can be shown to have bounded velocity. In §5 we describe our procedure for constructing a solution as a limit of solutions in domains with smooth boundary, $\Omega_j \nearrow \Omega$, and we establish some uniform estimates in the interior. In §6 we show how a formula of Kadlec gives global estimates on second order derivatives of the velocity, sufficient to construct a limiting solution. In §7 we consider initial data with vorticity in L^p, for some range of p (namely for $p > 12/7$), extending some of the results of [8] to this situation.

Part I: Navier-Stokes Equations on Lipschitz Domains

1. General set-up

In this section we let Ω be a bounded region in $\mathbb{R}^n$, and consider the Navier-Stokes equations on $\mathbb{R}^+ \times \Omega$, with external force f:

$$\frac{\partial u}{\partial t} + \nabla_u u = \nu \Delta u - \operatorname{grad} p + f, \quad \operatorname{div} u = 0. \tag{1.1}$$

We impose the "no slip" boundary condition

$$u = 0 \text{ for } x \in \partial\Omega. \tag{1.2}$$

We also set an initial condition

$$u(0) = u_0. \tag{1.3}$$

We consider the following spaces of vector fields on Ω. First, set

$$\mathcal{V} = \{u \in C_0^\infty(\Omega, T\Omega) : \text{div } u = 0\}. \tag{1.4}$$

Then set

$$W^k = \text{ closure of } \mathcal{V} \text{ in } H^k(\Omega, T), \quad k = 0, 1. \tag{1.5}$$

We can rewrite (1.1) as

$$\frac{\partial u}{\partial t} + P\,\nabla_u u = -\nu A u + Pf. \tag{1.6}$$

Here, P is the orthogonal projection of $L^2(\Omega, T)$ onto W^0. The operator A is an unbounded positive self adjoint operator on W^0, defined via the Friedrichs extension method, as follows. We have

$$A_0 : W^1 \to (W^1)^*, \tag{1.7}$$

given by

$$\langle A_0 u, v\rangle = (u, v)_{W^1} = (du, dv)_{L^2}, \tag{1.8}$$

the last identity holding because div $u =$ div $v = 0$. Then set

$$\mathcal{D}(A) = \{u \in W^1 : A_0 u \in W^0\}, \quad A = A_0\big|_{\mathcal{D}(A)}, \tag{1.9}$$

using $W^1 \subset W^0 \subset (W^1)^*$. Automatically, $\mathcal{D}(A^{1/2}) = W^1$. The operator A is called the *Stokes operator.*

We will define

$$W^s = \mathcal{D}(A^{s/2}), \quad s \geq 0. \tag{1.10}$$

Note that this is consistent with (1.5), for $s = k = 0$ or 1.

We try to construct a local solution to the initial value problem for the Navier-Stokes equation, by converting (1.6) into an integral equation:

$$u(t) = e^{-t\nu A}\, u_0 - \int_0^t e^{(s-t)\nu A}\, P\big[\text{div}\big(u(s) \otimes u(s)\big) + f(s)\big]\, ds = \Psi u(t). \tag{1.11}$$

We want to find a fixed point of Ψ on $C(I, X)$, for $I = [0, T_0]$, with some $T_0 > 0$, and X an appropriate Banach space. We take X to be of the form

$$X = W^s = \mathcal{D}(A^{s/2}), \tag{1.12}$$

for a value of s to be specified below. For the standard contraction mapping argument to work, we need a Banach space Y such that

$$\Phi : X \to Y \text{ is Lipschitz, uniformly on bounded sets,} \tag{1.13}$$

where

$$\Phi(u) = P \operatorname{div}(u \otimes u), \tag{1.14}$$

and such that, for some $\gamma < 1$,

$$\|e^{-tA}\|_{\mathcal{L}(Y,X)} \leq Ct^{-\gamma}, \tag{1.15}$$

for $t \in (0, 1]$. Also assume $Pf(t)$ is continuous in t with values in Y. Under such conditions, the verification that Ψ is a contraction on $C(I, X)$ for T_0 sufficiently small is straightforward.

We take

$$Y = W^0. \tag{1.16}$$

As $\|e^{-tA}\|_{\mathcal{L}(W^0,W^s)} \sim Ct^{-s/2}$ for $t \leq 1$, the condition (1.15) requires $s \in (0, 2)$, in (1.12). We need to verify (1.13). Indeed, (1.13) will hold provided

$$M : W^s \to H^1(\Omega, T \otimes T), \text{ with } M(u) = u \otimes u. \tag{1.17}$$

Since $W^1 \subset H_0^1(\Omega)$, we see that (1.17) holds provided $s \in [1, 2)$ and

$$W^s \subset L^\infty(\Omega). \tag{1.18}$$

Thus we have the following existence result:

Proposition 1.1. *Suppose that $s \in [1,2)$ and (1.18) holds. Let $u_0 \in W^s$. Assume $f(t)$ is a continuous function of t with values in $L^2(\Omega, T)$. Then there exists $T_0 > 0$ such that (1.11) has a unique solution*

$$u \in C([0, T_0], W^s). \tag{1.19}$$

Note that, by interpolation, (1.18) holds, for s sufficiently close to 2, whenever

$$\mathcal{D}(A) \subset C^r(\overline{\Omega}), \quad r > 0. \tag{1.20}$$

2. Solutions in planar Lipschitz domains

Here we will show that, if Ω is a bounded Lipschitz domain (with connected boundary) in $\mathbb{R}^2$, then the condition (1.20) is satisfied, so Proposition 1.1 applies to yield a weak solution to the Navier-Stokes equations (1.1)–(1.3). That (1.20) holds will be seen to follow from work of [9].

If $u \in \mathcal{D}(A)$, then $u \in H_0^1(\Omega)$, div $u = 0$, and

$$\Delta u = g + \nabla p, \quad g \in L^2(\Omega, T), \ p \in L^2(\Omega). \tag{2.1}$$

Suppose that $\overline{\Omega} \subset\subset \mathbb{R}^n$, and let $\mathcal{O}$ be a bounded open neighborhood of $\overline{\Omega}$, with smooth boundary. Extend g to $\mathcal{O}$, e.g., set $g = 0$ on $\mathcal{O} \setminus \Omega$. Then solve

$$\Delta v = g + \nabla p_1, \ \text{ div } v = 0 \ \text{ on } \ \mathcal{O}, \quad v\big|_{\partial\mathcal{O}} = 0. \tag{2.2}$$

Since $\partial\mathcal{O}$ is smooth, standard results imply $v \in H^2(\mathcal{O})$, and $p_1 \in H^1(\mathcal{O})$. Now write

$$u = v + w \tag{2.3}$$

on Ω, where

$$\Delta w = \nabla(p - p_1), \ \text{ div } w = 0 \ \text{ on } \ \Omega, \quad w = -v \ \text{ on } \ \partial\Omega. \tag{2.4}$$

Note that, assuming $\partial\Omega$ is Lipschitz, since

$$\nabla v \in H^1(\Omega) \Longrightarrow \nabla v\big|_{\partial\Omega} \in H^{1/2}(\partial\Omega) \subset L^2(\partial\Omega),$$

$$h = -v\big|_{\partial\Omega} \in H^1(\partial\Omega). \tag{2.5}$$

It follows from Theorem 4.15 of [9] that

$$\|\mathcal{M}(w)\|_{L^2(\partial\Omega)} + \|\mathcal{M}(\nabla w)\|_{L^2(\partial\Omega)} \le C\|h\|_{H^1(\partial\Omega)}, \tag{2.6}$$

where $\mathcal{M}(w)$ is a nontangential maximal function. It is easy to see that, if Ω is a bounded Lipschitz domain in $\mathbb{R}^2$, then

$$\|w\|_{C^{1/2}(\overline{\Omega})} \le C\|\mathcal{M}(\nabla w)\|_{L^2(\partial\Omega)} + C\|w\|_{L^2(\Omega)}. \tag{2.7}$$

We deduce that, for such planar domains,

$$\mathcal{D}(A) \subset C^{1/2}(\overline{\Omega}). \tag{2.8}$$

Thus we have the following:

Proposition 2.1. *Let Ω be a bounded Lipschitz domain in $\mathbb{R}^2$ with connected boundary. Let $f \in C(\mathbb{R}^+, L^2(\Omega, T))$ be given. Assume $s \in (4/3, 2)$, $u_0 \in$*

$\mathcal{D}(A^{s/2})$. Then there exists $T_0 > 0$ such that the Navier-Stokes equation (1.6) has a unique solution satisfying

$$u \in C([0,T_0], \mathcal{D}(A^{s/2})), \quad u(0) = u_0. \tag{2.9}$$

Proof. Just note that

$$[L^2(\Omega), C^{1/2}(\overline{\Omega})]_\theta \subset H^{\theta/2-\varepsilon,q(\theta)}(\Omega), \quad q(\theta) = \frac{2}{1-\theta}, \tag{2.10}$$

as can be seen by applying a convenient extension operator to extend functions from Ω to $\mathbb{R}^2$, and applying the holomorphic family of operators $(1-\Delta)^{(1/2-\varepsilon)z}$. Alternatively, more precise results (involving Besov spaces) can be found in [4]. When dim $\Omega = 2$, then as long as $\theta q(\theta)/2 > 2$, i.e., $\theta > 2/3$, the right side of (2.10) is contained in $L^\infty(\Omega)$.

3. Solutions in Lipschitz domains in $\mathbb{R}^3$

We now establish the main result of this part.

Proposition 3.1. *Let Ω be a bounded Lipschitz domain in $\mathbb{R}^3$, with connected boundary. Let $f \in C(\mathbb{R}^+, L^2(\Omega, T))$ be given. There exists $s_0 = s_0(\Omega)$ such that, if $s \in (s_0, 2)$ and $u_0 \in \mathcal{D}(A^{s/2})$, then there exists $T_0 > 0$ such that the Navier-Stokes equation (1.6) has a unique solution satisfying*

$$u \in C([0,T_0], \mathcal{D}(A^{s/2})), \quad u(0) = u_0. \tag{3.1}$$

Proof. As for the case considered in §2, it suffices to verify (1.20). In this case, this follows from the result (0.2) of [5], since

$$p = \frac{3}{1-r}, \ 0 < r < 1 \Longrightarrow H^{1,p}(\Omega) \subset C^r(\overline{\Omega}). \tag{3.2}$$

We mention that M. Dauge [6] showed that, if Ω is a convex, piecewise smooth domain (with edges and corners) in $\mathbb{R}^3$, then $\mathcal{D}(A) \subset H^2(\Omega)$. The planar case of such a result had been done by [14]. Thus the argument of §1 applies to this class of domains, without requiring the results of [5]. It is natural to conjecture that this result of [6] extends to arbitrary bounded convex domains in $\mathbb{R}^3$.

4. Further estimates and conjectures

Let Ω be a bounded Lipschitz domain in $\mathbb{R}^3$. We want to apply the set-up of §1, with

$$Y = W^0_p = \text{ closure of } \mathcal{V} \text{ in } L^p(\Omega, T). \tag{4.1}$$

We believe that e^{-tA} acts as a strongly continuous semigroup on W^0_p, at least for some $p > 2$. Denote the closed generator by A_p. We take X to be the interpolation space

$$X = W^s_p = [W^0_p, \mathcal{D}(A_p)]_{s/2}, \tag{4.2}$$

for some $s \in [1,2)$, which will be discussed below. (The terminology here perhaps conflicts with some common conventions; here W^s_p is *not* a Sobolev space.) Then we should have

$$\|e^{-tA_p}\|_{\mathcal{L}(Y,X)} \le Ct^{-s/2}, \tag{4.3}$$

for $t \in (0,1]$. In fact, we make the following

Conjecture. *For a given Lipschitz domain $\Omega \subset \mathbb{R}^3$ there exists $\varepsilon = \varepsilon(\Omega) > 0$ such that e^{-tA_p} extends to be a holomorphic semigroup, provided $3/2 - \varepsilon < p < 3 + \varepsilon$.*

When $\partial\Omega$ is smooth, such a result is established for all $p \in (1,\infty)$ in [12]. The restriction on p here is related to the following result. According to [10], there exists $\varepsilon = \varepsilon(\Omega) > 0$ such that

$$P : L^p(\Omega, T) \longrightarrow W^0_p, \quad \frac{3}{2} - \varepsilon < p < 3 + \varepsilon, \tag{4.4}$$

at least if $\partial\Omega$ is connected. In [20] there is a more general result, removing this connectivity hypothesis.

Now, we can verify (1.13) provided

$$M : W^s_p \longrightarrow H^{1,p}(\Omega, T \otimes T), \tag{4.5}$$

where, as in (1.17), $M(u) = u \otimes u$.

Thus, we are in a favorable position to solve (1.1), provided there exists $s \in [1,2)$ such that

$$W^s_p \subset H^{1,p}(\Omega) \cap L^\infty(\Omega). \tag{4.6}$$

On the other hand, (4.6) holds whenever

$$\mathcal{D}(A_p) \subset H^{1+\delta,p}(\Omega) \cap C^\delta(\overline{\Omega}), \quad \text{for some } \delta > 0. \tag{4.7}$$

We also conjecture that such a result is valid. In connection with this, we mention the following result of [5] (which also implies (0.2)).

Proposition. *Given a Lipschitz domain $\Omega \subset \mathbb{R}^3$ with connected boundary, there exists $\varepsilon = \varepsilon(\Omega)$ such that the solution u to (2.1) (with $g \in H^{-1,p}(\Omega)$ and $u|_{\partial\Omega} = 0$) satisfies*

$$\|u\|_{H^{1,p}(\Omega)} \le C\|g\|_{H^{-1,p}(\Omega)}, \quad \frac{3}{2} - \varepsilon < p < 3 + \varepsilon. \tag{4.8}$$

Part II: Euler Flows on Rough Planar Domains

5. Euler equations in planar domains

Let Ω be a bounded open set in $\mathbb{R}^2$. The Euler equation for ideal incompressible flow in Ω is

$$\partial_t u + \nabla_u u = -\nabla p, \quad \text{div } u = 0. \tag{5.1}$$

Here u is the fluid velocity and p the pressure. The boundary condition is that u be tangent to the boundary. There is a good existence theory when $\partial\Omega$ is smooth, which can be found for example in Chapter 17 of [25]. Here we want to allow $\partial\Omega$ to be rough.

Our basic strategy will be to use an approximation procedure. Let Ω_j have smooth boundary and satisfy

$$\Omega_1 \subset\subset \Omega_2 \subset\subset \cdots \subset\subset \Omega_j \nearrow \Omega.$$

For convenience, we assume each domain has connected boundary. Assume $v_j \in C^\infty(\overline{\Omega}_j)$ are vector fields, tangent to $\partial\Omega_j$, satisfying div $v_j = 0$ and

$$\|v_j\|_{C^1(\overline{\Omega}_j)} \le K < \infty, \quad v_j \to v \in C^1(\Omega). \tag{5.2}$$

We want to construct a weak solution to the Euler equation for incompressible flow on Ω, with initial data v, as a limit of (a subsequence of) solutions $u_j \in C^\infty(\mathbb{R} \times \overline{\Omega}_j)$ to

$$\partial_t u_j + \nabla_{u_j} u_j = -\nabla p_j, \quad \text{div } u_j = 0, \quad u_j(0) = v_j. \tag{5.3}$$

We require u_j to be vector fields on Ω_j, tangent to $\partial\Omega_j$. The functions p_j give the fluid pressure.

We will obtain certain uniform bounds on the u_j. First, there is the well-known conservation law

$$\|u_j(t)\|_{L^2} = \|v_j\|_{L^2} \le C. \tag{5.4}$$

Next, we have a vorticity estimate. Recall that, if $\tilde{u}_j$ denotes the 1-form on Ω_j corresponding to u_j via the Riemannian metric, then the vorticity is given by $w_j = *d\tilde{u}_j$. As a consequence of the 2D vorticity equation,

$$\frac{\partial w_j}{\partial t} + \nabla_{u_j} w_j = 0,$$

we have

$$\|w_j(t)\|_{L^\infty} = \|dv_j\|_{L^\infty} \le K. \tag{5.5}$$

Now div $u_j = 0$, $*d\tilde{u}_j = w_j$ is an elliptic system. In fact, we can write

$$\tilde{u}_j(t) = *df_j(t), \tag{5.6}$$

where $f_j \in C^\infty(\mathbb{R} \times \overline{\Omega}_j)$ satisfies, for each t,

$$\Delta f_j = w_j, \quad f_j\big|_{\partial\Omega_j} = 0. \tag{5.7}$$

Thus, if $\mathcal{O} \subset\subset \Omega_N$, we have

$$\|u_j(t)\|_{H^{1,\infty}(\mathcal{O})} \le C(\mathcal{O}), \quad \forall\, j \ge N, \tag{5.8}$$

where we set

$$H^{k,\infty}(\mathcal{O}) = \{u : D^\alpha u \in \mathrm{BMO}(\mathcal{O}), |\alpha| \le k\}. \tag{5.9}$$

Hence

$$\|\nabla_{u_j(t)} u_j(t)\|_{\mathrm{BMO}(\mathcal{O})} \le C(\mathcal{O}), \quad \forall\, j \ge N. \tag{5.10}$$

In the analysis of solutions to (5.3), it is convenient to rewrite the system as

$$\partial_t u_j + P_j(\nabla_{u_j} u_j) = 0, \quad u_j(0) = v_j, \tag{5.11}$$

where P_j is the Helmholtz projection onto the space of divergence-free vector fields on Ω_j tangent to the boundary. Also, it is useful to have the identity

$$\nabla_{u_j} u_j = \operatorname{div}(u_j \otimes u_j), \quad \text{if} \quad \operatorname{div} u_j = 0. \tag{5.12}$$

The pressure p_j is eliminated from (5.11), but it can be recovered as the solution (well defined modulo a constant) to the boundary problem

$$\Delta p_j = -\mathrm{Tr}\big((\nabla u_j)^2\big) \quad \text{on} \quad \Omega_j, \qquad \frac{\partial p_j}{\partial \nu} = -\kappa_j\, |u_j|^2 \quad \text{on} \quad \partial\Omega_j, \tag{5.13}$$

where $\kappa_j(x)$ is the curvature of $\partial\Omega_j$ at x.

6. Flows in convex planar domains

We now assume that Ω and all the Ω_j are *convex* domains in $\mathbb{R}^2$. We can then use Kadlec's formula,

$$\sum_{k,\ell} \int_{\Omega_j} |\partial_k \partial_\ell u|^2\, dx = \int_{\Omega_j} |\Delta u|^2\, dx + (n-1) \int_{\partial\Omega_j} |\partial_\nu u|^2\, H(x)\, dS(x),$$

(in n dimensions, where $H(x)$ is the mean curvature of $\partial\Omega_j$, cf. [13], p. 139), to obtain a uniform bound $\|f_j\|_{H^2} \le C\|w_j\|_{L^2}$, for solutions to (5.7), and hence, by (5.5),

$$\|f_j(t)\|_{H^2(\Omega_j)} \le C_2 < \infty. \tag{6.1}$$

This in turn implies

$$\|u_j(t)\|_{H^1(\Omega_j)} \le C. \tag{6.2}$$

Furthermore, by Proposition 3.5 in Chapter 3 of [26], given this convexity we have a uniform bound

$$\|f_j\|_{\mathrm{Lip}} \le C\|w_j\|_{L^p}, \quad p > \dim \Omega_j,$$

for solutions to (5.7), and hence the L^∞ bounds (5.5) give

$$\|f_j(t)\|_{\mathrm{Lip}(\overline{\Omega}_j)} \le C_3 < \infty, \tag{6.3}$$

which implies

$$\|u_j(t)\|_{L^\infty(\Omega_j)} \le C. \tag{6.4}$$

We have

$$\|u_j(t) \otimes u_j(t)\|_{H^1(\Omega_j)} \le C' \|u_j(t)\|_{L^\infty} \|u_j(t)\|_{H^1} \le C; \tag{6.5}$$

hence

$$\big\|\operatorname{div}\big(u_j(t) \otimes u_j(t)\big)\big\|_{L^2(\Omega_j)} \le C, \tag{6.6}$$

so

$$\big\|P_j \operatorname{div}\big(u_j(t) \otimes u_j(t)\big)\big\|_{L^2(\Omega_j)} \le C. \tag{6.7}$$

Thus we have

$$\|\partial_t u_j(t)\|_{L^2(\Omega_j)} \le C. \tag{6.8}$$

Interpolation with the bound (6.2) yields

$$E_j u_j \ \text{ compact in } \ C\big([-T,T], H^s(\Omega)\big), \quad \forall\, T < \infty,\ s < 1, \tag{6.9}$$

where E_j is a sequence of extension operators:

$$E_j : H^1(\Omega_j) \to H^1(\Omega), \quad E_j : L^2(\Omega_j) \to L^2(\Omega), \tag{6.10}$$

with uniform operator norm bounds. Also, interpolation of (6.8) and (5.8) yields, for any $\mathcal{O} \subset\subset \Omega_N$,

$$\{u_j\big|_{\mathcal{O}} : j \ge N\} \ \text{ compact in } \ C\big([-T,T], H^{s,p}(\mathcal{O})\big), \tag{6.11}$$

for any $T < \infty,\ s < 1,\ p < \infty$. Hence

$$\{u_j\big|_{\mathcal{O}} : j \ge N\} \ \text{ compact in } \ C\big([-T,T] \times \overline{\mathcal{O}}\big). \tag{6.12}$$

Now, perhaps passing to a subsequence, we obtain

$$E_j u_j \to u \in L^\infty\big(\mathbb{R}, H^1(\Omega)\big) \cap L^\infty(\mathbb{R} \times \Omega), \tag{6.13}$$

with convergence in the weak* topology. Furthermore, (6.9) implies

$$E_j u_j \to u \ \text{ in norm, in } \ C\big([-T,T], H^s(\Omega)\big), \tag{6.14}$$

for any given $T < \infty,\ s < 1$. Hence

$$E_j u_j \otimes E_j u_j \to u \otimes u \ \text{ weak* in } \ L^\infty\big([-T,T], L^2(\Omega)\big), \tag{6.15}$$

for any $T < \infty$. Boundedness results above then imply

$$E_j u_j \otimes E_j u_j \to u \otimes u \ \text{ weak* in } \ L^\infty\big(\mathbb{R}, H^1(\Omega)\big) \cap L^\infty(\mathbb{R} \times \Omega), \tag{6.16}$$

and hence

$$\operatorname{div}(E_j u_j \otimes E_j u_j) \to \operatorname{div}(u \otimes u) \ \text{ weak* in } \ L^\infty\big(\mathbb{R}, L^2(\Omega)\big). \tag{6.17}$$

To see that u solves the Euler equation, we rewrite (5.3) as

$$\partial_t u_j + \operatorname{div}(u_j \otimes u_j) = -\nabla p_j, \quad \operatorname{div} u_j = 0, \tag{6.18}$$

where

$$\nabla p_j = -(I - P_j) \operatorname{div}(u_j \otimes u_j) \tag{6.19}$$

satisfies

$$\|\nabla p_j\|_{L^2(\Omega_j)} \le C, \tag{6.20}$$

as a consequence of (6.6). Normalizing p_j so that $\int_{\Omega_j} p_j(t,x)\,dx = 0$, we have

$$E_j p_j \ \text{ bounded in } \ L^\infty(\mathbb{R}, H^1(\Omega)), \tag{6.21}$$

so (perhaps passing to a subsequence), we have

$$E_j p_j \to p \ \text{ weak}^* \text{ in } \ L^\infty(\mathbb{R}, H^1(\Omega)), \tag{6.22}$$

and then results above yield

$$\partial_t u + \operatorname{div}(u \otimes u) = -\nabla p, \quad \operatorname{div} u = 0. \tag{6.23}$$

We next show that, for each t,

$$u(t) \in V^0(\Omega) = L^2\text{-closure of } \{v \in C_0^\infty(\Omega, T) : \operatorname{div} v = 0\}. \tag{6.24}$$

In fact, for each $j < \infty$, $u_j(t)$ belongs to the analogous space $V^0(\Omega_j)$, by the standard theory for smooth boundaries. Hence, for each t, there exist $v_j(t) \in C_0^\infty(\Omega_j, T) \subset C_0^\infty(\Omega, T)$ such that

$$\operatorname{div} v_j(t) = 0, \quad \|u_j(t) - v_j(t)\|_{L^2} \le 2^{-j}.$$

It follows from (6.14) that $v_j(t) \to u(t)$ in norm in $L^2(\Omega)$ as $j \to \infty$, so (6.24) is established.

Regarding other properties of the limit, we have from (6.12)

$$u \in C(\mathbb{R} \times \Omega) \tag{6.25}$$

and, from (5.8),

$$u \in L^\infty(\mathbb{R}, H^{1,\infty}(\mathcal{O})), \quad \forall\, \mathcal{O} \subset\subset \Omega. \tag{6.26}$$

We derive some further local bounds, starting with bounds on the pressures p_j, which solve (5.13). Given $\mathcal{O} \subset\subset \Omega_N$, the right side of the first equation of (5.13) is bounded in $L^p(\mathcal{O})$ for $j \ge N$, for any $p < \infty$, as a consequence of (5.8). The global bound (6.20) plus local elliptic regularity yields

$$\|\nabla p_j\|_{H^{1,p}(\mathcal{O})} \le C_p(\mathcal{O}), \tag{6.27}$$

for $j \ge N$. Hence, if we use the equation

$$\partial_t u_j + \nabla_{u_j} u_j = -\nabla p_j \tag{6.28}$$

plus the bound (5.10), we get

$$\|\partial_t u_j\|_{\mathrm{BMO}(\mathcal{O})} \leq C(\mathcal{O}). \tag{6.29}$$

In the limit, these yield associated bounds on u.

We summarize what has been established:

Proposition 6.1 *Let Ω be a bounded convex domain in $\mathbb{R}^2$. Let*

$$v \in C^1(\Omega) \cap Lip(\overline{\Omega}) \cap V^0(\Omega)$$

be given. Then there exists a solution u to the Euler equation

$$u_t + \nabla_u u = -\nabla p, \quad div\, u = 0 \ \ in \ \ \mathbb{R} \times \Omega, \quad u(0) = v, \tag{6.30}$$

satisfying

$$\begin{aligned} u &\in L^\infty(\mathbb{R}, H^1(\Omega)) \cap L^\infty(\mathbb{R} \times \Omega) \cap Lip\,(\mathbb{R}, V^0(\Omega)), \\ u_t &\in L^\infty(\mathbb{R}, L^2(\Omega)), \\ \nabla p &\in L^\infty(\mathbb{R}, L^2(\Omega)). \end{aligned} \tag{6.31}$$

Furthermore, given any $\mathcal{O} \subset\subset \Omega$, we have

$$\begin{aligned} u &\in L^\infty(\mathbb{R}, H^{1,\infty}(\mathcal{O})), \\ u_t &\in L^\infty(\mathbb{R}, BMO(\mathcal{O})), \\ \nabla p &\in L^\infty(\mathbb{R}, H^{1,p}(\mathcal{O})), \quad \forall\, p < \infty. \end{aligned} \tag{6.32}$$

7. Flows with L^p initial vorticity

We can extend Proposition 6.1, obtaining some results of a flavor such as obtained for flows in all of $\mathbb{R}^2$ in [8]. That is, we can obtain weak solutions to the Euler equations (6.30), with initial vorticity in $L^p(\Omega)$, for some range of p. Some of these results follow from the analysis done above. Indeed, given

$$\|w_j(0)\|_{L^p} \leq C, \tag{7.1}$$

we have $\|w_j(t)\|_{L^p} \leq C$. As long as $p \geq 2$, the bound (6.1) still holds. If $p = 2$, we do not quite get (6.3)–(6.4), but if $p > 2$ we have $\|f_j\|_{\mathrm{Lip}} \leq C\|w_j\|_{L^p}$, so we do get (6.3)–(6.4) when (7.1) holds and $p > 2$. The rest of the arguments involving (6.5)–(6.24) then go through (except we get weaker bounds on $\mathcal{O} \subset\subset \Omega_N$), and thus Proposition 6.1 extends to the case of initial data satisfying

$$v \in V^0(\Omega) \cap H^{1,p}(\Omega), \tag{7.2}$$

provided $p > 2$.

We can go further, obtaining weak solutions for smaller p. To do this, we need a couple of more tools. One is the following result of V. Adolfsson [Ad], extending the consequence of Kadlec's formula described above. Namely, for our sequence of convex domains Ω_j, and for f_j satisfying (5.7), there is for each $p \in (1,2]$ a uniform estimate

$$\|f_j(t)\|_{H^{2,p}(\Omega_j)} \leq C_p \, \|w_j\|_{L^p(\Omega_j)}. \tag{7.3}$$

This gives, in place of (6.2),

$$\|u_j(t)\|_{H^{1,p}(\Omega_j)} \leq C. \tag{7.4}$$

If $p < 2$ here, we cannot expect the L^∞-estimate (6.4). Sobolev's theorem yields

$$\|u_j(t)\|_{L^q(\Omega_j)} \leq C, \quad q = \frac{2p}{2-p}, \tag{7.5}$$

if $p \in (1,2)$. Hence, in place of (6.6), we have

$$\|\nabla_{u_j} u_j(t)\|_{L^r(\Omega_j)} \leq C, \tag{7.6}$$

with

$$r = \frac{2p}{4-p} \in (1,2], \quad \text{provided} \quad p \in \Big(\frac{4}{3}, 2\Big]. \tag{7.7}$$

The second additional tool we need is an L^r-operator norm estimate on P_j. In [10] it is proven that P_j acts on $L^r(\Omega_j)$, for $3/2 \leq r \leq 3$, with uniform bounds $C = C_r$, as long as $\partial\Omega_j$ satisfy uniform Lipschitz estimates. That situation can be assumed to hold for our family of convex domains, so we have, in place of (6.20),

$$\|P_j \operatorname{div}(u_j(t) \otimes u_j(t))\|_{L^r(\Omega_j)} \leq C, \tag{7.8}$$

provided (7.5)–(7.7) hold with $r \in (3/2, 2]$. Note that

$$p \in \Big(\frac{12}{7}, 2\Big] \Longrightarrow r \in \Big(\frac{3}{2}, 2\Big]. \tag{7.9}$$

As before, we can push through variants of (6.8)–(6.24), and obtain the following.

Proposition 7.1 *Let Ω be a bounded convex domain in $\mathbb{R}^2$. Let*

$$v \in V^0(\Omega) \cap H^{1,p}(\Omega) \tag{7.10}$$

be given. Assume $p \in (12/7, 2)$. Then there exists a solution u to the Euler equation (6.30), satisfying

$$\begin{aligned} u &\in L^\infty(\mathbb{R}, H^{1,p}(\Omega) \cap V^0(\Omega)), \\ u_t &\in L^\infty(\mathbb{R}, L^r(\Omega)), \\ \nabla p &\in L^\infty(\mathbb{R}, L^r(\Omega)), \end{aligned} \tag{7.11}$$

with r given by (7.7).

8. Concluding remarks

We end this part with a remark on planar Lipschitz domains. Namely, the Lipschitz estimate (6.3) and consequent L^∞-estimate (6.4) on the fluid velocity can definitely fail. To see this, start with $f(x) = \log |x|$ on the annulus $\Omega = \{x \in \mathbb{R}^2 : 1/2 \le |x| \le 1\}$. It is harmonic, and $\tilde{u} = *df$ defines a vector field u on Ω which gives a steady, irrotational solution to the Euler equation. Now one can easily construct a conformal mapping $\Phi : \Omega \to \mathcal{O}$ of Ω onto a planar region $\mathcal{O}$ with a corner whose internal angle lies in $(\pi, 2\pi)$, such that $g = f \circ \Phi^{-1}$ is not Lipschitz, and $\tilde{v} = *dg$ is not bounded on $\mathcal{O}$, nor does it belong to $H^1(\mathcal{O})$, though it does still generate a steady irrotational flow. Furthermore, one can arrange that v not be in $H^{1,p}(\Omega)$ for p in the range considered in Proposition 7.1. This illustrates that different methods will be required to prove existence theorems for Euler flows on such Lipschitz domains.

Acknowledgement. Partially supported by NSF grant DMS-9877077.

References

[1] V. Adolfsson, L^p-integrability of the second derivative of Green potentials in convex domains, *Pacific J. Math.* 159 (1993), 201–225.

[2] V. Adolfsson and D. Jerison, L^p-integrability of the second order derivatives for the Neumann problem in convex domains, Preprint, 1994.

[3] H. Bahouri and J. Chemin, Equations de transport relatives a des champs de vecteurs non-lipschitzeans et mecanique des fluides, *Arch. Rat. Mech. Anal.* 127 (1994), 159–181.

[4] J. Bergh and J. Löfström, *Interpolation Spaces, an Introduction*, Springer-Verlag, Berlin, 1976.

[5] R. Brown and Z. Shen, Estimates for the Stokes operator in Lipschitz domains, *Indiana Univ. Math. J.* 44 (1995), 1183–1206.

[6] M. Dauge, Stationary Stokes and Navier-Stokes systems on two- and three-dimensional domains with corners. Part I: linearized equations, *SIAM J. Math. Anal.* 20 (1989), 74–97.

[7] P. Deuring and W. von Wahl, Strong solutions of the Navier-Stokes system in Lipschitz bounded domains, *Math. Nachr.* 171 (1995), 111–148.

[8] R. DiPerna and A. Majda, Concentrations in regularizations for 2-D incompressible flow, *Comm. Pure Appl. Math.* 40 (1987), 301–345.

[9] E. Fabes, C. Kenig, and G. Verchota, The Dirichlet problem for the Stokes system on Lipschitz domains, *Duke Math. J.* 57 (1988), 769–793.

[10] E. Fabes, O. Mendez, and M. Mitrea, Boundary layers on Sobolev spaces and Poisson's equation for the Laplacian in Lipschitz domains, Preprint, 1997.

[11] H. Fujita and T. Kato, On the Navier-Stokes initial value problem, *Arch. Rat. Mech. Anal.* 16 (1964), 269–315.

[12] Y. Giga, Analyticity of the semigroup generated by the Stokes operator in L_r spaces, *Math. Zeit.* 178 (1981), 297–329.

[13] P. Grisvard, *Elliptic Problems in Nonsmooth Domains*, Pitman, Boston, 1985.

[14] R. Kellog and J. Osborn, A regularity for the Stokes problem in a convex polygon, *J. Funct. Anal.* 21 (1976), 397–431.

[15] O. Ladyzhenskaya, *The Mathematical Theory of Viscous Incompressible Flow*, Gordon and Breach, New York, 1969.

[16] C. Marchioro and M. Pulvirenti, *Mathematical Theory of Incompressible Nonviscous Fluids*, Springer-Verlag, New York, 1994.

[17] D. Mitrea, M. Mitrea, and J. Pipher, Vector potential theory on nonsmooth domains in $\mathbb{R}^3$ and applications to electromagnetic scattering, *J. Fourier Anal. Appl.* 3 (1997), 131–192.

[18] D. Mitrea, M. Mitrea, and M. Taylor, Layer potentials, the Hodge Laplacian, and global boundary problems in nonsmooth Riemannian manifolds, *Memoirs AMS*, to appear.

[19] M. Mitrea and M. Taylor, Boundary layer methods for Lipschitz domains in Riemannian manifolds, *J. Funct. Anal.* 163 (1999), 181–251.

[20] M. Mitrea and M. Taylor, Potential theory on Lipschitz domains in Riemannian manifolds: L^p, Hardy, and Hölder space results, Preprint, 1998.

[21] M. Mitrea and M. Taylor, Potential theory on Lipschitz domains in Riemannian manifolds: Sobolev-Besov space results and the Poisson problem, Preprint, 1999.

[22] M. Mitrea and M. Taylor, Potential theory on Lipschitz domains in Riemannian manifolds: Hölder continuous metric tensors, Preprint, 1999.

[23] Z. Shen, Boundary value problems for parabolic Lamé systems and a nonstationary linearized system of Navier-Stokes equations in Lipschitz cylinders, *Amer. J. Math.* 113 (1991), 293–373.

[24] Z. Shen, A note on the Dirichlet problem for the Stokes system in Lipschitz domains, *Proc. AMS* 123 (1995), 801–811.

[25] M. Taylor, *Partial Differential Equations*, Vols. 1–3, Springer-Verlag, New York, 1996.

[26] M. Taylor, *Tools for PDE: Pseudodifferential Operators, Paradifferential Operators, and Layer Potentials*, Preprint, 1999.

MICHAEL E. TAYLOR, DEPARTMENT OF MATHEMATICS, UNIVERSITY OF NORTH CAROLINA, CHAPEL HILL, NORTH CAROLINA 27599, U.S.A.

E-mail address: met@math.unc.edu

Progress in Nonlinear Differential Equations
and Their Applications, Vol. 42

EXACT BOUNDARY CONTROLLABILITY OF THERMO-ELASTIC PLATES WITH VARIABLE COEFFICIENTS

M. ELLER
TENNESSEE TECHNOLOGICAL UNIVERSITY

I. LASIECKA
R. TRIGGIANI
UNIVERSITY OF VIRGINIA, CHARLOTTESVILLE

Abstract. We study a controllability problem (exact in the mechanical variables $\{w, w_t\}$ and, simultaneously, approximate in the thermal variable θ) of thermo-elastic plates by means of boundary controls when the 'thermal expansion' term is variable in space (and, possibly, in time).

1. Thermo-elastic systems. Boundary controllability problem

Let $\Omega \subset \mathbb{R}^2$ be an open bounded domain with smooth boundary Γ. We shall here consider the following thermo-elastic plate [La.1], [L-L.1] on a finite time interval in the unknown $w(t, x)$ [vertical displacement] and $\theta(t, x)$ [relative temperature about the stress-free state $\theta = 0$]:

$$\begin{cases} w_{tt} - \gamma\Delta w_{tt} + \Delta^2 w + \operatorname{div}(\alpha(x)\nabla\theta) = 0 & \text{in } Q \equiv (0, T] \times \Omega; \quad (1.1a) \\ \theta_t - \Delta\theta - \operatorname{div}(\alpha(x)\nabla w_t) = 0 & \text{in } Q; \quad (1.1b) \\ w(0, \cdot\,) = w_0;\ w_t(0, \cdot\,) = w_1;\ \theta(0, \cdot\,) = \theta_0 & \text{in } \Omega, \quad (1.1c) \end{cases}$$

to be augmented by boundary conditions on $\partial\Omega = \Gamma$. Throughout this paper the constant γ is positive, $\gamma > 0$, so that the model accounts for moments of inertia (rotational forces). The coefficient γ is proportional to the square of the thickness of the plate. The thermal coefficient $\alpha(x)$ represents 'thermal expansion' of the plate material and is assumed to be a function of $x \in \Omega$, with regularity to be specified below. We shall consider two cases of *boundary* controls associated with system (1.1): the case where the boundary controls act in the *hinged mechanical/Dirichlet thermal boundary conditions (B.C.):*

$$w|_\Sigma = u_1; \qquad \Delta w|_\Sigma = u_2; \qquad \theta|_\Sigma = u_3 \quad \text{on } \Sigma = (0, T] \times \Gamma, \tag{1.2}$$

and the case where the boundary controls act in the *clamped mechanical/Dirichlet thermal B.C.*:

$$w|_\Sigma = u_1; \qquad \left.\frac{\partial w}{\partial \nu}\right|_\Sigma = u_2; \qquad \theta|_\Sigma = u_3 \quad \text{on } \Sigma. \tag{1.3}$$

The regularity assumption on $\alpha(x)$ is as follows: $\alpha \in C^2(\bar{\Omega})$. Well-posedness of the two mixed problems above: (1.1),(1.2), and (1.1),(1.3) is discussed below.

Boundary controllability problem. Qualitatively, the boundary controllability problem studied in this note is as follows. Let $T > 0$ be sufficiently large, depending on the geometry of Ω. Given any initial condition $\{w_0, w_1, \theta_0\}$ and any preassigned target condition $\{w_{0,T}, w_{1,T}, \theta_T\}$ in specified Sobolev spaces, seek boundary controls $\{u_1, u_2, u_3\}$ in specified function spaces (compatible with the regularity of the underlying dynamics) that steer the solution of the corresponding mixed problem (1.1), (1.2), or (1.3) to a state $\{w(T), w_t(T), \theta(T)\}$ at time T, such that: $w(T) = w_{0,T}, w_t(T) = w_{1,T}$, while $\theta(T)$ is arbitrarily 'close' to θ_T in the relevant topology. Thus, the above is a problem of *exact controllability* in the mechanical variable and, simultaneously, of *approximate controllability* in the thermal variable. A more precise statement is given in the following theorems.

Main results. Our main results on the above boundary exact/approximate controllability problem follow next.

Theorem 1.1. (hinged B.C. (1.2)) Let $\Gamma_1, \Gamma_2, \Gamma_3 \subset \Gamma$ be subsets of the boundary Γ, with non-empty intersection of positive measure. (We think of Γ_1 and Γ_3 as being arbitrarily small.) Moreover, regarding Γ_2, we assume that: there exists a point $x_0 \in \mathbb{R}^2$, such that

$$(x - x_0) \cdot \nu(x) \leq 0 \quad \text{for } x \in \Gamma \setminus \Gamma_2, \tag{1.4}$$

where here and throughout the paper $\nu(x)$ denotes the unit outward normal at $x \in \Gamma$. Let

$$T_0 \equiv 2\sqrt{\gamma} \max_i \sup_{x\in\Omega} \operatorname{dist}(x, \Gamma_i), \qquad i = 1, 2, 3. \tag{1.5}$$

Let $\alpha \in C^2(\bar{\Omega})$. Finally, let $\{w_0, w_1, \theta_0\}$ and $\{w_{0,T}, w_{1,T}, \theta_T\}$ be pre-assigned initial and target states, with:

$$\{w_0, w_1\} \text{ and } \{w_{0,T}, w_{1,T}\} \in [H^2(\Omega) \cap H_0^1(\Omega)] \times H_0^1(\Omega); \;\; \theta_0, \theta_T \in H_0^1(\Omega). \tag{1.6}$$

Then, for any $T > T_0$ and any $\delta > 0$ arbitrarily small, there exist boundary control functions

$$u_1 \equiv \begin{cases} \hat{u}_1 \in C_0^\infty(\Sigma_1) \\ 0 \text{ on } \Sigma - \Sigma_1 \end{cases}; \quad u_2 \equiv \begin{cases} \hat{u}_2 \in L_2(\Sigma_2) \\ 0 \text{ on } \Sigma - \Sigma_2 \end{cases}; \quad u_3 \equiv \begin{cases} \hat{u}_3 \in C_0^\infty(\Sigma_3) \\ 0 \text{ on } \Sigma - \Sigma_3 \end{cases}, \tag{1.7}$$

such that the corresponding solution of the mixed problem (1.1), (1.2) satisfies the terminal condition at T:

$$w(T) = w_{0,T}; \qquad w_t(T) = w_{1,T}; \qquad \|\theta(T) - \theta_T\|_{H_0^1(\Omega)} \leq \delta. \tag{1.8}$$ □

Theorem 1.2. (clamped B.C. (1.3)) Let $\Gamma_1, \Gamma_3 \subset \Gamma$ be, as in Theorem 1.1, subsets of the boundary Γ, with non-empty intersection of positive measure, possibly arbitrarily small. Let $T_0 > 0$ be defined by (1.5), and let $\alpha \in C^2(\bar{\Omega})$. Finally, let $\{w_0, w_1, \theta_0\}$ and $\{w_{0,T}, w_{1,T}, \theta_T\}$ be preassigned initial and target states, with

(1.9) $\{w_0, w_1\}$ and $\{w_{0,T}, w_{1,T}\} \in H_0^1(\Omega) \times L_2(\Omega);\quad \theta_0, \theta_T \in H^{-\epsilon}(\Omega),\quad \epsilon > 0.$

Then, for any $T > T_0$ and any $\delta > 0$ arbitrarily small, there exist boundary control functions

$$(1.10)\quad u_1 \equiv \begin{cases} \hat{u}_1 \in C_0^\infty(\Sigma_1) \\ 0 \text{ on } \Sigma - \Sigma_1 \end{cases};\quad u_2 \in L_2(\Sigma);\quad u_3 \equiv \begin{cases} \hat{u}_3 \in C_0^\infty(\Sigma_3) \\ 0 \text{ on } \Sigma - \Sigma_3 \end{cases},$$

such that the corresponding solution of the mixed problem (1.1), (1.3) satisfies the terminal condition at T:

$$(1.11)\quad w(T) = w_{0,T};\quad w_t(T) = w_{1,T};\quad \|\theta(T) - \theta_T\|_{H^{-\epsilon}(\Omega)} \leq \delta. \qquad \square$$

Remark 1.1. We note that, in the above two theorems, the essential, critical control mechanism is provided by the control u_2 in the highest mechanical B.C., with $u_2 \in L_2(\Sigma_2)$ on a suitable portion of the boundary in the hinged case, and $u_2 \in L_2(\Sigma)$ on the full boundary in the clamped case. The addition of infinitely smooth controllers u_1 and u_3 in the lowest mechanical B.C. and in the thermal B.C., compactly supported on $\Sigma_1 = (0,T] \times \Gamma_1$ and $\Sigma_3 = (0,T] \times \Gamma_3$, with Γ_1, Γ_3 *arbitrarily small* portions of the boundary and non-empty intersection with Γ_1, is only for the purpose of obtaining the property of 'approximate controllability' of the overall thermo-elastic plate. By duality (Hahn-Banach theorem), this latter property is equivalent to the property of *unique continuation* across the boundary of a corresponding over-determined dual or adjoint problem, see Theorem 4.2.1. At present, the results available in the literature [I.1] require that the dual problem be over-determined with *all* four boundary conditions, in order to assert that, then, the corresponding solution is identically zero. See statement of Theorem 4.2.1. This is the reason why we assume three active controls in (1.2), or (1.3), in lieu of just u_2. However, any progress in the area of the unique continuation property for thermo-elastic plates will imply corresponding improvements of our results, by allowing us to drop unnecessary controls. $\square$

Remark 1.2. Theorems 1.1 and 1.2 continue to hold true in case the thermal expansion coefficient α depends (smoothly) not only on space x, but also on time t, provided that, in this case, all control functions u_1, u_2, u_3 are supported (active) over the *entire* lateral boundary $\Sigma \equiv (0,T] \times \Gamma$. This limitation results, as in the case of Remark 1.1, by the lack, at present, of appropriate unique continuation theorems which can handle also time dependence of α, in the presence of over-determined boundary data active only on a portion of the boundary. Indeed, in the case of α being also time dependent and with controls u_1, u_2, u_3 being active (supported) on the *entire* boundary $\Sigma = (0,T] \times \Gamma$, we may invoke [I.1, Theorem 3.5] rather than Theorem 4.2.1. Applicability of Isakov's theorem requires that the over-determined boundary information be given on the entire boundary $\Sigma = (0,T] \times \Gamma$. □

Literature. To our knowledge, the only papers in the literature in the area of *boundary* controllability of *thermo-elastic plates*—to which we must restrict—are [La.2] and [A-L.1]. In [La.2], thermo-elastic plates with *constant* coefficients and *boundary* controls in the free B.C. were considered. Here, the main theorem established is an exact controllability result in the mechanical variables $\{w, w_t\}$ *only*, with no information about the thermal variable θ, subject, moreover, to the condition that the (constant) thermal parameters be sufficiently *small.* Such smallness of the thermal connection plays an essential role in the arguments of [La.2]. The first result on exact/approximate boundary controllability of mechanical/thermal variables for the same thermo-elastic plate with controls in the free B.C., which *does not* require any smallness hypothesis on the model, is [A-L.1]. Here, however, the coefficients, in particular α, are constant. This assumption is also critical to the arguments of [A-L.1], as it allows for the introduction of a certain transformation of variables, to make the problem more tractable. On the other hand, it is known that observability/controllability estimates are sensitive with respect to variable perturbations of the "energy level" terms in the equations. In fact, even for simple *plate* equations, standard energy method (multipliers) used to obtain the right continuous observability estimates are not adequate in the presence of variable coefficients at the energy level (as the one represented by α in (1.1). Thus, it is expected that similar difficulties will recur in the study of thermo-elasticity. Thus, the main contribution of this paper is the presence of a variable thermal coefficient α without *any smallness* requirement.

2. Well-posedness. Dual problems

Well-posedness. Homogeneous problem. Let, at first, $u_1 = u_2 = u_3 = 0$ in both cases of B.C. (1.2) or (1.3).

Hinged B.C. (1.2). We introduce the following operators and spaces (with equivalent norms),

$$\mathcal{A}f = -\Delta f,\ \mathcal{A}_\gamma = (I + \gamma\mathcal{A}),\ \mathcal{D}(\mathcal{A}_\gamma) = \mathcal{D}(\mathcal{A}) = H^2(\Omega) \cap H_0^1(\Omega); \tag{2.1a}$$

$$\mathcal{D}(\mathcal{A}^{\frac{3}{2}}) = \{f \in H^3(\Omega) :\ f|_\Gamma = \Delta f|_\Gamma = 0\}; \tag{2.1b}$$

$$\mathcal{D}(\mathcal{A}_\gamma^{\frac{1}{2}}) = \mathcal{D}(\mathcal{A}^{\frac{1}{2}}) = H_0^1(\Omega),\ (x_1, x_2)_{\mathcal{D}(\mathcal{A}_\gamma^{\frac{1}{2}})} = ((I + \gamma\mathcal{A})x_1, x_2)_H, \quad H = L_2(\Omega) \tag{2.2}$$

$$Y_\gamma \equiv \mathcal{D}(\mathcal{A}) \times \mathcal{D}(\mathcal{A}_\gamma^{\frac{1}{2}}) \times H \equiv [H^2(\Omega) \cap H_0^1(\Omega)] \times H_0^1(\Omega) \times L_2(\Omega). \tag{2.3}$$

Clamped B.C. (1.3). In this case, we introduce

$$Af = \Delta^2 f,\ \mathcal{D}(A) = \{f \in H^4(\Omega):\ f|_\Gamma = \left.\frac{\partial f}{\partial \nu}\right|_\Gamma = 0\}; \tag{2.4a}$$

$$\mathcal{D}(A^{\frac{1}{2}}) = H_0^2(\Omega); \qquad \mathcal{D}(A^{\frac{1}{4}}) = \mathcal{D}(\mathcal{A}^{\frac{1}{2}}) = H_0^1(\Omega); \tag{2.4b}$$

$$Y_\gamma \equiv \mathcal{D}(A^{\frac{1}{2}}) \times \mathcal{D}(\mathcal{A}_\gamma^{\frac{1}{2}}) \times H \equiv H_0^2(\Omega) \times H_0^1(\Omega) \times L_2(\Omega). \tag{2.5}$$

Lemma 2.1. Let $u_1 = u_2 = u_3 = 0$ in (1.2) or (1.3). Then:

(i) Problem (1.1), (1.2), or (1.1), (1.3), defines a s.c. semigroup denoted by $e^{\mathbb{A}_\gamma t}$: $y_0 = \{w_0, w_1, \theta_0\} \to e^{\mathbb{A}_\gamma t} y_0 = \{w(t), w_t(t), \theta(t)\}$ in the space Y_γ given by (2.3), or (2.5), respectively. In the case of *hinged* B.C. (1.2), the generator $\mathbb{A}_\gamma$ is given by

$$\mathbb{A}_\gamma = \begin{bmatrix} 0 & I & 0 \\ -\mathcal{A}_\gamma^{-1}\mathcal{A}^2 & 0 & -\mathcal{A}_\gamma^{-1}\mathrm{div}(\alpha\nabla(\ \cdot\ )) \\ 0 & \mathrm{div}(\alpha\nabla(\ \cdot\ )) & -\mathcal{A} \end{bmatrix}; \tag{2.6a}$$

$$\mathcal{D}(\mathbb{A}_\gamma) = \mathcal{D}(\mathcal{A}^{\frac{3}{2}}) \times \mathcal{D}(\mathcal{A}) \times \mathcal{D}(\mathcal{A}); \tag{2.6b}$$

(ii) The Y_γ-adjoint $\mathbb{A}_\gamma^*$ of $\mathbb{A}_\gamma$ is given by

$$\mathbb{A}_\gamma^* = \begin{bmatrix} 0 & -I & 0 \\ \mathcal{A}_\gamma^{-1}\mathcal{A}^2 & 0 & \mathcal{A}_\gamma^{-1}\mathrm{div}(\alpha\nabla(\ \cdot\ )) \\ 0 & -\mathrm{div}(\alpha\nabla(\ \cdot\ )) & -\mathcal{A} \end{bmatrix}, \tag{2.7}$$

$$\mathcal{D}(\mathbb{A}_\gamma^*) = \mathcal{D}(\mathbb{A}_\gamma).$$

Similar expressions, which are omitted, hold true in the case of *clamped* B.C. (1.3). The dual semigroup dynamics $\bar{y}_0 = \{\phi_0, \phi_1, \chi_0\} \to e^{\mathbb{A}_\gamma^* t}\bar{y}_0 = \{\phi(t), \phi_t(t), \chi(t)\}$ is given by the following thermo-elastic (dual) problem:

$$\begin{cases} \phi_{tt} - \gamma\Delta\phi_{tt} + \Delta^2\phi + \mathrm{div}(\alpha(x)\nabla\chi) \equiv 0 & \text{in } Q; \quad (2.8a) \\ \chi_t - \Delta\chi - \mathrm{div}(\alpha(x)\nabla\phi_t) \equiv 0 & \text{in } Q; \quad (2.8b) \\ \phi(0,\ \cdot\ ) = \phi_0;\ \phi_t(0,\ \cdot\ ) = \phi_1;\ \chi(0,\ \cdot\ ) = \chi_0 & \text{in } \Omega, \quad (2.8c) \end{cases}$$

with the corresponding homogeneous B.C.

$$\text{hinged } \phi = \Delta\phi \equiv 0 \text{ on } \Sigma; \text{ or else clamped } \phi \equiv \frac{\partial\phi}{\partial\nu} \equiv 0 \text{ on } \Sigma; \tag{2.8d}$$

$$\chi = 0 \text{ on } \Sigma.$$

Proof. Part (i) is an application of the Lumer-Phillips theorem.

(ii) To compute the Y_γ-adjoint, we use the $\mathcal{D}(\mathcal{A}_\gamma^{\frac{1}{2}})$-inner product in (2.2), as well as the identity

$$(2.9)\quad (\text{div}(\alpha\nabla\theta_3), y_2)_{L_2(\Omega)} = -\int_\Omega \alpha\nabla y_2\cdot\nabla\theta_3 d\Omega = (\theta_3, \text{div}(\alpha\nabla y_2))_{L_2(\Omega)}, \qquad \theta_3, y_2 \in \mathcal{D}(\mathcal{A}),$$

so that $\theta_3|_\Gamma = 0$, $y_2|_\Gamma = 0$. Eqn. (2.9) is obtained by using the standard (divergence theorem) identity

$$(2.10)\qquad \int_\Omega \psi \text{div} h\, d\Omega = \int_\Gamma \psi h\cdot\nu\, d\Gamma - \int_\Omega \nabla\psi\cdot h\, d\Omega,$$

first with $\psi = y_2$, $h = \alpha\nabla\theta_3$ to get the middle term in (2.9), and next with $\psi = \theta_3$ and $h = \alpha\nabla y_2$ to get the final right-hand side term in (2.9). □

Well-posedness. Non-homogeneous problems. The following results are relevant [Tr.1], Theorem 1.1, Theorem 4.1]. See also [H-Z.1] in 1-D.

Proposition 2.2. With reference to problem (1.1), (1.2), or (1.3), let $u_1 \equiv u_3 \equiv 0$, $w_0 = w_1 = \theta_0 = 0$. Then

hinged case, (1.2): the following maps are continuous

(2.11a)

$$u_2 \in L_2(0,T;L_2(\Gamma)) \Rightarrow \{w, w_t, \theta\} \in C\left([0,T]; \mathcal{D}(\mathcal{A}) \times \mathcal{D}(\mathcal{A}^{\frac{1}{2}}) \times \mathcal{D}(\mathcal{A}^{\frac{1}{2}})\right),$$

$$(2.11b)\quad w_{tt} \in L_2(0,T;L_2(\Omega)); \quad \theta_t \in L_2(0,T;L_2(\Omega)) \cap C\left([0,T]; [\mathcal{D}(\mathcal{A}^{\frac{1}{2}})]'\right).$$

clamped case, (1.3): the map

$$(2.12a)\qquad u_2 \in L_2(0,T;L_2(\Gamma)) \Rightarrow \{w, w_t\} \in C\left([0,T]; \mathcal{D}(\mathcal{A}^{\frac{1}{2}}) \times L_2(\Omega)\right);$$

$$(2.12b)\qquad \theta \in L_p(0,T;L_2(\Omega)) \cap C\left([0,T]; H^{-\epsilon}(\Omega)\right),$$

for *all* $1 \le p < \infty$, and for all $\frac{1}{2} > \epsilon > 0$ is continuous. □

The controllability results of Theorems 1.1 and 1.2 are consistent with the regularity results of Proposition 2.2, particularly since $u_1(T) = u_3(T) = 0$ in those theorems.

3. Associated Kirchoff-equation. Structural decomposition of the s.c. semigroup $e^{\mathbb{A}_\gamma t}$, $\gamma > 0$

Associated Kirchoff equation. When $\gamma > 0$, the thermo-elastic plate has a hyperbolic-dominated character, in the sense of the next result. Write $\text{div}(\alpha(x)\nabla\theta) = \alpha(x)\Delta\theta + \nabla\alpha \cdot \nabla\theta$ in the first equation (1.1a), and substitute $\Delta\theta$ from the second Eqn. (1.1b) to obtain

$$(3.1)\quad w_{tt} - \gamma\Delta w_{tt} + \Delta^2 w - \alpha\ \text{div}(\alpha\nabla w_t) = -\alpha\theta_t - \nabla\alpha\cdot\nabla\theta \quad \text{in } Q.$$

This, then, induces one to introduce the purely mechanical Kirchoff equation

$$\begin{cases} v_{tt} - \gamma\Delta v_{tt} + \Delta^2 v - \alpha\text{div}(\alpha\nabla v_t) \equiv 0 \text{ in } Q; & (3.2a)\\ \text{either } v|_\Sigma = u_1,\ \Delta w|_\Sigma = u_2; \text{ or else } v|_\Sigma = u_1,\ \left.\dfrac{\partial v}{\partial \nu}\right|_\Sigma = u_2. & (3.2b)\end{cases}$$

Homogeneous case. $u_1 = u_2 = 0$ in (3.2b). The abstract version of problem (3.2) with $u_1 = u_2 = 0$ in (3.2b) is given by

$$(3.3)\quad \frac{d}{dt}\begin{bmatrix} v \\ v_t\end{bmatrix} = \mathbb{A}_{1,\gamma}\begin{bmatrix} v \\ v_t\end{bmatrix} \quad \text{on the space } Y_{1,\gamma},$$

where, in the case of *hinged* B.C. (3.2b) (left), with $u_1 = u_2 = 0$, we have

$$(3.4)\quad \mathbb{A}_{1,\gamma} = \begin{bmatrix} 0 & I \\ -\mathcal{A}_\gamma^{-1}\mathcal{A}^2 & \mathcal{A}_\gamma^{-1}[\alpha\text{div}(\alpha\nabla(\ \cdot\ ))] \end{bmatrix};$$

$$\mathcal{D}(\mathbb{A}_{1,\gamma}) = \mathcal{D}(\mathcal{A}^{\frac{3}{2}}) \times \mathcal{D}(\mathcal{A});$$

$$(3.5)\quad Y_{1,\gamma} \equiv \mathcal{D}(\mathcal{A}) \times \mathcal{D}(\mathcal{A}_\gamma^{\frac{1}{2}}) \equiv [H^2(\Omega)\cap H_0^1(\Omega)] \times H_0^1(\Omega),$$

The $Y_{1,\gamma}$-adjoint $\mathbb{A}_{1,\gamma}^*$ of $\mathbb{A}_{1,\gamma}$ in (3.4) is, in the case of *hinged B.C.*, given by

$$(3.6)\quad \mathbb{A}_{1,\gamma}^* = \begin{bmatrix} 0 & -I \\ \mathcal{A}_\gamma^{-1}\mathcal{A}^2 & \mathcal{A}_\gamma^{-1}\text{div}(\alpha\nabla(\alpha\ \cdot\ )) \end{bmatrix}, \quad \mathcal{D}(\mathbb{A}_{1,\gamma}^*) = \mathcal{D}(\mathbb{A}_{1,\gamma}),$$

with an analogous expression in the case of *clamped* B.C. Because of the topology on the second component space $\mathcal{D}(\mathcal{A}_\gamma^{\frac{1}{2}})$ given by (2.2), the lower right corner element in (3.6) is a bounded perturbation on $\mathcal{D}(\mathcal{A}_\gamma^{\frac{1}{2}})$.

Lemma 3.1. (i) The operator $\mathbb{A}_{1,\gamma}$ in (3.4) generates a s.c. group $e^{\mathbb{A}_{1,\gamma}t}$ on $Y_{1,\gamma}$: $\tilde{v}_0 = \{v_0, v_1\} \Rightarrow e^{\mathbb{A}_{1,\gamma}t}\tilde{v}_0 = \{v(t), v_t(t)\}$ where v solves (3.2) with $u_1 = u_2 = 0$.

(ii) The adjoint semigroup $e^{\mathbb{A}_{1,\gamma}^*t}$ describes the following dynamics: $\tilde{\psi}_0 = \{\psi_0, \psi_1\} \to e^{\mathbb{A}_{1,\gamma}^*t}\tilde{\psi}_0 = \{\psi(t), \psi_t(t)\}$, where $\hat{\psi}$ solves the (dual) Kirchoff problem

$$\begin{cases} \hat{\psi}_{tt} - \gamma\Delta\hat{\psi}_{tt} + \Delta^2\hat{\psi} - \text{div}(\alpha\nabla(\alpha\hat{\psi}_t)) = 0; & (3.7a)\\ \text{either } \hat{\psi} = \Delta\hat{\psi} \equiv 0 \text{ on } \Sigma; \text{ or else } \hat{\psi} = \dfrac{\partial\hat{\psi}}{\partial\nu} \equiv 0 \text{ on } \Sigma. & (3.7b)\end{cases}$$

Structural decomposition of $e^{\mathbb{A}_\gamma t}$. The following result, critical for our present development, holds true.

Theorem 3.2. Consider the thermo-elastic semigroup $e^{\mathbb{A}_\gamma t}$ on Y_γ of Lemma 2.1 and the associated Kirchoff-group $e^{\mathbb{A}_{1,\gamma} t}$ on $Y_{1,\gamma}$ of Lemma 3.1. The following structural decomposition holds true for any $t > 0$; and $[w_0, w_1, \theta_0] \in Y_\gamma$:

$$\begin{bmatrix} w(t) \\ w_t(t) \\ \theta(t) \end{bmatrix} = e^{\mathbb{A}_\gamma t} \begin{bmatrix} w_0 \\ w_1 \\ \theta_0 \end{bmatrix} = \begin{bmatrix} e^{\mathbb{A}_{1,\gamma} t} \begin{bmatrix} w_0 \\ w_1 \end{bmatrix} \\ 0 \end{bmatrix} + \mathcal{K}_\gamma(t) \begin{bmatrix} w_0 \\ w_1 \\ \theta_0 \end{bmatrix} \tag{3.8}$$

where $\mathcal{K}_\gamma(t)$ is a compact operator on Y_γ. □

The above is a special case of a more general structural decomposition result given in [L-T.4, Theorems 1.2.1 and 1.2.2]. [[L-T.4] considered explicitly the case of constant α. A variable α in space produces the additional contribution $\nabla\alpha \cdot \nabla\theta$ in (3.1) [not present in [L-T.4]], which, however, still yields a compact additional contribution to the operator denoted by L_t in [L-T.4, Eqn. (3.5)]. Thus, [L-T.4, Theorem 1.2.2] applies to the present case as well.]

4. Consequence of the structural decomposition: A strategy for the controllability problem

Section 6 of [L-T.4] presents a strategy, essentially already used in [L-T.3, p. 119–120], to obtain an exact controllability (surjectivity) result. In the present case of thermo-elastic plates, its applicability is based on the structural decomposition Theorem 3.2, combined with a soft argument as in [A-L.1] (R.T.'s private notes). This is amply elaborated in [L-T.4, Section 6.2]. With reference to the thermo-elastic plate (1.1), we take henceforth zero initial condition $y_0 = [w_0, w_1, \theta_0] = 0$ and boundary controls $u = [u_1, u_2, u_3]$ in (1.2), or (1.3), of the same class as specified in Theorem 1.1 and Theorem 1.2, respectively. We then define the input-solution operator L_T at the terminal time T by

$$L_T u = \{w(T), w_t(T), \theta(T)\} = \int_0^T e^{\mathbb{A}_\gamma (T-t)} \mathcal{B} u(t) dt, \tag{4.1}$$

which is continuous (mostly) by Proposition 2.2. Let Π_m be the orthogonal projection $Y_\gamma \to Y_{1,\gamma} : [v_1, v_2, v_3] \to [v_1, v_2]$ onto the mechanical state space and let $\Pi_m^* : [v_1, v_2] \to [v_1, v_2, 0]$ be its adjoint $Y_{1,\gamma} \to Y_\gamma$. The strategy for controllability hinges on the following two steps.

Step 1. Show *exact controllability* from the origin at time $t = T$ of the *thermo-elastic plate problem* (1.1), (1.2), or (1.3) in *the mechanical variable*; in symbols, with reference to (4.1), show that

$$\Pi_m L_T : \text{ surjective onto } Y_{1,\gamma}. \tag{4.2}$$

Step 2. Show *approximate controllability* from the origin at time $t = T$ of the *thermo-elastic plate* (1.1), (1.2), or (1.3): in symbols, show that the range of L_T is dense in $H_\gamma = Y_{1,\gamma} \times Y_2$, $Y_2 = H_0^1(\Omega)$ (hinged), on $Y_2 = H^{-\epsilon}(\Omega)$ (clamped):

$$\overline{\mathcal{R}(L_T)} = H_\gamma; \qquad \mathcal{R} = \text{range}. \tag{4.3}$$

Once Step 1 and Step 2 are accomplished, a soft argument [A-L.1] (R.T.'s private notes), since L_T is continuous, then shows the following.

Desired conclusion. Step 1 and Step 2 imply *exact controllability* from the origin at time $t = T$ of the thermo-elastic plate (1.1), (1.2), or (1.3) in the *mechanical variable* and, simultaneously, *approximate controllability* from the origin at time $t = T$ *in the thermal variable*; i.e., precisely, the statements of Theorem 1.1 or, respectively, Theorem 1.2.

4.1. Implementation of Step 1

As explained in [L-T.4, Section 6], it is at the level of implementing Step 1 that the structural decomposition of the thermo-elastic semigroup as in Theorem 3.2 is critically used. The key is the following simple result, essentially already used in [L-T.3, pp.119–120], from approximate to exact controllability.

Proposition 4.1.1. [L-T.4, Proposition 6.1.1]. Let $J = S + Q$, where:

(i) J is a closed operator $\mathcal{U} \supset \mathcal{D}(J) \to X$ with dense range $\overline{\mathcal{R}(J)} = X$ (approximate controllability); equivalently, with trivial null space of the adjoint $J^* : \mathcal{N}(J^*) = \{0\}$;

(ii) S is a closed, surjective operator: $\mathcal{U} \supset \mathcal{D}(S)$ onto X, where $\mathcal{D}(S) = \mathcal{D}(J)$;

(iii) Q is a compact operator: $\mathcal{U} \to X$.

Then, J is surjective $\mathcal{U} \supset \mathcal{D}(J)$ onto X (exact controllability). □

To implement Step 1, and with reference to the decomposition (3.8) of Theorem 3.2, we return to (4.1) and take

$$Ju \equiv \Pi_m L_T u \equiv \Pi_m \int_0^T e^{\mathbb{A}_\gamma (T-t)} \mathcal{B} u(t)dt; \tag{4.4}$$

$$Su \equiv \int_0^T e^{\mathbb{A}_{1,\gamma}(T-t)} \Pi_m \mathcal{B} u(t)dt \overset{\text{def}}{\equiv} L_{m,T} u; \tag{4.5}$$

$$Qu \equiv \Pi_m \int_0^T \mathcal{K}_\gamma (T-t) \mathcal{B} u(t)dt. \tag{4.6}$$

Assumption (i) of Proposition 4.1.1 then means that $\Pi_m L_T$ has dense range in $Y_{1,\gamma} : \overline{\mathcal{R}(\Pi_m L_T)} = Y_{1,\gamma}$: but this is, *a fortiori*, assured by the more demanding condition of Step 2.

Assumption (iii) of Proposition 4.1.1 was verified in the case of the Kirchoff elastic plate (3.2) with controls in the hinged or clamped B.C. in [L-T.4, Proposition 6.2.1]. For these boundary controls the abstract trace condition

$$\mathcal{B}_m^* e^{\mathbb{A}_{1,\gamma} t} : \text{ continuous } Y_{1,\gamma} \to L_2(0,T;U),$$

$U = L_2(\Gamma)$ holds true. As a matter of fact, the proof in [L-T.4, Proposition 6.2.1] referred explicitly to the case of constant α. If α is space dependent, then it produces an additional contribution $\nabla\alpha \cdot \nabla\theta$ in (3.1), but the conclusion is the same. Finally, one needs to verify Assumption (ii) of Proposition 4.1.1. More precisely, one needs to establish the following *exact controllability results* of the Kirchoff equation (3.2).

Theorem 4.1.2. (hinged B.C.) Let $\Gamma_1, \Gamma_2 \subset \Gamma$ be subsets of the boundary Γ, with non-empty intersection of positive measure (we think of Γ_1 as arbitrarily small). Moreover, regarding Γ_2, we assume that: there exists a point $x_0 \in \mathbb{R}^2$, such that

$$(x - x_0) \cdot \nu(x) \leq 0 \quad \text{for } x \in \Gamma \setminus \Gamma_2, \tag{4.7}$$

with $\nu(x)$ the unit outward normal at $x \in \Gamma$. Let

$$T_{0,h} = 2\sqrt{\gamma} \max_i \sup_{x\in\Omega} \operatorname{dist}(x, \Gamma_i), \quad i = 1,2 \tag{4.8}$$

(h stands for 'hinged'). Let $\alpha \in C^2(\overline{\Omega})$. Finally, let $\{v_0, v_1\}$ and $\{v_{0,T}, v_{1,T}\}$ be pre-assigned initial and target states of the (mechanical) v-problem (3.2), with

$$\{v_0, v_1\} \text{ and } \{v_{0,T}, v_{1,T}\} \in [H^2(\Omega) \cap H_0^1(\Omega)] \times H_0^1(\Omega). \tag{4.9}$$

Then, for any $T > T_{0,h}$, there exist control functions

$$u_1 = \begin{cases} \bar{u}_1 \in C_0^\infty(\Sigma_1) \\ 0 \text{ on } \Sigma - \Sigma_1 \end{cases}; \quad u_2 = \begin{cases} \bar{u}_2 \in L_2(\Sigma_2) \\ 0 \text{ on } \Sigma - \Sigma_2 \end{cases}, \tag{4.10}$$

such that the corresponding solution to Eqn. (3.2a) with controls $\{u_1, u_2\}$ in (4.10) in the *hinged B.C.* (3.2b) (left), satisfies the terminal condition

$$v(T) = v_{0,T}, \qquad v_t(T) = v_{1,T}. \quad \square \tag{4.11}$$

The proof of Theorem 4.1.2 is given, by duality, in Section 5.

Theorem 4.1.3. (clamped B.C.) Let $\Gamma_1 \subset \Gamma$ be a non-empty subset of the boundary Γ with positive measure (we think of Γ_1 as arbitrarily small). Let

$$T_{0,c} \geq 2\sqrt{\gamma} \sup_{x\in\Omega} \operatorname{dist}(x, \Gamma_1) \tag{4.12}$$

(c stands for 'clamped'). Let $\alpha \in C^2(\bar{\Omega})$. Finally, let $\{v_0, v_1\}$ and $\{v_{0,T}, v_{1,T}\}$ be pre-assigned initial and target states of the Kirchoff v-problem (3.2), with

$$\{v_0, v_1\} \text{ and } \{v_{0,T}, v_{1,T}\} \in H_0^1(\Omega) \times L_2(\Omega). \tag{4.13}$$

Then, for any $T > T_{0,c}$, there exist control functions

(4.14) $$u_1 = \begin{cases} \hat{u}_1 \in C_0^\infty(\Sigma_1) \\ 0 \text{ on } \Sigma - \Sigma_1 \end{cases} ; \qquad u_2 \in L_2(\Sigma),$$

such that the corresponding solution to Eqn. (3.2a) with controls $\{u_1, u_2\}$ in (4.14) in the *clamped* B.C. (3.2b) (right), satisfies the terminal condition

(4.15) $$v(T) = v_{0,T}, \qquad v_t(T) = v_{1,T}. \qquad \square$$

Remark 4.1.1. We note that Theorems 4.1.2 and 4.1.3 do *not* follow from known results [La.1], [L-T.5], [L-L.1], [Li.1], [K.1], and references therein. Indeed, the two main novelties of Theorems 4.1.2, 4.1.3 over known literature are: (i) the coefficient α is space variable dependent; and, consequently, (ii) the control function u_1 has arbitrarily small support on Γ. These two factors contribute additional technical difficulties and the techniques/methods in the quoted literature are no longer directly applicable. □

4.2. Implementation of Step 2 by duality: Unique continuation of over-determined dual thermo-elastic plates

By duality, the denseness of the range of L_T in (4.3) is equivalent to the null space condition $\mathcal{N}(L_T^*) = \{0\}$, where L_T^* is defined via (2.7) or (2.8) in Lemma 2.1. This leads to the following equivalent unique continuation result.

Theorem 4.2.1. (unique continuation) Let Γ_3 be a subset of Γ of positive measure. Let $\{w, \theta\}$ be a solution of Eqn. (1.1a) possessing the regularity $\{w, \theta\} \in H^3(Q) \times H^1(Q)$, and satisfying the following over-determined B.C.

$$w = \Delta w = \theta \equiv 0 \quad \text{on } \Sigma = (0, T] \times \Gamma;$$

$$\frac{\partial w}{\partial \nu} = \frac{\partial \Delta w}{\partial \nu} = \frac{\partial \theta}{\partial \nu} \equiv 0 \quad \text{on } \Sigma_3 = (0, T] \times \Gamma_3.$$

Let $T > T_3$, where $T_3 \equiv 2\sqrt{\gamma} \sup_{x \in \Omega} \operatorname{dist}(x, \Gamma_3)$. Then, in fact, we have $w \equiv 0$, $\theta \equiv 0$ in $Q = (0, T] \times \Omega$. □

The proof is difficult and is given in [E.1] on the basis of Carleman estimates in [H.1], [I.1], [Ta.1], [Ta.2].

5. Exact controllability of Kirchoff problem by duality: Hinged case

An argument by duality, which we have to omit for lack of space, shows that: the exact controllability result stated in Theorem 4.1.2 for the Kirchoff equation (3.2) and controls u_1, u_2 in the hinged case is equivalent to the following continuous observability inequality for the (dual) Kirchoff problem (3.7), as modified by the addition of suitable lower order terms. For $\alpha \in C^2(\bar{\Omega})$ and $F(\psi) = \mathcal{A}^{-1}\text{div}\{[\alpha, \nabla]\alpha\Delta\psi_t\}$, write

$$\begin{cases} \psi_{tt} - \gamma\Delta\psi_{tt} + \Delta^2\psi + \alpha^2\Delta\psi_t + F(\psi) = 0 & \text{in } Q; \quad (5.1a) \\ \psi(0, \cdot) = \psi_0; \ \psi_t(0, \cdot) = \psi_1 & \text{in } \Omega; \quad (5.1b) \\ \psi|_\Sigma \equiv 0, \ \Delta\psi|_\Sigma \equiv 0 & \text{in } \Sigma. \quad (5.1c) \end{cases}$$

Abstractly, via (2.1), problem (5.1) may be rewritten as

$$\psi_{tt} + \gamma\mathcal{A}\psi_{tt} + \mathcal{A}^2\psi - \alpha^2 Aw_t + F(\psi) \equiv 0, \tag{5.2}$$

where, via the generation Lemma 3.1, we have the *a-priori* regularity result

$$\begin{cases} \{\psi_0, \psi_1\} \in \mathcal{D}(\mathbb{A}_\gamma^*) = \mathcal{D}(\mathcal{A}^{\frac{3}{2}}) \times \mathcal{D}(\mathcal{A}) \Rightarrow \\ \{\psi, \psi_t\} \in C([0,T]; \mathcal{D}(\mathcal{A}^{\frac{3}{2}}) \times \mathcal{D}(\mathcal{A})). \end{cases} \tag{5.3}$$

Theorem 5.1. Let $\Gamma_1, \Gamma_2 \subset \Gamma$ be subsets of the boundary Γ, with non-empty intersection of positive measure, as in Theorem 4.1.2. Thus, we think of Γ_1 as arbitrarily small, and there exists a point $x_0 \in \mathbb{R}^2$, such that the geometrical condition (4.7) holds true. Let

$$\{\psi_0, \psi_1\} \in \mathcal{D}(\mathbb{A}_\gamma^*) = \mathcal{D}(\mathcal{A}^{\frac{3}{2}}) \times \mathcal{D}(\mathcal{A}^{\frac{1}{2}}) \tag{5.4a}$$

so that by (2.1b), (2.2)

$$\psi_0 \in H^3(\Omega), \ \psi_0|_\Gamma = 0, \ \Delta\psi_0|_\Gamma = 0; \ \psi_1 \in H^2(\Omega) \cap H_0^1(\Omega). \tag{5.4b}$$

Let $T > T_{0,h}$, defined by (4.8). Let k be an arbitrary positive real number.

Then: there exists a constant $C_{T,\gamma,k} > 0$ ($\gamma > 0$ fixed) such that the following inequality holds true:

$$\|\{\psi_0, \psi_1\}\|^2_{\mathcal{D}(\mathcal{A}^{\frac{3}{2}}) \times \mathcal{D}(\mathcal{A})} \le C_{T,\gamma,k} \int_0^T \int_{\Gamma_2} \left(\frac{\partial\Delta\psi}{\partial\nu}\right)^2 d\Sigma_2 + \left\|\frac{\partial\psi}{\partial\nu}\right\|^2_{H^{-k}(\Sigma_1)}. \tag{5.5}$$

Proof of Theorem 5.1. Step 1. We first establish the desired inequality (5.5), polluted, however, by lower-order terms.

Proposition 5.2. Under the same assumptions and setting of Theorem 5.1, we have: there is an (explicit) constant $C_{T,\gamma} > 0$ ($\gamma > 0$ fixed) such that

$$\begin{aligned} C_T\|\{\psi_0, \psi_1\}\|^2_{\mathcal{D}(\mathcal{A}^{\frac{3}{2}}) \times \mathcal{D}(\mathcal{A})} \le & \int_0^T \int_{\Gamma_2} \left(\frac{\partial\Delta\psi}{\partial\nu}\right)^2 d\Sigma_2 \\ & + \frac{C_T}{\gamma^2} \int_0^T \left[\|\psi\|^2_{H^2(\Omega)} + \|\psi_t\|^2_{H_0^1(\Omega)}\right] dt. \end{aligned} \tag{5.6}$$

Proof of Proposition 5.2. Step (i). As in [L-T.6], we reduce the problem to the wave equation with variable coefficient $\alpha^2(x)$ in the damping term. We then apply known estimates to this wave equation problem. To do this, with the *a-priori* regularity of ψ given by (5.3), we introduce a new variable z

$$z = -\Delta\psi = \mathcal{A}\psi \in C([0,T]; \mathcal{D}(\mathcal{A}^{\frac{1}{2}}) = H_0^1(\Omega)) \text{ in } Q. \tag{5.7}$$

Then, using (5.7) in (5.1a) where readily

$$\operatorname{div}(\alpha\nabla(\alpha\psi_t)) = \alpha^2\Delta\psi_t + \operatorname{div}(\alpha\psi_t\nabla\alpha) + \nabla(\alpha^2)\cdot\nabla\psi_t, \tag{5.8}$$

we then obtain the following wave equation problem in z:

$$\begin{cases} \gamma z_{tt} = \Delta z - \alpha^2 z_t + f & \text{in } Q; \quad (5.9a) \\ z|_\Sigma \equiv 0 & \text{in } \Sigma; \quad (5.9b) \end{cases}$$

$$f \equiv -\psi_{tt} + \operatorname{div}(\alpha\psi_t\nabla\alpha) + \nabla(\alpha^2)\cdot\nabla\psi_t + F(\psi), \tag{5.10}$$

which has the variable coefficient $\alpha^2(x)$ in front of the damping term z_t.

Step (ii). We invoke [L-T.2, Theorem 2.1.2(ii), Eqn. (2.1.10b) (which applies for a general wave equation with variable coefficients in the space variable in the first-order terms (energy level). We obtain, in the notation of [L-T.2]:

$$(\overline{BT})|_\Sigma + \frac{C_T}{\gamma^2}\int_Q f^2 dQ + TC_{T,\gamma}\|z\|^2_{L_2(Q)} \geq k_{\gamma,T}E_z(0), \tag{5.11}$$

where in equivalent norm

$$E_z(0) = \|\{z(0), z_t(0)\}\|^2_{\mathcal{D}(\mathcal{A}^{\frac{1}{2}})\times L_2(\Omega)} \tag{5.12}$$

(the grad-norm is equivalent to the $H_0^1(\Omega)$-norm for the z-variable) and where via [L-T.2, (2.1.11) and (2.1.9)] we have by (5.9b) [so that $h\cdot\nabla z = \frac{\partial z}{\partial\nu}h\cdot\nu$, $|\nabla z|^2 = \left(\frac{\partial z}{\partial\nu}\right)^2$, $h = (x - x_0)$]:

$$(\overline{BT})|_\Sigma = \int_\Sigma e^{\tau\phi}\frac{\partial z}{\partial\nu}h\cdot\nabla z\, d\Sigma - \frac{1}{2}\int_\Sigma e^{\tau\phi}|\nabla z|^2 h\cdot\nu\, d\Sigma \tag{5.13}$$

$$= \frac{1}{2}\int_\Sigma e^{\tau\phi}\left(\frac{\partial z}{\partial\nu}\right)^2 h\cdot\nu\, d\Sigma \leq C_\phi\int_0^T\int_{\Gamma_2}\left(\frac{\partial z}{\partial\nu}\right)^2 d\Sigma_2 \tag{5.14}$$

(ϕ is the pseudo-convex function in [L-T.2]). In the last step in (5.14), we have invoked assumption (4.7) on Γ_2.

Step (iii). From the definition of f in (5.7), we readily obtain with $\alpha \in C^1(\bar{\Omega})$,

$$\|f\|^2_{L_2(\Omega)} \leq C_\alpha\left\{\|\psi_{tt}\|^2_{L_2(\Omega)} + \|\psi_t\|^2_{H_0^1(\Omega)} + \|\Delta\psi_t\|^2_{[H^1(\Omega)]'}\right\}. \tag{5.15}$$

We now estimate the terms on the right side of (5.15).

Claim. With reference to problem (5.1), or (5.2), we have

$$\begin{aligned} \|\psi_{tt}\|^2_{L_2(\Omega)} &\le C_\alpha \left\{ \|\psi\|^2_{\mathcal{D}(\mathcal{A})} + \|\psi_t\|^2_{\mathcal{D}(\mathcal{A}^{\frac{1}{2}})} \right\} \\ &= C_\alpha \left\{ \|\psi\|^2_{H^2(\Omega)} + \|\psi_t\|^2_{H^1_0(\Omega)} \right\}. \end{aligned} \tag{5.16}$$

Proof of Claim. We return to the abstract equation (5.2) with *a-priori* regularity (5.3) and obtain by recalling, via $\mathcal{A}_\gamma$ in (2.1), that $\mathcal{A}_\gamma^{-1}\mathcal{A}^2$: continuous $\mathcal{D}(\mathcal{A}) \to L_2(\Omega)$:

$$\|\psi_{tt}\|_{L_2(\Omega)} \le c\|\mathcal{A}\psi\|_{L_2(\Omega)} = \|\psi\|_{\mathcal{D}(\mathcal{A})} = \|\psi\|_{H^2(\Omega)}. \tag{5.17}$$

Moreover, since $\alpha \in C^2(\bar{\Omega})$,

$$\operatorname{div}(\alpha\nabla(\alpha\,\cdot)) : \text{ continuous } H^1_0(\Omega) = \mathcal{D}(\mathcal{A}^{\frac{1}{2}}) \to H^{-1}(\Omega) = [\mathcal{D}(\mathcal{A}^{\frac{1}{2}})]', \tag{5.18}$$

duality with respect to $L_2(\Omega)$ as a pivot space. Hence, by (5.18),

$$\mathcal{A}_\gamma^{-1}\operatorname{div}(\alpha\nabla(\alpha\,\cdot)) : \text{ continuous } H^1_0(\Omega) = \mathcal{D}(\mathcal{A}^{\frac{1}{2}}) \to \text{itself}. \tag{5.19}$$

Hence, from (5.18) and (5.3) on ψ_t:

$$\|\mathcal{A}_\gamma^{-1}\operatorname{div}(\alpha\nabla(\alpha\psi_t))\|_{H^1_0(\Omega)} \le C\|\psi_t\|_{H^1_0(\Omega)}. \tag{5.20}$$

Using (5.17) and *a-fortiori* (5.20) in (5.2) yields (5.16) as desired.

Step (iv). Using (5.16) in (5.15) yields

$$\|f\|^2_{L_2(\Omega)} \le C_\alpha \left\{ \|\psi\|^2_{H^2(\Omega)} + \|\psi_t\|^2_{H^1_0(\Omega)} + \|\Delta\psi_t\|^2_{[H^1(\Omega)]'} \right\}. \tag{5.21}$$

Next, we use (5.21) on the right side of (5.11), and finally substitute (5.14) for $(\overline{BT})|_\Sigma$ in (5.11). We thus arrive at

$$\begin{aligned} \int_0^T \int_{\Gamma_2} \left(\frac{\partial z}{\partial \nu}\right)^2 d\Sigma_2 + \frac{C_T}{\gamma^2} \int_0^T \left[\|\psi\|^2_{H^2(\Omega)} + \|\psi_t\|^2_{H^1_0(\Omega)} + \|\Delta\psi_t\|^2_{[H^1(\Omega)]'} \right] dt \\ \ge k_{\gamma,T} \|\{z(0), z_t(0)\}\|^2_{\mathcal{D}(\mathcal{A}^{\frac{1}{2}}) \times L_2(\Omega)}, \end{aligned} \tag{5.22}$$

after recalling (5.7): $z = -\Delta\psi$ for the z-term in (5.11), and (5.12). Finally, again by (5.7) we have

$$\begin{cases} z(0) = & -\Delta\psi(0) = \mathcal{A}\psi(0), \text{ so } z(0) \in \mathcal{D}(\mathcal{A}^{\frac{1}{2}}) \iff \psi(0) \in \mathcal{D}(\mathcal{A}^{\frac{3}{2}}); \\ z_t(0) = & -\Delta\psi_t(0) = \mathcal{A}\psi_t(0), \text{ so } z_t(0) \in L_2(\Omega) \iff \psi_t(0) \in \mathcal{D}(\mathcal{A}). \end{cases} \tag{5.23}$$

Using (5.23) in the right side of (5.22) yields (5.6), as desired. Proposition 5.2 is proved. □

Step 2. We now absorb the lower-order terms (l.o.t.) in the interior, on the right side of inequality (5.6), by a compactness/uniqueness argument, as usual.

Proposition 5.3. Let ψ be a solution of (5.1) satisfying inequality (5.6). Let $T > T_{0,h}$ defined by (4.8). Let k be any positive number. Then: there exists a constant $C_T \equiv C_{T,\gamma,k}$ ($\gamma > 0$ fixed) such that the l.o.t. in (5.6) satisfies

$$\int_0^T \left[\|\psi\|^2_{H^2(\Omega)} + \|\psi_t\|^2_{H_0^1(\Omega)} + \|\Delta\psi_t\|^2_{[H^1(\Omega)]'}\right] dt$$

$$\leq C_T \left\{ \int_0^T \int_{\Gamma_2} \left(\frac{\partial \Delta\psi}{\partial \nu}\right)^2 d\Sigma_2 + \left\|\frac{\partial\psi}{\partial\nu}\right\|^2_{H^{-k}(\Sigma_1)} \right\}. \tag{5.24}$$

Proof. By contraction, suppose that inequality (5.24) is false. Then, there exists a sequence

$$\begin{cases} \{\psi^{(n)}, \psi_t^{(n)}\} \in C([0,T]; \mathcal{D}(\mathcal{A}^{\frac{3}{2}}) \times \mathcal{D}(\mathcal{A})), \ n = 1, 2, \ldots \\ \text{continuous in the initial data } \{\psi_0^{(n)}, \psi_1^{(n)}\} \in \mathcal{D}(\mathcal{A}^{\frac{3}{2}}) \times \mathcal{D}(\mathcal{A}), \end{cases} \tag{5.25}$$

of solutions of (5.1) such that

$$\begin{cases} \int_0^T \left[\|\psi^{(n)}\|^2_{\mathcal{D}(\mathcal{A})} + \|\psi_t^{(n)}\|^2_{\mathcal{D}(\mathcal{A}^{\frac{1}{2}})} + \|\Delta\psi_t\|^2_{[H^1(\Omega)]'}\right] dt \equiv 1, & (5.26) \\ \int_0^T \int_{\Gamma_2} \left(\frac{\partial\Delta\psi^{(n)}}{\partial\nu}\right)^2 d\Sigma_2 + \left\|\frac{\partial\psi^{(n)}}{\partial\nu}\right\|^2_{H^{-k}(\Sigma_1)} \to 0 \text{ as } n \to \infty. & (5.27) \end{cases}$$

The sequence $\{\psi^{(n)}\}$ satisfies inequality (5.6). Thus, the corresponding Initial Conditions are uniformly bounded,

$$\|\{\psi_0^{(n)}, \psi_1^{(n)}\}\|_{\mathcal{D}(\mathcal{A}^{\frac{3}{2}}) \times \mathcal{D}(\mathcal{A})} \leq \text{const}, \quad \forall n. \tag{5.28}$$

Then, by (5.25),

$$\|\{\psi^{(n)}, \psi_t^{(n)}\}\|_{C([0,T]; \mathcal{D}(\mathcal{A}^{\frac{3}{2}}) \times \mathcal{D}(\mathcal{A}))} \leq \text{const}, \quad \forall n. \tag{5.29}$$

By (5.2) applied to $\psi^{(n)}$, and by invoking (5.16) we obtain via (5.29),

$$\|\psi_{tt}^{(n)}\|_{C([0,T]; L_2(\Omega))} \leq C, \quad \forall n. \tag{5.30}$$

Then, *a fortiori* from (5.29) and (5.30), we can apply Aubin's Lemma, since $\mathcal{A}^{-1}$ is compact in $L_2(\Omega)$, and obtain

$$\{\psi^{(n)}, \psi_t^{(n)}\} \to \text{ some } \{\tilde{\psi}, \tilde{\psi}_t\} \text{ strongly in } L_2(0,T; \mathcal{D}(\mathcal{A}^{\frac{1}{2}+\theta}) \times \mathcal{D}(\mathcal{A}^{\theta})). \tag{5.31}$$

$\theta < 1$, since $H^{-\frac{1}{2}+2\epsilon}(\Omega) = [\mathcal{D}(\mathcal{A}^{\frac{1}{4}-\epsilon})]' \to [H^1(\Omega)]'$ is continuous, hence

$$\|\Delta\psi_t\|_{[H^1(\Omega)]} \le c\|\psi_t\|_{H^{\frac{3}{2}+2\epsilon}(\Omega)} \le c\|\psi_t\|_{\mathcal{D}(\mathcal{A}^\theta)}. \tag{5.32}$$

Then (5.31) and (5.26) imply that the limit $\tilde{\psi}$ satisfies

$$\int_0^T \left[\|\tilde{\psi}\|^2_{\mathcal{D}(\mathcal{A})} + \|\tilde{\psi}_t\|^2 + \|\Delta\tilde{\psi}_t\|^2_{[H^1(\Omega)]'}\right] dt = 1. \tag{5.33}$$

The limit $\tilde{\psi} \in L_2(0,T;\mathcal{D}(\mathcal{A}))$ satisfies Eqn. (5.1a–c) in particular the B.C.

$$\tilde{\psi}|_\Sigma = 0 \text{ in } L_2(0,T;H^{\frac{3}{2}}(\Gamma)) \text{ and } \Delta\tilde{\psi}|_\Sigma = 0 \text{ in } L_2(0,T;H^{-\frac{1}{2}}(\Gamma)). \tag{5.34}$$

Moreover, by (5.27),

$$\left.\frac{\partial\Delta\tilde{\psi}}{\partial\nu}\right|_{\Sigma_2} = 0 \text{ on } \Sigma_2; \quad \left.\frac{\partial\tilde{\psi}}{\partial\nu}\right|_{\Sigma_1} = 0 \text{ on } \Sigma_1. \tag{5.35}$$

Thus, $\tilde{\psi}$ satisfies the homogeneous Kirchoff equation (5.1a) with over-determined homogeneous B.C. (5.34), (5.35).

Recalling that $T > T_{0,h}$, we can invoke Theorem 4.2.1 and include that, in fact, $\tilde{\psi} \equiv 0$ is Q. But this contradicts (5.32). Thus, (5.24) holds true. □

Acknowledgement. Research partially supported by the National Science Foundation under Grant DMS-9804056 and by the Army Research Office under Grand DAAH04-96-1-0059.

References

[A-L.1] G. Avalos and I. Lasiecka, Boundary controllability of thermoelastic plates with free boundary conditions, *SIAM J. Control*, to appear.

[E.1] M. Eller, Uniqueness of continuation theorems, *Proceedings of the ISAAC'97*, to appear.

[H-Z.1] S. W. Hansen and B. Zhang, Boundary control of a linear thermo-elastic beam, *J. Math. Anal. & Appl.* 210 (1997), 182–205.

[H.1] L. Hörmander, On the uniqueness of the Cauchy problem under partial analyticity assumptions, in *Geometric Optics and Related Topics*, F. Colombini & N. Lerner, eds., 1997.

[I.1] V. Isakov, On the uniqueness of continuation for a thermoelasticity system, preprint, 1998.

[La.1] J. Lagnese, *Boundary Stabilization of Thin Plates*, SIAM, Philadelphia, 1989.

[La.2] J. Lagnese, The reachability problem for thermoelastic plates, *Arch. Rational Mechanics and Analysis* 112 (1990), 223-267.

[L-L.1] J. Lagnese and J. .L. Lions, *Modeling, Analysis and Control of Thin Plates*, Masson, 1988.

[Li.1] J. L. Lions, *Controllabilite exacte, stabilization de systemes distrbues*, Vol. 1 and 2, Masson, Paris, 1988.

[K.1] V. Komornik, *Exact controllability and stabilization*, Masson, Paris, 1994.

[L-T.1] I. Lasiecka and R. Triggiani, Uniform boundary stabilization of the wave equation with Dirichlet and Neumann feedback control without geometric conditions, *Appl. Math. Optimiz.* 25 (1992), 189–224.

[L-T.2] I. Lasiecka and R. Triggiani, Carleman's estimates and boundary controllability for a system of coupled, nonconservative second-order hyperbolic equations, *Lecture Notes in Pure and Applied Mathematics*, Marcel Dekker, Vol. 188, pp. 215–245, 1997.

[L-T.3] I. Lasiecka and R. Triggiani, Exact controllability of semilinear abstract systems with application to waves and plates boundary control problems, *Appl. Math. & Optimiz.* 3 (1991), 109–154.

[L-T.4] I. Lasiecka and R. Triggiani, Structural decomposition of thermoelastic semigroups with rotational forces, *Semigroup Forum*, to appear.

[L-T.5] I. Lasiecka and R. Triggiani, Exact controllability and uniform stabilization of Kirchoff plates with boundary control only on $\Delta w|_\Sigma$, *J. Diff. Eqns.* 93 (1991), 62–101.

[L-T.6] I. Lasiecka and R. Triggiani, A sharp trace result of a thermo-elastic plate equation with hinged/Neumann coupled B.C., *Discrete & Continuous Dynamical Systems* 3 (1999), 585–598.

[Ta.1] D. Tataru, Unique continuation for solutions to PDE;s between Hörmander's theorem and Holmgren's theorem, *Comm. Part. Diff. Eqn.* 20 (1995), 855–884.

[Ta.2] D. Tataru, Unique continuation for operators with partially analytic coefficients, preprint, 1998.

[Tr.1] R. Triggiani, Sharp regularity theory for thermo-elastic mixed problems, *Applicable Analysis*, to appear.

M. Eller, Department of Mathematics, Tennessee Technological University, Cookeville, TN 38501, U.S.A.

I. Lasiecka, R. Triggiani, Department of Mathematics, Kerchof Hall, University of Virginia, Charlottesville, VA 22903, U.S.A.

Progress in Nonlinear Differential Equations
and Their Applications, Vol. 42

THE OPERATOR EQUATION $AX - X\mathcal{D}^2 = -\delta_0$ AND SECOND ORDER DIFFERENTIAL EQUATIONS IN BANACH SPACES

ELENA SCHÜLER
VU QUOC PHONG
OHIO UNIVERSITY

1. Introduction

Consider the differential equation

$$u''(t) = Au(t) + f(t), \tag{1}$$

where A is a closed linear operator on a Banach space E, with a dense range $D(A)$, and $f(t)$ is a continuous function from $\mathbf{R}$ to E. We consider $Y = D(A)$ as a Banach space with the graph norm defined by

$$|||y||| = \|y\| + \|Ay\|, \quad y \in Y.$$

Let $C^{(n)}(\mathbf{R}, E)$, $n \geq 0$, be the space which consists of bounded continuous functions u with bounded continuous derivatives $u', ..., u^{(n)}$, and $BUC(\mathbf{R}, E)$ be the space of bounded uniformly continuous functions (with values in E). Further, we denote by C_c^∞ the class of functions $\widehat{\psi}$ such that $\psi \in L^1(\mathbf{R})$ and $supp(\widehat{\psi})$ is compact, where $\widehat{\psi}$ is the Fourier transform of ψ.

A function $u \in C(\mathbf{R}, E)$ is called a *strong solution* of Eq.(1) if $u \in C(\mathbf{R}, Y) \cap C^{(2)}(\mathbf{R}, E)$ and (1) holds for all $t \in \mathbf{R}$. A bounded function $u \in C(\mathbf{R}, E)$ is called a *mild solution* of Eq.(1) (with bounded f) if for every $\widehat{\psi} \in C_c^\infty$ the function $v = u * \psi$ is a strong solution of the equation $v'' = Av(t) + (f * \psi)(t)$, $t \in \mathbf{R}$. Note that a mild solution u is a strong solution if (and only if) $u \in C^{(2)}(\mathbf{R}, E)$.

Suppose that $\mathcal{M}$ is a closed translation-invariant subspace of $BUC(\mathbf{R}, E)$. The subspace $\mathcal{M}$ is said to be *regularly admissible* for Eq.(1), if for every $f \in \mathcal{M}$ there exists a unique mild solution $u \in \mathcal{M}$ of Eq.(1). Many statements about stability and asymptotic behavior of solutions of Eq.(1) can be formulated in terms of the regular admissibility of corresponding function spaces $\mathcal{M}$. In this paper we apply the method of the operator equation of Lyapunov's type to the study of the regular admissibility of $\mathcal{M}$. Analogous questions for equations of the first order

$$u'(t) = Au(t) + f(t) \tag{2}$$

have been investigated in our previous publications [10,11]. Namely, we have related the regular admissiblity of Eq.(2) to solution of an operator equation of Lyapunov type

$$AX - XB = C, \tag{3}$$

where A is the generator of a C_0-semigroup $T(t)$ and $(-B)$ is the generator of an isometric group $S(t)$. In fact, it was shown that the space $\mathcal{M}$ is regularly admissible if and only if the operator equation

$$AX - X\mathcal{D}_{\mathcal{M}} = -\delta_0^{\mathcal{M}} \tag{4}$$

has a unique bounded solution. Here, $\mathcal{D}_{\mathcal{M}}$ is the restriction of the operator of differentiation $D := \frac{d}{dt}$ onto the subspace $\mathcal{M}$, $\delta_0 : BUC(\mathbf{R}, E) \to E$ is the Dirac operator defined by $\delta_0 f \equiv f(0)$, and $\delta_0^{\mathcal{M}} \equiv \delta_0 \mid_{\mathcal{M}}$ - the restriction of δ_0 to $\mathcal{M}$.

Using properties of isometric groups, we obtained in [11] conditions for the unique solvability of Eq.(4), which in turn give criteria for the regular admissibility of $\mathcal{M}$ with respect to Eq.(2).

In the present article analogous results are proved for equations of second order. Namely, we show that the subspace $\mathcal{M}$ is regularly admissible if and only if the operator equation

$$AX - X\mathcal{D}^2_{\mathcal{M}} = -\delta_0^{\mathcal{M}} \tag{5}$$

has a unique bounded solution. However, since the operator $-\mathcal{D}^2_{\mathcal{M}}$ is not a generator of an isometric group, the method in [11] is not directly applicable for solving Eq.(5). Nevertheless, using specific properties of the operator $\mathcal{D}$, we obtain a necessary and sufficient condition for the unique solvability of Eq.(5) which is analogous to the condition in [11] for equations of the first order.

2. The equation $AX - X\mathcal{D}^2_{\mathcal{M}} = -\delta_0^{\mathcal{M}}$

Let A and B be closed, generally unbounded, linear operators on Banach spaces E and F, with dense domains $D(A)$ and $D(B)$ respectively, and let C be a bounded linear operator from F to E. A bounded linear operator $X : F \to E$ is called a *solution* of equation

$$AX - XB = C \tag{3}$$

if, for every $f \in D(B)$, we have $Xf \in D(A)$ and $AXf - XBf = Cf$.

We will need the following result which belongs to Krein-Rosenblum [2, 7] (see also [8, Lemma 22]).

Theorem 1. *Assume that B is bounded and $\sigma(A) \cap \sigma(B) = \emptyset$. Then for every bounded operator C equation (3) has a unique bounded solution X which has the following integral form*

$$X = \frac{1}{2\pi i} \int_\Gamma (\lambda I - A)^{-1} C (\lambda I - B)^{-1} d\lambda, \tag{6}$$

where Γ is a closed Cauchy contour around $\sigma(B)$ and separated from $\sigma(A)$.

Theorem 1 and formula (6) have been generalized to the case of unbounded operators A and B in [8]. Recall that an operator A is the generator of an analytic semigroup if and only if there exists θ, $(0 < \theta < \pi/2)$ and a real number ω such that

$$\sigma(A) \subset \Sigma_{\omega,\theta} = \{\lambda \in \mathbf{C} : |\arg(\omega - \lambda)| < \theta\} \cup \{\omega\} \tag{7}$$

and

$$\sup_{\lambda \in \mathbf{C}\setminus\Sigma_{\omega,\theta}} \|\lambda(\lambda I - A)^{-1}\| < \infty. \tag{8}$$

For convenience, we will call the corresponding analytic semigroup generated by an operator A with conditions (7)-(8) a $\Sigma_{\omega,\theta}$-*analytic semigroup.*

Further, let B be a closed operator. Suppose there exists a real number α such that

$$\sigma(B) \subset \mathbf{C}_\alpha^+ := \{\lambda \in \mathbf{C} : Re\,\lambda \geq \alpha\} \tag{9}$$

and $\|\lambda(\lambda I - B)^{-1}\|$ are uniformly bounded when $\lambda \in \Sigma_{\omega,\theta} \cap (\mathbf{C} \setminus \mathbf{C}_\alpha^+)$, i.e

$$\sup\{\|\lambda(\lambda I - B)^{-1}\| : \ \lambda \in \Sigma_{\omega,\theta} \cap (\mathbf{C} \setminus \mathbf{C}_\alpha^+)\} < \infty. \tag{10}$$

For convenience, an operator B satisfying condition (9)-(10) will be called an *operator of class* (ω, θ, α).

Assume that $\sigma(A) \cap \sigma(B) = \emptyset$. Then under the conditions (7) and (9) we can construct an infinite contour $\Gamma = \Gamma_1 \cup \Gamma_2 \cup \Gamma_3$ with positive orientation, where

$$\Gamma_1 = \{\lambda \in \mathbf{C} : \omega - \lambda = re^{\theta i}, \ \ r\cos\theta > -\alpha\},$$
$$\Gamma_2 = \{\lambda \in \mathbf{C} : \omega - \lambda = re^{-\theta i}, \ \ r\cos\theta > -\alpha\},$$

and Γ_3 is a piecewise smooth contour which connects Γ_1 and Γ_2 and separates $\sigma(A) \cap (\mathbf{C}_\alpha^+ \cap \Sigma_{\omega,\theta})$ from $\sigma(B) \cap (\mathbf{C}_\alpha^+ \cap \Sigma_{\omega,\theta})$. Since $\mathbf{C}_\alpha^+ \cap \Sigma_{\omega,\theta}$ is compact, such contour Γ_3 is easily constructed.

Theorem 2. *(See [8, Theorem 15]). Assume that A is the generator of a $\Sigma_{\omega,\theta}$-analytic semigroup and B is an operator of class (ω, θ, α) such that $\sigma(A) \cap \sigma(B) = \emptyset$, and Γ is the contour described above. Then Eq.(3) has a unique solution which is expressed by*

$$X = \frac{1}{2\pi i}\int_\Gamma (\lambda I - A)^{-1}C(\lambda I - B)^{-1}d\lambda. \tag{11}$$

It follows from the Hille-Yosida Theorem (see [4,6]) that the condition on B is satisfied if $(-B)$ is the generator of a C_0-semigroup. Hence Theorem 2 implies that if A is the generator of an analytic semigroup and $-B$ is the generator of a C_0-semigroup such that $\sigma(A) \cap \sigma(B) = \emptyset$, then for every bounded operator

C equation (3) has a unique bounded solution and the solution is given by formula (11).

The situation when B is the generator of an analytic semigroup and A is a closed operator such that $\sigma(A) \cap \sigma(B) = \emptyset$ can be treated analogously. We state the corresponding result below since it will be used in the sequel.

Theorem 3. *Assume that B is the generator of a $\Sigma_{\omega,\theta}$-analytic semigroup and A is an operator of class (ω, θ, α) such that $\sigma(A) \cap \sigma(B) = \emptyset$, and Γ is the contour described above. Then equation (3) has a unique solution which is expressed by formula (11). In particular, the above statement holds if B is the generator of an analytic semigroup and $-A$ is the generator of a C_0-semigroup, such that $\sigma(A) \cap \sigma(B) = \emptyset$.*

Next we consider another situation which occurs in the connection with the regular admissibility problem, namely when $B = Q^2$, where Q is the generator of an isometric group. We recall some well known properties of generators of isometric groups which will be needed in the sequel.

Assume that Q is the generator of a C_0-group of isometries $S(t), t \in \mathbf{R}$, on a Banach space F. Then the following properties hold (see [3]).

(P1) The spectrum of Q is a subset of $i\mathbf{R}$. For every compact subset $\Lambda \subset i\mathbf{R}$, there is a maximal closed invariant subspace L such that $Q|L$ is bounded, $\sigma(Q|L) \subset \Lambda$ and if K is another closed invariant subspace of Q such that $\sigma(Q|K) \subset \Lambda$, then $K \subset L$. The space L is called the *spectral maximal subspace* corresponding to Λ and denoted by $F(\Lambda)$.

(P2) For each vector $x \in F$, the local resolvent $R(\lambda; x) := (\lambda I - Q)^{-1}x$ has the so called *single valued extension property*, i.e. if f and g are analytic extension of $R(\lambda; x)$, with domains $D(f)$ and $D(g)$, respectively, then $f(\xi) = g(\xi)$ for all $\xi \in D(f) \cap D(g)$. The union of the sets $D(f)$ as f varies over all analytic extensions of $R(\lambda; x)$ is called the *local resolvent* of x and denoted by $\rho(x)$. The set $\sigma(x) := \mathbf{C} \setminus \rho(x)$ is called the *local spectrum* of x (with respect to Q).

(P3) For every closed subset $\Lambda \subset i\mathbf{R}$, we have $F(\Lambda) = \{x \in F : \sigma(x) \subset \Lambda\}$.

(P4) $\sigma(Q) = \cup\{\sigma(x) : x \in F\}$.

Assume now that $Q = \mathcal{D}_{\mathcal{M}}$ - the restriction of the operator $D = d/dt$ to the translation-invariant subspace $\mathcal{M}$. For each function $f \in BUC(\mathbf{R}, E)$ we define its Carleman transform by

$$\tilde{f}(\lambda) = \begin{cases} \int_0^\infty e^{-\lambda t} f(t)dt, & \text{if } Re\,\lambda > 0 \\ -\int_{-\infty}^0 e^{-\lambda t} f(t)dt, & \text{if } Re\,\lambda < 0. \end{cases}$$

A point $\lambda \in i\mathbf{R}$ is called a *regular point* of f if there exists a neighborhood $\mathcal{U}$ of λ such that $\tilde{f}$ admits an analytic extension into $\mathcal{U}$. The complement in $i\mathbf{R}$ of the set of regular points is called the *Carleman spectrum* of the function f and denoted by $Sp(f)$.

Since $\tilde{f}(\lambda) = (\lambda I - \mathcal{D})^{-1} f$ for $Re\,\lambda \neq 0$, it follows that for each element $f \in \mathcal{M}$, the local spectrum $\sigma(f)$ with respect to $Q = \mathcal{D}_{\mathcal{M}}$ coincides with the Carleman spectrum $Sp(f)$. Let $\mathcal{M}_n := \{f \in \mathcal{M} : \sigma(f) \subset [-in, in]\}$, $n \geq 1$. Clearly, $\mathcal{M}_n$ are translation-invariant subspaces and $\mathcal{M}_n \subset \mathcal{M}_{n+1}, n = 1, 2, \ldots$.

Lemma 4. $\sigma(\mathcal{D}_{\mathcal{M}}) = \cup_{n=1}^{\infty} \{\sigma(\mathcal{D}_{\mathcal{M}_n})\}$.

Proof. We show that

$$\sigma(\mathcal{D}_{\mathcal{M}}) \subset \cup_{n=1}^{\infty} \{\sigma(\mathcal{D}_{\mathcal{M}_n})\}. \tag{12}$$

Let $\lambda \in \sigma(\mathcal{D}_{\mathcal{M}})$. Since $\mathcal{D}_{\mathcal{M}}$ generates an isometric group, it has the single valued extension property and (P4) holds. Hence, there exists a function $f \in \mathcal{M}$ such that $\lambda \in \sigma(f)$. Let $a > 0$, $i\mathcal{U} = (\lambda - ia, \lambda + ia)$ and $k \in L^1(\mathbf{R})$ be such that $\widehat{k}|\mathcal{U} \equiv 1$. Then $\lambda \in \sigma(k * f)$. In fact, if $\lambda \in \rho(k * f)$, then there exists a neighborhood $\mathcal{V}$ of λ such that $\psi * k * f \equiv 0$ whenever $\psi \in L^1(\mathbf{R})$, $supp(\widehat{\psi}) \subset \mathcal{V}$. Hence, for every $\varphi \in L^1(\mathbf{R})$, such that $supp(\widehat{\varphi}) \subset \mathcal{V}$, there exists $\psi \in L^1(\mathbf{R})$ such that $\varphi = \psi * k$. Since $\widehat{\psi} = \widehat{\varphi}/\widehat{k}$, it follows that $supp(\widehat{\psi}) \subset \mathcal{V}$, and $\varphi * f = \psi * k * f = 0$, which implies that $\lambda \in \rho(f)$, a contradiction.

Since $\mathcal{M}$ is translation-invariant, we have $k * f \in \mathcal{M}$, that is the function $g = k * f \in \mathcal{M}$ has a compact spectrum and $\lambda \in \sigma(g)$. Let n be such that $\sigma(g) \subset [-in, in]$. Then $\lambda \in \sigma(\mathcal{D}_{\mathcal{M}_n})$ and the inclusion (12) follows. Since the inverse inclusion to (12) is obvious, the lemma is proved. □

From Lemma 4 we obtain the following

Lemma 5. $\sigma(\mathcal{D}^2_{\mathcal{M}}) = \cup_{n=1}^{\infty} \{\sigma(\mathcal{D}^2_{\mathcal{M}_n})\}$.

Proof. Again, we need only to prove the inclusion

$$\sigma(\mathcal{D}^2_{\mathcal{M}}) \subset \cup_{n=1}^{\infty} \{\sigma(\mathcal{D}^2_{\mathcal{M}_n}). \tag{13}$$

We note that $\sigma(\mathcal{D}^2) = (-\infty, 0]$, therefore $\sigma(\mathcal{D}^2_{\mathcal{M}}) \subset (-\infty, 0]$. Assume that $-\lambda^2 \in \sigma(\mathcal{D}^2_{\mathcal{M}}), \lambda \in \mathbf{R}$. Then there exists a sequence of elements $(f_n) \in \mathcal{M}$ such that $\|f_n\| = 1$, $f_n \in D(\mathcal{D}^2_{\mathcal{M}}), n = 1, 2 \ldots$, and

$$\lim_{n \to \infty} \|\mathcal{D}^2_{\mathcal{M}} f_n + \lambda^2 f_n\| = 0.$$

Consider $g_n = (\mathcal{D}_{\mathcal{M}} - i\lambda I) f_n$. Then either $\inf_{n \geq 1} \|g_n\| = 0$, in which case $i\lambda \in \sigma(\mathcal{D}_{\mathcal{M}})$, or $\inf_{n \geq 1} \|g_n\| > 0$, in which case $(\mathcal{D}_{\mathcal{M}} + i\lambda I) g_n \to 0$, so $-i\lambda \in \sigma(\mathcal{D}_{\mathcal{M}})$. Assume, for definiteness, that $i\lambda \in \sigma(\mathcal{D}_{\mathcal{M}})$. By Lemma 4, there exists n such that $i\lambda \in \sigma(\mathcal{D}_{\mathcal{M}_n})$. Since $\mathcal{D}_{\mathcal{M}_n}$ is a bounded operator, it follows that $-\lambda^2 \in \sigma(\mathcal{D}^2_{\mathcal{M}_n})$. The inclusion (13) is proved. □

From Lemmas 4 and 5 we have

Lemma 6 $\sigma(\mathcal{D}^2_{\mathcal{M}}) = \{\lambda^2 : \lambda \in \sigma(\mathcal{D}_{\mathcal{M}})\}$.

Let us now return to the operator equation (5). Assume that

$$\sigma(A)\cap\{\lambda^2:\lambda\in\sigma(\mathcal{D}_{\mathcal{M}})\}=\emptyset. \tag{14}$$

From Lemma 6 it follows that (14) is equivalent to

$$\sigma(A)\cap\sigma(\mathcal{D}^2_{\mathcal{M}})=\emptyset. \tag{15}$$

Therefore

$$\sigma(A)\cap\sigma(\mathcal{D}^2_{\mathcal{M}_n})=\emptyset,\ n=1,2,.... \tag{16}$$

By Theorem 1, there exists a unique bounded solution X_n to the operator equation

$$AX-X\mathcal{D}^2_{\mathcal{M}_n}=-\delta_0^{\mathcal{M}_n},$$

which has the integral form

$$X_n=-\frac{1}{2\pi i}\int_{\Gamma_n}(\lambda I-A)^{-1}\delta_0^{\mathcal{M}_n}(\lambda I-\mathcal{D}^2_{\mathcal{M}_n})^{-1}d\lambda, \tag{17}$$

where Γ_n is a suitable contour around $\sigma(\mathcal{D}^2_{\mathcal{M}_n})$ and separated from $\sigma(A)$. Moreover, the uniqueness of solutions X_n implies

$$X_n|\mathcal{M}_k=X_k,\quad k<n. \tag{18}$$

Theorem 7. *Assume that the condition (14) holds. Then the operator equation (5) has a unique bounded solution if and only if*

$$\sup_{n\geq 1}\|X_n\|<\infty, \tag{19}$$

where X_n are defined by (17).

Proof. Assume that X is a bounded solution to Eq.(5). Then we have $X|\mathcal{M}_n=X_n$, which implies $\|X_n\|\leq\|X\|$, i.e. (19) holds.

Conversely, assume that (19) holds. Define an operator X on $\mathcal{M}_\infty:=\cup_{n\geq 1}\mathcal{M}_n$ by $Xf=X_nf$ if $f\in\mathcal{M}_n$. Observe that $\overline{\mathcal{M}_\infty}=\mathcal{M}$ (see the first equality in (21)). Hence, from (17)-(19) it follows that X is correctly defined and can be extended by continuity to a bounded operator from $\mathcal{M}$ to E, which is also denoted by X. We show that X is a bounded solution of Eq.(5).

Suppose $f\in D(\mathcal{D}^2_{\mathcal{M}})$. We show that $Xf\in D(A)$ and

$$AXf-X\mathcal{D}^2_{\mathcal{M}}f=-\delta_0^{\mathcal{M}}f. \tag{20}$$

Let k_n be a sequence of functions in $L^1(\mathbf{R})$ such that

(i) $\int_{\mathbf{R}}k_n(t)dt=1$;

(ii) $\sup_{n\geq 1}\|k_n\|_{L^1(\mathbf{R})}:=C<\infty$;

(iii) $\lim_{n\to\infty} \int_{|t|>\delta} |k_n(t)|dt = 0$ for all $\delta > 0$.

(iv) $isupp(\widehat{k}_n) \subset [-in, in]$.

Such a sequence k_n is called an *approximate unity* in $L^1(\mathbf{R})$, see [5].
Let $f_n = k_n * f,\ n = 1, 2, \ldots$ Since $f \in D(\mathcal{D}^2_{\mathcal{M}})$, it follows that

$$\lim_{n\to\infty} f_n = f,\ \lim_{n\to\infty} \mathcal{D}f_n = \mathcal{D}f,\ \lim_{n\to\infty} \mathcal{D}^2 f_n = \mathcal{D}^2 f, \tag{21}$$

in the uniform topology in $BUC(\mathbf{R}, E)$. In fact, for every $\varepsilon > 0$ there exists $\delta > 0$ such that

$$|f(t-s) - f(t)| < \varepsilon/(2C), \quad \text{if } |s| < \delta.$$

By (iii), there exists $N > 0$ such that

$$\int_{|s|\geq\delta} |k_n(s)|ds < \varepsilon/(4\|f\|_\infty), \quad \text{if } n \geq N.$$

Therefore

$$|f_n(t) - f(t)| \leq \int_{|s|<\delta} |f(t-s) - f(t)||k_n(s)|ds$$

$$+2\|f\|_\infty \int_{|s|\geq\delta} |k_n(s)|ds < \varepsilon, \quad t \in \mathbf{R},\ n \geq N.$$

Hence the first identity in (21) is proved. The other identities are proved analogously.

Since $f \in \mathcal{M}$ and $\mathcal{M}$ is translation-invariant, we have $f_n \in \mathcal{M}, n \in \mathbf{N}$, and the property (iv) implies that $\sigma(f_n) \subset [-in, in]$. Hence $f_n \in D(\mathcal{D}^2_{\mathcal{M}_n})$. Therefore, we have

$$AXf_n = X\mathcal{D}^2_{\mathcal{M}} f_n - \delta_0^{\mathcal{M}} f_n.$$

Thus, by (21), $Xf_n \to Xf$, $AXf_n \to X\mathcal{D}^2_{\mathcal{M}} f - \delta_0^{\mathcal{M}} f$, and the closedness of A implies that $Xf \in D(A)$ and (20) holds. The proof is complete. □

3. The regular admissibility.

If a closed translation-invariant subspace $\mathcal{M} \subseteq BUC(\mathbf{R}, E)$ is regularly admissible for Eq. (1), then a standard argument involving the Closed Graph Theorem shows that there exists a bounded linear operator $G : \mathcal{M} \to \mathcal{M}$ such that for each $f \in \mathcal{M}$, $u = Gf$ is the unique mild solution of Eq.(1) in $\mathcal{M}$. In what follows, we assume that $\mathcal{M}$ satisfies the following additional assumption:

(22) *For all $C \in \mathcal{L}(\mathcal{M}, E)$ and $f \in \mathcal{M}$,
the function $\phi(t) = CS(t)f$ belongs to $\mathcal{M}$.*

Theorem 8. *Assume that $\mathcal{M} \subseteq BUC(\mathbf{R}, E)$ is a closed translation-invariant subspace which satisfies (22). Then the following are equivalent:*

(i) $\mathcal{M}$ is regularly admissible for Eq. (1).

(ii) The operator equation

$$AX - X\mathcal{D}^2_{\mathcal{M}} = -\delta_0^{\mathcal{M}} \tag{5}$$

has a unique bounded solution.

(iii) For every bounded linear operator $C : \mathcal{M} \to E$ the operator equation

$$AX - X\mathcal{D}^2_{\mathcal{M}} = C \tag{23}$$

has a unique bounded solution.

Proof. (i) $\Longrightarrow$ (ii). Let $\mathcal{M} \subseteq BUC(\mathbf{R}, E)$ be regularly admissible for Eq.(1). Let $G : \mathcal{M} \longrightarrow \mathcal{M}$ be the bounded operator defined by $Gf = u$, where u is the unique mild solution in $\mathcal{M}$ of Eq.(1). Because Eq.(1) is autonomous, it is not hard to see that G commutes with the translations $\mathcal{S}(t)$, which in turn implies that G commutes with $\mathcal{D}^2_{\mathcal{M}}$. Define a bounded linear operator $X : \mathcal{M} \to E$ by $Xf := (Gf)(0)$. Note that

$$X\mathcal{S}(t)f = (G\mathcal{S}(t)f)(0) = (Gf)(t). \tag{24}$$

It follows from (24) that if $f \in D(\mathcal{D}^2_{\mathcal{M}})$, then $(Gf)(t) \in C^{(2)}(\mathbf{R})$. Therefore, Gf is a strong solution of Eq.(1), i.e.

$$(Gf)''(t) = A(Gf)(t) + f(t), \quad \forall t \in \mathbf{R}.$$

Hence, $(Gf)''(0) = A(Gf)(0) + f(0)$, or $X\mathcal{D}^2_{\mathcal{M}}f = AXf + \delta_0^{\mathcal{M}}f$, so that

$$AXf - X\mathcal{D}^2_{\mathcal{M}}f = -\delta_0^{\mathcal{M}}f. \tag{25}$$

By repeating the final paragraph in the proof of Theorem 7, one can show that for every $f \in \mathcal{D}^2_{\mathcal{M}}$, we have $Xf \in D(A)$ and (25) holds. Thus, X is a bounded solution of Eq.(5).

Next we show that the solution of Eq.(5) is unique. We note that, if X is a bounded solution of Eq.(5), then for every $f \in D(\mathcal{D}^2_{\mathcal{M}})$ the function u defined by $u(t) = X\mathcal{S}(t)f$ is a strong solution of Eq.(1), since

$$u''(t) = X\mathcal{D}^2_{\mathcal{M}}\mathcal{S}(t)f = AX\mathcal{S}(t)f + \delta_0^{\mathcal{M}}\mathcal{S}(t)f = Au(t) + f(t).$$

If $f \in \mathcal{M}$ is arbitrary, then $u(t) = X\mathcal{S}(t)f$ is a mild solution of Eq.(1). Indeed, let $\widehat{\phi} \in C_c^\infty$ be given, $u_\phi = \phi * u$, $f_\phi = \phi * f$. Then $u_\phi(x) = X\mathcal{S}(t)(f_\phi)(x)$, and $\mathcal{S}(t)(f * \phi) \in D(\mathcal{D}^2_{\mathcal{M}})$. Hence, we have that

$$\begin{aligned} u''_\phi(t) &= X\mathcal{D}^2_{\mathcal{M}}\mathcal{S}(t)(f_\phi) = AX\mathcal{S}(t)(f_\phi) + \delta_0^{\mathcal{M}}\mathcal{S}(t)(f_\phi) \\ &= Au_\phi(t) + f_\phi(t), \end{aligned}$$

which shows that $u(t)$ is a mild solution of Eq.(1). Assume now that X_1 and X_2 are two solutions of Eq.(5). Then, for every $f \in \mathcal{M}$, $u(t) = (X_1 - X_2)\mathcal{S}(t)f$ is a mild solution of equation $u''(t) = Au(t)$. By uniqueness of the mild solution we have that $u(t) = 0$, which implies $X_1 = X_2$.

(ii)$\Longrightarrow$(iii). The uniqueness of the solution to equation equation $AX - X\mathcal{D}^2_{\mathcal{M}} = -\delta_0^{\mathcal{M}}$ implies that the homogeneous equation $AX - X\mathcal{D}^2_{\mathcal{M}} = 0$ has only the trivial solution $X = 0$, which, in turn, implies the uniqueness of the solution to equation $AX - X\mathcal{D}^2_{\mathcal{M}} = C$.

Let Y be the solution of Eq.(5). Define an operator $X : \mathcal{M} \to E$ by

$$Xf = Y(-C\mathcal{S}(t)f), \quad f \in \mathcal{M}.$$

Note that, by (22), $\phi : t \mapsto C\mathcal{S}(t)f$ is a function in $\mathcal{M}$. Since Y is a bounded linear operator from $\mathcal{M}$ to E, X is correctly defined as a bounded operator from $\mathcal{M}$ into E. We have, for every $f \in D(\mathcal{D}^2_{\mathcal{M}})$,

$$\begin{aligned} AXf - X\mathcal{D}^2_{\mathcal{M}}f &= AY(-C\mathcal{S}(t)f) - Y\mathcal{D}^2_{\mathcal{M}}(-C\mathcal{S}(t)f) \\ &= -\delta_0^{\mathcal{M}}(-C\mathcal{S}(t)f) = Cf. \end{aligned}$$

Therefore, X is a bounded solution of Eq.(23).

We remark that from (iii) of Theorem 8 it follows that $\sigma(A) \cap \sigma(\mathcal{D}^2_{\mathcal{M}}) = \emptyset$ (see [1, Theorem 2.1]).

(iii) $\Longrightarrow$ (i). We have shown above that, if X is a bounded solution of Eq.(5), then $u(t) = X\mathcal{S}(t)f$ is a mild solution of Eq.(1). Hence, the operator G defined by $(Gf)(t) := X\mathcal{S}(t)f, f \in \mathcal{M}$, is a bounded linear operator from $\mathcal{M}$ to $\mathcal{M}$, such that Gf is the mild solution in $\mathcal{M}$ of Eq.(1). It remains to show that this solution is unique. Assume, on the contrary, that there are two mild solutions u_1 and u_2 of Eq.(1). Then $v = u_1 - u_2$ is a non-trivial mild solution of $u'' = Au$. It follows from [9], Lemma 4.2 (iii), that if $i\lambda \in \sigma(v)$ then $-\lambda^2 \in \sigma(A)$, i.e. $(\sigma(v))^2 \subset \sigma(A)$. On the other hand, since $v \in \mathcal{M}$ we have $\sigma(v) \subseteq \sigma(D_{\mathcal{M}})$, hence $(\sigma(v))^2 \subset \sigma(\mathcal{D}^2_{\mathcal{M}})$, by Lemma 6. Since $\sigma(A) \cap \sigma(\mathcal{D}^2_{\mathcal{M}}) = \emptyset$, we have $\sigma(v) = \emptyset$, so that $v \equiv 0$. Theorem 8 is proved. □

By Theorem 10, the question of regular admissibility for a subspace $\mathcal{M}$ has been reduced to the question of solvability of the operator equation (5).

The following result follows from Theorem 3, the well known fact that $\mathcal{D}^2_{\mathcal{M}}$ is the generator of an analytic semigroup (see [6, p. 34]), and Theorem 2.1 in [1].

Theorem 9. *Assume that A is an operator of class (ω, θ, α), and $\mathcal{M}$ is a translation-invariant subspace satisfying condition (22). Then $\mathcal{M}$ is regularly admissible if and only if $\sigma(A) \cap \sigma(\mathcal{D}^2_{\mathcal{M}}) = \emptyset$.*

Corollary 10. *Assume that $-A$ is the generator of a C_0-semigroup. Then $\mathcal{M}$ is regularly admissible if and only if $\sigma(A) \cap \sigma(\mathcal{D}^2_{\mathcal{M}}) = \emptyset$.*

The following theorem follows from Theorems 7, 8 and Theorem 2.1 in [1].

Theorem 11. *The space $\mathcal{M}$ is regularly admissible for Eq.(1) if and only if $\sigma(A) \cap \sigma(\mathcal{D}^2_{\mathcal{M}}) = \emptyset$ and*

$$\sup_{n\geq 1} \|X_n\| < \infty,$$

where X_n are defined by formula (17).

Below we apply Theorem 11 to a class of subspaces $\mathcal{M}(\Lambda)$ defined by

$$\mathcal{M}(\Lambda) := \{f \in BUC(\mathbf{R}, E) : Sp(f) \subset \Lambda\},$$

where Λ is a closed subset of $i\mathbf{R}$. It is easy to see that $\mathcal{M}(\Lambda)$ is a translation-invariant subspace which satisfies condition (22). Moreover $\sigma(\mathcal{D}_{\mathcal{M}(\Lambda)}) = \Lambda$.

The following are cases of $\mathcal{M}(\Lambda)$ of particular interest in the study of the asymptotic behavior of Eq.(1).

(i) $\Lambda = \{2k\pi i/\omega : k \in \mathbf{Z}\}$, then $\mathcal{M}(\Lambda) := \mathcal{P}(\omega)$ coincides with the class of all (continuous) ω-periodic functions.

(ii) $\Lambda = \{i(2k+1)\pi/\omega : k \in \mathbf{Z}\}$, then $\mathcal{M}(\Lambda) := \mathcal{AP}(\omega)$ coincides with the class of all (continuous) ω-anti-periodic functions, i.e. all functions f such that $f(t+\omega) = -f(t), \forall t$.

(iii) Λ is a closed countable subset of $i\mathbf{R}$ and E does not contain an isomorphic copy of c_0 (the Banach space of numerical sequences which converge to 0). Then every function in $\mathcal{M}(\Lambda)$ is almost periodic.

(iv) Λ is a closed countable subset of $i\mathbf{R}$ and

$$\mathcal{M}_e(\Lambda) = \Big\{ f \in BUC(\mathbf{R}) : Sp(f) \subset \Lambda, \text{ and}$$

$$\lim_{t\to\infty} \frac{1}{2T} \int_{-T+h}^{T+h} e^{-\lambda t} f(t)dt \text{ exists uniformly in } h \ \forall \lambda \in i\Lambda \Big\}.$$

Then $\mathcal{M}_e(\Lambda)$ also is translation-invariant, satisfies (22), and consists of almost periodic functions (see [9]).

Theorem 11 can be applied to obtain conditions for the regular admissibility of $\mathcal{M}$ in all the special cases (i)-(iv). We restrict ourselves by stating the result for periodic functions.

Corollary 12. *The space $\mathcal{P}(\omega)$ is regularly admissible if and only if $-4k^2\pi^2/(\omega^2) \in \rho(A)$ for all $k = 0, 1, ...$ and there exists a constant C such that*

$$\Big\| \sum_{k=-n}^{n} \Big(A + \frac{4k^2\pi^2}{\omega^2}\Big)^{-1} v_k \Big\| \leq C \sup_{t\in\mathbf{R}} \Big\| \sum_{k=-n}^{n} e^{(2k\pi i/\omega)t} v_k \Big\|,$$

for any finite sequence $\{v_k\}_{k=-n}^{n}$ in E.

Proof. Consider the space $\mathcal{P}_n(\omega) = span\{e^{(2k\pi i/\omega)t}v_k : |k| \leq n, v_k \in E\}$. Let $\mathcal{D}_n := \mathcal{D}|\mathcal{P}_n(\omega)$, $\delta_0^{(n)} = \delta_0|\mathcal{P}_n(\omega)$, and X_n be the solution of $AX - X\mathcal{D}_n^2 = -\delta_0^{(n)}$. By Theorem 11, $\mathcal{P}(\omega)$ is regularly admissible if and only if $\sup_{n\geq 0}\|X_n\| < \infty$, where X_n are defined by (17) with suitable contours Γ.

Let ϕ be an arbitrary function in $\mathcal{P}_n(\omega)$, i.e. $\phi(t) = \sum_{k=-n}^{n} e^{(2k\pi i/\omega)t}v_k, v_k \in E$. We show that the solution X_n satisfies

$$X_n\phi = \sum_{k=-n}^{n}\left(A + \frac{4k^2\pi^2}{\omega^2}\right)^{-1}v_k,$$

from which the statement of the corollary follows.

Choose a contour $\Gamma = \cup_{k=0}^{n}\Gamma_k$, where Γ_k is a small circle around $\frac{4k^2\pi^2}{\omega^2}$ such that Γ_k and its interior are contained in $\rho(A)$. By (17), we have

$$X_n\phi = -\frac{1}{2\pi i}\int_\Gamma(\lambda I - A)^{-1}\delta_0^{(n)}(\lambda I - \mathcal{D}_n^2)^{-1}\phi d\lambda.$$

Since

$$\begin{aligned}\delta_0^{(n)}(\lambda I - \mathcal{D}_n^2)^{-1}\phi &= \delta_0^{(n)}\Big[\sum_{k=-n}^{n}\left(\lambda + \frac{4k^2\pi^2}{\omega^2}\right)^{-1}e^{(2k\pi i/\omega)t}v_k\Big]\\ &= \sum_{k=-n}^{n}\left(\lambda + \frac{4k^2\pi^2}{\omega^2}\right)^{-1}v_k,\end{aligned}$$

we have

$$\begin{aligned}X_n\phi &= -\frac{1}{2\pi i}\int_\Gamma(\lambda I - A)^{-1}\sum_{k=-n}^{n}\left(\lambda + \frac{4k^2\pi^2}{\omega^2}\right)^{-1}v_k d\lambda\\ &= -\frac{1}{2\pi i}\sum_{j=0}^{n}\sum_{k=-n}^{n}\int_{\Gamma_j}(\lambda I - A)^{-1}\left(\lambda + \frac{4k^2\pi^2}{\omega^2}\right)^{-1}v_k d\lambda.\end{aligned}$$

Since, by Cauchy's Theorem,

$$\frac{1}{2\pi i}\int_{\Gamma_j}(\lambda I - A)^{-1}\left(\lambda + \frac{4k^2\pi^2}{\omega^2}\right)^{-1}v_k d\lambda = \begin{cases} 0, & \text{if } j \neq k,\\ -\left(A + \frac{4k^2\pi^2}{\omega^2}I\right)^{-1}v_k, & \text{if } j = k,\end{cases}$$

it is easily seen that $X_n\phi = \sum_{k=-n}^{n}\left(A + \frac{4k^2\pi^2}{\omega^2}I\right)^{-1}v_k$. □

References

[1] W. Arendt, F. Räbiger, A. Sourour, *Spectral Properties of the operator equation $AX + XB = Y$*, Quart. J. Math. Oxford **45:2** (1994), 133–149.

[2] Ju.L. Daleckii and M.G. Krein, *Stability of Solutions of Differential Equations in Banach Spaces*, Amer. Math. Soc., Providence, RI, 1974.

[3] I. Erdelyi and S.W. Wang, *A local spectral theory for closed operators*, Cambridge Univ. Press, London, 1985.

[4] J. Goldstein, *Semigroups of linear Operators and Applications*, Oxford Univ. Press, New York, 1985.

[5] Y. Katznelson, *An Introduction to Harmonic Analysis*, Dover Pub. Inc., New York, 1976.

[6] R. Nagel *et al.*, *One-Parameter Semigroups of Positive Operators*, Springer, Berlin, 1986.

[7] M. Rosenblum, *On the operator equation $BX - XA = Q$*, Duke Math. J. **23** (1956), 263–269.

[8] Vu Quoc Phong, *The operator equation $AX - XB = C$ with unbounded operators A and B and related abstract Cauchy problem*, Mathematische Zeitschrift **208** (1991), 567–588.

[9] Vu Quoc Phong, *Almost Periodic Solutions of Volterra Equations*, Diff. Int. Eq. **7:4** (1994), 1083–1093.

[10] Vu Quoc Phong, *On the exponential stability and dichotomy of C_0-semigroups*, Studia Mathematica, to appear.

[11] Vu Quoc Phong and E.Schüler, *The operator equation $AX - XB = C$, admissibility, and asymptotic behavior of diff. equations*, J. Differential Equations **145** (1998), 394–419.

[12] E. Schüler, *On the spectrum of cosine functions*, J. Math. Anal. Appl., to appear.

Elena Schüler, Vu Quoc Phong, Department of Mathematics, Ohio University, Athens, OH 45701, U.S.A.

E-mail address: qvu@bing.math.ohiou.edu

Progress in Nonlinear Differential Equations
and Their Applications, Vol. 42

DILATION TO UNILATERAL SHIFTS

PEI YUAN WU
NATIONAL CHIAO TUNG UNIVERSITY

KATSUTOSHI TAKAHASHI
UNIVERSITY OF HOKKAIDO

Which bounded linear operator on a complex separable Hilbert space can be dilated to a unilateral shift? This is the problem we are going to address here. Recall that an operator A on space H is said to *dilate* (resp. *power dilate*) to operator B on K if there is an isometry V from H to K such that $A = V^*BV$ (resp. $A^n = V^*B^nV$ for all $n \geq 1$) or, equivalently, if B is unitarily equivalent to a 2-by-2 operator matrix $\begin{bmatrix} A & * \\ * & * \end{bmatrix}$ with A in its upper left corner (resp. B^n is unitarily equivalent to $\begin{bmatrix} A^n & * \\ * & * \end{bmatrix}$ under the same unitary operator for all $n \geq 1$). The *unilateral shift* S_k of multiplicity k $(1 \leq k \leq \infty)$ is the operator $S_k(x_0, x_1, x_2, \ldots) = (0, x_0, x_1, \ldots)$ on $\sum_{n=0}^{\infty} \oplus H$ with dim $H = k$.

The classical result of Foias (cf. [3, Problem 152]) settles the corresponding power dilation problem completely: an operator T power dilates to some unilateral shift S_k if and only if T is a contraction of class $C_{\cdot 0}$, that is, $\|T\| \leq 1$ and T satisfies $T^{*n} \to 0$ in the strong operator topology, and, moreover, in this case the minimum value of k is dim $\overline{\text{ran }(I - T^*T)^{1/2}}$. Here we consider the dilation problem for various classes of operators (contractions with at least one defect index finite, C_0 contractions, strict contractions, normal contractions and compact contractions) and show that in all such cases, the two classes, one consisting of those which dilate to a unilateral shift and the other those which power dilate, coincide. The only difference is on the minimum multiplicity of the unilateral shift which can be (power) dilated to. In the dilation case, this multiplicity can be as small as 1 if the two defect indices $d_T =$ dim $\overline{\text{ran }(I - T^*T)^{1/2}}$ and $d_{T^*} =$ dim $\overline{\text{ran }(I - TT^*)^{1/2}}$ of the contraction T under consideration are equal, and $d_{T^*} - d_T$ if otherwise. This is in contrast to the (minimum) multiplicity d_{T^*} of the unilateral shift to which T can be power dilated.

The monograph [6] by Sz.-Nagy and Foias is our standard reference for the terminology and results of their contraction theory. For Fredholm theory, the reader can consult [2, Chapter XI].

We say that operator A on H *dilates* to operator B on K *by n-dimension* $(0 \leq n \leq \infty)$ if $A = V^*BV$ for some isometry : $H \to K$ with dim $(K \ominus VH) = n$. The next theorem characterizes operators which dilate to a unilateral shift by finite dimension.

Theorem 1. *An operator T dilates to S_k by n-dimension* $(1 \leq k \leq \infty$, $0 < n < \infty)$ *if and only if T is a $C_{\cdot 0}$ contraction with $d_T < \infty$ and $d_T \neq d_{T^*}$. In this case, $k = d_{T^*} - d_T$ and the minimum value for n is d_T.*

Note that if T is a $C_{\cdot 0}$ contraction, then $d_T \leq d_{T^*}$ always holds (cf. [6, Proposition VI.3.5]).

The necessity part depends on the following proposition and some basic Fredholm theory. Recall that a contraction T is *of class* C_{11} if the sequences $\{T^n x\}_{n=1}^{\infty}$ and $\{T^{*n} x\}_{n=1}^{\infty}$ do not converge to 0 in norm for any nonzero vector x. A C_{11} contraction must have equal defect indices (cf. [6, Proposition VI.3.5]).

Proposition 2. *If T is a C_{11} contraction with finite defect indices or a contraction with $d_{T^*} < d_T$, then it has no unilateral shift dilation.*

The proposition can be proved based on Carey's result [1, Proposition] on the finite-rank perturbations of isometries. The proof techniques are adapted from [7, Proposition 3.5].

The sufficiency part of Theorem 1 is proved via a series of lemmas. The first of these concerns rank-one perturbations of $C_{\cdot 0}$ contractions; it is proved by Nakamura [4, Corollary 3].

Lemma 3. *If T is a $C_{\cdot 0}$ contraction with $d_T = 1$ and $d_{T^*} > 1$, then there is a rank-one operator F with* $\operatorname{ran} F \subseteq \ker T^*$ *such that $T + F$ is unitarily equivalent to S_k, where $k = d_{T^*} - 1$.*

This is transferred into a result on dilations to the unilateral shift by one dimension.

Lemma 4. *If T is a $C_{\cdot 0}$ contraction with $d_T = 1$ and $d_{T^*} > 1$, then T dilates to the unilateral shift S_k by one dimension, where $k = d_{T^*} - 1$.*

By judiciously applying the preceding lemma repeatedly, we can have it generalized to $C_{\cdot 0}$ contractions T with $d_T < \infty$.

Lemma 5. *If T is a $C_{\cdot 0}$ contraction with $1 \leq d_T < \infty$ and $d_{T^*} > d_T$, then T dilates by one dimension to another $C_{\cdot 0}$ contraction $\tilde{T}$ with $d_{\tilde{T}} = d_T - 1$ and $d_{\tilde{T}^*} = d_{T^*} - 1$.*

Our main result is the next theorem, which shows that among contractions with at least one defect index finite, dilation and power dilation to a unilateral shift amount to the same thing. The only difference is on the minimum multiplicity of the unilateral shift to which the contraction is (power) dilated.

Theorem 6. *Let T be a contraction with at least one defect index finite. Then the following conditions are equivalent:*

(a) *T dilates to some S_k, $1 \leq k \leq \infty$;*

(b) *T power dilates to some S_l, $1 \leq l \leq \infty$;*

(c) *T is of class $C_{\cdot 0}$.*

Moreover, in this case, the minimum value of k in (a) is 1 if $d_T = d_{T^*}$, and $d_{T^*} - d_T$ if otherwise, and the minimum value of l in (b) is d_{T^*}.

In view of the results stated before, to complete the proof we need show that (1) every $C_{\cdot 0}$ contraction with finite equal defect indices dilates to S, the unilateral shift with multiplicity 1, and (2) if the $C_{\cdot 0}$ contraction T with $d_T < d_{T^*}$ dilates to S_k by infinite dimension, then $k \geq d_{T^*} - d_T$.

(1) is a consequence of the following more general lemma since every such contraction is of class C_0. Recall that a contraction T is *of class* C_0 if T is completely nonunitary (c.n.u.), that is, T has no nontrivial reducing subspace on which T is unitary, and there is a nonzero function f in H^∞ such that $f(T) = 0$, where $f(T)$ is formed according to the Sz-Nagy-Foias functional calculus. In this case, there is a (nonconstant) inner function ϕ in H^∞ such that $\phi(T) = 0$ which divides any function f in H^∞ with $f(T) = 0$. This uniquely determined (up to a scalar multiple with modulus one) ϕ is called the *minimal function* of T.

Lemma 7. *Every C_0 contraction dilates to the unilateral shift S.*

Since every C_0 contraction with minimal function ϕ can be extended to $S(\phi) \oplus S(\phi) \oplus \ldots$ (cf. [5, Lemma 4]), where $S(\phi)$ denotes the operator $S(\phi)f = P(zf(z))$ on $H^2 \ominus \phi H^2$, with P being the (orthogonal) projection from H^2 onto $H^2 \ominus \phi H^2$, the proof of the lemma is reduced to that for operators of the form $S(\phi) \oplus S(\phi) \oplus \ldots$, which is achieved by exploiting the structure of the minimal isometric dilation of $S(\phi)$.

On the other hand, (2) is proved via Carey's result [1, Proposition] and some Fredholm theory.

The following are easy consequences of Theorem 6.

Corollary 8. *Let T be contraction with at least one defect index finite. Then T dilates to S if and only if T is of class $C_{\cdot 0}$ and $d_{T^*} - d_T = 0$ or 1.*

An operator T is *algebraic* if $p(T) = 0$ for some polynomial p.

Corollary 9. *The following conditions are equivalent for an algebraic operator T:*

(a) *T dilates to S;*
(b) *T dilates to some S_k, $1 \leq k \leq \infty$;*
(c) *T power dilates to some S_l, $1 \leq l \leq \infty$;*
(d) *T is a C_0 contraction;*
(e) *T is a contraction with spectrum contained in $\mathbf{D}$, the open unit disc.*

Corollary 10. *Every strict contraction ($\|T\| < 1$) dilates to S.*

Corollary 11. *Every operator dilates to a scalar multiple of S.*

Corollary 12. *The following conditions are equivalent for a normal operator T:*

(a) *T dilates to S;*
(b) *T dilates to some S_k, $1 \leq k \leq \infty$;*

(c) *T power dilates to some S_l, $1 \leq l \leq \infty$;*

(d) *T is a C_{00} contraction, that is, $T^n \to 0$ and $T^{*n} \to 0$ in the strong operator topology;*

(e) *T is a c.n.u. contraction.*

Corollary 13. *The following conditions are equivalent for a compact operator T:*

(a) *T dilates to some S_k, $1 \leq k \leq \infty$;*

(b) *T power dilates to some $S_l, 1 \leq l \leq \infty$;*

(c) *T is a C_{00} contraction;*

(d) *T is a contraction with spectrum contained in* $\mathbf{D}$.

We conclude by asking whether contractions which dilate to a unilateral shift can always power dilate to one. In all the cases we considered here, this is indeed true. The general case remains open. Another question worth investigating is whether a contraction with equal defect indices always dilates to S. In particular, this is unknown for compact contractions.

Acknowledgements. This is a joint work with Katsutoshi Takahashi, formerly of University of Hokkaido, who passed away in November, 1996. The work was partially supported by the National Science Council of the Republic of China and the full paper has appeared in *Integral Equations Operator Theory*, 32 (1998), 101–113. The author would like to thank the organizers, among them C. Kubrusly and N. Levan, for inviting him to take part in this Semigroup Conference and for the hospitality extended to him during his stay at Newport Beach.

References

[1] R.W. Carey, Trace class perturbations of isometries and unitary dilations, *Proc. Amer. Math. Soc.*, 45 (1974), 229–234.

[2] J.B. Conway, *A course in functional analysis*, 2nd ed., Springer-Verlag, New York, 1990.

[3] P.R. Halmos, *A Hilbert space problem book*, 2nd ed., Springer-Verlag, New York, 1982.

[4] Y. Nakamura, One-dimensional perturbations of the shift, *Integral Equations Operator Theory*, 17 (1993), 373–493.

[5] T. Nakazi and K. Takahashi, Two-dimenional representations of uniform algebras, *Proc. Amer. Math. Soc.*, 123 (1995), 2777–2784.

[6] B. Sz.-Nagy and C. Foias, *Harmonic analysis of operators on Hilbert space*, North Holland, Amsterdam, 1970.

[7] P.Y. Wu and K. Takahashi, Singular unitary dilations, *Integral Equations Operator Theory*, 33 (1999), 231–247.

PEI YUAN WU, DEPARTMENT OF APPLIED MATHEMATICS, NATIONAL CHIAO TUNG UNIVERSITY, HSINCHU 300, TAIWAN